HARTMUT WALRAVENS (HG.)

JOSEPH FRANZ ROCK
(1884–1962)

TAGEBUCH DER REISE VON CHIENG MAI NACH YÜNNAN, 1921–1922

BRIEFWECHSEL MIT C. S. SARGENT, UNIVERSITY OF WASHINGTON, JOHANNES SCHUBERT UND ROBERT KOC

ÖSTERREICHISCHE AKADEMIE DER WISSENSCHAFTEN
PHILOSOPHISCH-HISTORISCHE KLASSE
SITZUNGSBERICHTE, 757. BAND

BEITRÄGE ZUR KULTUR- UND GEISTESGESCHICHTE ASIENS

Nr. 55

ÖSTERREICHISCHE AKADEMIE DER WISSENSCHAFTEN
PHILOSOPHISCH-HISTORISCHE KLASSE
SITZUNGSBERICHTE, 757. BAND

Hartmut Walravens (Hg.)

Joseph Franz Rock

(1884–1962)

Tagebuch der Reise von Chieng Mai nach Yünnan, 1921–1922

Briefwechsel mit C. S. Sargent, University of Washington, Johannes Schubert und Robert Koc

Verlag der
Österreichischen Akademie
der Wissenschaften

ÖAW

Wien 2007

Vorgelegt von w. M. ERNST STEINKELLNER
in der Sitzung am 15. Dezember 2006

Die verwendete Papiersorte ist aus chlorfrei gebleichtem Zellstoff hergestellt, frei von säurebildenden Bestandteilen und alterungsbeständig.

ISBN 978-3-7001-3830-3

Druck: Börsedruck Ges.m.b.H., A-1230 Wien

http://hw.oeaw.ac.at/3830-3
http://verlag.oeaw.ac.at

INHALT

ABKÜRZUNGEN

CAJ	Central Asiatic journal
HJAS	Harvard journal of Asiatic Studies
MIO	Mitteilungen des Instituts für Orientforschung
MS	Monumenta serica
NGM	National Geographic Magazine
NOAG	Nachrichten der Gesellschaft für Natur- und Völkerkunde Ostasiens
OE	Oriens extremus
OLZ	Orientalistische Literaturzeitung
ZAS	Zentralasiatische Studien

EINLEITUNG

Joseph F. Rock (1884-1962) hat in den letzten Jahren verstärkte Aufmerksamkeit gefunden. Dies hängt teils mit einem wachsenden Interesse an den Na-hsi (Pinyin-Transkription: Naxi, Rock schreibt: Na-khi) und ihren Bilderhandschriften zusammen, aber auch mit der beeindruckenden Qualität und Quantität von Rocks photographischen Aufnahmen von seinen Reisen. Für die Botaniker und Ornithologen war Rocks Name ohnehin ein Begriff. Etwa 300 Pflanzenarten sind nach ihm benannt. Neuerdings wird Rock auch an seinem langjährigen Wirkungsort Li-chiang (Pinyin: Lijiang) wieder geschätzt, und es dürfte dort kaum einen Touristen geben, der seinen Namen nicht gehört und sein ehemaliges Wohnhaus nicht gesehen hat. Es mag daher hier genügen, auf die wichtigste Literatur hinzuweisen:

Sutton, Stephanne B.: *In China's border provinces: The turbulent career of Joseph Rock, botanist-explorer.* New York: Hastings House 1974. 334 S.

Walravens, Hartmut: Joseph Franz Rock (1884-1962). Sammler und Forscher. Eine Übersicht. *Oriens extremus.* 38.1995, 209-237

Aris, Michael; with the assistance of Patrick Booz and contributions by S. B. Sutton and Jeffrey Wagner: *Lamas, princes, and brigands. Joseph Rock's photographs of the Tibetan borderlands of China.* New York: China House Gallery 1992. 141 S.

Joseph Franz Rock (1884–1962): *Berichte, Briefe und Dokumente des Botanikers, Sinologen und Nakhi-Forschers.* Mit einem Schriftenverzeichnis. Hrsg. von H. Walravens. Stuttgart: Franz Steiner Verlag 2002. 452 S. (VOHD Supplement. 36.)

Joseph Franz Rock: *Expedition zum Amnye Machhen in Südwest-China im Jahre 1926.* Im Spiegel von Briefen und Tagebüchern. Hrsg. von H. Walravens. Wiesbaden: Harrassowitz 2003. 237 S. (Orientalistik Bibliographien und Dokumentationen.19.)

Edwards, Mike
Our man in China [1922–1935]: Joseph Rock.
NGM 191.1997:1, 62-82
Mit wichtigem Bildmaterial und einer Kartenskizze (S. 69) der Reisen Rocks in China.

Todd, A. L.
Joseph Francis Charles Rock (1884-1962).
Bulletin, Hunt Institute for Botanical Documentation 14.2002:1, S. 4-5, 11

Mayhew, Bradley; Korina Miller, Alex English
South-West China. [Reiseführer.]
Melbourne, Oakland, London, Paris: Lonely Planet Publ. 2002, 362

Harris, Paul B.
A King in China.
Köln: People and Places 2003. Video

Kleinhaus, G.
Moderne Pflanzenjäger in China.
Garten Praxis 2003:11, S. 20-26

Wang Ta-wei 王大衛
Hsün-chao t'ien-t'ang 尋找天堂 / A way to paradise. Retracing Joseph F. Rock's 27-year adventurous journey in China
Peking: Chung-kuo wen-hua ch'u-pan-she 2003. 287 S.
Persönlich gefärbte Darstellung auf der Basis insbesondere der NGM-Artikel Rocks. Mit zahlreichen Fotos von Rock, aber auch anderen Quellen.

Palmer, B.
Joseph Rock: some musings on a renaissance man.
Journal of the American Rhododendron Society 58.2004, 71-73

Wei Chien 為劍 (Hrsg.):
Lo-k'o yü hsiang-ko-li-la 洛克與香格里拉 / The legend of Shangri-la.
Shen-chen: Hai-t'ien ch'u-pan-she 2004. 283 S.
Das Buch handelt großenteils von Rock und seiner Rolle bei der Entstehung des Shangri-la-Mythus. Mit Übersetzung von Texten von Rock und P. Goullart.

Selections from the Naxi Manuscript Collection: Joseph Rock
http://international.loc.gov/intldl/naxihtml/rock.html
2 Seiten

Joseph Francis Rock Collection, 1922-1929
Record Unit 7287, Smithsonian Institution Archives
http://www.si.edu/archives/archives/findingaids/FARU7287.htm

Rock, Joseph Francis Charles, 1884-1962. Papers of Joseph Francis Charles Rock, 1922-1962. A finding aid
Arnold Arboretum Archives, Jamaica Plain, Harvard University, Cambridge, MA 02138
http://oasis.harvard.edu/html.ajp00007.html
26 S.

Joseph Franz Rock: *Briefwechsel mit Egbert H. Walker, 1938-1961.* Hrsg. von . Walravens.
Wien: Österreichische Akademie der Wissenschaften 2005. 328 S. (Österreichische Akademie der Wissenschaften. Philosophisch-historische Klasse. Sitzungsberichte. 738.)

Der vorliegende Band dokumentiert in unterschiedlicher Weise Rocks Leben und Werk:

– Die Zeitungsberichte zeigen die Rezeption seiner Arbeit in den USA
– Der Briefwechsel mit dem Gelehrtenkreis an der University of Washington belegt eine spätere Schaffensphase Rocks, die insbesondere der Ausarbeitung des Na-hsi-Wörterbuchs galt. Zugleich wird deutlich, daß Rock, entgegen einer Äußerung seiner Biographin, durchaus Kollegen mit dem Vornamen anredete, so Franz Michael, Sinologen und Historiker. Auch zu dem Tibetologen Turrell Wylie scheint das Verhältnis etwas persönlicher gewesen zu sein. Vielleicht wäre Rock in Seattle heimisch geworden, aber das Wetter verleidete ihm längere Aufenthalte. Lägen Rocks Briefe an die Adressaten vor, erführen wir mehr über Rock selbst; so beleuchten die meisten Details naturgemäß Rocks Korrespondenten.
– Für die Kenntnis seines persönlichen Handelns und Fühlens ist die Korrespondenz mit seinem Neffen Robert Koc aufschlußreich. Rock scheint sich damit abgefunden zu haben, daß sein Neffe Hans ein schwieriger Charakter war, über den er sich verschiedentlich ärgerte, und daß Robert, den er gern in seine Arbeit eingebunden hätte, dieser Aufgabe nicht gewachsen war, jedenfalls dem ungeduldigen Onkel nur schwer Dinge recht machen konnte. So verkörperte Robert das ihm verbliebene Stückchen Familie und die Bindung an Wien. Er nahm Robert mit in seine Ferienquartiere, wo dieser ihm als Gesellschafter und Kopist diente. Er war verständnisvoll Roberts dichterischen Versuchen gegenüber, die er im Falle einer deutschen poetischen Wiedergabe der Yu-vu-Gesänge aktiv unterstützte, mußte allerdings auch hier erleben, daß Robert nicht die erwartete Tatkraft besaß.
Immerhin bietet der Briefwechsel eine Menge Details, die Rock mit seinen ganz persönlichen Gefühlen zeigen.

– Der Briefwechsel mit C. S. Sargent ergänzt die schon in der Darstellung der Amnye Machhen-Expedition 1926 gebotenen Proben und gibt weitere Details zum Verlauf der Expedition.
– Die Korrepondenz mit dem Bibliothekar und Tibetologen Schubert begann, als Rocks eigene Kenntnisse der tibetischen Sprache noch sehr gering waren. Sie zeigt Rocks Bereitschaft zur Teamarbeit, beleuchtet die schwierige Situation in Deutschland zur Nachkriegszeit und kommt schließlich zum Ende, da Schuberts Übersetzungen von dem italienischen Tibetologen Giuseppe Tucci als nicht adäquat angesehen wurden und Schubert inzwischen als Professor Karriere in der DDR gemacht hatte und mit anderen Arbeiten überhäuft war.
– Das Tagebuch 1921-1923 berichtet über Rocks Reise von Chieng Mai nach Li-chiang, eine Reise, die insbesondere der Suche nach einer gegen gewisse Schädlinge resistenten Kastanienart (blight-resistant chestnut) diente. Der Bericht wird in einer von Rock selbst überarbeiteten und gekürzten, allerdings auch mit korrigierten Namenschreibungen versehenen Fassung geboten. Von besonderem Interesse ist der Bericht, da Rock teils der selben Route folgte wie Major Davies[1], so daß hier zwei Beschreibungen in nicht großem zeitlichem Abstand verglichen werden können.

Einige Einzelheiten verdienen schon hier herausgestellt zu werden, die im Briefwechsel mit dem Neffen Robert Koc erwähnt werden und für Rocks Biographie und seine Einstellung von Bedeutung sind.

Biographisches

Rock hat sich selten über seine Jugend geäußert und war mit Mitteilungen über sein Leben in Österreich sparsam, da er seine autodidaktische Bildung nicht offenkundig machen wollte. Hier sind zwei Stellen, die seine eigene Einschätzung zeigen:

> Meine Jugend war nicht erfreulich wenn ich da zurückdenke graut es mich. Wenn man mit 5 Jahren keine Mutter mehr hat ist das schlecht bestellt mit der Erziehung eines Kindes. Ich könnte Dir viel erzählen von dem Du nichts weisst. (An Koc, 15. 9. 1962)
> Der Anfang war nicht gut. Ich erinnere mich auf das erste Begegnen mit ihm [i.e. dem Stiefbruder]. Es war knapp nach dem Tode oder war es noch vor dem Tode meiner Mutter, ja es war vor dem Tode der Mutter denn ich war 5 Jahre alt. Mein Vater und ich fuhren nach Brunn a/S und besuchten einen grossen jungen Mann der damals 24 Jahre alt war. Wir waren in

1 H. R. Davies: *Yünnan. The link between India and the Yangtze.* Cambridge Univ. Press 1909. XII,431 S., Klapptab., Kt.

> einem Garten wo er als Gärtner angestellt war und ich ging zwischen ihm und Vater den Garten auf und ab. Wer er war oder was mein Vater mit ihm sprach wusste ich nicht. Nur nach einigen Jahren nachher erfuhr ich dass er mein Halbbruder war. Er kam einmal am 19ten März nach Wien und dekorierte den kleinen Altar mit weissen Kamellien zu meinem Namenstag. Dann später kamen andere Zeiten die besser vergessen sind und von denen ich Dir einmal erzählte. So ist dieses Kapitel geschlossen, es hat lange gedauert. Für Euch ist es nicht das Letzte und jetzt bin ich an der Reihe und hoffentlich wird es nicht solange dauern. (An Koc, 7. 6. 1961)

Persönliches Leben

Rock galt als guter Erzähler, und nach Monaten des einfachen Lebens auf Forschungsreisen oder in Li-chiang lebte er in Gesellschaft auf:

> Alle meine Freunde fordern mich auf eine Biographie zu schreiben, und jetzt werde ich es auch thun sobald ich nach Honolulu komme, ich nehme mir keine andere Arbeit mit. (An Koc, 26. 12. 1952)

Was hier *Biographie* genannt wird, war in seinen Augen allerdings eher ein *funny book*, wie er es bei anderen Gelegenheiten nannte – eine populäre Beschreibung seiner Reisen mit dem Akzent auf dem Anekdotischen. Sicherlich entbehrt die folgende Feststellung nicht der Begründung:

> Ich glaube am besten wird es sein ich widme mich nicht mehr wissenschaftlichen Arbeiten die nichts einbringen, sondern schreibe für das Publikum. Ich habe manches erzählt von meinen Erfahrungen und alles sagt „ja warum schreiben Sie das nicht wieder, man würde sich reissen um das alles zu veröffentlichen, sie könnten damit viel Geld verdienen". Und nun will ich das thun, aber nicht hier in diesem Lande. Alles ist furchtbar teuer. (An Koc, 11. 1. 1953)

Daß in der Praxis nichts aus diesem verschiedentlich ganz konkret angesprochenen Projekt wurde, hat sicherlich mit der Absicht zu tun, seine wissenschaftlich relevanten Materialien mit Priorität zu bearbeiten und zu veröffentlichen. So locker Rock allerdings in Gesellschaft plaudern konnte, wir erfahren aus der früheren Korrespondenz mit der National Geographic Society, daß es Rock nicht leicht fiel, nach dem Publikumsgeschmack zu schreiben, und nach Feststellung seiner Biographin brachte Rock die Redakteure des *National Geographic Magazine* zum Schwitzen.

Besonders aufschlußreich ist Rocks Mitteilung über seine Reiselust, die schon eher zwanghaft war:

> Ich bin Tod müde. Ich muss aufrichtig sagen, es gefällt mir nicht. Oder bin ich rastlos?, weiter, immer wieder weiter, Zerstreuung ist das Einzige dass [!] mich am Leben hält. Aber wohin? Ich möchte am liebsten morgen nach Zürich fliegen und von dort nach Meran gehen um in Frieden mit Dir zu sein und zu schreiben. (An Koc, 11. 1. 1953)

Er konnte sich in der Tat einen Weltbürger nennen, und es ist erstaunlich, wie er, häufig von seiner Bibliothek getrennt und nur mit den eigenen Notizen, so vieles schaffen konnte.

> Ich bin traurig und doch fühle ich dass ich fort muss. Es ist ein Drang wie schon aus alter Zeit. In San Francisco erwarten mich Freunde auch in British Columbia, überall wo ich hingehe. Es ist komisch, ich komme mir wirklich wie ein Weltbürger vor und bin es auch. Alles sagt schreiben sie nur ihre Erlebnisse es würde ein grosser Erfolg werden. (An Koc, 12. 3. 1953)

Die traurigen Gedanken, die ihm nach dem Tode einer alten Bekannten kamen, sind leicht nachzuvollziehen:

> Miss Eagle, the 81 year old lady who always used to cook puddings for me in the evening, died suddenly of heart failure on Nov. 13th. She was a nice good old soul, worked hard all her life up to the last. She loved flowers and when she was through cooking, instead of lying down and resting, went on her knees and worked in the garden with her plants and flowers. It makes me feel very sad indeed. It shows how ephemeral human life is, here today, gone tomorrow for good. I wonder where my last day will find me? Time is getting short and really I often think, if I only could go soon, so that I could leave you both some money. (An Koc, 21. 11. 1956)

Auch die gefühlvollen Gedanken beim Sonnenaufgang sind für einen alternden Mann nicht ungewöhnlich:

> Heute morgen glühten die Schneeberge rosa von den Strahlen der aufgehenden Sonne. Ich verfiel in tiefe Gedanken über das Wesshalb [!] und Warum und Woher! Alles und wir selbst sind ein Rätsel dass [!] Niemand wird lösen können. (Darjeeling, an Koc, 8. 12. 1960)

Bemerkenswert sind seine sozialkritischen Äußerungen, gerade auch in Hinsicht auf seine guten Freunde, denen er ja zu großem Dank verpflichtet war. Insofern blieb ihm nur die resignative Schlußfolgerung, daß es in der Welt nun mal so ist:

> Meine Freunde die vor Geld strotzen, thun mir doch leid, die Frau lässt nichts wie Kleider machen, sie denkt nichts, 20 paar Schuhe machen zu lassen. Wenn ich die Armut dieser Flüchtlinge sehe, kleine Kinder in Fetzen gekleidet, nichts zu essen, und wenn ich dann die Frau ansehe, die sich im Spiegel schaut ob jener Stoff besser steht wie ein Anderer, da werde ich traurig. Nun Schluss mit diesen Betrachtungen, das Leben ist halt so. (An Koc, 8. 12. 1960)

Neffen

Rock fühlte sich als Vermittler zwischen seinen ungleichen Neffen, auch wenn er sich durch Hans' Benehmen mehrfach verletzt sah. Er führte indes zahlreiche Entschuldigungsgründe an, wie Militär- und Kriegserlebnisse, den Verlust der Frau und schließlich auch ein familiäres Element – das Erbe seiner Schwester Lina erkannte er in Hans' Verhalten.

> Mein lieber Robert es tut mir leid dass Du mit Deinem Bruder Auseinandersetzungen hast. Er scheint mir doch eine grobe Natur zu haben so wie sein Vater, der Krieg hat ihn wohl auch nicht freundlicher gemacht, und dazu noch der Verlust seiner Frau, so musst Du halt ein Auge zudrücken. Reize ihn nicht das hat keinen Zweck. (An Koc, 12. 10. 1952)
>
> Auch gestern kam ein trauriger Brief von Hans, er spricht von lebensmüde und allem möglichen. Jetzt hat er doch eine „Frau". Er schreibt jetzt hat der Robert Ruhe, als ob du Dich immer beschwert hattest, es war er der immer sagte ich kann mit Robert nicht leben, mit ihm in einem Raum sein. Er ist die ganze Lina, sie hatte immer etwas und war nie zufrieden. Ich werde nie vergessen als [ich] ihren Brief bekam nachdem ich ihr 1000 Dollar schickte; kein Wort des Dankes, sonder[n] „Ja wie lange soll ich denn damit auskommen." Er ist ein Bild (physisch) von ihr und hat auch ihr Gemüt. Solchen Brief kann man nicht, und ich werde ihn auch nicht beantworten. (An Koc, 5. 11. 1958)

Demgegenüber hielt er sich an den weicheren Robert, dessen Briefe ihm einen familiären Bezug gaben. Nachdem er sich damit abgefunden hatte, daß Robert nicht über seinen Schatten springen konnte, akzeptierte er dies und konstatierte, daß man sich jetzt verstehe:

> Deine Briefe erfreuen mich sehr und ich hoffe dass wir bald wieder zusammen sein werden. Hoffentlich kommt wegen der jetzigen Weltlage und Vorfällen nichts dazwischen. (An Koc, 5. 3. 1953)
> Ich fühle ich gehöre nicht in die jetzige Zeit, ich lebe viel in der Vergangenheit. Nach China sehne ich mich nicht mehr, nur hie und da bekomme ich Heimweh. Ich bin ja ganz allein und so sehne ich mich nach Dir. Wir verstehen uns jetzt. (An Koc, 12. 3. 1953)

Amerika

Seiner Wahlheimat Amerika gegenüber war Rock kritisch. Eigentlich hatte er gehofft, seinen Lebensabend in Li-chiang *bei seinen Na-hsi* zu verbringen. Da die politische Entwicklung dies nicht zuließ, schwankte er zwischen Amerika und Europa, hauptsächlich der Lebenskosten wegen. Bei aller Negativentwicklung, die er nicht müde wurde festzustellen, erkannte er doch die positiven Seiten an.

> Man hat mich sehr freundlich empfangen hier [Seattle], aber mir gefällt es nicht. Ich bin allein und traurig. Soll ich nach Hawaii fliegen oder zurück nach Europa? (An Koc, 11. 1. 1953)
> Das Tempo hier ist fürchterlich! Es graut mir, Licht Reklame drehen sich wie im Ringelspiel automatische Menschen öffnen Thüre, laden dich ein! Alles ist künstlich, aber keine *Kunst.* Luxus überall, speziell in den Prachtzügen, herrliche Speisewagen, so etwas gibt es in Europa nicht, alles Diesel Züge, keine Kohlen mehr, alles ist rein und sauber. Kein Fenster braucht man öffnen, alles ist *air conditioned.* 3 Tage und Nächte war ich im Zug und mein Hemd war weiß wie Schnee. Aber ich fliege doch lieber, es ist so viel schneller und man spürt nichts. (An Koc, 1. 1. 1953)

Was ihn in Seattle störte, war freilich das feuchte, kühle Klima, das ihn depressiv machte. Da fühlte er sich in Hawaii wie im Paradies, trotz der wenig inspirierenden Touristenscharen:

> Für mich hat das Leben keinen Zweck mehr. Mein Heim in China ist weg. Honolulu hat sich sehr verändert und ist nicht mehr was es war. Voll von Touristen welche halb nackt mit übergrossen Bäuchen am Strande spazieren gehen, in unterhosen [!]. Es ist abscheulich. (An Koc, 3. 3. 1953)
> Gestern abends um 8 Uhr flog ich mit schwerem Herzen von Honolulu ab und kam heute morgen um 6.30 in San Francisco an. Was für ein Unterschied. In Honolulu schöner blauer Himmel, blühende Bäume, Palmen, Berge die in Pastellfarben wie verschleiert aussahen, das saftige Grün der Wälder, die bunten Vögel die jeden morgen [!] vor meinem Fenster auf *Cassia* Bäumen in voller Blüte sangen, und hier ein grauer

> Himmel, keine Bäume, nur mit Wiesen bedeckte Hügel, Häuser die wie Bienenkisten aussehen so nahe aneinander gereiht. Mir wurde buchstäblich übel als ich das heute morgens sah! Bin vom Paradies auf die Erde gekommen! Meine alten Freunde in Honolulu wollten alle dass ich dableibe, mir war das Scheiden sehr schwer und ich musste schlucken um das Weh zu erdrucken. (An Koc, 18. 3. 1953)

Aber auch aus dem *Paradies* trieb es ihn fort:

> Ich hatte sollen von Hawaii nicht fort, aber dieses Jahr war es später wie letztes, 15 Tage später. Auch das Paradies wird einem langweilig, aber ich arbeitete doch und beschrieb eine neue Lobelia nahezu während der letzten Minuten meines Aufenthaltes. (An Koc, 14. 4. 1957)

Dank Loy und Lester Marks, seiner Freunde in Honolulu, konnte er schließlich seine letzten Jahre von Finanzsorgen unbeschwert, in Hawaii leben:

> Meine Freunde werden mir eine Einsiedelei am Bergabhang bauen und da will ich dann wohnen. (An Koc, Febr. 1957)
> Ich habe meinen Freunden in Boston geschrieben und ihnen definitive gesagt dass ich ihr Angebot bei ihnen zu wohnen angenommen habe und es mich freuen wird meine letzten Tage in Frieden und Arbeit verbringen zu können. (An Koc, 21. 4. 1957)

Deutschland
Deutschland hat Rock nur gelegentlich besucht, z.B. München auf der Durchreise. Erst der Verkauf von Na-hsi-Handschriften an die Staatsbibliothek in Marburg und seine Einbindung in das Projekt der Katalogisierung der orientalischen Handschriften, brachte ihn etwas länger nach Deutschland. Die ersten Eindrücke vom Kurhotel in Bad Nauheim waren freilich niederschmetternd:

> Hier sieht man hässliche sehr fette dicke Frauen die Körper wie Fässer haben und auf entweder sehr dicken or dünnen Stelzen gehen. Die Männer sehen grob und dick aus mit furchtbaren Bäuchen vom Saufen und Überessen. Sie machen hier die Kur durch. Es würde mir auf die Dauer unmöglich sein hier zu leben und ich möchte nur raus und sobald wie möglich. (An Koc, 9. 5. 1960)

Freundlichere Worte fand er indes über das hübsche Marburg, über Dr. Wolfgang Voigt, den Leiter des Katalogisierungsprojektes, den Neurochirurgen in Gießen, der ihn operierte und manche andere.

Veröffentlichungen

Rock versuchte Robert Koc zu einer schönen Veröffentlichung zu verhelfen – etwas ganz Wichtiges für jemanden, der mit dem Gedanken umgeht, als Schriftsteller zu arbeiten. Aber alle gebotene Hilfestellung erwies sich als nicht ausreichend, und von Koc sind aus späterer Zeit nur ganz wenige Seiten umfassende Gedichtpublikationen bekannt (in der Österreichischen Nationalbibliothek):

> After thinking the matter over I have come to the conclusion that it would be a very good idea to publish a comprehensive work on the Na-khi and their literature religion etc. A kleines Büchlein does not seem to satisfy me, „wenn schon denn schon". Why not get up a historic account from my *Na-khi Kingdom*, a description of the country (a geographic account) types of the people, their religion with special to their poetic comparisons with nature, etc. I shall be very glad to so supply you with fine photographs, unpublished ones, original Na-khi texts to show their methods of writing, their way of combining symbols, their ideographic and phonetic uses.
> This would make a book about at least 350 pages with say 50 Plates, and a map of the country. The Li-chiang map from my Nakhi kingdom could be copied. I think Ullstein might be the proper *Verleger*. I would write a preface and all in all a very decent and instructive book could be the result. If you want I could write to Ullstein when the time comes or to some other *Verleger* you may have in mind, but I don't believe in the *Büchlein* business. (An Koc, 13. 2. 55)

Wissenschaftliche Beziehungen

Rocks Beziehungen zu Johannes Schubert wurden schon kommentiert:

> Von Schubert höhre [!] ich nichts mehr, er scheniert sich wahrscheinlich denn Tucci wollte seine Übersetzungen nicht veröffentlichen. (An Koc, 28. 8. 1958)

Rock hatte gehofft, daß das Istituto italiano per il Medio ed Estremo Oriente die ihm geschenkten Na-hsi-Handschriften behalten würde, und fühlte sich etwas düpiert, als Tucci sie gleich weiterverkaufte. Freilich konnte er nichts dagegen sagen, da ja so immerhin die Veröffentlichung des Wörterbuchs gesichert wurde:

> Ich schickte Tucci 500 Na-khi mss. um die Kosten teilweise zu bestreiten. Er verkaufte sie um $ 10,000 Dollar. Wenn ich das gewusst hätte, hätte ich sie selbst verkauft und ihm einen Teil davon gegeben. Ich besitze noch

> 229 Na-khi mss. sie sind die Besten, schön illuminiert. So geht es im Leben. (An Koc, 24. 2. 1962)

Bemerkenswert ist der Hinweis, daß die Transkription der Tagebücher, wovon Rock einige Bände selbst noch korrigiert und überprüft hat, auf Staatskosten erfolgte (vermutlich des Außenministeriums, vielleicht auch des CIA). Bislang war der Hintergrund dieser Aktion, die durch Rocks Tod dann auch nicht zuende geführt wurde, unklar:

> Ich ediere jetzt den 8ten Band meiner Tagebücher welche eine Sekretärin (auf Kosten der Regierung) tipiert. Sie werden in fünf Kopien tipiert, drei davon für mich, mit elektrischer Schreibmaschine getiept [!]. (An Koc, Febr. 1962?)

Bibliothek

Rock war sehr stolz auf seine Bibliothek, die für ihn in der Tat eine ideale Arbeitsbibliothek war, und benutzte sie als Einsatz, um sich damit gewissermaßen in Forschungspositionen „einzukaufen" – er stellte den Institutionen seine Bibliothek letztwillig in Aussicht und bekräftigte dies durch Kodizille zu seinem Testament. Er nahm es der Harvard-Universität aber ausgesprochen übel, daß sie wegen einer größeren Duplikationsrate nur an Einzelstücken seiner Sammlung interessiert war:

> As to the library, as I hear nothing more from Harvard, and as they only want to buy the very cream I am not interested, and they want everything at prices that prevailed in 1920 in Peking. Try and get anything now. The damn silly missionaries feeling so sorry for the poor Chinese students who were red long before the country ever went red, insisted supplying funds for Yenching University on condition that the money wont be used for subversive activities, and inspite of stating that their curriculum included Marxism, Leninism and Stalinism, the trustees sent them the money regularly till the State Department stepped in and prohibited them to send any more money to the various anti American Red Universities. Yes I find I did not finish my sentence as to my library. I have arranged to loan it for the time being to the Far Eastern Institute of the University of Washington in Seattle. They also want to buy it. (An Merrill, 20. 4. 1951)

Die Bibliothek wurde schließlich von der University of Washington erworben, der Rock sie vorab als Leihgabe zur Verfügung gestellt hatte:

> Meine Bücher sind alle schön angekommen und sehr gut aufgehoben. Alle bewundern diese Bibliothek denn sie besteht aus Büchern die nirgends zu haben sind. Man hat mir ein schönes office eingeräumt wo ich arbeiten kann. Alle sind sehr liebenswürdig und thun alles was sie können um mich hier zu behalten.
> Doch ich bin traurig und als ich alles auspackte und durchsah da kamen alle die guten Zeiten vor meinen müden Augen und ich wurde sehr traurig. (An Koc, 11. 1. 1953)

Dank gebührt allen Institutionen, die Material für diesen Band zur Verfügung gestellt haben: der Staatsbibliothek zu Berlin Preußischer Kulturbesitz (Reisebericht 1921-1923 sowie Briefe von Johannes Schubert), der Hamilton Library, University of Hawaii at Manoa (Zeitungsberichte), dem Hunt Institute of Botanical Documentation, Carnegie-Mellon University, Pittsburgh (Briefwechsel mit University of Washington, Robert Koc und Johannes Schubert) sowie dem Arnold Arboretum, Harvard University (Briefwechsel mit C. S. Sargent).

Nach Abschluß des Manuskripts hatte der Herausgeber Gelegenheit, K'un-ming, Ta-li, Li-chiang, den Yü-lung hsüeh-shan und Rocks ehemaliges Wohnhaus (inzwischen Museum) in Yü-hu zu besuchen und eine (teils kommerziell motivierte) Renaissance des Na-hsi-Schamanismus (Lehre der dto-mba, chin.: tung-pa) wie auch eien Hochschätzung Rocks zu beobachten.

ZEITUNGSBERICHTE ÜBER ROCK

EXPLORING FOR PLANTS

After an eleven months' exploring trip through the jungle of Siam, Burma, Assam and Bengal, Professor J. F. Rock, plant explorer for the United States Department of Agriculture, exhibited recently in a stereopticon lecture before the Botanical Society of Washington the first photographs of the tree *Taraktogenos Kurzii*, which is the source of the chaulmoogra oil, known to a greater or lesser extent for a long time as a remedy for leprosy in its native habitat.

The immediate result of Professor Rock's enterprise was the sending of enough of the seeds to the Department of Agriculture to assure the establishment of a considerable plantation under American jurisdiction. The Hawaiian Government has set aside 100 acres of ground for the purpose. The trees, while they attain great size and age, come into bearing, it is believed, within eight years after the planting of the seed.

Professor Rock, while in Asia, discovered in old Buddhist histories a legend of a Burmese king voluntarily exiled for leprosy about 1,000 years ago who cured himself with the oil and likewise effected the cure of a beautiful young woman whom he afterwards married, founding a dynasty. Because of the inaccessibility of the forests producing the oil, however, no European people made any effort to study and apply the remedy until 1856, and then the British scientists who undertook the work distributed seeds of an entirely different tree, which have not the same curative properties, and it was not until 1899 that the mistake was discovered.

In 1902 investigations were begun by Dr. Frederick B. Power, then Director of the Wellcome Chemical Research Laboratories of London and now engaged in research work in the Bureau of Chemistry, Department of Agriculture. Dr. Power and his co-workers isolated some new physiologically active acids from chaulmoogra oil, and the same acids were obtained from the oil of certain species of *Hydnocarpus*. He also prepared the ethyl ester of these acids, in which form the remedy can be injected into the muscles, and a sufficient quantity can thus be introduced into the system to effect a cure. It is in this form that the oil has been used with the Hawaiian lepers, 200 of whom have apparently been cured.

Professor Rock said that the news of these cures had had the result that persons who had previously concealed their disease came forward and acknowledged being leprosy victims in order to receive treatment. Realizing that worldwide circulation of the fact of these cures would result in a heavy demand for the meagre supply of *Taraktogenos Kurzii* seeds, certain interests in Hawaii arranged with Professor Bock [!] to obtain seeds, from their native source for propagation. The first effort was made in 1919, while Professor Rock was on his way to Java regarding reforestation problems for the Hawaiian Sugar Planters' Experiment Station. Because of travel difficulties he was unable to penetrate to the actual source of supply at that time, but sent seeds of the Siamese species.[1]

AGRICULTURAL EXPLORER REPORTS FINDING CHAUL-MOOGRA TREE IN CHINA

Seeds of the tree producing chaulmoogra oil, which recently has been found to cure leprosy, are being brought to Washington by J. F. Rock, agricultural explorer, who is expected to arrive next week. Cables have been received by officials at the Department of Agriculture, stating Mr. Rock has sailed from Honolulu for San Francisco.

Mr. Rock has been in the northern part of China three years, looking for the seeds of the tree. Lately, word was received here of his success. He will also bring specimens of rare plants and flowers to the horticultural bureau.

Chaulmoogra oil has worked many apparently miraculous cures and has been used with signal success in the fight against leprosy in the West Indies and the Canal Zone. Hence officials here await Mr. Rock's return with much interest. It is believed that the tree may be planted in government reservations and the oil used to check the spread of leprosy.[2]

---

1 *New York Times* 14.8.1921

2 *The Washington Post* 18,.6.1924

SCIENTIST HAS CLOSE CALLS IN WILD CHINA

Joseph F. Rock, of Washington, writes of experiences in Tibet seeking reason for American corn there prior to 1492 – also in quest of blightless chestnut tree for importation.

Joseph F. Rock, Washington scientist, who found the chaulmoogra tree, whose oil is used to treat leprosy, now figures in a series of dangerous adventures and narrow escapes from bandits in China.

Mr. Rock went to the remote Burma-Tibet frontiers of China as head of a National Geographic Society expedition which hoped to bring back numerous plants for American introduction, study the mystery of how American native corn was transplanted there before 1492 and to seek a blight-resisting chestnut tree.

From Likiangfu, China, he writes, under date of June 11:

„ I just returned from an exciting trip to the Lachiming salt mines.

Money is dreadfully scarce; paper is useless, as no one will take it, especially up here with all the tribespeople.

I decided to risk it and make a dash for Lachiming, in the heart of the wildest mountain ranges imaginable, to get funds to carry on the work and to see me through to wherever God knows where I will emerge.

I had four Chinese drafts on the Lachiming salt mines from the Tali salt office.

At first the Likiang official refused to give me an escort and said he feared for my life and would not let me go. Finally I said I must have money and I would risk it. He gave me two opium-sodden disreputable wretches of soldiers as escorts. They were simply a nuisance and an expense. They were, of course, only to go as far as the next magistrate, which is at Chien Chican [!].

There I got four soldiers as far as Lauping. Lauping is not on any map, as it is in unsurveyed territory. At Lauping, which is five days west from Likiang, I got five soldiers, and thence we started out through the wildest country imaginable to Lachiming. Dense forests, deep ravines, high mountain passes, the loveliest and weirdest imaginable. I was not afraid going to Lachiming, but I feared the return trip, with about $ 4,000 in silver in the trunks.

I took seven of my Moso men along and the five soldiers from Lauping besides myself, so we were thirteen persons in all. We reached Lachiming safely and had no trouble to get the money. I had to go at the end of the month or arrange it so that I would reach the mines on the 1st, which I did.

It is especially at that time that the roads are watched by the brigands, as usually on such dates money is dispatched. I was told at Talifu to arrange to be at the mines about the 1st of the month. It is exceedingly difficult to keep things secret. We had no sooner arrived in Lachiming than everybody knew why we came. I did not stay longer than I had to to prevent the news traveling faster than we did, so I left the next morning with an additional escort from Lachiming. Thanks to the gods I reached Likiang without mishap.

I put four scoundrels into the jail in Lauping, where they got a beating, as they tried to foul me, but I got ahead of them, and so here I am, safe with the cash with which to carry on.

On my trip up the Yangtze and the Yundodye river toward Chungtien I met the grand lama on the road, who informed me that not a house has been left standing in Chungtien. The whole place has been burned and the Chinese official fled to Likiang, where he is at present. Two thousand Tibetan bandits are at Hsiao Chuntien, five days north of Likiang. What will happen no one knows.

I had a letter from the customs commissioner. He tells of the Tengyueh magistrate having been captured by brigands and a ransom is asked of $ 20,000. It is certainly a problem how to get out of this country without losing one's life or all his belongings. However, our collecting goes on apace as if it were the most peaceful place in the world.“[3]

CREATION PICTURE STORY IN MANUSCRIPTS BROUGHT HERE FROM CHINA BY EXPLORER

Curious religious books of Nashti [!] tribes tell version similar to that found in present Bible. Big collections of bird and animal life made; rust-resisting chestnut will aid industry.

Returning from two years of life with the little known aboriginal Nashti [!] tribe of the Yun-nan province of China, Joseph F. Rock, National Geographic Society explorer, has just brought here original copies, as well as the first known translations, of this tribe’s religious

3 *The Washington Post* 1.9.1923

books, which relates a story of the creation similar in many respects to that told in the Bible.

The Nashti is one of about 30 aboriginal non-Chinese tribes living in the province with many of which he came in contact, Mr. Rock said yesterday. He was the only white man on the expedition sent by the society into the province, and had 27 Nashti assistants. One or two of these acted as the explorer's interpreters, conversing with him in Chinese.

The religious book which tells the story of the creation is about 10 inches long and three inches wide, has a red cover with a painted design, and contains several pages of manuscript inscribed in picture writing resembling Egyptian hieroglyphics. It tells of the first man and woman, of the creation of the lands, oceans, and rivers, and relates other events of the beginning of the world, as known to this tribe.

This book and the others he has are believed by Mr. Rock to have been handed down in the tribe from generation to generation, though he does not know their ages.

„Only the priests can read these manuscripts, and as I could not understand the Nashti language, a priest read each section of the picture writing to my interpreter, who, in turn, translated it for me into Chinese“, the explorer said. „The tribes also has another form of writing, which appears to be more advanced, and has some Chinese characters in it, but the priest would not read for me anything in this other writing.“

In addition to these manuscripts Mr. Rock has brought with him hundreds of photographs of the weird religious ceremonies and singular customs of the Nashti tribe, whose mode of life has heretofore been virtually unknown to the civilized world. These are to be published in the National Geographic Society's magazine. Until 270 years ago, when Chinese for a time invaded the territory of the Nashti and neighboring tribes, these aborigines had never had contact with the rest of China, Mr. Rock asserted. He found the Nashti men and women to be black-haired, with brownish skin and some of them rosy-cheeked.

Thirty seven boxes of plant and bird specimens were also brought back by the explorer. In this collection are 500 mammal and 1,700 bird specimens, some of them believed by Mr. Rock to be unknown here. He has also obtained in Yun-nan the seeds of 493 different kinds of rhododendron, unknown in this country, and has distributed many of these seeds in various parts of the United States and England.

Mr. Rock, who returned to Washington Wednesday, after being away nearly three years, was already widely known before he undertook his last expedition, for his finding in northwestern Burma in 1920 and 1921 of the Chaulmoogra tree, which yields an oil used as a leprosy cure. This tree has since been transplanted in Hawaii.

He left in August, 1921, for his expedition to Yun-nan, in the course of which he made his fourth round-the-world trip. He has been active in foreign scientific study since 1913 and in addition to his work for the National Geographic Society is connected with the bureau of plant industry of the Department of Agriculture.

The department is now engaged in cultivating at its nurseries in Bell, Md., seeds of what may prove to be Dr. Rock's most important discovery in Yun-nan, a chestnut tree which resists blight. Thousands of seedlings of such healthy trees, found by the explorer, are now being grown by the bureau of plant industry, and though they are not nut-bearing trees, will provide excellent timber and bark for tanning.

Discovery of these chestnut trees in China is regarded as of great importance to the United States, which, according to Mr. Rock, may be entirely denuded of their native chestnut trees in 20 years as a result of the blight, for which no remedy has been found. It is believed to be caused by a parasite carried from tree to tree by woodpeckers, and once a tree is affected its branches begin to die, and when the trunk of a tree is attacked it dies within a year.[4]

EXPLORER FROM CHINA BRINGS 60,000 PLANTS

J. F. Rock has 2,200 specimens of animals – found blight-resisting chestnut tree

Washington, July 13 (Associated Press). Thrilling stories of adventure and interesting scientific discoveries have been reported to the National Geographic Society by Joseph F. Rock, discoverer of the chaulmoogra tree, that yields a leprosy-combating oil, and leader of the society's expedition into Yunnan, a vast and little explored province in Southwestern China.

---

4 *The Washington Post* 29.6.1924

Mr. Rock, a wellknown plant authority, has just returned to Washington, and in narrating his experinces tells of finding in Yunnan aboriginal tribes who still practice religious ceremonies that pre-date the introduction of Buddhism, and the discovery of a blight-resisting chestnut tree, which, it is hoped, will aid in restoring the diseased chestnut timber crop in the United States.

The explorer brought back from the wilds of the Chinese province some 1,700 specimens of birds, 500 mammal specimens and more than 60,000 plants. He also came intro possession of books used by the Nashi, or Moso, priests, containing ancient writing in picture form. The books also antedate Buddhism in China and Tibet and include a story of a great flood.

During his adventures Mr. Rock traversed a hitherto unexplored gorge of the Yangtze River, described as much deeper than the Grand Canyon of the Colorado.

The crossing of streams presented one of the most difficult problems of the expedition. Mr. Rock and his twenty-three native assistants and pack animals having to be swung across them on single-strand rope bridges. The party also had narrow escapes from bandits along the China-Tibet frontier.[5]

RIVAL OF GRAND CANYON DISCOVERED IN CHINA

Natives still beyond pale of civilization practice rites antedating Buddhism. 60,000 plant specimens

Thrilling stories of adventure and interesting scientific discoveries have been reported to the National Geographic Society by Joseph F. Rock, discoverer of the chaulmoogra tree, that yields a leprosy-combating oil, and leader of the society's expedition into Yunnan, a vast and little explored province in Southwestern China.

Mr. Rock, a wellknown plant authority, has just returned to this city, and in narrating his experiences tells of finding in Yunnan aboriginal tribes who still practice religious ceremonies that pre-date the introduction of Buddhism, and the discovery of a blight-resisting chestnut tree, which, it is hoped, will aid in restoring the diseased chestnut timber crop in the United States.

---

5 *New York Times* 14.7.1924

The explorer brought back from the wilds of the Chinese province some 1,700 specimens of birds, 500 mammal specimens and more than 60,000 plants. He also came intro possession of books used by the Nashi, or Moso, priests, containing ancient writing in picture form. The books also antedate Buddhism in China and Tibet and include a story of a great flood.

Mr. Rock traversed a hitherto unexplored gorge of the Yangtze River, described as much deeper than the Grand Canyon of the Colorado. The crossing of streams presented one of the most difficult problems of the expedition. Mr. Rock and his 23 native assistants and pack animals having to be swung across them on single-strand rope bridges. The party also had narrow escapes from bandits along the China-Tibet frontier.[6]

PHOTOGRAPHS RITES IN REMOTE CHINA

Dr. Rock takes color pictures of Lamaian ceremonies rarely seen by white men. Aided by Chinese Prince. National Geographic Society sends five other parties to film Old and New World scenes

Special to the New York Times.
Washington, March 14. – Color photographs of the religious ceremonies of Lamaism, which activities have rarely been seen, much less pictured, by white man, are now being made in China for the National Geographic Society, by Dr. Joseph F. Rock, plant explorer.

Dr. Rock's former explorations in Burma resulted in the discovery of the tree that produces chaulmoogra oil, which is used in the treatment of leprosy, and his more recent expedition into Western Yun-nan and Tibet brought to America the blight-proof chestnut tree and discovered the black rhododendron.

The fact that he had been sent on still another photographic expedition into a remote province of China was announced by the society tonight along with the fact that it had sent five other important photographic expeditions into the field.

Dr. Rock is now the guest of a prince in China in the far Chinese hinterland. A letter received from the explorer by the society says he is enjoying the hospitality of this prince, that the latter is cooperating

---

6 *The Washington Post* 20.7.1924

with the photographic expedition in every possible manner and that they have become fast friends.

Prince Choni has placed at Mr. Rock's disposal one of the buildings in the lamasery at his capital, and he is experiencing unusual winter comfort due to the loan by the Prince of a European stove. The explorer has become so well established in this faraway community in the Chinese interior, where few white men have visited, that he is being admitted to the religious ceremonies of Lamaism and is taking photographs.

Because of the unusually colorful costumes used by these Lamas in their ceremonies and the extraordinary opportunity Dr. Rock has to photograph them, he cabled for equipment to make natural color plates of these unusual subjects. With the co-operation of the Chinese Legation in Washington, a consignment of plates with color screens and special lenses was dispatched to him via Peking.

The equipment was specially packed in small boxes within larger ones, so that when the end of the railroad was reached and the boxes were to be transported by coolies or camels, these smaller boxes could be adapted to their packs.

Five other photographic expeditions which the National Geographic Society stated tonight it had sent into the field, are taking pictures in the West Indies, the Balkans, the Russian trans-Caspian sector, Guatemala and the Rocky Mountains. [...][7]

EXPLORER RETURNS WITH RARE PLANTS

Dr. Joseph F. Rock, leader of expedition to Tibetan lands, is home. Valuable data gained

Dr. Joseph F. Rock, famous plant explorer, leader of the National Geographic Society's 1927–1930 expedition to unexplored areas of the China-Tibetan borderland, has just returned to Washington with a noteworthy collection of plant and zoological specimens, and remarkable photographs of lofty, snow-covered mountains, which have been the subjects of much geographic speculation.

Dr. Rock is the man who braved tigers and wild elephants to locate the trees which yield chaulmoogra oil, used in the treatment of leprosy. He found aboriginal tribes in west China practicing pre-

---

7 *New York Times* 15.3.1926

Buddhist religious rites, and obtained for the Library of Congress a complete set of the priceless Tibetan classics from the Buddhist monastery at Choni before it was destroyed last year.

On his last expedition Dr. Rock devoted his first year to exploration of the majestic Konkaling snow peaks northwest of the semi-independent Llam [Lama] kingdom of Nuli [Muli] in the Chinese procince of Sacchwan [Szechwan].

These peaks had not previously been visited by any foreigner. They are situated in the territory infested by Tibetan bandits, and the menace of these bandits has made the area too perilous for exploration. The expedition was conducted with the aid of the Muli king who had friendly relations with the head bandit of this lofty stronghold.

Among the mass of valuable scientific data, and photographs of high geographic interest brought back by Dr. Rock, are about 20,000 sheets of plants he collected. Many of these are rhododendrons not previously collected.

While he was in the field he shipped back seeds of 173 species of rhododendron not previously introduced in America; in addition, many other ornamental plants, both herbaceous and wooded.[8]

„MYSTERY" PEAKS EXPLORED BY ROCK

Strange plants and animals are brought back by leader of Tibet Border Expedition

Special to The New York Times.
Washington, April 9. – Dr. Joseph F. Rock, leader of the National Geographic Society's expedition to unexplored areas of the China-Tibetan borderland, has returned to Washington with a noteworthy collection of plant and zoologcal specimens and remarkable photographs of lofty snow-covered mountains which have been the subjects of much geographic speculation.

Dr. Rock is the man who braved tigers and wild elephants to locate the trees which yield chaulmoogra oil, used in the treatment of leprosy; later found aboriginal tribes in West China practicing pre-Buddhist religious rites, and obtained for the Library of Congress a

8 *The Washington Post* 10.4.1930

set of the Tibetan classics from the Buddhist monastery at Choni before it was destroyed last year.

On his latest expedition Dr. Rock devoted his first year to exploration of the majestic Konkaling snow peaks, northwest of the semi-independent Lama kingdom of Muli in the Chinese Province of Szechwan.

„These peaks had not previously been visited by any foreigner", the National Geographic Society stated today.

„In the Konkaling area he traversed the gorges of the Shou Chu, a river whose mouth only had hitherto been known. For eleven days the expedition followed this river toward its source.

Last year Dr. Rock explored another region of Southwestern Szechwan, where he was the first civilized man to see at close range, and the first to photograph the lofty Minya Konka peak, which has so long been a geographic mystery.

Dr. Rock also explored territory between Muli and Tatsienlu, a large stretch of country hitherto unexplored either from a botanical or geographical standpoint. He followed the gorges of the Yatung River which, in one place, are 12,000 feet deep."[9]

BOTANIST TO LIVE WITH NASHI TRIBE

Hong Kong – Dr. Joseph F. Rock, internationally known botanist, working under the auspices of Harvard University and the United States Department of Agriculture, arrived here from America on July 22 en route to the remote city of Lichiang where he plans to work five years among the Nashi tribe.

Dr. Rock will proceed to Hanoi, Tongking, and thence by train to Yunnanfu, Yunnan Province, where Nashi tribesmen will be awaiting him.

From Yunnanfu he will go to Lichiang, which is 23 days distant by horseback.

Lichiang is situated at the headwaters of the Yangtse River in the extreme northwest corner of Yunnan. Dr. Rock, who visited the Nashi country twice previously, intends to spend the first six months writing two volumes for the National Geographic Magazine, describing his travels and researches in Tibet.

---

9 *The New York Times* 10.4.1930

The remainder of his stay will be devoted to translation into English of interesting Nashi literature and also to the compilation of English and Nashi dictionaries from ancient Nashi writings.

He also plans to conduct a research of plants of the region for introduction into the United States under supervision of the Department of Agriculture.[10]

2 AMERICANS FIRST WHITES TO SCALE HIGH TIBETAN PEAK

25,000-foot Minya Konka is described as pyramid mountain of granite of superb beauty, located in extremely isolated section of China

Minya Konka, Tibetan mountain, which is one of the world's highest peaks outside the Himalayas, is reported to have been scaled by two American explorers, according to news dispatches which have just reached the United States from China. The mountain, which has been seen by only a few white men, was first definitely located, mapped and photographed in 1929 by an expedition of the National Geographic Society under the leadership of Joseph F. Rock.

Many travelers from the East have penetrated the mountain districts of western China and visited Tatsienlu, Szechwan Province, great border gateway to Tibet, but only three or four have ventured farther into an almost unknown region of sky-piercing mountain peaks.

Minya Konka is the dominant peak of the Minya Konka range. It had been seen by only one white man previous to Dr. Rock's visit. Dr. Rock photographed every peak of the range, mapped its jagged slopes and explored them for plants, birds and animals.

He found many peaks in the range rising to an altitude of more than 20,000 feet, and reported to the National Geographic Society that in his opinion Minya Konka itself was close to 25,000 feet high.

Dr. Rock was assisted by guides furnished by a *Living Buddha* of a nearby Lama monastery. The region is so effectively surrounded by lofty mountains that it is cut off for six months of the year by snow-filled passes.

---

10 *The Washington Post* 27.11.1932

In his report to the National Geographic Society, Dr. Rock wrote of his first view of Minya Konka:

„The scenery was superb. Indeed, words fail to describe this marvelous panorama, which until I came no white man had viewed from such close range. Seen from my vantage point, Minya Konka appears as a triangular peak not unlike one of the pyramids of Egypt. A light, transparent veil of clouds here enshrouded the very apex of the peak, allowing the contours to be seen.

The walls of Minya Konka are of a blackish gray rock and are apparently granite, as is the entire Djesi Valley, which extends from the northern end of the range to Tatsienlu.

One of the finest peaks overlooking the head of Riuchi Valley and also the Djesi Pass is an enormous snowy mass of great beauty, Reddomain Solo, 23,000 feet in height. This had been mistaken by the missionaries of Tatsienlu for the Minya Konka, which is, however, visible from the Djesi Pass, but only on perfectly clear days, and even then only a small part of its apex proper can be seen."[11]

SEEKS GRASSES IN TIBET

American botanist prepares for expedition in November

Special Correspondence, The New York Times.

Shanghai, China, June 3. – Dr. Joseph F. Rock, American explorer and botanist, will leave for Northeastern Tibet next November on an hunt for the seeds of wild grasses which will assist in increasing the fodder production of cattle ranges in Utah and other Western States in the United States. He will also search for fodder grasses that will help in stopping soil erosion.

This next trip, which will require at least a year, is being undertaken at the request of the Bureau of Plant Industry of the Department of Agriculture in Washington. The exploring party will consist of Dr. Rock, one other American, an agronomist to be sent from Washington, and from twelve to twenty natives of the grasslands west of the upper reaches of the Yellow River.

Dr. Rock once before traversed the grasslands of Northeastern Tibet. He hurried over them during an expedition lasting from early

11 *The Washington Post* 27.11.1932

1925 to late 1927. On that trip he was in quest of new trees and paid no attention to grasses.

The Northeastern Tibetan grassland areas are inhabited by savage tribesmen, known as the Ngoloks, said to number about 90,000.[12]

AMNE MACHIN

To the editor:
Speaking of the discovery and exploration of the great mountain Amne Machin, on the China-Tibet frontier, in his review of Leonard Clark's *The Marching Wind,* your reviewer Roy Chapman Andrews writes that the undersigned „saw it from a distance of seventy-five miles". As a matter of fact I explored the Amne Machin range for a whole summer in 1926 for the Arnold Arboretum of Harvard. It was our expedition which published the first pictures of Amne Machin and of the Ngolok, the Tibetan tribes that surround it.

The mountain range was not discovered by General Pereira in the Twenties, as the reviewer said. It had been visited by the Russian explorer Roborowski in the Eighteen Seventies, who published a map of the northwestern part of it. The picture taken from the book and reproduced in the review showing a Ngolok chieftain is inaccurate. It is of an ordinary Tibetan nomad. The Ngolok look very different. And as for Amne Machin being higher than Everest, that is a myth.
Joseph F. Rock, Seattle, Wash.[13]

ANCIENT YUNNAN, TUCKED AWAY IN INNER CHINA, BECKONS SAVANT FOR ANOTHER PERIOD OF EXILE

Dr. Joseph Rock leaves Saturday for distant and strange land. Plans to continue studies of language and pictographs

Have you ever desired to „get away from it all?" Yes? Then give a thought to Yunnan, which is in southwest China and is the primitive retreat to which Dr. Joseph F. Rock will return for another period of self-exile.

12 *New York Times* 1.7.1934

13 *New York Times* 16. 1. 1955

Yunnan, and Northwest Yunnan, where the people live like they did when Christ left Nazareth, where they write by drawing pictures like the Egyptians did – in short, a place which Kipling might have had in mind when he begged to be shipped east of Suez.

Dr. Rock should have a Chinese, or, more specifically, a Nakhi accent, after 12 years of studying the religion and language of the Nakhi (pronounced Nashi) peoples, but instead he still betrays his Viennese nativity when talking. And he talks enthusiastically of the fascinating land of the Nakhis and of his plans to write four volumes on their history, religion, literature and written language.

TO LEAVE SATURDAY

The Austro-Chinese-American has been in Washington only a short time. He returned from China last fall and is to leave Saturday for his adopted East. It will be his fifth trip to China and Tibet since 1923.

An explorer, botanist, zoologist and photographer, Dr. Rock has found time during his years of research to master the written Chinese language and translate the difficult Nakhi pictographs. Or perhaps it is more correct to say that the Tibetan scholar has found time to continue his work as an explorer, botanist, ornithologist, zoologist and photographer.

One of the pictographs tells the story of a couple of lovers of long ago whose sad fate is reminiscent of Romeo and Juliet. Its crude pictures and phonetic markings are translated into the activities of a Chinese maiden who is urged by demons to climb a mountain and hang herself.

TALE STILL LIVES

Some time later, her sweetheart finds her lifeless body hanging from a limb. Her soul, though, is still on the scene and tries to convince him he should join her in death. When he demurs, the soul of his departed lover compels him to end his life.

The Nakhis, reading this bitter tale, believe Kamegyumigkian (the woman) and Ndgibuejulapar (the man) live in the clouds above the mountains to this day, and when lovers end their lives, their souls go to join them.

Dr. Rock again will go back to Yunnan as a $ 1-a-year-man in the Department of Agriculture, for which he performs botanical research. On his last visit, which began in 1930 and ended last October, he led a botanical expedition for the University of California and an ornithological expedition for Harvard University.

Although only 50 years old, Dr. Rock doubts whether he will do any more serious exploring in Tibet but intends to locate in Peiping and write his history of the Nakhis, which will be published in London. It is a formidable assignment which he has given himself because the natives do not date their writings. It is fairly certain, however, that their pictographics are upward of 1,500 [!] or 200 years old.

600 VOLUMES IN LIBRARY

The forthcoming volumes also will describe the geography of Eastern Tibet and Western China from Siam to Southwest Mongolia. Jaded Washingtonians looking for new places to go and who are interested in this exotic section of the globe may get further ideas of its character by visiting the Library of Congress and browsing through the 600 volumes of pictographic literature contributed by Dr. Rock when he returned from an earlier expedition several years ago. But they will have to do their own translating. A little practice first on deciphering Egyptian hieroglyphics may be helpful.

That Yunnan has its virtues is demonstrated by the fact that two Nakhi boys who accompanied Dr. Rock back to the United States in 1927 were willing enough to return, after a two-month dose of automobiles, subways, pie a la mode and elevators.

They came to study taxidermy, the one, and photography, the other. Now the boys are Dr. Rock's assistants.[14]

---

14 *Washington Post* 21. 3. 1934

BRIEFWECHSEL MIT GELEHRTEN AN DER UNIVERSITY OF WASHINGTON, SEATTLE

Übersicht

1.	27. 10. 1950	G. Taylor
2.	18. 1. 1951	Rock an G. Taylor
3.	28. 3. 1951	Rock an H. Wilhelm
4.	10. 4. 1951	H. Wilhelm
5.	12. 4. 1951	Rock an G. Taylor
6.	19. 4. 1951	Rock an H. Wilhelm
7.	20. 4. 1951	Rock an Brew
8.	20. 4. 1951	Rock an Brew
9.	20. 4. 1951	Rock an Merrill
10.	20. 4. 1951	Rock an Merrill
11.	20. 4. 1951	Rock an G. Taylor
12.	22. 5. 1951	Rock an G. Taylor
13.	20. 6. 1951	F. Michael
14.	18. 7. 1951	Annie Perry
15.	18. 7. 1951	H. Wilhelm
16.	21. 7. 1951	G. Taylor
17.	17. 8. 1951	H. Wilhelm
18.	18. 10. 1951	H. Wilhelm
19.	13. 2. 1952	R. Krader
20.	3. 3. 1952	F. Michael
21.	20. 5. 1952	F. Michael
22.	11. 12. 1952	Rock an L. Eisenhardt
23.	21. 7. 1953	F. Michael
24.	26. 4. 1955	G. Taylor
25.	26. 4. 1955	G. Taylor an Tucci
26.	26. 4. 1956	F. Michael
27.	6. 4. 1956	F. Michael an Belden
28.	30. 11. 1956	Rock an F. Michael
29.	3. 12. 1956	F. Michael
30.	24. 4. 1957	G. Taylor
31.	17. 10. 1957	F. Michael
32.	13. 11. 1957	C. Leo Hitchcock
33.	25. 4. 1958	F. Michael
34.	3. 6. 1958	F. Michael
35.	2. 7. 1958	F. Michael
36.	24. 8. 1958	T. Wylie
37.	12. 9. 1958	F. Michael
38.	16. 2. 1959	T. Wylie

39.	13. 3. 1959	F. Michael
40.	3. 6. 1959	F. Michael
41.	10. 7. 1959	F. Michael
42.	10. 8. 1959	T. Wylie
43.	11. 8. 1959	F. Michael
44.	4. 9. 1959	T. Wylie
45.	21. 10. 1959	F. Michael
46.	28. 10. [?1959]	T. Wylie
47.	13. 11. 1959	F. Michael
48.	30. 11. 1959	F. Michael
49.	11. 1. 1960	F. + Anita Michael
50.	22. 1. 1960	T. Wylie
51.	4. 2. 1960	T. Wylie
52.	19. 2. 1960	F. Michael
53.	18. 11. 1960	Sherman
54.	10. 3. 1961	T. Wylie
55.	6. 4. 1961	Sherman
56.	19. 4. 1961	F. Michael
57.	5. 5. 1961	F. Michael
58.	9. 5. 1961	Sherman
59.	13. 7. 1961	Sherman
60.	4. 8. 1961	Sherman
61.	8. 8. 1961	Anita Michael
62.	8. 9. 1961	Sherman
63.	1. 10. 1961	G. Smith
64.	10. 10. 1961	Anita Michael
65.	28. 10. 1961	Anita Michael
66.	28. 11. 1961	Anita Michael
67.	3. 12. 1961	F. Michael
68.	19. 12. 1961	F. Michael
69.	12. 1. 1962	Sherman
70.	15. 1. 1962	F. Michael
71.	21. 1. 1962	G. Smith
72.	3. 2. 1962	T. Wylie
73.	4. 3. 1962	Anita Michael
74.	2. 5. 1962	T. Wylie
75.	1. 6. 1962	F. Michael
76.	1. 6. 1962	Anita Michael
77.	6. 6. 1962	Sherman
78.	26. 11. 1962	F. Michael
79.	27. 11. 1962	F. Michael
80.	27. 11. 1962	W. W. Ritchie
81.	29. 11. ?	W. W. Ritchie
82.	o. D.	H. v. Wissmann
83.	o. D.	F. Michael

1

University of Washington
Far Eastern and Russian Institute
Seattle 5, Washington

October 27, 1950

Mr. J. F. Rock
Himalayan Hotel
Kalimpong, West Bengal
India

Dear Mr. Rock:
Thank you for your good letter written from Darjeeling. We are, of course, awaiting your catalog in order to form a definite opinion about the library and its value to us. We are most definitely interested. There might, however, be some duplication, both in the Chinese and the Western material. You have not yet quoted an overall price. We would be glad to hear from you on that subject.

If you could acquire for us the Tibetan Kanjur[1] as well as the set of 63 volumes of the compendium of the ritzogs chen pa sect[2], you would be obliging us very much indeed. We would also like other Tibetan books that can be acquired without too heavy a financial obligation.

We do not at the present time have any Tibetan collection but we definitely plan to expand our interests in that direction. We already have one student studying spoken and written Tibetan. We do not expect to acquire any funds, unfortunately, for Tibetan art.

Your plan to settle down here in Seattle after your return next October is greeted here with great enthusiasm. You know, of course,

---

1 Rock hatte früher bereits einen tibetischen Kandjur für die Library of Congress beschafft. Zum Kandjur vgl. die Literatur bei Günter Grönbold: *Der buddhistische Kanon.* Eine Bibliographie. Wiesbaden: Harrassowitz 1984. 70 S.

2 Gemeint is wohl das Rin-chhen gter-mdzod im folgenden Brief.

that Prof. Nicholas Poppe[3] is on our staff and that Dr. Li Fang-kuei[4] heads our Chinese language work. Our Mongolian and Inner Asian studies have found a great response among the students and we are just now making plans to expand our course work into Tibet and India. You would, I think, feel very much at home here, and we can most certainly provide you with working space.

Sincerely yours,
George E. Taylor
Director

2

Kalimpong, January 18th/51

Dr. George E. Taylor
Director, Far Eastern Institute
University of Washington
Seattle 5, Washington

Dear Dr. Taylor:
In continuation of my last letter I would like to state that I have been able to purchase the following Tibetan books:

The Rin-chhen gter-mdzod 63 volumes, block print, it is a compendium of all the unreformed sects, it comes from mTshur-phu north of Lhasa U.S.$ price 350.00
The bKah-thang sder-lnga, in 5 volumes 50.00
The dPag-bsam-khri-shing, one volume 10.00

The hDzam-gling rgyas-bshad, an old manuscript copy from 1830,

---

3 Nikolaus Poppe (1897-1991), seinerzeit bedeutendster Mongolist, ursprünglich Professor in Leningrad, dann wissenschaftlicher Mitarbeiter in Deutschland und schließlich Professor für Mongolistik an der University of Washington. Vgl. seine *Reminiscences.* Edited by Henry G. Schwarz. (Bellingham:) Western Washington University (1983). XIII, 331 S. (Studies on Asia 16.)

4 Li Fang-kuei (1902-1988), Linguist, 1950-1969 Professor der chinesischen Sprachwissenschaft an der University of Washington. Vgl. A list of writings of Dr. Li Fang-kuei published up to 1966. *Monumenta Serica* 56.1967, 1-5, Widmung, Porträt; *International journal of American linguistics* 34.1968,212-214; Ch'ing-chu Li Fang-kuei hsien-sheng liu-shih-wu sui lun-wen chi. *Chung-yang yen-chiu-yüan, Li-shih yü-yen yen-chiu-so chi-k'an* 39.1969. 438, 457 S.; Li Fang-kuei chi-nien lun-wen chi. *Chung-yang yen-chiu-yüan, Li-shih yü-yen yen-chiu-so chi-k'an* 59.1988. 576 S.

written in *U-med.*[5] It is a famous geographical work, very well written and in excellent condition. 75.00

The Chhos-skyong-bi rgyal-po bstan-skam-poi-bkah-hbum-las. Bla-ma brgyud-pai-gsol-hdebs, one vol., 331 fol. It is the Ma-ni-bkah-hbum or Life story of King Srong-tsan-gam-po 25.00

The rNam-tar of the main teachers of the Kar-gyu-pa Sect, as Marpa, Naropa, Tilopa, the 100.000 songs of Milaraspa, etc. One volume of 482 folios 50.00

The Drang-srong dwang-po-nyin zhag-la-spyod-tshul-gyi-riu-rgya-mtsho, this is a Manuscript written in beautiful *U-chen*[6], about 100 leaves; a book on astrology 30.00

The rJe-btsun-thams-chan-mkhan-pa Vai-ro-tsa-nai rnam-thar, Life story of the Lama Vairocana, 130 folios 10.00

The Deb-ther-sngon-po, about 300 folios 35.00

The rNam-thar rin-phreng. In two volumes of 328 and 420 leaves respectively, it is the life story of the 13th Dalai Lama, now difficult to obtain 50.00

$	685.00
–	410.00
	275.00
	150.00
	425.00

This is the cost delivered at Kalimpong and includes transportation by mule from mTshur-pu and Lhasa respectively. Sometime ago I bought a very rare Gesar epic[7] in manuscript form, written in U-med, it is over 100 years old and comes from Kham, it consists of 500

5 u-med, d.i. dbu-med, die tibetische Kursive ("ohne Kopf").

6 u-chen, d.i. dbu-chan, die tibetische Normalschrift ("mit Kopf").

7 Die Gesar-Sage ist in Europa insbesondere durch Isaak Jakob Schmidts Textausgabe und Übersetzung bekannt geworden: *Die Thaten des Vertilgers der zehn Übel in den zehn Gegenden, des verdienstvollen Helden Bogda Gesser Chan.* Eine mongolische Heldensage, nach einem in Peking gedruckten Exemplar aufs Neue abgedruckt unter der Aufsicht des Akademikers J. J. Schmidt. St. Petersburg 1836. 191 S. – *Die Thaten Bogda Gesser Chans, des Vertilgers der Wurzel der zehn Übel in den zehn Gegenden.* Eine ostasiatische Heldensage, aus dem Mongolischen übersetzt von I. J. Schmidt. St. Petersburg: W. Gräff, Leipzig: Leopold Voss 1839. XIV, 287 S.

folios. It is one of the rarest versions namely the sTag-gzig one, the war between Gesar and Persia. Tucci has a copy but a printed one of much fewer leaves. You can have it for 100.00 Dollars. If you wish the books please wire me accept, and I will forward them to the American Consulate General Calcutta for despatch to Seattle. To the cost of the books will be added packing, and transportation to Calcutta. At any rate please wire me yes or no. I have given you first call, as there are other institutions after me who want me to furnish books for them, as the Sinological Institute in Leiden (Dr. Duyvendak[8]), Lessing[9], the Congressional Library, etc.

As regards the Kangyur, I have not yet heard from the party whom I had asked to secure it for me, the party has arrived with the Dalai Lama at Yatung, and he is expected here in the near future. If you want the books please cable the amount in U.S. Dollars to the American Express Company Incorporated, Calcutta, West Bengal, in my name.

I have decided to remain here till next summer and am working now on the Historical Geography of North-west China and North-east Tibet. It is much easier as I have an opportunity to ask questions as to the proper spelling of place names, etc. There are now Tibetans here from Amdo, Kham, Lhasa, etc.

If there are other books you want, I will inform you what is available, at any rate I will give you first choice.

Hoping to hear from you by wire, address: Rock Himalayan Kalimpong India, I remain with best regards, very sincerely yours,

---

8 Jan Julius Lodewijk Duyvendak (1889-1954), Professor für Sinologie an der Universität Leiden, Direktor des Sinologisch Instituut, Mitherausgeber der Fachzeitschrift *T'oung Pao* (Paris, Leiden). Vgl. Erich Haenisch: J. J. L. Duyvendak in memoriam, 1889-1954. *Oriens extremus* 2.1955, 121-122.

9 Lessing, ursprünglich Professor am Seminar für Orientalische Sprachen ind Berlin und Oberkustos am Völkerkundemuseum, war seit 1935 Professor an der University of California, Berkeley. Vgl. H. Walravens: *Ferdinand Lessing (1882-1961), Sinologe, Mongolist, Kenner des Lamaismus.* Materialien zu Leben und Werk, mit dem Briefwechsel mit Sven Hedin. Osnabrück: Zeller Verlag 2000. 425 S.

3

Kalimpong, March 28th/51

Prof. Helmut Wilhelm[10]
University of Washington
Seattle 5, Washington, U.S.A.

Dear Prof. Wilhelm:
Yesterday I received your wire saying *Yes* which indicates that your Institute is buying the 27 items on the second list, the price for the 27 items being $ 150.00.

As regards the bKah-thang sde-lnga I may say that I have been able to secure another copy in lieu of the one the lama refused to sell me after having agreed to do so. There are five volumes, but of the first volume there are three (the first three) folios missing, the price is just half of that of the copy the lama refused to let me have, namely $ 25.00. As it is very difficult to get any books now as nothing comes through to Kalimpong, all Tibetans having gone or are going back with the approach of the hot season, I bought it and am sending it with the other books. I packed the four volumes separately also the first volume with a note attached saying that the first three folios are missing, both packages are marked no. 11 in pencil on the first page of the books. All the different books are tied with string at each end, those of the first list have the numbers written on the title page in pencil, those of the second list have a tag tied to each item numbered from 1 to 27. As regards the Kesar sTag-gzig you will receive it from Prof. Rolf A. Stein[11] of Paris to whom I had loaned it, as I have written. As the first three pages or so of the bKah-thang sde-lnga consist mostly of invocations not much is lost, but as Tucci has a

---

10 Hellmut Wilhelm (1905-1990), Sohn des bekannten Sinologen Richard Wilhelm; seit 1948 Professor der Sinologie an der University of Washington. Vgl. die Würdigungen: Memories and bibliography, von David R. Knechtges, George E. Taylor, Donald W. Treadgold, Frederick W. Mote, und Herbert Franke in *Oriens extremus* 35.1992, 5-34.

11 Rolf Alfred Stein (1911-1999), Sinologe und Tibetologe, Professor am Collège de France. Rock und Stein kannten sich aus Indochina und standen im Briefwechsel. Vgl. *Tantric and Tibetan studies in honour of R. A. Stein.* 1-3. Bruxelles 1981. (Mélanges chinois et bouddhiques 20.); Kuo Liying: In memoriam: Rolf Alfred Stein (1911-1999). *Cahiers d'Extrême-Asie* 11.1999/2000, XI-XXX.

copy which he secured in Lhasa, he will I am sure, let you have photostat copies of the first three folios.

There is in the collection also an Item marked no. 12, it is entitled: Legs-par bshad-pa bka-gdams rin-po-chhei gsung, 1 vol. with about 200 folios, it is the life-story of the above rin-po-chhe, and his wise sayings; others not on the first list are no. 13, the rNam-thar of the Fifth Dalai Lama, an abbreviated version of 246 folios; no. 14 is rGyal-po lugs kyi bstan bchos gzhi skyong bai rgyan zhes bya ba bzhugs so, of 157 folios. No. 15 is Byang chub lam gyi rim pa chhen po las byung bai brda bkrol nyer mkho bsdus pa bzhugs so, of 55 folios. No. 16 is Bya bral ba kun dgah rang grol dang sprang byang chhub seng ges gtsos chhos kyi rgyal pos bkai hphrin etc., about 200 folios, it gives descriptions of the hells, Bardo, the Lord of the Dead, etc. As these items were not included, I do not know if you want them, however I am sending you items 14, 15, and 16 with my compliments for your Institute.

Item no. 12 costs $ 15.00, and no. 13 $ 20.00, presuming that you will want these I will deduct the price from the amount at the American Express Co. and you will find these items also in the box.

I am packing all the books in one strong box and am shipping them to Siliguri by car, thence by air to Calcutta, airfreight is not expensive and much safer. The Railway passes through part of Pakistan and there may be a detention of goods by the Pakistan Customs, and it may take the box a month to reach Calcutta, and furthermore goods are often rifled in transit. The box is addressed to the American Express Co., Inc. of Calcutta for transshipment to the States (either to San Francisco or direct to Seattle if steamer available). The box will be insured for $ 600.00.

Enclosed find statement which is self explanatory. With best regards,

Very sincerely yours,

Itemized Statement

Items as per letter of January 18th, 1951. (Item 1, 2, and 3 not available)

hDzam-gling rgyas-bshad	$ 75.00
Chhos-skyong-bi rgyal-po etc.	$ 25.00
rNam-thar of the main teachers of the Kar-gyu-pa Sect	$ 50.00
Drang-srong dwang-po nyin etc.	$ 30.00
rJe-btsun-thams-chan-mkhen-pa etc.	$ 10.00

Debter (Deb-gter) sngon-po	$ 35.00
rNam-thar rin-phreng (Biography of 13th Dalai Lama)	$ 50.00
Gesar sTag-gzig	$ 100.00
Incomplete (three folios missing) bKah-thang sde-lnga	$ 25.00
Total	$ 400.00

Items of second list from 1 to 27	$ 150.00
Other items not on list:	
No. 10 rNam-thar and mGur-hbum (Hundred Thousand Songs) of Mi-la-ras-pa	$ 30.00
No. 12 Life-story and sayings of bKah-gdams-rin-po-chhe	$ 15.00
No. 13 rNam-thar of Fifth Dalai Lama	$ 20.00
Items nos. 14, 14, and 16 personal gift to Far Eastern Institute	
	$ 65.00
Total	$ 615.00

Balance on hand at American Express Co., Calcutta $ 170.00

4

University of Washington
Far Eastern Institute
Seattle 5, Washington

April 10, 1951

Dr. J. F. Rock
Himalayan Hotel
Kalimpong, West Bengal
India

Dear Dr. Rock:
Many thanks for your good letter of March 28th. I am glad you received our wire in time and have been in the position to bring the 27 items of your second list on the way to us, together with the items of the first list. We are especially grateful that you included some of the items as a personal gift.

It is with great appreciation that we take note of your willingness of making a loan of your library to our institute. We will see to it that it is properly housed and taken care of. A member of our staff, probably George Taylor[12] himself, will go to Boston to take delivery of the library in the way described by you. Thus, we would be very grateful if you could notify Dr. Merrill accordingly and send us a letter, on the presentation of which Dr. Merrill would deliver the books.

The things you say with regard to the closing of the open door to Tibet has been of greatest interest to us. I hope that this will be only temporary and that opportunity will present itself again to draw on the treasures of Tibetan libraries. For that reason we would like to ask you to keep the balance of the amount that was remitted to you for the time being. We still hope that some good use might be made of it.

Professor Stein of Paris is a very close friend of mine and I am very glad you loaned him the copy of the Kesar manuscript. I am sure that Professor Stein will find it of great value and we are not in a particular hurry to have the book here. Thank you very much for all the troubles you went through for us. My best regards.

Sincerely yours,
Hellmut Wilhelm

5

Kalimpong, April 12th/51

Dr. George E. Taylor
Director, Far Eastern Institute
University of Washington
Seattle 5, Wash.

Dear Dr. Taylor:
On the 15th of this month I am leaving for Calcutta and I shall take the box of books with me by air from Bagdogra to Calcutta. They will be shipped by the American Express Company of that City, and as requested will be fully insured.

Yesterday I had a letter from Mr. Evan M. Wilson, American Consul General at Calcutta to whom I had written several weeks ago about the books, and whom I had requested to assist in getting them through the Customs in Calcutta. He writes that he was away and on

---

12 George E. Taylor (1905-2000) war Direktor des Far Eastern (and Russian) Institute. Er veröffentlichte u.a. *The struggle for North China.* New York: Institute for Pacific Relations 1940. XIV, 252 S.; *The Philippines and the United States.* New York: Praeger 1964. X, 325 S.

his return he found that my various letters to him had not been acknowledged. In regard to the shipping of the books he writes: *We shall be glad to help you with the forwarding of the books to the University of Washington at Seattle; we have also received a letter from Mr. Harry C. Bauer, Director of Libraries at that University in the same connection.* So I presume there will be no trouble getting them through the Customs.

In the meantime, since writing you or rather Dr. Wilhelm, I bought another lot of Tibetan books, and as the box was not full I included them in the shipment, and enclose a list of them. They are numbered from 28 to 46, although numbers 38 and 39 contain many different items which I could not all enumerate, they being too numerous. If you want them the price can be taken out of the funds still on hand. There are some rare items among them as no. 28 the life story of Rwa lo tsa ba, no. 35 the rGyal-rabs, 33 on the Lhasa Cathedral, 34 on Tibetan Prosody, etc. The cost of the lot is $ 120.00. If you do not want them please send them to Professor F. D. Lessing at the University of California. They are all marked with tags on the inside of the package and the numbers of the items are on the outside written in pencil on the wrapping paper.

I know of a large Ms. of a Gesar version other than the sTag-gzig, but to buy books in this country is very slow business, endless talk and putting off, they want too much, knowing that it is difficult to get these things with Tibet being in the process of becoming hermetically sealed. I was very fortunate to get what I did, for to order them now from Lhasa is hopeless, and may take a year or more, especially as the Dalai is not in Lhasa.

As soon as the box leaves Calcutta, invoices will be sent as requested. Hoping that you have fully recovered, and that the box of books will arrive safely, I remain with best regards, very sincerely yours,

If I can get any other books I will give you first choice, provided you want them, if not there are other institutions who will take them, so please let me know if you are interested in case I should get other individual works, these can be sent by parcels post. Enclosed list of new lot of books and letter from American Express.

6

Kalimpong, April 19th, 1951

Far Eastern Institute
University of Washington
Seattle 5, Wash.

Dear Dr. Wilhelm:
Your good letter of April 10th reached me today on my return from Darjeeling. As I wrote to Dr. Taylor, I had intended to go to Calcutta and take the box of Tibetan books with me, but I have not been very well, and the heat in Calcutta went up to 105 Fahr., furthermore the plague broke out there plus cholera and smallpox, all of which decided me not to go. Instead I went to Darjeeling, took the box with me and sent it by airfreight from Bagdogra to Calcutta, and by car transport from Darjeeling to Bagdogra. The road from Darjeeling to the plains is now open, while the road from Kalimpong to Siliguri is still interrupted and transport is by coolie in part. The box was sent to the American Express Co., Calcutta, who will forward it to the States insured, I also wrote to Mr. Wilson the American Consul General in Calcutta advising him of the shipment and asked him to assist in passing the box through Customs. I had a special box made for the long journey, it is so nice that you can have a cabinet made from it.

I am sending you the last lists of the foreign books in my library (seven lists), this is all but there are still a few volumes not listed and some I have here with me. There are also trunks containing negatives and photographs and photostat copies of rare Na-khi manuscripts, all with title of ceremony and name of book, there are no duplicates among them, also one trunk containing original Na-khi manuscripts, labelled and numbered. At the Peabody Museum are trunks containing Tibetan Art Objects also photographs, etc. All these would have to be brought with the books because I don't want to keep one part in Cambridge and another in Seattle. An exhibition could be made of the Tibetan Collection which I am sure would draw quite a crowd. You did not mention having received the second instalment of the lists of Chinese books I sent you, I hope it has arrived. I have decided to make a temporary loan of my library to your Institute till such a time when I come back to the States and a sale can be arranged, your Institute will be given preference and first option.

I had a letter from Dr. Merrill today who has just returned from Central America, he writes that he is leaving shortly for England for the summer and may stay away a year, so it would be imperative that the library be fetched as soon as possible, please write to Dr. Merrill and ask him when you could take over the library; I am also writing him by this mail, and am sending him copies of the letter of introduction on the presentation of which he will turn over the library to you. As I stated in my last letter to Dr. Taylor or to you I have now forgotten, the books are packed in cartons, they should be repacked in boxes, especially those cartons which are not full, for the trunks I will send you the keys, or rather one key of the trunk containing all the keys to the other trunks. I hope Dr. Taylor or your good self will go to Boston, preferably the Director of the Institute, for the library is a very valuable one and all possible care should be taken in its transfer to Seattle. Each book is wrapped in strong Manila paper, and as I said, some are in trunks and boxes and the bulk is packed in cartons.

The trunks both those at the Peabody Museum in Cambridge and at the Arnold Arboretum in Jamaica Plain, Mass., should be strapped as some of them are not very strong, and of one as I remember at the Peabody Museum one hinge is broken. I am very glad that the books will find a home and will be made use of, rather than being packed in boxes and stored. I hope you or Dr. Taylor will be able to go to Boston as soon as possible and take delivery of the books.

With best regards and all good wishes and success in bringing the library to Seattle, I remain, very sincerely yours, J. F. R.
Enclosure: Letter of Introduction to Dr. Merrill.

7

Kalimpong, April 20th/1951

Dr. J. O. Brew[13]
Director,
Peabody Museum of Archaeology and Ethnology
Harvard University
Cambridge Mass.

Dear Dr. Brew:
Five years have passed since I have seen or even written to you.

---

13 John Otis Brew (1906-), Promotion in Archäologie 1941; Direktor des Peabody Museum 1948-1967. Vgl. auch sein *One hundred years of anthropology*. Cambridge, Mass.: Harvard University Press 1968. 276 S.

Much have I also experienced since my departure from Boston in 1946. I was one month under the Communist in China, and I am glad I left when I did. I am now in the Himalayas continuing my studies; this is a very interesting place and one meets many interesting people, but it is not China.

I have decided to turn over my library to the Far Eastern Institute of the University of Washington, Seattle Washington, and also my Tibetan Art Collection now stored at the Peabody Museum. I have given Letters of Introduction to Dr. George E. Taylor, Director of the above Institute, or his representative, Dr. Hellmut Wilhelm of the same Institute. There are two boxes and six trunks in storage at the museum, which according to our agreement and which in part reads: *Received from Dr. J. F. Rock, two wooden boxes and seven (actually there are only six) trunks, to be held at his risk and returned at any time to Dr. J. F. Rock, on demand of him, his heirs, assigns, or authorized representative.*

Either Dr. Taylor or Dr. Wilhelm will call at the Peabody Museum and will take delivery of the two boxes and six trunks belonging to me, containing Tibetan Art treasures, on presentation of a letter of introduction, authorizing them to take delivery of the above mentioned boxes and six trunks.

Please be so good and turn over to either of the above mentioned gentlemen the two boxes and six trunks on presentation of the letter of introduction a copy of which is here enclosed. Please allow me to thank you for the storing of the Art Collection at your Museum. With kindest regards and best wishes, I remain, very sincerely yours,

Enclosure: Letter of Introduction

8

Kalimpong, April 20th/1951

Dr. J. O. Brew
Director,
Peabody Museum of Archaeology and Ethnology
Harvard University
Cambridge Mass.

Dear Dr. Brew:
This will introduce to you either Dr. George E. Taylor, Director of the Far Eastern Institute of the University of Washington, Seattle, Wash.,

or Dr. Hellmut Wilhelm of the same Institute.

On presentation of this letter please be so good and turn over to either one of the above mentioned gentlemen the two wooden boxes and six trunks belonging to me and now in storage at the Peabody Museum in Cambridge, Mass., obtaining receipts in duplicate on delivery.

Thanking for your kindness, I remain with best regards,
very sincerely yours,
J. F. Rock

9

Kalimpong April 20th/51

My dear Merrill:[14]
Your good letter of April 11th reached me last night. I was very glad to know that you and Mrs. Merrill had a nice and interesting time in Central America, but I am surprised that you are again about to fly from the nest, in your old days you are it seems getting the wanderlust, are you trying to imitate me? So you are going to England the food wont [!] be much, one lambchop a week, but you don't seem to mind what you eat, judging hm ..... no disparaging remarks on Mrs. Merrill. I must say the food here is nothing to brag about nothing like Italy, but of course it is also dearer there.

As to the library, as I hear nothing more from Harvard, and as they only want to buy the very cream I am not interested, and they want everything at prices that prevailed in 1920 in Peking. Try and get anything now. The damn silly missionaries feeling so sorry for the poor Chinese students who were red long before the country ever went red, insisted supplying funds for Yenching University on condition that the money wont be used for subversive activities, and inspite of stating that their curriculum included Marxism, Leninism and Stalinism, the trustees sent them the money regularly till the State Department stepped in and prohibited them to send any more money to the various anti American Red Universities. Yes I find I did not finish my sentence as to my library. I have arranged to loan it for the time being to the Far Eastern Institute of the University of Washington in Seattle. They also want to buy it. So either Dr. George

---

14 Elmer Drew Merrill (1876-1956), Taxonom, seit 1902 Professor der Botanik in Manila, seit 1935 an der Harvard-Universität. Vgl. *Dictionary of American Biography* Suppl. 6.1980, 449-450.

E. Taylor, the director of the Institute, or one of their Professors will come to Boston to take delivery of the library, all the carton boxes, trunks and wooden boxes, all there is that belongs to me. I enclose copy of letter sent to the director of the Institute above mentioned. Please turn over my belongings to who brings the original of the letter, either Dr. Taylor or Dr. Hellmut Wilhelm, it might be best to ask them also for their identity cards, etc., so that in case the letter gets lost and some theological student finds it will present it and collect the library. I have had experience enough with Harvard theological students. It is possible that I will see you in England or Scotland for I cannot do all the work on the Plant Geography of West China and East Tibet here, there is a lot that I need, I am sorry now I gave you the little specimens which I had kept for just the purpose I want them now, but what is done is done. I thought I was finished as no one in America took any interest, it takes the British to appreciate things, I am as much interested in live plants as in herbarium specimens, and also in photographs, good ones, old Prof. Sargent[15] had the right idea. I am sorry the A[rnold] A[rboretum] has gone to pot, I told you at the time that I did not trust that Dean of yours, I have forgotten his name, still waters run deep, I sized him up the first day I saw him. I hope however that matters will turn out in favor of the AA, right will after all prevail.

There is little news here, I gave up my trip to Nepal for I could not find any one to go with me without my paying his way, and alone I did not care to go especially to the wilds where the government has little control. The Tibetans are returning to Tibet but the Dalai Lama is still in Yatung, and may remain there till something has been settled with the Chinese reds. I probably will stay here till autumn when travelling becomes a little more comfortable, the heat in Calcutta is 105 Fahr., besides the plague, cholera and smallpox are raging there, a good place to stay away from.

Now let me thank you for all your kindness in allowing me to store my library at the Arboretum, I had intended either giving it or selling it to H[arvard]-Y[enching], but they are as ... well I won't say, they are at a paar [!] with the whole H[arvard] outfit that you are so well acquainted with. The Far Eastern Institute of Washington University

---

15 Charles Sprague Sargent (1841-1927), seit Direktor des Arnold Arboretum, der eine Expedition Rocks sponserte. Vgl. *Dictionary of National Biography* 16.1935, 354-355.

is a new concern and they would find it very difficult to get such a collection together, and they appreciate what I have, and know that it can never again be collected.

Hoping that we will meet in England, I remain with kindest regards to your good self and Mrs. Merrill, cordially yours, J. F. Rock
Enclosure: Letter of Introduction

10

Kalimpong, April 20th/1951

Dr. Elmer D. Merrill
Emeritus Director
The Arnold Arboretum
Jamaica Plains, Mass.

Dear Dr. Merrill:
This will introduce to you either Dr. George E. Taylor, Director of the Far Eastern Institute of the University of Washington, Seattle, Washington, or Dr. Hellmut Wilhelm of the same Institute. Either one of these Gentlemen will come to take delivery of my library, now in storage at the Arnold Arboretum, and I would be obliged if you would kindly turn over to either one of them all the cartons, trunks and boxes on presentation of this letter.

I have written to them and asked them to repack certain cartons which are not full and have the trunks strapped, I have also asked them to write to you when they could come to take delivery of the library in view of your early departure for England.

Anything you will be able to help them with will be highly appreciated. Again thanking you I remain, very sincerely yours,
Joseph F. Rock

11

Kalimpong, April 20th, 1951

Dr. George E. Taylor, Director
Far Eastern Institute
University of Washington
Seattle, Washington

Dear Dr. Taylor:
I have just heard from Dr. Wilhelm that you would like to have the loan of the Library belonging to me, and now stored in the Arnold

Arboretum of Harvard University at Jamaica Plain Mass. I have written to Dr. Merrill as requested and enclose letter of introduction to him, a duplicate of the letter has been sent to Dr. Merrill.

I enclose letter to Dr. Wilhelm, I hope you have fully recovered and that [you] will be able to go in person to take delivery of the Library.

With kindest regards, very sincerely yours,

P.S. Please read enclosed letter to Dr. Wilhelm.

[wrote on June 13th that owing to absence of reply, will be forced to cancel letters of introduction.]

12

Kalimpong, May 22/51

Dr. George E. Taylor
Director Far Eastern Institute
University of Washington
Seattle 5, Washington

Dear Dr. Taylor:

I have just received the Shipping charges, etc., for the box of books, from the American Express Co., Inc., Calcutta, and enclose same with bills for expenditures in connection with the shipping etc., of the books from here to Calcutta.

You will notice that there is due me U.S. $ 12.75 (Twelve Dollars seventy-five cents). I would be obliged if the amount would be deposited to my account with the National City Bank of New York, Farmers Branch, 22 William Street, New York, this will avoid trouble of forwarding the amount to me by international Money Order.

I hope that by this time the box will have arrived in good condition.

Sincerely yours,
J. F. Rock

13

University of Washington
Far Eastern and Russian Institute
Seattle 5, Washington

June 20, 1951

Dr. J. F. Rock
c/o Professor G. Tucci
Piazza Vescovio 21
Rome, Italy

Dear Dr. Rock:
In the absence of Professor Taylor who has just left for New York I want to write you about your library. We have just received your letter of the 13th of June in which you express your concern about the handling of the library loan and indicate that you have not received any answer to your letter of April 20. We sent you last night the following night-letter: *Letter thirteenth received. Regret your not receiving reply to April letter. Prof. Taylor now enroute to New York. He called on Dr. Merrill and Dr. Brew and presented letters on previous trip. Had very pleasant visit with them. Appreciate loan of books. Arrangements now being made for shipment to Seattle. Letter follows.* I want to repeat how sorry I am that you have not received George Taylor's letter of May 7, a copy of which is attached. As our night-letter indicated, it was Professor Taylor himself who visited Dr. Merrill and Dr. Brew and discussed the library loan with them. He handed your introductory letter to Dr. Brew and showed the other to Dr. Merrill who returned it to him. You can be assured that no unauthorized use has been made of them.

In the meantime, we have learned that the cost of the packing and transportation of your library and art collection is almost $ 1,000.00. As I understand it, a good part of this cost is taken up by packing and handling of the art objects in a safe way. You have indicated in your letter to Dr. Wilhelm that you want the library and the art objects to remain together. We can perfectly understand your wish, but our difficulty is that we do not have sufficient funds immediately available to us to meet the high cost of transportation of both the books and the art objects. As Dr. Merrill is leaving on the 25th, we will have, therefore, to postpone any action as much as we regret having to do so.

The problem of transportation cost for both the library and the art objects has only come to my notice after I sent you the night-letter yesterday in which I indicated we were making arrangements for the shipment of the books. I had not realized then that you wanted to keep both library and art objects together.

We are still very much interested in acquiring your library. The books would be of great value to our research work in our Inner Asian and Chinese History projects, and we would be interested in taking the whole collection, although there are naturally a good many duplications with our own holdings. We would, however, have no way of handling the art collection at our University.

We are sending now the list of your library to Miss Huff[16] in Berkeley in compliance with your wishes as expressed to Dr. Wilhelm.

We hope that you might see some way in which you could separate the art collection and the valuable Na-khi manuscripts from your library. If such a separation is possible we would be very grateful for a loan of your library in the way which you indicated and in the hope of acquiring it eventually with the preference and first option which you so kindly offered.

Sincerely yours,
Franz Michael[17] Acting Director

---

16 Elizabeth Huff (1912-1988), Sinologin, 1947-1967 Leiterin der ostasiatischen Bibliothek der University of California, Berkeley. Sie promovierte 1947 am Radcliffe College (Harvard) mit der Arbeit: *Shih hsüeh. Translation of a study of Chinese poetry by Huang Chieh* [1874-1935]. 229 Bl. Sie gab den chinesischen Katalog der Bibliothek heraus. *East Asiatic Library. Author-title catalog. Subject catalog.* 19 Bde. Boston: G. K. Hall 1968. 2°, und edierte auch Fang Chao-ying's Katalog *The Asami Library* (Berkeley: Univ. of California Press 1969. X, 424 S.).

17 Franz Michael (1907-1992), Jurist, Politologe und Sinologe, Professor of Chinese History an der University of Washington. Er promovierte 1933 in Freiburg in Jura, emigrierte 1934 nach China und war 1934-1938 Professor an der National Chekiang University, Hangchow; 1939 emigrierte er in die USA und war seit 1942 an der University of Washington tätig. Vgl. Marie-Luise Näth in *The China Quarterly* 138.1994, 513-516. Zu seinen größeren Arbeiten gehören: *The origin of Manchu rule in China.* Baltimore: Johns Hopkins Press 1942. VIII, 127 S.; *The Far East in the modern world.* Rev. ed. New York: Holt, Rinehart & Winston 1964. XII, 850 S.; *Rule by incarnation.* Boulder, CO: Westview Press 1982. XII, 227 S.; *China and the crisis of Marxism-Leninism*. Boulder, CO: Westview Pr. 1990. X, 214 S.

14

Himalayan Hotel
Kalimpong

18th July 1951

Dear Dr. Rock,
We were all so sorry to hear you had been so ill in Calcutta, but glad to know you had got to Switzerland and are in good hands. You must remember you are a very tired man, after so many years of strenuous mental work, and please make up your mind to give yourself a holiday. I am sure it will help to build up your health. Needless to say, we are missing you very, very much.

In the short time since your departure many things have happened in Kalimpong. First and foremost, the Communists are at Yatung now. There were many parties during their brief stay in Kalimpong, and Shakapba was the only man who returned a toast to the Dalai Lama and Tibet while they were toasting their Mao Tse-tung. Their arrival at Gangtok was met by the Maharaja's Private Secretary and Dayal's assistant. They did not bother to return the call, but went straight through without having anything to do with them whatsoever. It so happened that they went over the NathuLa and twenty-three White Russians came over the JelepLa. The latter are expected in Kalimpong today. Very sad. Mary and Jigme left day before yesterday, and Shakapba yesterday. The Chinese seem to have a very full programme with regard to Tibet. To begin with, the idea is to increase the population and open out schools. No need for me to enlarge on this – who knows better than you do?

The Dalai Lama leaves for Lhasa via Sakya and Shigatse on the 22nd of this month. The troops from Chamdo are fast approaching Lhasa. Reggie Fox, from whom I had a letter, expects to be in Kalimpong very soon. I quote you the following from his letter:

„Yatung, 12th July 1951.

Is Dr. Rock still with you? I feel awfully guilty after having promised I would do what I could for him in Lhasa. As a matter of fact I was hardly out of bed all the time I was there in Lhasa and the sudden departure of His Holiness for Yatung made the whole commission impossible.

If he is there, please explain, Annie, and when I see him personally I can do so properly. In any case it is still not too late. D.

L. is returning and I have many friends in Lhasa, but it will take time …"

Robert Lynn has just been and gone, and now Gibson, also of the Consular Service, is with us. It is good to see them again.

The zinnias have come and gone, but I could not save any seed due to too much rain. The daliahs are just coming on. The solo papyias too are doing well and the cassias are just germinating. The gloxynias have not arrived yet.

30th July 1951
As we were typing the letter the White Russians arrived and we had to leave it. As they had nowhere to stay we let them have Daddy's little cottage at the bottom of the garden, where they were very happy.

Twenty of the men have gone down to Calcutta, and they are being cared for by the Government of West Bengal in a camp and they hope to be able to make arrangements to send them elsewhere. The father, mother and daughther are still in Kalimpong, but they, too, will go when final arrangements are made for them.

The Dalai Lama has gone by the direct route to Lhasa. He will be spending some days at Gyantse. Apparently Fox is still at Yatung.

We have a woman here, Miss Barclay from France, who tells me that you had written saying Kalimpong was very expensive to live in, and it cost one two hundred dollars to live for a month. I wonder where they get these ideas. Another remark I heard (at a party) was *Why is Mrs. Perry always siding with Dr. Rock, when he didn't have very nice things to say about the Hotel?* which is not true. I did my best for the twelve rupees daily which I charge people for long stays, so of course the food won't always be what one would expect in a private home, or where the charges are much more. I know you understood this. I have to allow for food, the room, service, and upkeep, and to make a little for myself – which I never seem to be able to do, and I am sure you understood my position.

I trust you will have received some of the mail which I have forwarded you from here. I re-addressed some before I got your Rs 10/-, and since then have sent everything by Airmail, including a Registered Letter. The exact amount I have used on postage to date is Rs 5-15-0.

We have a houseful even now, and among them a very nice family from Singapore.

Vera [Macdonald] and Vicky [Williams] are not here just now, as

they are working, but I am sure they would wish to join me in sending you our very best wishes, and to hope you keep better health.
Yours sincerely,
Annie Perry[18]

15

University of Washington
Far Eastern and Russian Institute
Seattle 5, Washington

July 18, 1951

Dr. J. F. Rock
c/o American Express Co.
Zürich, Switzerland

Dear Dr. Rock:
I hasten to answer your two letters of July 1st and 6th, and want to express my deepest concern about the heavy price in health you had to pay for your stay in the tropics. I sincerely hope that the people in Zurich will be in a position to take care of the situation, but if I may venture a suggestion I would advise calling in a specialist on tropical diseases. Even highly trained specialists in Europe sometimes have not enough experience with tropical diseases and are liable to trot along the well-established path. It was due to this fact that my father's illness was not found out in time, and when the specialist for tropical diseases was called in it was too late.

We are delighted that you have our needs of Tibetan books and manuscripts at heart and we feel privileged that you offer us the very valuable manuscript of the rNam-thar of the Sixth Dalai Lama, which we would very much like to procure for our library, at the price mentioned by you ($ 300). As far as I remember, we still have a small amount on our account with you. As of the date of your letter of March 28th, it was $ 170, but we would like you to check on this figure and let us know how large it actually is, and also let us know to which address we should transfer the rest.

---

18 Annie Perrie war eine der drei älteren Töchter David Macdonalds (†1962); sie führte das Himalayan Hotel, das Haus der Familie, in Kalimpong, das auch heute noch in Familienbesitz ist. Ihre Schwestern Vera und Vicky sind im Brief genannt.

Amongst the things in the possession of Mr. McDonald[19] are also several that interest us intensely. You were kind enough to offer to act as a go-between and if you could find out whether McDonald is willing to sell his books and manuscripts and at what price, we would be extremely grateful.

We have taken up your suggestion to get in touch with Jigmed Tharing and I'm confident that this connection will be very helpful to us.

It might interest you to know that the brother of the Dalai Lama is just now touring this country. How much this mission is scholarly and how much political, I do not dare decide.

With my best wishes regarding your health, I remain,
Sincerely yours,
Hellmut Wilhelm

16

University of Washington
Far Eastern and Russian Institute
Seattle 5, Washington
c/o Dr. K. A. Wittfogel[20]. R.D. 1 Highland Mills, N.Y.

July 21, 1951

Dear Dr. Rock,

---

19 David MacDonald (†1962) war Sohn eines schottischen Vaters und einer sikkimesischen Mutter und wuchs in Darjeeling auf. Über seine Tätigkeit als British Trade Agent 1905-1925 in Tibet hat er in seinen Memoiren berichtet: *Twenty years in Tibet. Intimate & personal experiences of the closed land among all classes of its people from the highest to the lowest.* London: Seeley, Service 1932. 318 S. Vgl. auch Macdonald: *The land of the lama. A description of a country of contrasts.* With a forword by the Earl of Ronaldshay. London: Seeley, Service 1929. 283 S.

20 Karl August Wittfogel (1896-1988), Schriftsteller, Sozialwissenschaftler und Sinologe; vgl. G. L. Ulmen: *The science of society:* Toward an understanding of the life and work of Karl August Wittfogel. The Hague, Paris: Mouton 1979; Karl H. Menges in *Central Asiatic Journal* 33.1989, 1-7. Zu seinen Hauptwerken gehören: *Wirtschaft und Gesellschaft Chinas.* 1. Teil. Leipzig: C. L. Hirschfeld 1931. XXIV, 767 S.; *History of Chinese society: Liao, 907-1125.* Philadelphia: American Philosophical Society 1949. XV, 752 S. (Transactions of the American Philosophical Society.39.1946)(mit Feng Chia-sheng); *Die orientalische Despotie.* Köln: Kiepenheuer & Witsch 1962. 625 S.

Your letters are being handled by the Assistant Director of our Institute, Dr. Franz Michael, but I get them too and am most interested in our finding some solution to the matter of accepting your very generous offer. But now I am writing merely to send my very kindest regards about your health and to wish you a speedy recovery. I was very distressed to hear in your letter of the ailments you had contracted. The drastic steps you had to take are sufficient evidence of their seriousness. Good luck.

I am putting in a pleasant summer's work out here in the country near West Point. It is useful to get away from one's usual haunts and have nothing to interrupt writing and reading.

Sincerely yours,
George E. Taylor

17

University of Washington
Far Eastern and Russian Institute
Seattle 5, Washington

August 17, 1951

Dr. F. J. Rock
c/o American Express Co.
Zurich, Switzerland

Dear Dr. Rock:

Thank you very much for your kind letters of July 24th and 25th. We are really very much looking forward to receiving the manuscripts including the ones of Tharanatha [!]. As far as the latter is concerned, I am very glad that you will not be too exacting with your bill. Now that the academic year is over, we first have to arrange our funds before we can meet our obligations.

Your letter to George Taylor in which you give a complete survey of your account, I haven't yet seen as George is absent. As soon as he is back I will see to it that your bill is paid. I must apologize that in the summer months all these matters can be settled only against great disadvantages.

Yours very sincerely,
Hellmut Wilhelm

18

University of Washington
Far Eastern and Russian Institute
Seattle 5, Washington

October 18, 1951

Dr. J. F. Rock
American Express Company
London, England

Dear Dr. Rock:
I am extremely sorry that your last letter had to wait so long until it was answered, but I was away for two months and as my mail did not reach me, it is only now that I can thank you for everything that you have done for us. You mentioned that the Taranatha manuscript is being mailed to you. Queerly enough, we received information from the American Express Company in Calcutta that this manuscript is on its way to our address. I wonder whether you could let us know your wishes in this respect.

Franz Michael, who is now in the hospital after undergoing an operation, asked me to thank you also for your good letter of October 12th. I am very glad that the affair of your library has been settled after all and that you have been kind enough to inform Dr. Sax[21] at the Arboretum in Boston. I do hope that the books will arrive soon. We are all very much looking forward to using them. One passage in your letter made me very happy, that is the indication of your decision to settle down in Seattle. I think that you will find the intellectual climate here, even though somewhat provincial, highly stimulating and very much alive.

Your letter from the Mirror Press is very interesting indeed. Our University Press has been contemplating buying type of Chinese, Japanese and Tibetan script for some time. I have forwarded your letter to them and I hope that they will act accordingly.
Very sincerely yours,
Hellmut Wilhelm

---

21 Karl Sax (1892-1973), 1936-1959 Professor für Pflanzenzytologie am Arnold Arboretum, 1947-1954 Direktor des Arboretums.

19

University of Washington
Far Eastern and Russian Institute
Seattle 5, Washington

February 13, 1952

Dr. J. F. Rock
Villa Igea
Palermo

Dear Dr. Rock:
It was brought to my attention that you had not yet heard about the arrival of your loan library. If, meanwhile, you did not receive Dr. Kirchhoff's communication mailed about one month ago, this is to acknowledge the receipt of 69 units (trunks, cases, boxes). We have begun to unpack the books, and as soon as we shall have finished checking, you will be furnished with a more detailed account.

Your collection is highly valued by us, and we are all very grateful to you for letting us use it.
Yours truly,
Dr. Ruth Krader[22]
Curator, Far Eastern Library

20

University of Washington
Far Eastern and Russian Institute
Seattle 5, Washington

March 3, 1952

Dr. Joseph F. Rock
Villa Igea, Palermo
Sicily

Dear Dr. Rock:
Your letter to the President has been handed to me. I fully understand your concern about the library. However, you have probably heard in the meantime from Professor Kirchhoff who wrote you in January

---

22 Ruth Krader (1911-1996), Sinologin aus Hamburg, Leiterin der East Asiatic Library der University of Washington. Vgl. H. Walravens: Ruth Krader in memoriam. *NOAG* 159/160.1996, 13-14, Porträt.

and from Dr. Krader who wrote you at the beginning of February, that your library has arrived safely. It will take some time to unpack it and then you will receive a detailed report as Dr. Krader promised.
My best regards,
Yours sincerely,
Franz Michael
Acting Director

cc: President Everest[23]

21

University of Washington
Far Eastern and Russian Institute
Seattle 5, Washington

May 20, 1952

Dr. J. F. Rock
c/o Royal Horticultural Society
Vincent Square
London, England

Dear Dr. Rock:
Thank you very much for your good letter of May 10. We are looking forward to meeting you and having you here with us. If you will tell me beforehand the date of your arrival, we will make a reservation for you in the Meany Hotel.[24] This is very close to the campus and comfortable and pleasant. I am sure you would be satisfied there until you have found something else. We will prepare a desk for you in our research room. The library is in the basement of our building and there are tables in a partitioned room, but you will probably prefer to have the desk upstairs.

As to the books we have your original list, we have checked the books against your list but did not find all of them – they may be in the trunks. We have not yet opened any of the twelve trunks and shall wait until your arrival before we open them. We found books which were not on your lists so we made a list of them which we will give

23 Everest, vermutlich: Everett?

24 Meany Hotel, das spätere University Tower Hotel in der Nähe des Campus.

you here. You will find that your books are well-stored and taken care of.

Our vacation begins and our spring quarter ends June 13. There is a Summer School from June 23 to August 22. The following quarter starts the end of September. I may not be here myself during the summer, but Doctor Taylor and Doctor Wilhelm will be here until August 18. I hope though to see you before I leave in July.

With best regards,
Very sincerely yours,
Franz Michael
Acting Director

22

University of Washington
Far Eastern and Russian Institute
Seattle 5, Washington

December 11, 1953

Dr. Luther Eisenhardt[25]
American Philosophical Society

Dear Dr. Eisenhardt:
At the suggestion of the Director of the Far Eastern and Russian Institute I am writing to you about the problem of completing my work on a dictionary of the Na-khi language.

From the year 1930 on I dedicated myself to the serious study of the voluminous Na-khi literature written in pictographs with annotations in syllabic characters, after I had been associated with members of the Na-khi tribe from May 1922 to December 1929 during which time I learned their language.

I started making collections of Na-khi manuscripts as early as 1923 and later from 1930 on they were collected systematically. Their religious ceremonies, which are shamanistic rites of Bön origin, and of which there are about 120, were studied and the major ones I had performed for me in Li-chiang. The ritual texts belonging to or chanted during these ceremonies were assiduously assembled from all over the Li-chiang district wherever dto-mbas or priests lived. I was thus enabled to amass the most comprehensive collection of original

---

25 Luther Eisenhardt (1876-1965), Mathematiker und Pädagoge. Vgl. *Biographical memoirs, National Academy of Science* 40.1969,69-90.

Na-khi pictographic manuscripts, many of them beautifully illuminated and dating back to the early days of the Ming dynasty. These manuscripts were carefully sorted according to ceremony, numbered, their titles transcribed and translated and arranged in the sequence in which they were chanted.

From 1930 to 1944, with minor interruptions, the work of translating hundreds of their manuscripts with the help of two dto-mbas or priests was carried out; all this I did at my own expense.

The interruptions were caused by communist invasions of Yunnan in 1935 and 1936 when foreigners were evacuated to Indo-China. In 1938 the Japanese bombed Kunming and it was again necessary to move to Indo-China; I assembled my servants and accompanied by the two dto-mbas moved to Dalat where we remained until the capitulation of France and when the Japanese moved into Indo-China. The work had to be interrupted, the Na-khi were sent home and I returned via Siam, Manila to Honolulu with my entire library and manuscripts.

During the time translation work was in progress every pictographic symbol encountered was written down, its origin explained, its Na-khi sound complex transcribed and the tones noted. As the pictographic nature of the script does not lend itself to express abstract ideas the dto-mbas employed the symbols phonetically for such words for which no symbols existed, selecting them according to sound complex and tone required for these abstract ideas.

Names of many hundreds of gods, spirits, Nagas, minor deities and demons are written in part phonetically in pictographic symbols, often accompanied by syllabic characters similar to the Japanese Kana employed alongside Chinese characters to ensure the correct reading of the name. This resulted in a collection of thousands of words, symbols and phonetic characters which filled four volumes of 300 pages each, plus a phrase book to explain the syntax and the grammar.

All the results of my labors were destroyed when the Japanese sunk the ship which carried my material to the States from Calcutta except the dictionary of which a photostat copy was kept in the United States. Fourteen volumes of notes and translations plus many precious Na-khi mss. and books were thus lost.

After the war I was determined to return to Li-chiang to start the work over again and repair what has been lost. As a Research Fellow of Harvard-Yenching I returned to Li-chiang late in 1946 and remained there until August 1949. The communists by that time had taken over the government in Li-chiang and it was not advisable to remain there. Furthermore most of the Na-khi priests were forbidden to practice their religion and I could get no one to work with me.

Since my return several volumes on the Na-khi literature have already been published with subvention from Harvard-Yenching. There remains however the crowning work of all my labors the Na-khi pictographic and phonetic dictionary, but it is not possible for me now to carry on the work at my own expense. The preparation of the dictionary will take a year. It will be arranged according to subject.

First: pictographs pertaining to heaven, sky, astronomy, etc.
Second: pictographs pertaining to land, earth, fields, topography, mountains, rivers, geographic terms and names of places in and outside of Li-chiang.
Third: pictographs pertaining to man, his activities, ancestors, historic characters and family relationships
Fourth: human anatomy
Fifth: animals, insects
Sixth: birds
Seventh: plants, economic, medicinal and wild, with their scientific names
Eighth: food and drink
Ninth: implements used by the Na-khi
Tenth: clothing and adornment
Eleventh: warriors, ancient weapons
Twelfth: houses, construction, tents
Thirteenth: numerals
Fourteenth: abstract ideas, qualities, sensations
Fifteenth: Tibetan loan words and their original equivalents
Sixteenth: Chinese loan words and their original equivalents
Seventeenth: gods
Eighteenth: spirits and protective tutelaries
Nineteenth: Nagas, ghosts, spirits, snake demons
Twenty: demons
Twenty-one: dto-mbas or priests (Bpö-mbö)
Twenty-two: ceremonial objects and paraphernalia
Twenty-three: religious ceremonials
Twenty-four: titles of books and their meanings

Although the Na-khi system of writing is a mnemonic one actual facts and descriptive matter is more or less fully written leaving allegoric matter, etc., unwritten, which missing symbols the dto-mbas, knowing the story by heart, read or supplied from memory.

By giving the names of all the gods, etc., as they occur written in the pictographs in the manuscripts, the symbols selected being used phonetically and preceding either the symbol for god, demon, spirit, etc., it will be possible for future investigations to sift the names from the text which can then be deciphered as far as written.

Side by side with the pictographic symbols I shall give the phonetic or syllabic characters or ggo-ba letters. At the end of the Lexicon I will add an alphabetic list of all words occurring in the same with reference to the page number where they are to be found.

Wherever possible reference to the particular text or manuscript in which a symbol occurs as given in the lexicon will be added.

In the phrase book samples of compositions giving both the pictographic text, transcription, translation and the explanation of untranslatable words namely prefixes, suffixes, etc., and comparable to the Chinese so-called empty words [虛字] will be given.

Sincerely yours,
Joseph Rock

23

University of Washington
Far Eastern and Russian Institute
Seattle 5, Washington

July 21, 1953

Dear Rock:

Forgive the typewriter but there is the usual rush. Thanks for all the postcards – the Wetterhorn looks as good as ever and for your good letter from Zürich. You seem to be in the right spirit and I am sure the funny book is already in the making.

Just got a letter from Paul who has sent your application to Sauer. He – that is Paul – likes it very much. But he says that the most important part is yet missing, the question of how much money you need and ask for. I have written to Paul asking him whether it would not be best if he would ask for the respective amount. What we have in mind would be the equivalent of a full professorial salary, that is, $ 7,500 a year or $ 15,000 for the two years. In addition, you could ask for $ 2,500 a year for typing assistance. That would make a total of $ 20,000. What do you think of this? And would it be all right if Paul asks for these figures?

Paul also writes that the Wenner Gren Foundation (the Viking fund) is closed until Sept. 15th, already since June 1st so that an application there will have to wait. If Guggenheim does not work, George thought also of writing for you to Rockefeller. So let's keep our fingers crossed.

The weather is fine all the time and we have had some more mountain hiking.

Many happy schnitzels and love from all your friends to whom I have given your greetings.

Herzlichst Franz[26]

I hope you got the previous letter.

24

University of Washington
Department of Far Eastern and Slavic Languages and Literature
Seattle 5, Washington

April 26, 1955

Dr. J. F. Rock
Hotel Pitscheider
Merano, Italy

Dear Dr. Rock:

I am sending you a letter which I have written to Professor Tucci.[27] It is self-explanatory. Our request for the money was held up, but we think that the request will go through if we have a letter of the sort that I have indicated in my note to Tucci. This follows pretty much the pattern set by Harrassowitz. I am sending Tucci's letter to you so that you can hold it up in case there is anything in it you feel would be unacceptable.

If the letter is alright would you be good enough to send it on to Tucci right away. If it is not, then perhaps you could write to me and suggest how it should be re-phrased.

---

26 Michael.

27 Giuseppe Tucci (1894-1984), Indologe und Tibetologe, Gründer des Istituto per il Medio ed Estremo Oriente in Rom, der Serie Orientale Roma und der Zeitschrift *East & West.* Vgl. Raniero Gnoli: *Ricordo di Giuseppe Tucci;* con contributi di Luciano Petech, Fabio Scalpi, Giovanna Galluppo Vallauri. Roma: IsMEO 1985. 79 S., Porträt (Serie Orientale Roma 55.)

The laws of the State of Washington are that public funds can be used only for the purchase of goods or services – hence the request for ten volumes. Harrassowitz gave us twenty-five but we paid him $ 1,000.

I am sorry to put you to this trouble, but we do want to get the money to you, and this should clear up the matter pretty quickly. At least we hope so.

We talk about you very often, and miss your genial presence. With kindest regards.

Sincerely yours,
George E. Taylor
Director

25

University of Washington
Department of Far Eastern and Slavic Languages and Literature
Seattle 5, Washington

April 26, 1955

Professor Tucci
c/o Dr. J. Rock
Hotel Pitscheider
Merano, Italy

Dear Professor Tucci:

Dr. Rock has informed me about the arrangements that you are making for the publication of this book. We are delighted that you are publishing this book which we consider to be a very important addition to scholarship in this field. As you know, the University of Washington wishes to make a contribution towards the cost of publication. We are able to suggest the sum of $ 400.

In order to assign this money for this purpose, however, it is necessary for us to have a letter of understanding from you which will describe the basis on which the money is accepted. I wonder whether you would be good enough to send me a letter, at your earliest convenience, in which the following information would be included:

1) the title of the book, and the exact name of the publisher,

2) the total estimated cost of the book, and the amounts of money contributed from all sources,

3) The credit shown on the title page for the University of

Washington (I suggest that the book could be listed as one of the University of Washington's Studies in Asia series. This is what Harrassowitz did with Professor Poppe's *Grammar of Written Mongolian*.[28] If this is not agreeable then I would suggest that mention be made that the University of Washington contributed toward the book and that Dr. Rock be identified as a research associate of the Far Eastern and Russian Institute.)

4) that the University of Washington would be presented with ten copies of the book.

Such a letter would make it possible for us to secure the money fairly rapidly. All these matters are necessary in order for us to comply with the laws of the State of Washington.

Sincerely yours,
George E. Taylor
Director

26

University of Washington
Far Eastern and Russian Institute
Seattle 5, Washington

April 26, 1956

Dr. J. F. Rock
Hotel Victoria
Kandersteg, Switzerland

Dear Rock:

I am sorry that you had such bad luck with the weather and some trouble too. I remember Kandersteg quite well from younger years when we hiked up to the Oeschinensee and did some mountain-climbing in the area of your postcard. You may have left the place by now, but I am writing you anyhow in the hope that it will be forwarded.

Thank you also for the letter from Tucci. I am very happy to hear that you have nailed him down on the publication, and I hope that you had some favorable answer from Harvard-Yenching on the subsidy. In the letter of Tucci, he speaks of a *subvention of at least 2000 or 3000 Dolls*. For that, I would suggest going to Hollywood; but it

---

28 Wiesbaden: Harrassowitz 1954. XII, 195 S. (Porta linguarum orientalium. NS 1.)

seems that even Tucci's Institute would have a hard time using them all ...

Here work is going on. George Taylor's and my book has at last come out, as I may have written you already. You will get a copy, either here or whenever I know of a somewhat more stable abode of yours than you seem to have at present.

At the moment, we have the visit of my brother and his wife and a cousin of Anita's,[29] so the house is full and things are somewhat hectic. The work here is going on, on the good old manuscripts. Take good care of yourself so that you will be in good shape when you come back. Best regards from Anita and your friends.

As ever, Franz

Just got your letter with the book outline; looks good. Let me know what Harvard says. Herzlichst Franz

27

University of Washington
Far Eastern and Russian Institute
Seattle 5, Washington

April 6, 1956

Mr. Allen Belden, Acting Director
Office of Geography – Office of the Secretary
Room 7134 – Interior Department
„C" Bet. 18th & 19th Sts. NW
Washington 25, D.C.

Dear Mr. Belden:

We received your letter of April 3 and are very much concerned about the disappearance of the maps. We have made a thorough search at our end and are certain that the maps have not arrived here.

As I told you, we are dependent on these maps for the coming publication of Dr. Rock's to which we have given our support. We therefore take a very grave view on this matter.

Please let me know immediately the result of your further attempts to locate the maps. They are invaluable documents of an area which has not otherwise been mapped, and we and Dr. Rock will have to hold the department responsible for them.

---

29 Anita Goldschmidt, Michaels Frau. Die Ehe wurde 1965 geschieden.

I expect to hear from you soon.
Sincerely yours
Franz Michael
Acting Director

28

Honolulu Nov. 30/56

Dear Franz:

I am sorry not to have been able to send you a sketch showing the entire length of territory in West China for which I have coverage plus photos, but I do so now. However it occurs to me that certain sketches as from Chieng-law (Keng-law of the maps) on the southwest corner of Yün-nan to Ta-li in West Central Yün-nan I have only a meager stretch up the Mekong Valley in part, then up the Black River Valley and to the source of the Red River to Meng-hua South of Hsia-kuan. We did not deviate from the route as our object was to get to Ta-li.

Only from Ta-li North, West and East have we better coverage. Then of course we broadened and covered the entire Na-khi country. The entire Mu-li area, Konkaling and Mi-nyag country to Ta-tsien-lu (K'ang-ting).

The area from K'un-ming to Huang-chiang on the Yangtze west of Suifu; from there we went up by houseboat to Chia-ting, and thence to Cheng-tu. For that we have no coverage as we were on a houseboat. Photography was not too good either due to the haze on the rivers and to our tying up on sandbanks where we were away from the towns which I disliked and stayed away from on purpose.

From Chengtu the route to Kansu covered a very narrow strip to Mien-ho, Lo-chiang, Chiang-yu, Chang-ming to Chung-pa and Ching-ch'uan the last town in Szechuan. Thence to Mo-tzu-p'ing, P'i-k'ou, Wu-tu partly up the Wan-hsien River and Pai-shui chiang, to Min-chou – T'ao-chou and Choni. That is a narrow strip up the gorges of the Pai-shui chiang, where there was no outlook. Photography was not too good owing to climatic condition, mostly snow storms, although I have some very good pictures of stretches.

So the routes to headquarters cover only a very narrow strip, a valley or valleys, and only where we had headquarters whence we explored the country, like the The-wu country where we spent months, collecting, making maps and carried on extensive photographic surveys, would appear to me worth while to reproduce maps of with photos; also the unsurveyed stretch from Cho-ni to Sung-pan, the extreme northwest of Szechuan would really be of interest.

Thence from Choni north to Lan-chou – Hsi-ning Ko-ko-Nor, to the Richthofen Range of which only the latter part would be of interest. In the densely populated Chinese areas I did not take pictures or only a few. The Chinese areas, all well mapped did not interest me. Then of course we have the maps now being published 5 sheets, for these we have many photos, except the loop within the Yellow River.

The best areas are the Northwest of Yünnan to Tsarong in Tibet, already published in my Na-khi Kingdom; these however can be augmented with areas extending to the Southwest to T'eng-ch'ang (T'eng-yüeh) up to Yün-lung on the Lo-ma Ho to Ch'ien-ch'uan – Ta-li and Ta-li to Li-chiang.

The second in Li-chiang – Yung-ning and the entire Mu-li Country from the Wa-liang Ho in the west to the Ya-lung in the East, then the Konkaling to the west of Mu-li, and the Chia-la Kingdom to the East and Northeast to K'ang-ting, for all these areas we have excellent coverage and photos.

The third would be the Kan-su and Ch'ing-hai areas mentioned.

We did not make maps of densely populated Chinese areas in Szechwan, first because those areas are already mapped, and second we did not wish to attract attention.

Therefore it would seem to me best to publish maps of such areas insufficiently known or poorly mapped or not at all mapped; this includes the areas where we had headquarters, and thence radiated in all directions exploring for years at a time. I would therefore suggest to leave out the areas we simply traversed and w[h]ere we do not have good photos. For example, the Black River (Pa-pien Ho) is uninteresting, a flat valley with forested banks.

Please discuss this with Dr. Sherman[30] and let me know. We are having southerly weather 100% humidity and downpours, the sky is black. I have a good notion to come back to Seattle as I don't accomplish anything here. It is a waste of time. I have been making an index to my diaries which will greatly help me in the geography

---

30 John Clinton Sherman (1916-), Geograph, lehrte seit 1942 an der University of Washington, 1963-1986 als Professor of Geography. Vgl. *Who's who* 1994, 3143.

work as it also gives the numbers of the negatives, detailed routes, etc. But that is a big job, for example vol. 20 of my diaries will have some 500 cards, and I have 14 volumes here.

Please write me soon and let me know what you and Sherman think about my prosposal publishing only maps of the more important areas where we spent years and explored rather than areas traversed. I will sketch the areas where we have better coverage and photos.
So I await your reply. In the meantime my kindest regards, cordially yours, Rock

P.S. I am flying to the big island today to get cooled off, will be back in a week from volcano.

29

University of Washington
Far Eastern and Russian Institute
Seattle 5, Washington

December 3, 1956

Dear Rock:
Thanks for your good letter. I am sorry that the climate of Hawaii hasn't kept its promises so far, but writing today with electric light and with miserable rain outside I still think that you will feel better amid the beauty of the islands than in this grey atmosphere.

Anita has written you that I got my grant. I am of course happy to go. Even if I am not the inveterate traveler you are, I have after all the same itch for it has been a long time that I have sat pat. I am going along, and Anita, who is staying here, will represent the Michaels to you in my absence. I will be back in July and I hope we will go to the mountains together then. You write that you might come over for Christmas. That would of course be awfully nice. But if not, we will have to wait till the summer.

As to your letter, Elaine will arrange for the plates and we shall send them to you as soon as they are done, or do you want us to keep them for you? I have spoken with Sherman who is drawing up the project. Have you yet had time to put the over-all area on a map? Elaine wants to know if you want the cards that are ready. She has finished the alphabet and is up to the gods and spirits of pleasant and nasty types.

Forgive the typing. I am quite pressed for time and this way I can say more and say it faster.

Take good care of yourself. There is lots to be done when you come back. Anita sends her love. All my very best.
Your old friend Franz

P.S. Just had your letter with the sketch and talked with Sherman. We feel in general that it would be very worth-while to use all the maps, and add only as much in pictures and in shading of the maps as seems good to all of us in each case. How many photos should be taken will of course depend on what you think of the pictures and also on the cost. As to your sketch, we were not quite sure what it included. You divided several areas called *strips* and referred to them in your letter, but there are also dividing lines without comment, and it is not quite clear whether all the area in the lines is included and how far it extends.

Then you have to consider that names very familiar to you cannot always be located by people not so familiar with the area. John Sherman is trying to draft an outline, and we will send it to you to check and complete it, but he asked whether he could have as many of the maps as possible now so that he could not only determine the whole area but also sketch in the area covered by each of the maps of the atlas. I believe you have the maps in the trunk here and I would like you to tell me what to do. As soon as possible, we want to apply for a grant to cover your expenses as well as the expenses here, and Sherman feels quite optimistic, so please let me know about the maps and add whatever words of advice you want to give us.
Herzlichst Franz
Ruth ist gerade hier und lässt sehr grüssen.

30

University of Washington
Far Eastern and Russian Institute
Seattle 5, Washington

April 24, 1957

Dr. J. F. Rock
University of Washington

Dear Dr. Rock:
I am sorry that you seem to have misunderstood my well-intentioned effort to assist your research. Obviously the main reason for the

seminar is, as you say, to submit the Dictionary for discussion and suggestions which would prove the value to its final execution. I cannot imagine where you got the idea that this was a matter of „approval“. What I suggested was that as we could not ourselves provide $ 2,000, we would have to try to raise it outside the University. For this purpose the seminar would be most useful if your colleagues got a clear understanding of the Dictionary and could, therefore, write well informed and enthusiastic letters in support of a subsidy. That is all that I have tried to suggest, and all that I have in mind. Obviously the Dictionary is of the greatest importance. If any one of us had been in doubt about this, we would never have suggested discussing the matter any further with you.

I hope that you will reconsider your reaction and that you will be willing to have a meeting so that we can make every effort to press to a speed conclusion. As I have told you before, we expect to have funds available within a few weeks which will enable us to provide more than adequate secretarial assistance. I think the amount of support that we have already provided you with should be sufficient indication of the pride that we take in your work and the pleasure we derive from your association with us.

Sincerely yours,
George E. Taylor
Director

31

University of Washington
Far Eastern and Russian Institute
Seattle 5, Washington

October 17, 1957

Dr. J. F. Rock
Far Eastern Department
University of Washington
Seattle 5, Washington

Dear Dr. Rock:

It is with great pleasure that I confirm to you in writing that we have been able to allocate $ 1,000 for the support of the publication of your book by Dr. Tucci in Rome. This amount is being given to you here as a matter of convenience, and it is our understanding that the amount will be handed by you to Dr. Tucci as part of the publication costs of the Na Khi dictionary.

I want to tell you again that we are very happy that we can give this amount of help toward the publication of such an important and unique piece of scholarschip and believe that our effort will indicate again our great interest in your work.

With best wishes,
Yours sincerely,
Franz Michael

32

University of Washington
Seattle 5, Washington

November 13, 1957

Dr. J. F. Rock
P.O. Box 497
Honolulu 9, Hawaii

Dear Dr. Rock:

I am sorry that I was unable to get over to your office before you left and to accept from you personally the 64 mounted sheets of *Rhododendron* from the Royal Botanical Garden at Edinburgh. I consider this accession a very valuable one and appreciate greatly your generosity in turning the material over to us. I hope that you will have the opportunity to convey my appreciation to Mr. Davidian also.

We shall keep the collection together as a unit and will be happy to lend it to you whenever you may have occasion to use it.

With most sincere personal regards, and with the hope that we may occasionally have the pleasure of seeing you if and when you come to Seattle, I am,

Very truly yours,
C. Leo Hitchcock[31]
Executive Officer
Department of Botany

---

31 Charles Leo Hitchcock (1902-), promovierte 1931 an der University of Washington. Verfasser (mit Arthur Cronquist Jr.) der *Flora of the Pacific Northwest*. Seattle: University of Washington Press 1973. XIX, 730 S.

33

April 25, 1958

Dear Rock:
Your letter from the ranch gave us a shock. I am sorry you had that attack and hope it was a light one. I am glad you take it easy, and your letter showed that you are in good spirits. In any case I wouldn't run for president yet.

I wrote you some time ago but my letter is probably waiting for you in Honolulu, and I am sending this one also to the old address since you say that you are flying back there. To this address I have sent also a copy of the report, which should reach you soon, and I am very curious to hear what you think of it.

You write about the duplication of the original cards of your book. You know I am all for it, as I wrote you before, but I mentioned that it will run quite a bit of money and that I should know roughly how much before I apply to the authorities in power for the requisite sum. I don't want to be between two fires – that is, the heavy artillery of the University and your 45 automatic Colt.

I'll stick my neck out for you as far as it reaches, but I have to ask the University for something specific and cannot give a blank promise. So if you can tell me at least roughly how many cards there are I'll do my best to rob the bank.

As to the English-Na-khi dictionary, I think I wrote you once before about the difficulties. Gladys has tried but has simply found it impossible to work out any system that would be helpful. This is a little difficult to explain on paper, but I'll try to give you at least a hint what it is all about. We take at random the cards under the letter „B". The first one is the „bone and meatless dried pig". Next: a lama handbell (which is phonetically written only.) The next: a „pig with white front legs or four white legs". Next: „a descending pig." Next: „to cause to congregate." Next: „the first turn." Next: „a woman is thus called when she gives birth to a child after her husband's death." Next: „a domestic pig killed and offered to the demons." Next: „to think (of a companion)." Next: „a girl born after her father's death." Next: „a boy born after his father's death." Next: „to look for a wife." Next: „to castrate a pig without cutting off its tail is equal to litigation without end." Next: „gone to look for a wife."

All these are the items under ^{1}bu. Some of them could be given under the English „pig" but it wouldn't be very enlightening to say:

„pig, descending“ or „pig, killed and offered to the demons“, or, „pig, to castrate without cutting off its tail is equal to litigation without end“, or, „pig, the bone and meatless dried.“ Not to speak of the difficulty of finding the lead word for the other items listed. And yet the meaning of these items is lost if they are not given in full. Your book is much more than a dictionary of words. It expresses the whole culture, and can therefore not be turned around in a pedestrian fashion. Gladys has been worried about this, and after we talked about it it seems to me that it wouldn't serve any real purpose and might not look very scientific if such a listing were attempted. This is at least my limited understanding of the matter. [Schluß fehlt.]

34

University of Washington
Far Eastern and Russian Institute
Seattle 5, Washington

June 3, 1958

Dear Rock,
This is just a sign of life and to tell you that we are well and that summer has come. You have heard from Anita about us. We hope very much that you are in good shape and take good care of yourself. A word of caution I know is hopeless, but I want to add it anyhow.

I got your telegram and your letter and gave them to Gladys. I still have questions on this but I think they are best discussed when you come. By the way, when do you come, and what is your present status of your plans for Italy?

We had a number of visiting firemen, some of whom you would have enjoyed. There was a British MP who had been to China and talked very interestingly about it -- his name is Hill. He had his eyes open and yet has some good and serious questions about what they are doing and what we are doing. There was a German Count and member of the Bundestag who was very interesting and reasonable. There is the usual quota of *scholaren* coming on their way to and from other places, so we hear what's going on in the big bad world.

Have you ever experienced any thing as preposterous as the French show? I always thought that the main crisis for the Western world would come in the Far East of the Middle East, but we should have known better.

Our own world is the same: Taiping documents[32], Gentry Income, and other manuscripts, and the progress is slow, much too slow. Chang Chung-li[33] is still with us, and the future is still undecided. I still need a vacation – don't we all – but I'm also still skiing on the week end on good old Firnschnee, and it is truly beautiful up above five thousand. Of course, for you that is no height at all.
Ihr alter, müder Freund Franz

35

University of Washington
Far Eastern and Russian Institute
Seattle 5, Washington

July 2, 1958

Dear Rock,
Thank your very much for your good letters. It was good to hear that you are well. I hope your back doesn't trouble you too much either.
The news here is that we are grandparents. Vera[34] has a little girl named Anita Elizabeth, weight seven pounds twelve ounce, twenty-one and a half inches long, and in the meantime she has gained in stature. Mother and child are well. Everybody is happy, life is going on. Since you do not smoke cigars, we reserve candy for you when you come.

I will have in August the visit of my brothers from New York[35] and Texas[36], and before that I may take a week's vacation to climb somewhere in the mountains. Otherwise work is going on.

As to your work, I was glad to hear that you can get the cards copied there. I have asked George Taylor about your request to help by providing the paper. We are buying $ 75.00 worth of paper for delivery there, which will save the expense of mailing it.

---

32 *The Taiping rebellion.* History and documents. 1-3. Seattle: University of Washington Press 1966, 1971. XII, 244; XXIX, 1815 S.

33 Chang Chung-li (1919-), Professor an der University of Washington, veröffentlichte u.a.: *The Chinese gentry.* Seattle: University of Washington Press 1955; *The income of the Chinese gentry.* Ebenda 1962.

34 Vera Ingrid Osterhang, geb. 1936.

35 Walther Michael (1913-), Prof. für Wirtschaft.

36 Wolfgang Michael (1909-), Professor für Deutsch, University of Texas, Austin.

I am returning the letters from Tucci and Schubert[37]. I don't see how you could do or could have done otherwise with his translation and just let it go. Tucci of course is traveling as ever. He is really an amazing man.

Take good care of yourself and don't forget us. We are going to keep some mountain-flowers for September and hope for good weather then.
As ever Ihr alter Freund Franz

P.S. Anita thanks very much for your letter and sends her love. She is busy with helping Vera over the first days of motherhood but will write soon herself.

36

University of Washington
Seattle 5, Washington

24 August, 1958

10666 Exeter N.E.
Seattle 55, Washington

Dear Dr. Rock:
Just received your letter of the 19th. Since you did not say anything to the contrary, I assume happily that you are feeling well.

First of all, about a copy of my thesis. I am truly sorry that I just haven't been able to have a copy made for you yet. I meant to do it in July, but couldn't. The only delay is due to my present financial situation. I have been unemployed since the 15th of June, and since the cost of having a copy reproduced for you is not really too expensive it is just enough so that I have had to put it off. You see, I also wish to have a copy made to send on to Tucci at the same time I have your's made. The reproduction is done by a new „light-developing" process called ozalid, and it will be better for all concerned if I have both copies done together. So – I anticipate going

37 Johannes Schubert (1896-1976), Bibliothekar an der Universitätsbibliothek Leipzig, dann Professor für Tibetisch und Mongolisch an der Universität Leipzig. Vgl. *Asienwissenschaftliche Beiträge Johannes Schubert in memoriam.* Hrsg. v. Eberhardt Richter und Manfred Taube. Berlin: Akademie Verlag 1978. 204 S., 44 Taf. (Veröffentlichungen des Museums für Völkerkunde zu Leipzig 32.)

back on the payroll the middle of next month, and shall have your copy made just as soon as I am solvent again. My wife and I have been living off our meager and fast diminishing savings since I went off the payroll in June, so I hope you'll understand and be patient. I shall get a copy to you as soon as possible.

Now, in regard to the occurrence of ཨ་མི་ a-mi in the Deb-ther-sngon-po. I check out the names beginning with A-mi and A-me in [George N.] Roerich's Blue Annals.[38] Those occurring in the first volume are used like those in the second volume, i.e., used as if A-mi was a clan (or place) name. A-mi is not used alone in any of its occurrences in the Blue Annals, but alway as the first part preceeding another name, as is the usual way for clan, or place, names.

I also checked the *Tibetan Painted Scrolls*[39] for A-mi. There are two instances.

1. – p. 634 A mi byang chub 'dre 'khol
2. – p. 632 A mi mu ji khri do

Here again, the usage of A-mi indicates a clan, or place, name. Both names belong to descendants in the genealogy of Phag-mo-gru.

Concerning the second one (A mi mu ji khri do), it occurs in Das' Tibetan Dictionary as follows:

A-mi mu-zi khri-do one of the earliest tribes of Tibet; acc. to some: an ancient dynasty (J. Zañ.).

[J. Zañ. representing: Pag-sam-jon-zang]

When you asked for the *exact reference for a-mi in Deb-ther-sngon-po*, precisely what is it you wish? Do you want the folio number in the Deb-ther original Tibetan xylograph where A-mi is mentioned? If so, which of the 7 A-mis do you want? Roerich gives the folio numbers beside his translation, so it would be a simple matter to supply the folio numbers for those names in the first volume which you do not have in Honolulu. But, if the A-mis listed in the BA refer to various people with a full name of which A-mi is only the first part ... would it serve the purpose of your work?

It is possible that A-mi, A-me, may occur in Nebesky's *Oracles and Demons of Tibet*[40], in a more favorable context, i.e., as the name

38 *The Blue Annals*. Part 1-2. Calcutta 1949, 1953. XXI, 1275 S. (The Royal Asiatic Society of Bengal, Monograph series.7.)

39 Tucci: *Tibetan painted scrolls*. Roma: Libreria dello Stato 1949. 3 Bde. 2°

40 René von Nebesky-Wojkowitz (1923-1959), Wiener Tibetologe, war

of a sorcerer or deity. Unfortunately, I don't own a copy of it (at $ 13.50 a copy, who can afford it?), but I shall check it out at the University Library and see if there is anything interesting there.

I checked out several of the works I have here at home, but didn't find any further A-mis. Only ཨ་མེས་ A-mes which is quite a common clan name.

Please let me know precisely what you wish as far as *exact reference*, and I shall do my best to help out.

Hoping that you are in the best of health and spirits and enjoying your work, (and please give our very best regards to the Marks[41])
I remain, yours sincerely
Turrell Wylie

37

University of Washington
Far Eastern and Russian Institute
Seattle 5, Washington

12 September 1958

Dr. J. F. Rock
P.O. Box 497
Honolulu, Hawaii

Dear Rock:
Thank you for your letter of September 7. I am sorry you were ill again and hope you have completely recovered. I took up a search for the films in question with the help of Mr. Roy Wright, the present director of the film center, who was however not yet in charge at the time when the films were made. He had only copies of the films which were given to the film center and are their property. It was my mistaken belief that you had taken another set of originals along, and also that you had the original positives from which the copies were made. We have here negatives of the films in the hands of the Institute. From this it would seem that there are two negatives and one positive rather than one negative and two positives. The company

---

zeitweise als Rocks Amanuensis tätig. Vgl. J. W. de Jong: René von Nebesky-Wojkowitz. *Indo-Iranian Journal* 3.1959, 306-309.

41 Rock hatte Lester und Loy Marks in China kennengelernt; sie luden ihn ein, bei ihnen in dem Anwesen auf der Old Pali Road zu wohnen, was er schließlich auch tat.

which made the films does no longer exist so that we have no way to find out what they did or hold them responsible. The director of the film center is sending to you, by airmail, the four films which he has so that you can show them to your friends. Would you please send us the negatives which you have so that we can compare them with the negatives here and perhaps find out what happened. Would you also please tell me whether you still have the original, brittle films from which the copies were made. This is all I can say at the moment but I will try to find out when we see your negatives and compare them with the negatives we have here.

Otherwise there is not much local news. My brothers have left and I am up to the ears in work. The weather is dry and beautiful. We do need more rain. The international situation is zum heulen. How we ever could maneuver ourselves into such a mess is unbelievable.
Take good care of yourself.
Herzlichst, Ihr alter Franz

38

University of Washington
Far Eastern and Russian Institute
Seattle 5, Washington

16 February, 1959

Dear Dr. Rock:
I have finally read through the Gzer-Myig[42] text by Francke. Incidentally, the years of the volumes are: 1924, 1926, 1927, 1928, and 1930. (You had 1939 instead of 1928). I'm sorry to report that I was unable to find any equivalent for your 1- Gkyi-ma-wu-ts'u, or for 2- Ssu-mi-muan-dta-k'aw-ssu-ma. I read through the entire five sections, but found only where Gshen-rab-myi-mo had one wife, and it is your si-zaw-gyi-man, the Hos-za-rgyal-myed-ma (ཧོས་ཟ་རྒྱལ་མྱེད་མ།) in Tibetan. The exact references for: Hos-za-rgyal-myed-ma are: Fol. 100a, line 5, Page 39; and Folio 100b, line 3, page 40, of Gzer Myig, part VI, *Asia Major,* vol. 5, 1928. Hos-za-rgyal-myed-ma was the daughter of the king named Dang-ba-yid-rings and the queen named Phya-za-agu-ling-ma-ti.

---

42 gZer-Myig. A book of the Tibetan Bonpos. Tibetan text according to the Berlin manuscript. [Books 1 to 7.] Edited and translated by A. H. Francke. *Asia Major* 1.1924, 243-346; 3.1926, 321-339; 4.1927, 161-239, 481-540; 5.1930, 1-40; 6.1931, 299-314; NS 1.1949/50, 163-188.

If there is some other work I could check through for you, please let me know.

In regard to Dr. Lokesh Chandra, who wrote you concerning the letter enclosed: I met him in India in 1957. He is the son of Dr. Raghu Vira,[43] and apparently a very learned fellow. Dr. Vira, Dr. Chandra, and Dr. Chandra's wife (who is also a Dr.) run the International Academy of Indian Culture (or some title like that) in New Delhi. Dr. Chandra is publishing a 20 volume Tibetan-Sanskrit dictionary, the first volume of which is already out and has received wide acclaim. I think you can feel perfectly safe in entering into any scholarly undertaking with Dr. Lokesh Chandra. You should find him, as well as his father Dr. Raghu Vira, to be fine scholars, industrious, friendly, and cooperative.

Are you still coming to Seattle soon? I hope so! I've almost finished the revisions on the 'Dzam-gling-rgyas-bshad thesis. I expect to turn it over to the photocopier after payday.

My wife and I are well, and looking forward to seeing you again soon.

Our best regards to the Marks. Take care of yourself.

Sincerely yours,
Turrell Wylie[44]

39

University of Washington
Far Eastern and Russian Institute
Seattle 5, Washington

13 March 1959

Dr. Joseph F. Rock
P. O. Box 497
Honolulu, Hawaii

---

43 Raghu Vira (1902-1963), bedeutender indischer Indologe, Gründer der International Academy of Indian Culture und einer umfassenden buddhologischen Publikationsreihe, Śata-piṭaka. Zahlreiche, teils von Lokesh Chandra posthum herausgegebene bzw. fortgeführte Werke.

44 Turrell V. Wylie (1927-1984), später Professor für Tibetisch an der University of Washington. Vgl. *Reflections on Tibetan culture.* Essays in memory of Turrell V. Wylie. Edited by Lawrence Epstein & Richard F. Sherburne. Lewiston, Queenstown, Lampeter: The Edwin Mellen Press 1990. XIII, 350 S. (Studies in Asian thought and religion 12.)

Dear Rock:
Thanks for your letter of March 9. I have taken note of your change of arrival and have informed the Wilsonian. The room will be ready on April 6 at 2:00, otherwise you would have to pay from April 5. If you want that let me know. And I will be at the airport with Ehrenjungfrauen.
Herzlichste Grüsse der Familie Ihr alternder Freund Franz

40

University of Washington
Far Eastern and Russian Institute
Seattle 5, Washington

3 June 1959

Dr. Joseph F. Rock
c/o American Express
6 Haymarket
London, England

Dear Rock:
Thank you very much for your letter from London with all the news about your social activities. Did you curtsy before the Queen or did you get confused by all the flowers?
On the same day that your letter arrived I got a bill from Jensen and I sent them today my check for the amount of $ 22.51. I called them up to find out what happened to the nine pieces of luggage which they picked up at the Wilsonian and our library, and I learned that they have been shipped to Honolulu and are being held under *Will Call* at the Matson terminal in Honolulu, but will be delivered to whatever address you want them sent. Since I was not sure where you wanted them, I am asking you to write either to Norman G. Jensen, Inc., 75 Marion Street, Seattle 4, Washington or to yours sincerely as to where they are to go.

I am impressed that you have already produced another lengthy article on Sorcerers. I have known for sometime that you practiced black magic and I am happy to learn that it is done on a scientific basis.

Here everything is topsy-turvey and we have to work lickety-split to get through but it might be worse. Udo[45] is leaving this week for Turkey and the short trip to Ulan Bator to a conference. For Wylie we have secured a job here with the help of the National Defense Education Act, which has also given us lots of fellowships and headaches. Peter[46] had a good time in England and must be by now in Germany. George is all over Europe and I am holding the bag. All I am going to do is go to conferences in Los Angeles and New York and battle it out with people called colleagues. But otherwise I am in fine fettle and would love to fly around with such world travelers as you.

Love from Anita and all the best. Herzlichst, Ihr alter Freund Franz

41

University of Washington
Far Eastern and Russian Institute
Seattle 5, Washington

10 July 1959

Dr. J. F. Rock
c/o American Express
Zurich, Switzerland

Dear Rock:

Thanks for your letter from Zurich with your hints about the profitable and interesting stay at Bonn and Marburg. As to the *Festschrift*, I am sorry to say that this was not an Institute venture and I never got a copy of it either. This seems to have been arranged by colleagues in the field, but I will try to get a copy for you after finding out where.

I have been through some trying times with very unpleasant conferences and all kinds of battles, about which I will tell you sometime. I have been in charge now of this show for the last six weeks and I have not had time for a stroke of honest work. But George will be back in a week and he can have it.

The family is fine. Anita is enjoying the cabin which by now has a

---

45 Udo Posch, österreichischer Altaist, der sich das Leben nahm, als hinsichtlich seiner Dissertation der Verdacht des Plagiats entstand. Vgl. Altaistic contributions by Udo Posch (1922-1965). *CAJ* 11.1966, 144-146.

46 Vermutlich der im folgenden mehrfach genannte Sohn Peter, geb. 1935.

usable outhouse and other improvements which will make it possible to take our friends there sometime when it doesn't rain.
All my best. As ever your old friend, Franz
P.S. The *Festschrift* has been published by Pritsak in the series of Ural-Altaische Bibliothek Studia Altaica.[47] I got it for you.

42

University of Washington
Far Eastern and Russian Institute
Seattle 5, Washington

10 August, 1959

My dear Rock:
It is difficult to realize how fast the time has flown since you were here. You have been everywhere. How I envy you having seen my beloved Tucci before he departed for India.

I don't know when I've had a busier period myself than since you left. I worked continuously the month of June on a research paper[48] which I presented to the Colloquium. I tried to show evidence from the biography of Milarepa, the hermit, that Marpa, the translator, was in fact a power-seeking hegemon during the 11th century in Tibetan history. The paper took far more time research-wise than its 27 pages would indicate. It was well received by members of the Colloquium, and Franz was quite pleased himself. I had to work on it night and day in order to finish it by my deadline.

Then, as soon as I finished that paper, I had to compile six maps which are to be published in Dr. Carrasco's book[49] on land economy in Tibet. His sketch maps were so inadequate, that I doubt I could have prepared the maps had I not already studied Tibetan geography for my 'Dzam-gling-rgyas-bshad manuscript. (Incident[al]ly, I had a letter from Gargano[50] and he said they hope to turn my 'Dzam-gling

---

47 *Studia altaica.* Festschrift für Nikolaus Poppe zum 60. Geburtstag am 8. August 1957. Wiesbaden: Harrassowitz 1957. 189 S. (Ural-altaische Bibliothek 5.)

48 Wohl veröffentlicht als: Mar-pa's tower: Notes on local hegemons in Tibet. *History of Religions Journal* 3.1964, 278-291.

49 Pedro Carrasco: *Land and polity in Tibet.* Seattle: University of Washington Press 1959. VII, 307 S. (Publications of the American Ethnological Society.)

50 Antonio Gargano (*1914), Direttore culturale dell'Istituto italiano per il Medio ed Estremo Oriente sowie Professor für Indologie.

manuscript over to the printer by November). The map-making was time consuming and the editors of the press here wanted major revisions on the first one. All in all, I put in over 115 hours doing them in little over two weeks. (I'm still trying to get my pay for doing the job.)

As soon as I finished the maps, my wife, Shirley, started her vacation. She had a week coming, but she is taking 3 weeks extra without pay, since we wanted to redecorate the upstairs of our home. The house is 5 years old and had never been painted after the builder's first paint job. It had dark forest green walls in the living room and dining room with atrocious flowered wallpaper on one wall in the living room. So far I've painted my bedroom, the two bathrooms, the living, dining room, and kitchen. The old paint was so old, that it has taken two coats to cover properly with the new paint. So far, I've been painting constantly for 17 days on the average of 12-to-15 hours a day, and the only time I've had is on Tuesdays, when they hold the Colloquium. I still want to paint another bedroom, my study, and the wash-room, before Shirley goes back to work next Monday.

Needless to say, I haven't taken care of my correspondence during the past two months as I should have. In fact, my mother wrote and wanted to know if I was sick, because she hadn't heard from me for so long. My trouble is I'm too tense and impatient. I have the tendency to plunge into a job and I want to get it done so badly that I neglect everything else. I finally had to go out and mow the lawn today. I let it go so long that I should have had a sickle to chop it down rather than a mower.

Shirley and I are very disappointed that you will not be passing through Seattle on your way back to Honolulu. We wanted you to see our home and even stay with us. I thought you were going to come back through here, and since you didn't have time to come and have dinner with us when you were here before, we wanted you to stay with us this time. Now you write and say your [!] going home via South America, and that we'll have to come to Honolulu to see you. Needless to say, we'd love to come to Honolulu and visit you, but that is a trip a little too expensive for my income right now.

I just don't think it is fair that you should run off to Honolulu so fast. Shirley is still disappointed that she didn't get to see you when you were here last spring. After all, I have very few friends whom I admire, love and respect, more than you and Tucci. You are a link with our wonderful nostalgic memories of Rome and Honolulu, and an integral part of our thoughts. I may not write very often, but you are always with us. It is well known in the Department and elsewhere how I feel about you ... I've never made any bones about how much I like you. You've made reference so often in your most welcome letters about the shortness of human life ... is not that the best argument for you to come to Seattle on your way home? I shall do my best to come see you in Honolulu as soon as it is possible .... but, who knows what will happen to my financial situation.

You travel so fast that this letter may have to be sent to you after you arrive in Honolulu. Your recent letter said to write to you at Zurich, but Franz told me to write in care of American Express in Rio. I think I'd better check with Franz's secretary. She'll know precisely when you'll be in Rio. If you don't make forwarding arrangement with American Express it could lay there forever.

Franz told me you wrote a complimentary letter in regard to my position next fall at the University. I want to thank you for taking the time to do so. Although the appointment will be for one year, the Department is hoping that the University will make it a permanent one. I shall have the title of Assistant Professor of Tibetan Language and History. I shall continue to conduct myself in such a manner that I hope will always reflect favorably on the judgement of Tucci and yourself.

I was shocked to learn of Nebesky-Woikowitz's premature death. You know, Udo Posch and Swami Bharati[51] were schoolmates of his in Vienna. Udo is in Turkey right now and probably has not heard the news, but the Swami was quite startled by the unexpected news.

Nebesky was in his thirties, as was Ferrari[52] (Tucci's promising Tibetan disciple) at the time of her unexpected death. It seems the

---

51 Swami Bharati = Agehananda Bharati (1923-1991), aus Wien stammender Ethnologe. 1957-1961 Research Associate, Far Eastern Studies, University of Washington; seit 1961 Prof. of Anthropology, Syracuse University. Vgl. seine Publikationen *Die Tantra-Tradition.* Freiburg i. Br.: Aurum Verlag 1977. 325 S.; *The ochre robe* [Autobiographie]. 2nd rev. ed. Santa Barbara: Ross-Erikson 1980. 300 S.

52 Alfonsa Ferrari (1918-1954), Tibetologin, begabte Schülerin Tuccis. Vgl. ihr *mK'yen brtse's Guide to the holy places of Central Tibet.* Completed and edited by Luciano Petech. With the collaboration of Hugh Richardson. Roma: IsMEO 1958. XXI, 198 S., 3 Kt. (Serie Orientale Roma 16.)

30's are a dangerous age for young Tibetanists ... and I shall be just 32 this month! So, I've a few more *dangerous* years to go.

I do hope that you're taking care of yourself. I was distressed to hear you've had more attacks. Do take it easy ... because when I can afford to come to Honolulu again, I want you there to meet Shirley and I with your usual smile and *to hell with the missionaries and government officials* bounce in your walk.

If you refuse to come through here and allow us to repay in small measure the generous hospitality you've given us in the past, at least remember, you're always a welcomed friend and much remembered „Roman". Come see us if you can. If you can't, think of us and do remember us to the Marks.

Aloha, Puhako[53] (is the spelling right for *Rock*?). With love
Turrell Wylie

43

University of Washington
Far Eastern and Russian Institute
Seattle 5, Washington

11 August 1959

Dr. J. F. Rock
c/o American Express Company
Rio de Janeiro
Brazil

Dear Rock:

Thanks for your last two letters from Zürich. I have shown the clippings to Turrell Wylie and told our Inner Asia people here about your friend Tarring. Ruth Krader is on vacation until the 24th so I thought I had better wait with the glass slides until she comes back, but will then have them sent to you and pay for it.

As to the reimbursement to you for the expenses of sending the manuscript, I checked and found that there is another bureaucratic angle which has held the thing up. According to state law, the University can only reimburse original receipts, not even thermofax copies. So I ask you please to send us the original if you can locate it and the money will be sent to you. This is what you might call force

53 Korrekt: Pohaku.

majeure.

I am under terrible pressure to meet some deadlines and to make up for all the work that was postponed while I was up on the mountain – by the way I climbed that hill.

As ever, Ihr alter Freund Franz
Und herzliche Grüsse, von Anita
P.S. I hope this reaches you in Brazil or will be forwarded.
Aber nächstes Jahr sehen wir Sie!

44

University of Washington
Seattle 5, Washington

4 Sept., 1959

My dear Rock:
I do hope you received the letter I sent you at Rio de Janeiro. I thanked you in it for writing a letter of recomme[n]dation for me. Perhaps you will be pleased to know that your efforts were not in vain. I've been appointed an Assistant Proffesor [!] of Tibetan Language an[d] History for the coming year.

My wife and I are fine, although still disappointed that you did not come thru Seattle on your way home.

School will start the 30 of Sept. I shall be teaching 2 Tibetan classes; beginners and two advanced students.

I have to finish a research paper before school starts. I am translating the Nepal section of the 'Dzam-gling. The translation is done, but I still have the notes to do.

I hope your [!] feeling better and staying well. Take care of yourself. If I can do anything for you here in Seattle, I'm at your service.

Give our regards to the kind and Gracious [!] Marks.
Affectionately
Terry Wylie

45

University of Washington
Far Eastern and Russian Institute
Seattle 5, Washington

21 October 1959

Dr. J. F. Rock
P. O. Box 497
Honolulu, Hawaii

Dear Rock:
Welcome back to Honolulu. We had a vague notion of your progress through the wilds of South America through your cards and letters, but I did not write you anywhere there since we did not know the stations of the *pilgrim's progress*. Now that it is over you seem not to be overly enthusiastic but I am sure you had to find out about it.

You ask about the business of the refund. I wrote you earlier in a letter which you obviously did not receive that the law requires the University to have the original Railway Express receipt and not only a copy, and I want to ask you please to send it to us. This, I repeat, is the law and not our local bureaucracy. As soon as we have it you will get the whole amount and then can give me my $ 22.00. The Festschrift was sent with the compliments of yours sincerely.

It is good to hear that you are back at the Nakhis in various ways. After all they are your first love. This has been a hell of a summer with several rather strenuous and unavoidable conferences, and we are back now at teaching and an attempt to write the things that were supposed to have been written during the summer.

Personally everybody is fine. The grandchild is growing and in the eyes of the grandparents delightful and most amusing. Anita sends her best and will write you herself.
With all the best wishes,
Your old friend,
Franz

46

University of Washington
Far Eastern and Russian Institute
Seattle 5, Washington

October 28th [1959?]

My dear Rock--
I am returning the letter you sent from L. Chandra. We are subscribing to the Śatapiṭaka both totally for the department and personally; (I am only taking those things relating to Tibetan.) Thanks for keeping an eye open for things of interest to us.

I, too, was quite disappointed when Rockefeller postponed my journey; however, under present plans, I am to depart shortly after the first of the year. I am to leave about the time Ekvall returns. I shall let you know as soon as my travel plans are confirmed; but, at present, it will be probably be [!] early December before I can give you a definite time of arrival. If I shall still be welcome to stay with you in January, I shall arrange for a short stop-over.

The postponement of my trip until Ekvall's return has certain advantages. One, I shall be able to travel a little more leisurely since I shall cancel my winter quarter classes and not have to adhere to a strict 60-day trip. Secondly, traveling alone can be more enjoyable in certain cases, particularly since I have friends and acquaintances enroute that I want to spend time with alone.

Take good care of yourself so we can *go wild* if and when I get to Honolulu. Please give my thanks and sincere gratitude to the Marks for extending me their warm hospitality again. I hope I shall be able to accept it soon. Shirley joins me in sending you and the Marks the best of wishes and warmest affection. Yours, Turrell Wylie

P.S. The Bpa'-bo-gtsug-lag is a valuable historical work. Tucci uses it constantly in his researches. I have ordered this work from Chandra both for myself and for the department. It will be of the same format as Chandra's other editions, i.e., Tibetan text only handwritten in u-can and reproduced in some photocopy process. His edition of Part III of Sum-pa's Pag Sam Jon Zang is well done and valuable. Wylie

47

University of Washington
Far Eastern and Russian Institute
Seattle 5, Washington

13 November 1959

Dr. J. F. Rock
P. O. Box 497
Honolulu 9, Hawaii

Dear Rock:
Thanks for your letter of October 26 and the receipt which was put to good use. The result is the enclosed check for $ 59.61. I am glad that you had such a good winter-summer, even without your overcoat.

You ask about whether I am working on anything particular. I am working on only about twenty-four particular things which all drive me crazy, but sometime I may get through. I am writing about two dozen encyclopedia articles, the Introduction to the T'ai-p'ing and a paper on Regionalism, and then there is the usual load of lectures and just managing things in as sweet a way as possible. I envy you about your page proofs and I am happy to hear that the work is coming out. I told Gladys about it and gave her your regards and she asked me to send you her very best.

Udo has returned from Mongolia enthusiastic about the wonderful time he had with linguistic scholarschip and roaming the steppes and fishing in rivers. As most other people, he came to be fond of the Mongols.

At home everything is fine. The grandchild is rapidly getting into the stage of language, so far of a not always recognizable variety. I will not even begin to discuss world events which I hope will at least not get worse until we meet again.

All my best to you. As ever, your old friend
Franz

48

University of Washington
Far Eastern and Russian Institute
Seattle 5, Washington

30 November 1959

Dr. J. F. Rock
P. O. Box 497
Honolulu 9, Hawaii

Dear Rock:
Thanks for the letter and the $ 22 which are herewith confirmed.
We have read with great fascination that your volcano has exploded again and are sure you have enjoyed the spectacle. Don't get too close.
All the best your old friend Franz

49

Seattle, January 11th 60

Dear Rock,
Now since it is legal again you will, no doubt, add another year to the actual impressive figure which you have assembled. We hope that this coming year will bring you good health and a continuation of this enterprising spirit that carried you during the last 12 months to new continents in addition to the familiar stomping grounds. Happy birthday and many happy returns!

It is a long time since we have seen you here and we add a personal wish that in 1960 you direct your itineraries to include good old Seattle. At the moment it is winterly. There was some snow yesterday here and more is expected tomorrow. And in the Cascades there is at last enough to ski.

The family is well. Peter has right now a cold but otherwise is allright [!]. He had moved out, as I think, we wrote you; and has taken a room at the YMCA in order to be independent. But he comes home off and on for better dinners and, as now, when he stays in bed with a cold and has to be nourished and looked after. Vera was for a while in bad health but seems to have gotten over it. And we carry on.

Politically I have given up. I feel that we live in a fools paradise and are led down the happy way to hell. But aside from playing Cassandra there is little we can do about it.

Udo is back from his Mongolian venture with many stories and pictures. Dori turned down her Fullbright grant for Berlin to lead her married life here and so we have them both. Wylie is probably going to India this month for a quarter to collect books and refugee scholars. There is now a great interest in Tibet and the Foundations all of a sudden compete with each other to give money for Tibetan studies. The barn door is locked after the horse is stolen. Ekvall is supposed to go to[o] and has been waiting for some time for his Indian visa. The most interesting thing is the change of the mood in India reflected in all letters from Sikkim and New Delhi. We are now friends against the Chinese of whom they tried to tell you good things. Now they wake up and we fall asleep. – What have you heard from all your friends?
Again all the best and auf Wiedersehen
Herzlichst Ihr alter Franz

My dear Rock.
First, the very best wishes for the new year of your life and many healthy and happy returns. We do hope to see you here in not too long a time. Thanks for your several post-cards. Were you also watching the eruptions of the Volcano?
Have you found new lobelias or other interesting plants?

We had nice holidays, only too many of them this year. Not much has happened that is worth writing about. I have seen little of the cabin, which is now rather finished, has many more windows and is now friendly and homy.

Our granddaughter talks now a blue streak and is a very amusing child. It goes so very fast, the growing up. – I feel again Wanderlust but have not yet decided where to wander to or into.

Do let us know how you are and what you are doing and when you are coming!
Again the very best wishes and greetings.
Very cordially, Yours Anita

50

University of Washington
Far Eastern and Russian Institute
Seattle 5, Washington

22 January, 1960

Dear Rock,

Just a short note to inform you of my present plans. The Indian government finally refused Ekvall's visa,[54] so he will not be going to India. Consequently, Rockefeller is now going to send me if I can get a visa. I applied for one about a week ago, and if I get one I should hear about it sometime next week. I'll let you know.

If I get the Indian visa, I shall be going by way of New York, in order to meet with the Rockefeller people. Then on to Rome to see Tucci. I shall spend the rest of the time in India and shall return to Seattle via Hong Kong and Honolulu. If all goes well, I shall be in Honolulu about the 1st of April. I would like very much to see you then.

Unfortunately, the Rockefeller grant doesn't allow for my wife to travel with me. She is disappointed in that she can not go to Rome again and India, and to see you in Honolulu. But, that's the way it goes!

I shall let you know in advance of my traveling dates, but I shall inform you right away when I hear of my visa. If the visa is rejected, then needless to say, I won't be going.

I hope this finds you in the best of health and happiness. Give my best regards to the Marks.

Yours cordially, Wylie

51

University of Washington
Far Eastern and Russian Institute
Seattle 5, Washington

4 February, 1960

Dear Rock,

I've received an Indian visa and will leave Seattle this Sunday, the 7th for New York to confer with Rockefeller officials. Then on to London, Munich, and then Rome. I'll be in Rome from Feb. 13-24th, then to New Delhi. As my travel itinerary now reads, I am scheduled

---

54 Robert Brainerd Ekvall (1898-1978), Missionar; arbeitete 1958-1974 für die University of Washington. Vgl. *Tibetan sky lines.* New York: Farrar, Straus & Young 1952. 240 S.; *Religious observances in Tibet.* Chicago: Univ. Press 1964. XIII, 313 S.; *Fields on the hoof.* New York: Holt, Rinehart & Winston 1968. XII, 100 S.

to be in Honolulu from about April 3rd to the 5th. However, your last letter says you'll be in Rome in April. What time in April will you leave Honolulu? I am scheduling my stop there only to visit you, so if you will not be in Honolulu, I shall cancel my stop-over. Please let me know. You wont [!] have time to write me here, so write to me c/o Tucci at the ISMEO (via merulana [!] 248). If you expect to leave Honolulu before I can get there, I will just extend my stay in India and then skip Honolulu on my way home.

Should you pass through Seattle (enroute to Rome) during my absence, my wife would be delighted to hear from you. Our home phone is Emerson 3-0948.

I shall await your letter in Rome. I do not hold confirmed reservations for the Tokyo-Honolulu-Seattle flight, so it will be no trouble to rearrange my itinerary in the event you are not going to be in Honolulu the 1st week in April. With best regards to you and the Marks, I remain
Cordially yours,
T. V. Wylie

52

University of Washington
Far Eastern and Russian Institute
Seattle 5, Washington

19 February 1960

Dr. Joseph Rock
P. O. Box 497
Honolulu, Hawaii

Dear Rock:
Thank you so much for your good letter of January 14. I was happy to learn that your Na-khi Chung-chia and Lolo manuscripts have found such a good home,[55] and I am also happy to hear of all your plans for work. Turrell Wylie received his Indian visa and left for a two month stay. I have shown him your letter about new books and publications and also about Jigme Taring. Let us see what we can do. We have

55 Die Handschriften gelangten in die Westdeutsche Bibliothek Marburg, und Rock wurde gebeten, die Katalogisierung im Rahmen des Projekts «Katalogisierung orientalischer Handschriften in Deutschland» zu übernehmen.

been swamped as usual with work and problems and have definitively undertaken too much.

Udo never received your letter from Hamburg, but I told him about yours to me and I believe you will hear from him. Eva has been rather ill and in a bad nervous state. This between us.

We are looking forward to having you here, I hope when the sun shines. Let us know your plans.

All the best from Anita and myself. Ihr alter Freund Franz

53

University of Washington
Department of Geography
Seattle 5

November 18, 1960

Dr. Joseph F. Rock
P. O. Box 497
Honolulu 9
Hawaii

Dear Dr. Rock:

I was unable to get an answer to your letter off and on the way to you before October 11 and, consequently, I decided to delay writing to you until I had received a bit more information from Washington, D.C. I thought that if the letter were waiting for you when you returned from the Far East, it could be more complete and specific than any that I could have written in early October. I have indicated my interest and willingness to work with you on the development of a pictorial atlas utilizing your materials to both Mr. Elmquist and to the Central Intelligence Agency. Mr. Elmquist stopped here on his way out to Hawaii last spring for a preliminary talk, and I in turn spent two days in Washington, D.C. in the latter part of September on my way back from Europe. The whole situation shapes up something like this.

1. The State Department, through CIA, is interested in bringing your material together in the form of a pictorial atlas of Western China. The format and content to be essentially as you and I and Franz Michael discussed it a number of years ago.

2. The State Department and CIA do not have funds available to cover this kind of an operation. They are also desirous of becoming an unidentified catalist in all this development. In other words, they are hopeful that other sources of funds and therefore identification

with the project can be found.

3. As a move in this direction, they carried on initial conversations with the Office of Naval Research and found that organization to be receptive to the idea of participating, in terms of funds, in the development of this atlas. As a result of this initial contact, I visited ONR when I was in Washington in September and found that there was a cordial atmosphere in regard to a possible project proposal and budget application.

4. I talked extensively with Elmquist and others at CIA to develop an initial idea of their point of view and their ideas as to what form the atlas might take. In their terms, the primary objective would be to develop an atlas utilizing your materials to present as much in the way of general geographic information which is in the main timeless or, in other words, to display information which will not change a great deal over time and therefore could be thought to represent, within limits, virtually a contemporary record of the area. Their point of view is particularly pertinent to the selection of photography and the captions which which would go with such photography.

5. I indicated to them that while my interest in this project continued, I was already committed solidly through the fall quarter and that I would not be able to start any real action in the development of a proposal, in cooperation with you, until after December 1960. That so far as I could see my schedule at that time, I would begin to work actively on the development of a research proposal about January 1, 1961. While they would have appreciated more rapid action, they did understand my situation and I am sure that you will too. However, as soon as I have the decks clear, and it looks more certain that this will be along towards the end of December, I will begin to develop a project proposal and will be corresponding with you in detail concerning it.

6. I also indicated to the people in Washington that I felt that it would be necessary for you and I to get together for several days so that we could achieve a meeting of the minds in regard to the final project proposal. This poses a financial problem of transportation costs but there was an indication on their part (absolutely no commitment) that they might be able to find funds to pay transportation costs for me to come to Hawaii to work with you for a few days. I will pursue such arrangements and hope that such funds are forthcoming. I assume that it would be more convenient for me to come to Hawaii, so far as you are concerned, than it would be for you

to come to Seattle. This assumption is based on the fact that you have all of your material there in Honolulu and if I travel to Hawaii it would remove the necessity of packing all this material together and bringing it to Seattle. In any case, I will work along in this direction and keep you posted on developments.

7. I had an oppportunity to examine some of the hundreds of photographic prints which the Agency has had made from your negatives. Mr. Elmquist has also indicated that he has received your letter regarding the photo coverage held by the Department of Agriculture and indicated that he was in the process of making whatever arrangements were possible to get hold of that series of photographs. There is, of course, an enormous volume here of very excellent material and even an initial selection, let alone the final decisions as to photographs to be included in the atlas, is going to be difficult and time-consuming.

This brings you up-to-date on the developments which have taken place so far. I am sorry that I am unable to plunge into all of this immediately but I must finish up present commitments before I can be free to do that. In any case, I will keep you informed of all developments including any statements regarding a project proposal and will also let you know what arrangements are possible for me to come to Honolulu to work with you. In connection with your letter, let me say that even though the ONR financing were not possible I would be more than happy to work with you on the four maps that you list. We could handle that in much the same way that we did previously as special projects in our advanced cartography class. On the other hand if a project proposal is accepted by ONR, we will be in a position to accomplish these as well as the atlas.

I hope this gives you sufficient information to see where we are right now and also to let you know that things are moving ahead even though at a very slow rate. I will keep you informed of all developments and in the meantime I hope that you are well and that you had a very interesting and profitable trip. Let me hear from you at your convenience.

Sincerely yours,
John C. Sherman
Associate Professor

54

10 March 1961

My dear Rock –
I was delighted to hear from you and learn that you are well. I am sorry you could not come to Seattle on your return to Honolulu. Maybe next time.

I shall write Mrs. Yuthok about education for her children. I do not see any help for her right now, but may be later. I assume this is the mother of Dundul & Tubchol Yuthok.

My life is a mess since the Tibetans arrived. I am trying to teach a full academic load, answer the mail they receive from all over the U.S., chaufeur [!] them to the dental clinic, hospital, etc., do research with them, and supervise a gov't language project, needless to say, I am going in six directions at once, and little is really accomplished. I hope you will understand!

Give our regards to the Marks. With affection and all the best to you.
I remain,
Yours truly,
Wylie

55

University of Washington
Seattle 5, Washington
Department of Geography

April 6, 1961

Dr. Joseph F. Rock
P. O. Box 497
Honolulu, Hawaii

Dear Dr. Rock:
Let me first apologize for having waited so long to answer your letters. I did receive your letter explaining why you had to go from San Francisco to Honolulu rather than coming to Seattle and I fairly understood your reasons. I hope that you are feeling better and were able to get rid of that cold. I had originally thought that I would complete all of the work on the Rand McNally maps by January but as it turned out I did not finish all of the work until well into early March. I have never underestimated the time required for a job as badly as I did on that one. However, now that it is out of the way,

final exams are over and the new quarter has started I should be able to get moving on our program for the atlas. In the meantime, since you are actively at work and need the five maps prepared and since the offer to prepare them for you still stands let us go ahead in the meantime with them.

Would you please send the materials to me as soon as you have worked out the romanization of the place names? I think that you can use the five maps that we prepared before as models more or less and thus make sure that we have the pertinent content information so that we can go ahead with the maps. Does it seem feasible to you, on the basis of costs, to plan for three colors on these maps as we did on the previous ones? Do you also see any argument against the preparation of these additional five in essentially the same way as the previous ones so that we would then have essentially fifteen maps (the AMS group, the group that we did earlier and this group) all to the same format and with essentially similar content. I have a class this quarter (started last week and runs until about the 5th of June) for which I would like to use these five maps as special projects. This means that we should have the material by around the 1st of May if it is possible for you to get the material together. I will be looking forward to hearing from you and can promise that I will answer your letters promptly.

Sincerely yours,
John C. Sherman
Associate Professor of Geography

56

University of Washington
Far Eastern and Russian Institute
Seattle 5, Washington

19 April 1961

Dr. Joseph Rock
P. O. Box 497
Honolulu 9, Hawaii

Dear Rock:
Thanks so much for your good letter. I know I should have written long before but now I will answer you right away. I had already heard from Anita how nice you have been to them, and I want to thank you for that too. I am sorry that I could not have come myself and be

together like in the old times.

As to your wish to have the article by Edgar[56] photostated, I have immediately asked the library to do so and I will keep check on it and send it to you as soon as it is done.*

I myself have been very busy and Anita may have told you that the bigger things get here, the less pleasant. I have at last decided to get out of as much as possible of the administrative work in order to have time for my own work. I have a sabbatical this quarter and hope to finish at last the study on the Taiping Rebellion, and then to turn to other things. Before I took a two weeks skiing vacation in Switzerland. This was a short trip across on the spur of the moment, somewhat à la Rock only that I did not combine it, as you do, with all the work that you produce. Mine was strictly pleasure in a condensed form. When I got back, however, I caught a very bad laryngitis and was quite ill. I lost my voice for a couple of days almost completely and though I have recovered it I still have to be careful and I am still feeding on antibiotics. Otherwise I am as good as new.

We were very sorry that you did not come through Seattle because writing never takes the place of good talks and we have been missing you.

With my warmest regards and good wishes.
Herzlichst Ihr alternder Freund Franz

*P.S. They promised it in a couple of days and I will send it right away.

57

University of Washington
Far Eastern and Russian Institute
Seattle 5, Washington

5 May 1961

Dr. J. F. Rock
P. O. Box 497
Honolulu 9, Hawaii

---

56 James Huston Edgar (1871/72-1936), Missionar in Westchina, Kenner des tibetischen Grenzgebiets. Vgl. D. S. Dye: James Huston Edgar: An appreciation and an interpretation. *Journal of the West China Border Research Society* 8.1936, 19-27.

Dear Rock:
Thanks for your good letter of April 29. I am happy that you got the pages and the gazetteer.

I am delighted to hear how well your work is going ahead, and I have admired the page proofs which you sent me and which look very impressive indeed. It certainly recalls the days at the Wilsonian where you reproduced all these fascinating word drawings. I return the two pages enclosed in the belief that you might want to have them. I also think that your new title for the dictionary is much better, and hope that this monumental work will soon appear, as well as your other piece.

I called John Sherman and he told me that he had the maps and is working on them.

If you fly to Europe and back, would you cut us in this time? We would love to see you again.

All the warmest regards from us all.
Your old friend (etwas abgekämpft aber unverzagt)
Franz
P.S. In looking over the proofs we discovered a printers [!] error on page 457 in the spelling of courageous.

58

University of Washington
Department of Geography
Seattle 5

May 9, 1961

Dr. Joseph F. Rock
P.O. Box 497
Honolulu 9, Hawaii

Dear Dr. Rock:
I am sorry that I have been remiss in dropping you a line to let you know that the maps arrived safely. I have been so busy getting my class organized, getting the maps organized and work started, plus other classes and graduate student theses, that I did not think to write to you. We are following the same general procedure which we followed for the other five maps, in fact using them as a model. Your compilations are certainly in good condition and we are being as careful as we possibly can with them. The students are finding them a

challenge and progress is going ahead very nicely.

Since these maps are to be in three colors we are preparing three separate drawings as before. When they are complete, what would you like to do? Shall I send all of the material to you for your examination and editing, certainly you should have a look at them before anything further is done, and how and where are they to be reproduced? Let me know your best wishes on this score and I will keep you informed about the progress we are making with the maps. I would guess that they will be completed in about two weeks.

Also, thank you for your letter regarding my staying with you when I come to Honolulu. I certainly appreciated and will certainly take advantage of your kind offer.

Sincerely yours,
John C. Sherman
Associate Professor of Geography

59

University of Colorado
Boulder, Colorado
Department of Geography

July 13, 1961

Dr. Joseph F. Rock
c/o American Express Co.
8 Haymarket
London, England

Dear Dr. Rock:

I'm not sure whether you are still in Europe or have returned to Honolulu so I'm sending a copy of this letter to both places.

Since your last letter we were able to finish the color separation drawings of the four maps. Except for your editing and probable minor corrections they are ready to go. Instead of trying to make a black and white print from the three overlays for each map I think it would be wiser for me to bring them along when I come to Honolulu at the end of August. Then you can take your time editing them and when that is done can send them back to me in Seattle for final finishing up.

I am scheduled to come to Honolulu at the end of August to attend and participate in the Tenth Pacific Science Conference which is to be held there. My paper is scheduled for the morning of the 31st of

August. I will be leaving from Seattle on the 28th or 29th (probably the 29th) and plan to return Sept. 5th. I thought that would give us at least two solid days for discussion of the atlas proposal or more. Is it possible, as you kindly indicated in your earlier letter, to take advantage of your kind offer to stay where you live. I do not want to take advantage of your hospitality or that of the people you are with. If there is the slightest doubt in your mind as to the advisability please do not hesitate to tell me so. I have to travel on a pretty slim budget (as you have had to and know about) so that it would be wonderful for me. It would also give us both more of an opportunity to talk together about the atlas and to plan accordingly. Please let me know as soon as possible if your kind invitation still holds, otherwise I must make arrangements for reservations in a hotel.

Paul Elmquist stopped by in Seattle on his way back to Washington. He just missed you in Honolulu. He and I had a very good visit and he assured both you and me of the continued interest and cooperation of his group. In fact I just had a call from Washington, D.C. this morning re-affirming that interest. I hope that my schedule for Honolulu fits in with yours. Let me know.

I am teaching this summer here in Boulder, Colorado. I have two classes and also much more time to get work done than I find I have in Seattle. The family is here with me and we are enjoying being here. Summer School is not over until the 25th of August so I'm going to be hard pressed to [go] back to Seattle and off to Honolulu. Write to me here, Department of Geography, University of Colorado, Boulder, Colorado.

Sincerely yours,
John C. Sherman

60

University of Colorado
Boulder, Colorado
Department of Geography

Aug. 4, 1961

Dr. Joseph F. Rock
P.O. Box 497
Honolulu 9, Hawaii

Dear Dr. Rock:
Thank you again for your kind invitation, I'm looking foreward to

being in Honolulu with you. I am presently scheduled to leave Seattle at 7:00 p.m. on Pan Am Flight #891, August the 29th. I do not know the estimated time of arrival in Honolulu. I would think it would be around 11:00 a.m. Aug. the 30th. If I can find out I will let you know, otherwise could you please call the airline and find out from them.

It is wonderful for you to offer to meet me – it will make the trip so much more pleasant to know you will be there when I get there. Thank you again – I will see you in a few weeks.
Sincerely,
John C. Sherman

61

Seattle, Aug. 8, 61

Dear Rock:
Thank you so very much for the Medinillas.[57] They arrived in perfect condition and seem very happy in their little pots. I would never have thought that one can send them that way. The seeds were rather dried out but I think they will come up in time. Again many thanks. –

I had to laugh about the description of your trip. It was just what I had expected. I don't know why these sport teams have always to be so misbehaving. I should think that their coach would not stand for it.

We had hot dry weather all along, but the nights are cool. I was three days at the cabin last week. We slept out on the porch. Nothing is more beautiful than a moonlit warm summer night.

Vera and family are back from their sailing trip, all braun [!] and happy. There is nothing new to relate from here. This Saturday is the yearly big department party at Gershefsky's. Franz works as usual, but he is not happy at his work as before. But there seems nothing one can do about it. I guess he has to try and rise above his immediate emotions if one can do it.

I hope you are not working too hard. Did you hear from Marcks [!]? Hows did you find all your plants at return? Was *Baby* happy to see you again? I guess on Hawaii is no bombscare as some people have here. I don't but Franzi has. I know it is silly, but somehow I don't believe that all the russian [!] achievements are true. I wonder if they could not fake that the man was in the capsule. All fits always so perfect, too perfect. – Give my regards to all I know on Old Pali

57 *Medinilla magnifica* Lindl. (Melastomataceae).

Road, to Oahu and Hawaii. We expect you back in September on your way to Butte, Montana.
In the meantime take very good care of yourself. The family sends best regards and also the Bauers.[58]
Sehr herzlich
Ihre Anita

62

University of Washington
Department of Geography
Seattle 5

Sept. 8, 1961

Dear Dr. Rock:
Just a short note to let you [k]now that I got here safe and sound. Had a very nice trip but did not sleep much and since I got to Seattle at 7:20 a.m. I lost a nights [!] sleep. Seems like I'm still trying to catch up. I'm already back at work, things have sure piled up here, but I'm starting to put the proposal together. I should have a statement to send you late this month. I'm getting prints of the index map made and will send you copies along with the tracing paper when they are ready.

I want to thank you again for the wonderful visit I had with you. I only wish Helen could sit with me and talk to you about China as we did those evenings. I will never forget this visit, for many reasons, but chiefly because of the opportunity it afforded to get to know you better and to learn more about your work. I never did see the Tibetan sword nor the armor but maybe, if I come again, you will show them to me.
Thank you again for everything and pet Baby for me, might even give him a tidbit for me. Sincerely yours,
John C. Sherman

63

1 October, 1961

Seattle, Wash.

Esteemed Dr. Rock:
Please excuse the long delay in replying to your kind letter of the 27th

---

58 Harry C. Bauer, der Bibliotheksdirektor?

of July. I have been away much of the last two months.

The list of gsung-'bum[59] was very welcome, indeed. The immediate task is to locate them. According to H. I. Poleman's Survey[60] (1961), the Library of Congress' holdings of Tibetan works are:

1. Nine manuscripts and 57 xylographs in the Rockhill collection.

2. 120 xylographs in the Laufer Collection. This collection is uncatalogued.

3. 50 miscellaneous manuscripts and xylographs. This collection, too, is uncatalogued.

4. The Sde-dge Kanjur, the Snar-thang Tanjur, and the Co-ne Kanjur + Tanjur.

I doubt very much whether the Gsung 'bums which you acquired for the library of Congress could be included in Item #3.
I intend to write to Poleman as soon as I can.

We have received a catalogue of the Collection in the Library of the University of Kentucky, Lexington, Ky. Surprisingly enough, this collection is rather exciting. There are, of course, the usual common works, however, there are several histories which are rather rare. There are also numerous illuminated manuscripts. The catalogue to this collecti[o]n has been prepared obviously by some Khalkha Mongol. The titles are all in modern Khalkha in Cyrillic script. Have you any information as to the origin of this collection?

Our catalogue is coming along slowly. It will probably run close to 400 pages in typescript. Our older lama (Sde-gzhung rin-po-che) is a gold mine of bibliographic knowledge.

I have read with much interest the two separata which you were so kind as to send. Particularly, the History of Sikkim excerpts are useful. The Rnam-rgyal family, although now adherents of the Rnying-ma-pa, in the 17th and 18th centuries, showed strong leanings toward the Sa-skya system. Mr. Richardson[61] has kindly written to

---

59 *gsuṅ-'bum* – gesammelte Werke.

60 Horace I. Poleman: *A census of Indic manuscripts in the United States and Canada.* New Haven, Conn.: American Oriental Society 1938. XXIX, 542 S. (American Oriental Series 12.)

61 Hugh Edward Richardson, 1905-2000, vgl. *Tibetan studies in honour of Hugh Richardson.* Proceedings of the International Seminar on Tibetan Studies Oxford 1979. Edited by Michael Aris. Warminster: Aris & Phillips 1980. XX, 348 S. Von seinen zahlreichen Publikationen über Tibet seien nur erwähnt: *A cultural history of Tibet.* Rev. ed. Boulder:

the Maharaj Kumar to attempt to obtain a copy of the Tibetan blockprint original. Should we succeed, would you like a microfilm.

With respect to the Ulan-bator edition of the 5th episode of a version of the Tibetan Gesariad: We do have this work in our library. Also, I have a copy of my own; but up to the present moment, I have not had the opportunity to read it. Right now, our Tibetan informants' English is hardly adequate to permit the undertaking of a translation, although the Sa-skya Rinpo-che's wife, Damola, is exceptionally quick and is learning English rapidly.

My Tibetan is, likewise, not adequate for attempting a scholarly translation of a work so complex as Gesar. Should Professor Stein be unable to produce a translation adequate for your needs, after the catalogue is finished, I would be pleased to summarize, with the help of our informants, this text and to prepare a list of place names, etc.

Have you seen Prof. Damdinsüren's book *The Historical Roots of the Gesar*?[62] The book is in Russian; the author, in it, presents the fruits of his many years of study on the various versions of the Gesar – Tibetan, Mongolian, Sarygh Uighur, Burushashki, etc. There is a copy in the library here.

While you were in the Ljang country, did you, by any chance, visit any Sa-skya monasteries? Supposedly up until the 18th century, the strongest religious sects were the Sa-skya-pa and the Karma-pa.

I have run afoul of another problem with which you could perhaps help me. In one of our chronicles (Rgya-bod-yig-tshang. 1434) it mentions a Sa-skya abbot and prince of the Sa-skya family as going to Sa-tham of Ljang to become the bla-mchod of one Hu-ar-che (ཧུ་ཨར་ཆེ་). I feel that this occurrence must be dated after the expulsion of the Yüan dynasty from China. Our Tibetan friends here insist that Hu-ar-che was not a Mongol. The name looks very Mongol. How long after the rise of the Ming were there Mongol overlords in

---

Prajñâ Press 1980. 307 S. (mit David Snellgrove); *Tibet and its history.* London: Oxford University Press 1962. VIII, 308 S.; *Ch'ing dynasty inscriptions at Lhasa.* Roma: IsMEO 1974. VIII, 104 S.; *Ceremonies of the Lhasa year.* London: Serindia 1993. 133 S.

62 C. Damdinsüren (1908-1988), Mongolist und Literaturwissenschaftler; vgl. *Zentralasiatische Studien* 2.1968, 179-182; A. Kajgorodov: *Cėndijn Damdinsurėn; bibliografičeskij ukazatel'.* Moskva: Kniga 1978. Das Buch ist: *Istoričeskie korny Geseriady.* Moskva: Akademija nauk 1957. 238 S.

Yünnan? Any help you can give me would be greatly appreciated.
Hoping to hear from you soon, I remain,
respectfully yours,
Gene Smith[63]
P.S. I would like to express my thanks again for the very enjoyable lunch to which you so kindly treated me.

64

Seattle, 10.10.61

My dear Rock.
Thank you so much for your letter. I am very sorry to hear that you are not feeling your best. I hope, when you receive this letter you will be quite on top again. My silence had mostly to do with our big news which is that we are going to move out to the country, near Bothell. Franzi got a bit worried about the world situation bombs etc. and felt we lived to[o] much in the center of the city. As I always wanted a smaller house, since the children are grown I took his mood to advantage and we found ourselves suddenly house hunting. Well, we found a nice roomy house on the corner of two country roads with 5 acres of land, cleared, but with some big trees still standing. Now I can plant all I want and go wild and have a greenhouse (when we have financially recovered) and build a fishpond and so on and on. Of course, in time, loads of Rhododendron! The ground is hard at places, but I shall manage. Franz will have about 1/2 hour drive to the Univ. It will also be nearer to my cabin. I do hope you will be our guest there soon, perhaps Christmas? Vera and John are also moving into a beautiful big new house more north, with view onto Puget Sound. It will be about 1/2 hour west of us.

---

63 E. Gene Smith (1936-); Tibetologe; von 1968 bis zu seiner Pensionierung war er für die Field Offices der Library of Congress in New Delhi, Jakarta und Kairo tätig. Vgl. E. Gene Smith: *Among Tibetan texts. History and literature of the Himalayan plateau.* Edited by Kurtis R. Schaeffer, with a foreword by Jeffrey Hopkins. Boston: Wisdom Publications 2001. 384 S. (Studies in Indian and Tibetan Buddhism.). S. ist ein exzellenter Kenner der tibetische Literatur und Bibliographie. Er ist Gründer des Tibetan Buddhist Resource Center, das sich die Digitalisierung traditioneller tibetischer Texte zur Aufgabe gemacht hat. Vgl. Barbara Stewart: War resister becomes savior of Tibet's literature. *New York Times* 15. 6. 2002. Ein Katalog der von ihm gesammelten tibetischen Texte auf Mikrofilm: *University of Washington. Tibetan catalogue.* 1969. 327 Bl.

I hope to move still in November and I hope we can sell our house soon and well. Peter is moving with us and I hope he stays with us, otherwise it would be very lonesome for us as Franz is gone so much. Guess I have to get a big dog. – How is your Baby pudel? I wish often that I could take a swimm [!] in Hawaii! My Europe plans are given up for the next year, but may be another short visit to your part of the world would be possible, or Alaska.

Franzi works too much as usual, keeps a lot to himself academically, i.e. he does not join very much in the planning of the show, others are busy there now. He says, he does not want to be president of a Uni, he wants to do academic work. He says besides one could not apply for that job, one has to be selected and the offer has to come from them. (Ask George, ha). He send[s] his greetings to you.

In the mountains it has snowed already quite a bit. Soon Franz can go skiing. I transplanted last week my large Rhodo to the new house and of course, the one from you is going to get a preferred place too. Also my big pink dogwood and the big Oxidendron[64] I shall take along. Plants are terribly high in price if they are a little bigger. Well, I have to stop this letter; a family with 5 children wants to look at the house.

Hope to have good news from you soon and in the meantime all the best. Cordially, Anita

I'll give you the new address when we know when we move.

I am drying some blackberry seeds for you. The salmon are now spawning in front of my cabin. Poor things, beat up and dying. Hard life!

65

Seattle, Oct. 28, 61

Dear Rock,

You must have been real mad with me. Hope you aren't any more. The reason I did not write the address was because John and Vera did not know yet when they would move. They still don't know, and we are in the same boat. First we have to sell our houses. I spoke with Johnny (Osterhang)[65] and he said the best would be to send the slides

---

64 *Oxydendron arboreum* DC (Ericaceae) – sourwood, sorrel tree.

65 Michaels Schwiegersohn, Veras Mann.

to the University to Franz. But he said there is absolutely no hurry. –

I hope you are feeling better now and enjoy again your gardening and your other work. We could use some Hawaiian sunshine at present.

We had yesterday the visit from Redfield who is now with his family in Nepal. Our friend the Dutchman, Marinus Meijer,[66] went from Tokyo via China to Holland. Was one week in China and described the dismal, hungry atmosphere. He was agast [!] to see how many of his former acquaintances took to drinking.

Udo had a bad car (?) accident, is in a hospital in Munich. We know nothing in detail, just the fact. If George knows more he did not tell us. Poor Dori, she must be terribly upset. – Otherwise I can not think of anything more to tell.

Keep well and take good care of yourself. At least you have now the company of Mr. Mark[s].
All good wishes und seien Sie von Herzen gegrüßt von Ihrer Anita

66
[Postcard: Cardinal flower, *Lobelia cardinalis*]

Dr. J. F. Rock
P. O. Box 497
Honolulu – Hawaii

Seattle, Nov. 28.61

Dear Rock
I am worried that we had no news from you in quite some time. Hope you are just too busy to write. Hope you had a nice Thanksgiving. We were all at the cabin, were we have now a new well and running water in the house. We intend to move the 11th of Dec. Address: Michael 3917-212th Bothell, Wash.
I bought many Rhododendrons. With best wishes from all of us. Cordially yours, Anita

---

66 Marinus Meijer, veröffentlichte u.a. *The introduction of modern criminal law in China.* Batavia: De Unie 1950. 214 S. (Sinica Indonesiana.2) (Dissertation); *Murder and adultery in late imperial China: a study of law and morality.* Leiden: Brill 1991. X, 137 S. (Sinica Leidensia 25.)

67

University of Washington
Far Eastern and Russian Institute

Seattle 5, Washington
December 3rd 1961

Dear Rock
Many, many thanks for your good letter. I am sorry I have not written for such a long time, but it has been a rough one and it has passed awfully fast. I am sorry to hear that your heart is making trouble but I know that you are made of good stuff and I [!] there is so much more that you have to give. Take good care of yourself.

The news that your dictionary is in the proofreading, final stage is wonderful. I am deeply touched that you want me to review it. As you know, I am a bloody laymen [!], but of course I have sort of seen the work grow and I'll love to do it.

I have given your slides to John, and he and Vera send their thanks and their very best regards. They are moving this month into a new house at Edmonds with a magnificent view of the Sound and the mountains which you will have to admire when you come next time, we hope soon. (No telephone wires in front.) We are, as Anita wrote to you, also moving, to a house near Bothel [!]. No view there, but lots of land, and Anita can have a greenhouse and plenty of rhododendron, and its [!] closer to the cabin which has been supplied with running water and made more homelike too. I, as you know, am not a cabin man, but the children like it very much. (that means the grandchildren)

You ask about the Institute. Well not much change. I continue as chairman of the Modern Chinese History Project, and we have had an excellent summer and fall of research. This fall at last I have had some time for my own work on the Taiping documentary history. I plan to work on it during Christmas vacation and enjoy it very much. Plenty of human foibles there – I mean among the Taipings. I have also expanded my article on *The Role of Law in traditional, nationalist and communist China* which is sort of a major theoretical piece and will appear in the January-March number of the *China Quarterly* in London. You will certainly get a reprint. – As for the rest of Thomson Hall there is peaceful coexistence at the moment. Our Tibetan project is flourishing. Poppe is back and wants now to increase the Mongolian program too. We had a sad accident. Leon

Hurvitz[67] smashed his car. He is for a week now in hospital, still more or less unconscious, and it looks serious.

This coming week I fly to Mills College to give an assembly speech and share in a seminar on introducing Far Eastern Courses there. And before Christmas I plan to go for a few days skiing at Sun Valley. Next March I'll go to Germany for a conference on Moscow-Peking, that fascinating new parlor game. All these doings will keep me busy if not out of mischief. –

We all had the flue, to make this report complete. It was a nuisance and hung on for quite some time. I am against it.

The world looks certainly black. This crisis seems to blow over, but in the long run I do not see how the final clash can be avoided. It wont [!] be the end either, but nobody can tell how things then will go on. Sehr herzlich alles Gute Ihr junger alter Freund Franz

P.S. Anita hopes you will soon be much better and will come hear [!] soon. She sends her love.

68

SUN VALLEY

Dec. 19th 1961

Dear Rock,

Thank you so much for your letters. I am happy to hear that work proceeds and that even if you feel sometimes weak you sound in your letter as good as ever. I have taken a few days skiing vacation here, to get away from it all, have taken no books and just ski, eat, drink and sleep. I will leave again on Saturday and will be home Saturday night, in time for Christmas.

I got your letter about the picture and its explanation just before I left and there was no time to handle it then, but I will, as soon as I am back. I have already looked up the plate in your book, but John and

---

67 Leon Nahum Hurvitz (1923-1985), Japanologe, Buddhologe; graduierte 1951 zum M.A., 1959 zum Ph.D. an der Columbia Universität, war 1955-1962 Acting Assoc. Professor of Far Eastern Languages an der University of Washington, 1962-1968 Associate Professor of Japanese, danach Professor of Asian Languages. Er übersetzte u.a. *Scripture of the Lotus Blossom of the Fine Dharma* (New York: Columbia University Press 1976. XXVIII,421 S.) und Tong-uk Kims *History of Korean literature* (Tôkyô: Centre for East Asian Cultural Studies 1980. XI, 321 S.).

Vera were just moving into their new house and everything was packed. Next week it will be unpacked and I will get hold of the Tibetans too.

We have moved too, and belong now to the landed gentry. So far the house is in a mess but eventually it may become livable. There is plenty of land (5 acres) and Anita can have a greenhouse and can garden after her hearts [!] delight.

The world situation is as dark as can be, but at least we begin to realize what we are up against. And may be there is still enough time to arms.

All my very warmest wishes for Christmas and the New Year. We hope you will soon come again.
Take good care of yourself.
Sehr herzlich Ihr Freund Franz

69

University of Washington
Department of Geography
Seattle 5

Jan. 12, 1962

Dear Dr. Rock:
I imagine that you have completely given up on ever hearing from me. I have been so busy that the whole fall is nothing but a blur. I had everything, so I thought, planned so well for this year and everything has gone by the boards. As soon as I got home from Hawaii I had to get busy on preparations for the fall quarter and I also found that the five graduate students who had done preliminary work on an atlas of Communist China had made a great deal of progress and if they were to continue we had to seek funds. So Prof. Murphey[68] (of our department), the students and I spent most of our spare time (and some which wasn't so spare) getting together a proposal for the National Science Foundation. I am particularly anxious for this work to go forward because it is so closely related to the work you and I

---

68 Rhoads Murphey (1919-), Historiker und Geograph, war 1952-1964 Professor of Geography an der University of Washinghton. Er veröffentlichte u.a. *Shanghai, key to modern China.* Cambridge: Harvard University Press 1953. XII, 232 S.; *The treaty ports and China's modernization: what went wrong?* Ann Arbor, MI: University of Michigan, Center for Chinese Studies 1970. 73 S.

will do. In addition to these things I also found that I had an evening class on top of my regular class load. It was a special class of men from city and county agencies here in Seattle who had requested work in Cartography. In a weak moment, in the spring I said I would teach it if they could get 15 people together. I didn't think it would work out but it did. That represented five hours a week in class in the evening plus preparation. Then the roof really fell in on me. I spent almost a year on illustrations for a book, elementary college, written by Rhoads Murphey and published by Rand McNally.[69] I finished the work last March and the book was published in May. Well, it sold better than they expected and they had to reprint. In September it became obvious that they were going to have to do a third printing by early spring of this year. Rand decided to spend the money to revise text, map illustrations and pictures. To make a long story short, I had to set to work and modify, expand and add to some fifty maps. I'm still working on them although I have almost finished them. I think you can see that the fall quarter for me has been chaotic.

I don't want to burden you with my troubles but this may also have some future influence on our work together. Last spring, during a routine physical, I discovered that I have diabetes and high blood pressure. The doctor started medication immediately and all during the summer I was on an experimental regime to see if it was going to do the trick. I had to go back again in early October and since then for further checks. Finally after a lot of modification we seem to have hit on the right combination of medicine. I've been really dressed down by Dr. McRae twice about the amount of work I'm doing. I don't know what I could have done about it this fall but I guess I'm going to have to reduce the load a bit. I've never been sick in my life, and I guess I've gotten to sort of take health for granted. I love to do what I'm doing (wouldn't want to do anything else) but I've also got four children and a wife to look after so I guess I'll slow down a bit. None of this changes your and my plans but it has helped to slow them down. I'm just not going to get involved with so many things from now on – concentrate on a few and do them as well as possible.

Please accept my apology for not writing and for being so slow on everything we planned. Thank you so much for the picture you sent of me – my wife really appreciated that and so did I. I just didn't have

---

69 Rhoads Murphey: *An introduction to geography.* Maps by John C. Sherman. 2nd ed. Chicago: Rand McNally 1966. XV, 734 S.

time to write. I am sending the tracing paper and the index next week. I did not think you would have time to use them in the last two months with all the work you were doing on the dictionary. The maps are coming slowly too. As you know, we do not have any money in the departmental budget for them and so I have to fit the work in my classes when it is appropriate. The students did not get the maps finished in the spring as you know, nevertheless they were sufficiently interested that those who were working on them in the spring came back in the fall and offered to carry on during their spare time. As a result the work has been slow but they are almost done now with the lettering. Also starting this quarter I also have a graduate student assistant who is going to wind them up. So don't give up, I said we would make the maps and we will. My Cartography class, during the fall quarter, was not the kind of a class where work on the maps fitted in.

The Christmas Holidays are a blurr in my mind. I worked on Rand maps, had to make a special map for the President of the University, finish preparing a paper and then go to Denver Christmas night to the AAAS meetings and give the paper. I got back the Saturday before New Years completely bushed. The weather was beautiful and sunny while I was there and I got a chance to see all of the people at the U of Colorado where I taught last summer. I stayed in Boulder and commuted every day to Denver for the meetings. Here in Seattle the weather has been mostly cloudy and rainy. We have had a lot of early snow in the Cascades and so the water prospect for next summer is looking good right now. Seattle is all hustle and bustle preparing for the Century Twenty-One World's Fair which is to open the first of April. It doesn't seem possible that they will be able to finish all the buildings and displays in time but I guess they will by working night and day on them.

All your pictures are safe and sound. I have done a little preliminary planning and selecting on them but as of now I do not have any way (money that is) to get half-tones made so I can return them to you. I hope you don't mind my hanging on to them until we get our project money and get going. I'm too tired tonight to think of any more to say — if I have left out any answers to questions please drop me a line and I will answer.

Again let me say how sorry I am for delaying so long and particularly not writing sooner. I won't be so slow again.

Sincerely yours (humbly too) John C. Sherman

70

University of Washington
Far Eastern and Russian Institute
Seattle 5, Washington

January 15, 1962

Dr. Joseph Rock
P. O. Box 497
Honolulu 9, Hawaii

Dear Rock:
Happy birthday and many happy returns. Our wire will have come by mistake a day too early and Anita's letter after the event, and so we spread our wishes to make sure the event is really properly celebrated. I did call Sherman and he is very sorry indeed. He just had written you a letter which you probably will have received by now, describing the whole situation, so I hope that he will now act.

I have also learned from Turrell Wylie that Gene Smith will have finished the work with the Tibetans for you by today, and so you will hear from him, too.

I was happy to hear that your heart trouble is better, and I am sorry for your nervous pain. It must be miserable, and I hope it is improving without any drastic action.

No, I do not worry about nuclear war, but just as a matter of logic, I think that eventually war is possible, to say the least. If so, it will be horrible, but it will not mean the end of the world, I think.

The new place has its advantages and disadvantages. It will take some time until we are used to it. There is certainly plenty of Lebensraum for Anita's plants, and she has already ordered a greenhouse.

I am terribly under pressure with all kinds of work and so you will have to forgive if I do not write a longer personal letter today. Nothing much has changed really, but I have made some progress with my own work on the Taiping history which fascinates me more end more.

All the best for your health, for good work and auf baldiges Wiedersehen
Ihr alter Freund Franz

71

21 January, 1962

6551 24th N.E.
Seattle, Washington

Dear Dr. Rock:
First, I suppose apologies are in order for having been so remiss in answering your letter of some months ago. I really am very sorry; and I feel especially embarassed in that I had intended to begin to summarize the Sa-tham episode of the Ge-sar for you. I have been laid up for some time with hepatitis. It is only now that I am beginning to feel like working again.

I understand that you have been somewhat ill. I hope you are feeling better now.

Your last letter contained some very interesting information. You may be interested to know what I have been able to find out from our old lama, Dezhung Rinpoche. You noted that the Ha-pa manuscripts which you obtained and are now in the possession of Madame de Zayas, were in a 'gos-lo འགོས་ལོ་ script. According to Dezhung Rinpo-che, a script for Tibetan was devised by a Rnying-ma-pa bla-ma 'Gos lo-tsa-ba (not the author of the Deb-ther-sngon-po). This script purportedly is phonetic and 'Gos lo-tsa-ba lived in the 13th or 14th century. If the script is phonetic and this old, it seems plausible that these manuscripts might be untampered with copies of originals which attempt to represent a colloquial, at least phonologically, Tibetan of the 13th or 14th century. According to Dezhung, 'Gos lo-tsa-ba was not the only one to try his hand at script devising; however, Dezhung Rinpoche has never seen any samples of these scripts but only has heard that they exist.

Thank you very much for the information on the last Mongol ruler of the Yüan of Yünnan. It has enabled us to pin down one of the Sa-skya hierarchs and revise our dating a little.

While I was in the hospital, I found time to read your work on the Ancient Na-khyi [!] Kingdom in Southwestern China and also the volume on the A-mnye Ma-chen [!] which I must confess I hadn't read before. I found a great deal of extremely interesting material in them.
I have a couple of questions that perhaps you can help me with.

1. Do you know where there are copies of the biographies of the

first three 'Jam-dbyangs-bzhed-pa?

2. Have you ever heard of peoples who are called the Dung-Reng(s) divided into two tribes – the Shar-Dung and Lho-Dung. I have found an extremely interesting and detailed history of the campaigns against these people by the combined armies of Dbus and Gtsang under the leadership of the Prince of Gyantse, 'Phags-pa-dpal (1318-1376). Tucci has mention of them in the Painted Scrolls, but this passage in the Rgya-bod-yig-tshang has detailed geographical information and strategical analyses, most of which I can't identify or understand. These people lived in Lho-kha but seem to be also located farther to the east in Yünnan.

I am enclosing some identifications on the Rma-chen-spom-ra tanka given to you by Shing-bza' Pandi-ta. We showed all of our Sa-skyas Dr. Michael's slide and I have tried to show the placement of the figures with numbers. They are rather doubtful on #1 and #4. the Sa-skya pantheon differs considerably from the Dge-lugs-pa and we have no Dge-lugs-pa iconographic manuals. The younger Rin-po-che 'Phrin-las seems to be positive in this [!] identifications, while the old Khams-pa lama insists that we say perhaps. I hope this is satisfactory for your purposes. If it is not, let us know.

I sincerely hope this finds you in good health and spirits and that this New Year proves happy and healthy for you.

Yours sincerely,
E. Gene Smith

72

University of Washington
Department of Far Eastern and Slavic Languages and Literature
Seattle 5, Washington

3 February, 1962

My dear Rock:

Franz told me about your impending operation. I am sorry to hear that it must be done and I hope you recover quickly. My love and prayers (to Buddha, naturally) are with you.

My 'Dzam-gling-rgyas-bshad is almost finished. The final proofs are in Rome. The map was sent last month as well as the Indexes (Tibetan, Chinese, General). I hope to see it on the market by June.

I shall be in Italy in June-July. Rockefeller is holding a Tibetan conference at Bellagio from July 2-9. Richardson with [will] chair the meeting. The leading scholars from the 8 Rockefeller centers will be there. Also Gene Smith has been invited! I'm looking forward to being in Italy again.

Tibetan New Year (Water-male-tiger year) begins Feb. 5th. The Tibetans are having a huge party – they've invited about 150 people – and, the Bdag-chen has chicken pox so bad today I must go and take him to doctor at 2:00. Some fun!

I am still too busy – with teaching, research, lectures, et al. I returned to oil painting last fall as a means to relax but haven't had time lately to do much. I am doing a portrait now.

Give my warmest regards to Loy & Lester.

བཀྲ་ཤིས་ [bkra-śis]

Always yours, Wylie

73

Bothell
3917-212th St.

March 4, 62

My dear Rock,

I am quite worried about your health and well being. You mentioned an operation. I hope you did not need it, or if, that you get over it quickly and it brings the expected relieve. Do let us know how things are. – I have been longing for Hawaii very much. It has been bitter cold here, and then over a foot of snow out here were [!] we live. All the traffic snarls possible and I was really more or less snow bound. – Slowly I am getting more order into the house and it is rather cozy in this weather. But we do miss the view!

On the whole not much is to relate from here. Franz leaves on Thursday for Europe, skiing in Switzerland and some work combined. He'll be gone for about a month. The children are fine, the little ones growing fast.

Helmuth[70] came here for a lecture, I think he'll be glad to be back in Seattle for good. I have had very little social life, especially since I live out here. If I can be outside gardening its [!] better, but it is without doubt rather lonesome for me. May be in time I shall get to

70 Hellmut Wilhelm?

know more people. I am afraid several of my plants won't survive the cold spell. It came so suddenly, that even in my greenhouse many froze, I am very unhappy about it, which does not help. I have seen few of our mutual friends. Last week Richardsons[71] were out here with Simon[72] (from England). All Seattle is in expectation and apprehension of the big worlds [!] fair.

Give my regards to Hawaii, its sunshine, bird songs and all the beauty. Vera's little girl said today she wanted to go with us to see you again. If I had the money I would do it but Johnny is not willing to let his family go again without him. Do let us know how you are and what your plans are and when we can expect you here again. All good wishes and regards from all of us. With love, Anita

I think this is the dullest letter I ever wrote. You see I am already am verbauern.

74

2 May, 1962

My dear Rock:

Just received your letter of April 29th. I am happy to get the report that your Na-khi dictionary is coming along. Such a slow process!

At present, my travel plans call for me to leave about the middle of June for Rome. Then, I shall be at the Villa Scorbolloni at Bellagio from July 2nd to 9th for the Rockefeller Conference on Tibetan Studies, then back to Seattle. You could write to me here until the middle of June. Or, leave word at the IsMEO if and when you will be in Rome.

So Tucci expects to leave for India in June. That is really surprising news! I thought he planned on attending the Conference in Bellagio in July before leaving for Asia. I am looking for a letter from him any day. I have begged permission to have some of his Sa-skya books microfilmed when I come to Rome. Our Tibetan project is bogged down without them. Gene Smith and I are working on Sa-

---

71 Wohl Hugh Richardson, 1905-2000, s. o.

72 Walter Simon (1893-1981), Professor für Sinologie an der Universität London, vorher Privatdozent an der Universität Berlin und Bibliothekar an der Universitätsbibliothek; Spezialist für indotibetische Linguistik. Vgl. C. R. Bawden: Ernst Julius Walter Simon. *Proceedings of the British Academy* 67.1981,459-477.

skya history, but we must have more Sa-skya books for reference. All other sources have been explored with no results, so if Tucci does not permit me to microfilm his materials, I guess we might as well give up the history study of Sa-skya. The Lamas return to India in just about 16 months from now.

Dr. Rock, it makes me sad to have to tell you this, but my wife and I have separated. We have never been very compatible emotionally or physically, but circumstances during the past year have made it wiser for us to get a divorce. She now has an apartment on Mercer Island across Lake Washington and is very happily settled there. We still have a great deal of affection for each other. After almost 13 years of marriage, we still treat each other like a brother and sister. Unfortunately, a brother-sister relationship is a poor excuse for a marriage, so we have separated by mutual agreement. Shirley asked me to convey her love to you and all good wishes for a happy trip to Europe.

Take good care of yourself. Give my best regards to Lester and Loy. Let me know your time schedule in Europe so we can raise a toast or two to auld lang syne ... eat raspberries and get sick together. Miss you!

Always yours,
Terry Wylie

75

1. Juni 1962

Dear Rock,

I am bad, I know, and I am really sorry for my bad behaviour – my long silence. But as you said: *There is just one damn thing after another.* All the more do we need your spirit and sense of humour. So, please, come and let us laugh at the latest stories and tell you all about this silly place and our own ventures abroad. Even the Fair is worth seeing – so they say. So be forgiving and let us know when you can be here. –

Anita has written and probably told you about this new gentry estate out here which is peaceful and quiet but a little too far away and too much for her to manage. –

I got one of these miserable infections first in Germany and then here again, three times in the last 3 months under anti-biotics! I lost my voice again (which may be good except that I have to teach) and it is all a little rough on one's mood. But otherwise there are hardly any

complaints. And about the world situation we cannot do too much. But all this is much better talked than written about so please, come by. Your old friend
Franz

76

June 1st, 1962

Dear Rock
Thanks for your grouchy letter. I liked it, it was real Rock when he is angry or disappointed. My, how could you think for a moment that we forsake you?! A friend is a friend, even if not writing often. But you are right, I should have thanked you at once. I do it herewith most cordially. But you wrote „I send you the *Medinilla magnifica* in another envelope“, and I waited to thank for both. The only weak excuse is that I have been very busy with Peter. His foot is not yet good (after 7 weeks) and needs 3 times special bathes (boil water 20 minutes etc.). He is still on crutches, can’t wear a shoe yet. Besides, much too big „garden“ takes too much time, I just cannot manage. Franz has been on trips and now has since some time a very nasty cold and laringitis which makes him miserable and hor[s] de combat. So its [!] all up to me. But we are getting better and we surely will be very much offended if you would cut us on your trips. At least stay a night or a day! Tracy can surely find a room for you in the Wilsonian, besides it is not at all full, the crowd has not really started. But the Fair is said to be really worth seeing.

We hope to hear from you soon dear friend. Only on the 7th of June we go to Aida, otherwise we are free. So long and in the meantime all the best, Anita
My rhodod. were beautifull [!] this year!
After all, you have to see our „estate“!

77

University of Washington
Department of Geography
Seattle 5

June 6, 1962

Dr. Joseph F. Rock
P. O. Box 497
Honolulu, Hawaii

Dear Dr. Rock:
It was wonderful to get your letter. I was sorry to hear that you have been ill and I think you have now discouraged me from ever trying to write a dictionary (not that I ever planned to). I know you will keep at it and create a very useful dictionary for oriental linguists.

It seems such a short time since I was with you in Honolulu, but you are right that it will soon be a year. I have not yet heard anything from the Office of Naval Research but hope to soon. If everything goes well I should be out to see you again some time this summer.

The maps are very nearly done. We have only one graduate assistant in cartography and he has to serve the needs of ten faculty people. He has had to fit work on the maps in here and there and in between other assignments. In any case we are now in the middle of final exams so I will get together the original maps late this week. I will get contact copies made and send the originals on to you.

As soon as exams are out of the way and also the Pacific Coast geographers meetings scheduled here next week, I will write to you more fully. I think there is only one map to finish up - my chief problem is how to get reproductions made so you can edit them.
More later!
Sincerely,
John C. Sherman

78

University of Washington
Far Eastern and Russian Institute
Seattle 5, Washington

November 26, 1962

Dr. Joseph Rock
P. O. Box 497
Honolulu 9, Hawaii

Dear Rock:
Thanks for your good letter. I wondered how you were and I am very happy to hear that you are safely back in Hawaii and seem to be better. I am sorry that you by-passed us, but I understand that you probably had had enough.

I certainly think that it would be silly to pay customs on your own books. I don't know whether George has done anything, but if I can help let me know to whom and what to write.

I have been terribly busy with my own work and have just completed an article with four more to go, letting alone the Taiping book on which I'm still working.

At home, the family is all right. Peter has slowly recovered from his foot infection. Every time I drive by the Wilsonian, I think of the good coffee.

All the best and love to you,

Herzlichst Franz

79

27-11-62

Dorrie is still in New York and highly successful in her professional education. Udo had a wonderful time there and is now back here. He had a stubborn flu and has still to be careful. He lives now in a houseboat which they acquired to escape the eternal moving around. Eva had some very successful appearances in California and other places. I gave them all your greetings.

I am trying to concentrate on working over the Taiping documents and writing the introduction. It is a loot of work, but I'm making some progress.

Take good care of yourself, don't work too hard. My best to your friends. Wie immer Ihr alter Freund Franz

80

W. W. Ritchie

2659 Currie Road, Victoria, B.C., Canada

November 27th, 1962

My dear Pohaku,

The best thing in your very welcome letter of Nov. 20th is *my health is fairly good.* You go on to mention angina and other effects, but the great thing is that you are well on the way to recovery, active in the work that interests you and enriches the world of people that matter. As for angina, I remember the story of a boy in my hometown. The doctor said to him „I hear you have been troubled with angina pectoris“ and the boy slyly answered: „Yes, I did have some trouble over a girl – no shot-gun stuff – but that wasn't her name.“ Not everyone has your vocabulary. I believe it is true that Procter and Gamble had to take the word concentrated out of their soap ads., because many good American women thought, with *consecrated* in mind, that it had something to do with the Pope.

I rejoice in the progress of your work, in your happiness at recognition by scholars and at your plans to move to where you will be in contact with men of culture and understanding. I hope there will be some clever and well-educated women around, too. Don't forget Ninon de l'Enclos[73], friend of many scholars, buried in (I think) Paris. Her epitaph contains this, in French of course: „She was renowned for her chastity during the last years of her life.“ She died at 85, after many amours, and I suppose that the chastity was say from 82 to 85. She was like a signpost, pointing the way, but not necessarily going there.

Your problem will be to find the scholarly atmosphere in a good climate. Marburg may be too cold in winter and Sicily or Majorca have men who, like an attic, have the least valuable possessions in the upper floor – the brain.

We are well and happy, and hope to see you once again before we all pass into the land that loveth silence, that old Egyptian expression that appeals much to me. It is at least 5,000 years old. Pam and Harry are well, and Ken is at his new post in Vernon, near enough for him to come over on long week-ends and for Daphne to run down to Vernon. They each have a car. Harry seems to me better than for years past.

I'm glad that you can arrange for your art treasures to be, when you no longer get inspiration from them, with people who value them and know what they stand for. Glad, too, that you have another big Historiography in hand and that your diaries are to be published. Don't talk about your *last work*. Gladstone was Prime Minister of the U.K. when he was well over 80, and some poets and painters have done outstanding work in the eighties and I think that I could instance fine work in the nineties.

Miss Forester at last has found a vacant bed in hospital. I conveyed your message to her.

Affectionate good wishes from us both. We miss you. We shall be with the family in Vancouver for Christmas. Will

81

Mrs. W. W. Ritchie

---

[73] Anne, genannt Ninon de L'Enclos, 1616-1706, berühmte Pariser Kurtisane, deren Salon Treffpunkt bedeutender Persönlichkeiten war. Ob die ihr zugeschriebenen Briefe von ihr stammen, ist unsicher.

2659 Currie Road
Victoria, B.C.
Tel.: Evergreen 5-3008

Nov. 29th

Pohaku dear

We were so glad to get your last letter of Nov. 20th. I quite understand how very busy you have been correcting all those proofs. 600 odd pages! You must have felt a weight off your chest. So sorry you have been bothered with angina pectoris. I know how very painful it is when it comes. Youve [!] had more than your share of pain + accomplished so much in spite of it.

Pam rang up to say how delighted she was to get a letter from you. We don't see them often – but always keep in touch. I meet her often at the Horticultural meetings. Tuesday night there was a very interesting meeting of the English Speaking Union. When Lady Maye [?] gave an account of the big ESU gathering in San Francisco which Prince Philip attended + then a lecture on „Britain Today" by a D-Pettit [?] of the U. BC I was glad to hear that he is not pessimistic + feels Britain is going ahead.

Miss Forester is back home again I heard yesterday – as there's not much they can do for her arthritis. Hospital beds are very precious I expect.

I am very glad you are warm anyway. Today is just wretched – high east wind – dull grey skies heavy with rain or maybe snow – I'm glad to stay put at home. We shall be thinking of you at Christmas. No we shall be in Vancouver with the family for two or three days. In May we shall see. Norah + Ronald in Chicago – Arch is today in Bahrein [!] hoping he may land a new hotel job. He + Lesley loved Rabat. Love + best Xmas wishes from us both Ella

82

Professor Dr. Hermann v. Wissmann
Tübingen, Wildermuthstrasse 24

Herrn Professor Josef F. Rock
Far Eastern Institute
University of Washington
Seattle, Washington
U.S.A.

Sehr verehrter Herr Professor Rock!
Erinnern Sie sich noch an den jungen Österreicher, der mit einem chinesischen Kollegen und drei chinesischen Assistenten in Kunming auftauchte, um dann 10 Monate in Süd-Yünnan zu reisen? In letzter Zeit habe ich mich viel mit Ihrer Arbeit, insbes. Nahsi-Kingdom und Ammye [!] Ma-chhen beschäftigt, einerseits schrieb ich nach Handel-Mazzettis[74] Nachlass eine Arbeit über die rezente und quartäre Vergletscherung des Yülungschan[75], andererseits ist jetzt gerade meine größere Abhandlung in der Mainzer Akademie (math.-nath. Kl.) über die heutige und letzteiszeitliche Vergletscherung Hochasien[s] erschienen. Ich möchte Ihnen dieses Buch und einen Sonderdruck der genannten Arbeit zusenden, möchte aber erst wissen, ob die Adresse richtig ist. Ich hoffe, daß Sie gesund sind. Kürzlich schrieb Kollege Weigold[76], der eine große Arbeit vorbereitet. Als der

---

74 Heinrich Freiherr von Handel-Mazzetti (1882-1940), Botaniker, seit 1903 am Botanischen Institut Wien, 1923 am Wiener Naturhistorischen Museum. 1914-1919 war er auf Reisen in China. Vgl. *Chronica Botanica* 7.1942, 88-89. Zu Handel-Mazzettis größeren Werken gehören *Naturbilder aus Südwest-China. Erlebnisse und Eindrücke eines österreichischen Forschers während des Weltkrieges.* 1927. XIV, 380 S.; *Symbolae Sinicae. Botanische Ergebnisse der Expedition der Akademie der Wissenschaften Wien nach Südwest-China, 1914-1918.* Bd 1-7. 1929-1937.

75 Hermann von Wissmann (1895-1979); Geograph an der Universität Tübingen. Vgl. *Hermann von Wissmann-Festschrift.* Zum 65. Geburtstag von seinen Freunden, Kollegen und Schülern. Hrsg. von Adolf Leidlmair. Tübingen: Geograph. Institut der Universität 1962. 384 S., 18 Taf., 2 Faltbl. (Tübinger geographische Studien. Sonderband.1.) – Die angesprochene Arbeit ist: *Die heutige Vergletscherung und Schneegrenze in Hochasien mit Hinweise auf die Vergletscherung der letzten Eiszeit.* Mit einem Beitrag: Bemerkungen zur Klimatologie von Hochasien. Aktuelle Schneegrenze und Sommerklima. Von Hermann Flohn. Mit 27 Abb. im Text, 4 Bildtaf. u. 3 Faltkt. Mainz: Akademie, Wiesbaden: Steiner in Komm. 1960. 331 S. (Akademie der Wissenschaften und der Literatur, Mainz. Math.-nat. Klasse. 1959, Nr 14.)

76 Hugo Weigold (1886-1973), Zoologe und Naturschützer. *Vgl. Natur und Jagd in Niedersachsen.* Festschrift zum 70. Geburtstage, Hannover: Arbeitsgemeinschaft für Zoologische Heimatforschung in Niedersachsen 1956. Weigolds Ostasieninteressen hängen mit seiner Teilnahme an Walther Stötzners Expedition zusammen; vgl. Weigold: *Zoologische Ergebnisse der Walter Stötznerschen Expedtionen nach Szetschwan, Osttibet und Tschili auf Grund der Sammlungen und Beobachtungen Dr. Hugo Weigolds.* Leipzig: B. G. Teubner 1922.

Amnye-Ma-chhen entstand, müssen Sie ja nahe von hier, in Zürich, gelebt haben. Hoffentlich erreicht Sie dieser Brief!
Mit den allerbesten Wünschen und herzlichen Grüßen bin ich Ihr ergebener Hermann Wißmann
Ich habe mir erlaubt, in beiden Arbeiten einige Bilder von Ihnen abzudrucken, selbstverständlich mit genauer Angabe, woher ich sie nahm.

83

University of Washington
Far Eastern Institute
Seattle 5, Washington

Dear Rock,
Thank you for your Gruss of February 25th with the news of your arrival on March 15th, Sunday morning. I called the Wilsonian and they will try to reserve 609 for you. We are all looking forward to have you and had been waiting for this news for some time.

Yes, I did not answer your last letter and apologize. But this was crazier as usual and you will find a tired and chastened friend. There are lots of things to tell, but I better wait till your arrival. Die Ehrenjungfrauen sind auf den Flugplatz bestellt. Auf bald und herzlichste Grüsse von Anita, den Kindern und mir. Ihr alter Freund Franz

---

(Abhandlungen und Berichte des Zoologischen und Anthropologisch-Ethnographischen Museums zu Dresden 15.)

BRIEFWECHSEL MIT ROBERT KOC

1.	11. 3. 1952	36.	10. 12. 1957
2.	12. 10. 1952	37.	8. 1. 1958
3.	26. 12. 1952	38.	20. 5. 1958
4.	3. 1. 1953	39.	o. D. 6.58?
5/6.	11. 1. 1953 / 16.1. 1953	40.	o. D.
7.	12. 2. 1953	41.	10. 8. 1958
8.	3. 3. 1953	42.	16. 8. 1958
9.	5. 3. 1953	43.	18. 8. 1958
10.	12. 3. 1953	44.	28. 8. 1958
11.	18. 3. 1953	45.	19. 10. 1958
12.	11. 5. 1953	46.	5. 11. 1958
13.	24. 5. 1953	47.	31. 12. 1958
14.	13. 2. 1955	48.	18. 2. 1959
15.	20. 2. 1955	49.	19. 2. 1959
16.	16. 4. 1956	50.	25. 2. 1959
17.	27. 4. 1956	51.	27. 3. 1959
18.	27. 5. 1956	52.	3. 7. 1959
19.	4. 9. 1956	53.	4. 7. 1959
20.	31. 10. 1956	54.	5. 2. 1960
21.	5. 11. 1956	55.	18. 2. 1960
22.	21. 11. 1956	56.	29. 2. 1960
23.	10. 1. 1957	57.	26. 4. 1960
24.	19. 1. 1957	58.	9. 5. 1960
25.	24. 1. 1957	59.	15. 5. 1960
26.	4. 2. 1957	60.	18. 5. 1960
27.	5. 2. 1957	61.	26. 5. 1960
28.	Febr. ?	62.	26. 5. 1960
29.	28. 2. 1957	63.	23. 10. 1960
30.	3. 4. 1957	64.	26. 6. 1960
31.	14. 4. 1957	65.	28. 10. 1960
32.	21. 4. 1957	66.	8. 12. 1960
33.	11. 8. 1967	67.	4. 2. 1961
34.	5. 9. 1957	68.	5. 2. 1961
35.	21. 10. 1957	69.	3. 6. 1961

70. 7. 6. 1961
71. 24. 2. 1962
72. o. D.
73. 23. 3. 1962
74. 6. 4. 1962
75. 3. 6. 1962
76. 15. 9. 1962
77. 19. 8. 1962
78. 23. 9. 1962
79. 27. 9. 1962
80. 1. 10. 1962
81. 5. 10. 1962
82. 20. 10. 1962
83. 26. 10. 1962
84. 30. 10. 1962
85. 4. 11. 1962
86. 2. 12. 1962

Briefe von Robert Koc und Giuseppe Tucci

87. 31. 1. 1958 Von Koc
88. 17. 4. 1958 Von Koc
89. 27. 7. [1958?] Von Koc
90. 1. 8. 1958 Von Koc
91. 25. 10. 1958 Von Koc
92. 22. 1. 1964 Tucci
93. 12. 6. 1967 Von Koc
94. 27. 6. 1967 Tucci
95. 30. 6. 1967 Von Koc

1
28 West Park Road, Kew Gardens, Surrey
Mar. 11/52

My dear Robert,

Thank you for all the postcards and your last letter. I am O.K. again. I have a new system now, am taking half of the salts in the evening and half in the morning, and this is the solution, the ear ringing has also stopped, that is it is not so loud as it was. I also don't eat at night except soup and black bread.

Alas I cannot come to the continent there is no time, I have got to get back to the States. Tonight I am flying by BEA to Edinburgh Scotland where I shall be till Tuesday next a week hence I will be here again. Then shall pack and I will be off for Boston, New York, San Franciso-Honolulu. May be I shall have an operation there my old Doctor is still there, he has a large clinic. I would rather be there among friends than among some strange place in Europe. And if I die, I rather die there than anywhere else. It was there I really started life.

The Corras[1] are flying to Hong Kong on Friday so I shall not see them again, they come back on the 8th but I was too ill to phone them. Did so today but they were out.

There is little new so will close, don't worry about me and don't spend your shillings on postage you need them for something else. Have not heard from Hans but wrote him two letters I also wrote that long letter to Anton, but have had no reply, probably because I have not sent him anything yet. One can't from here, I should have done so from Zürich, but I could not find his address. And so it goes. fare well dear boy. Will write again soon. As ever yours, J. F. Rock

---

1 Henry Corra, ein alter Freund Rocks, war in den dreißiger Jahren als Vertreter einer texanischen Ölfirma durch China gereist. Er lebte offenbar später in Hong Kong. Vgl. S. B. Sutton: *In China's border provinces.* New York 1974, 87.

2

Sunday evening Oct. 12th/1952

Mein lieber Robert:
Deine zwei Briefe mit Dank erhalten. Es wunderte mich dass Du meinen langen Brief den ich Dir schon vor nahe zwei Wochen schrieb nicht erhalten hast. Ich schrieb Dir Du sollst zum Englischen Konsul gehen und fragen was nötig wäre für Dich ein Visum für England zu bekommen oder was ich hier thun müsste um für Dich gut zustehen etc. Hasst [!] Du denn diesen Brief nicht erhalten? Ich wartete auf Deine Antwort. Inzwischen war ich wirklich diesmal sehr krank, ich fürchtete ich hatte eine Perforation, Ich wachte auf um 4 h morgens mit heftigem Schüttelfrost der immer anhielt den folgenden Tag, ich hatte eine Temperatur nahe 103° Fahr. Ich glaubte wirklich das Ende war gekommen, mein abdomen war wie eine Trommel und ich erbrach fortwährend doch hatte ich wenig im Magen. Ich hatte solche mehrere male früher in China, aber nie mit so hohem Fieber. Vor dem Erbrechen kam wie ein Stück rohe [!] Fleisch mit Blut dran, so muss doch irgendwo ein Geschwür sein. Heute geht es mir wieder gut. Wie Du weisst ich komme schnell wieder weg über solche Anfälle. Aber diesmal hat es mehrere Tage nahezu eine Woche gebraucht und ich fühle mich geschwächt. Ich bin froh dass ich in keinem Hotel allein war. Die alten Damen hier die früher Krankenschwestern waren waren sehr gut zu mir.

Mein lieber Robert es tut mir leid dass Du mit Deinem Bruder Auseinandersetzungen hast. Er scheint mir doch eine grobe Natur zu haben so wie sein Vater, der Krieg hat ihn wohl auch nicht freundlicher gemacht, und dazu noch der Verlust seiner Frau, so musst Du halt ein Auge zudrücken. Reize ihn nicht das hat keinen Zweck. Wer ist denn der Neffe der da bei euch wohnt? Ich glaubte er kam nur auf Besuch. Wird er ständig bei euch wohnen? Wenn so, und er verdient, da soll er doch etwas bezahlen.

Ich hätte beinahe vergessen Dir zu danken für die Vanilla Kipferln, sie kamen gerade am Tag wo ich so sehr krank war. Ich wurde alle paar Minuten ohnmächtig und bewußtlos. Es war schrecklich. Und doch hatte ich den Tag vorher gefasstet [!]. Seit meiner Ankunft hier habe ich kein Fleisch gegessen und nur zweimal Fisch, sonst immer Grieskoch [?] und Reis in Milch.

Du hättest nicht Dir Auslagen machen sollen wegen den Kipferln das ist überflüssig. Du brauchst dass [!] bischen Geld das Du noch hast.

Ich hatte der Frau Ciani geschrieben und ihr gedankt für die Kipferln, ich schrieb ihr auch dass ich wegen meiner Krankheit meine Abreise aufgegeben habe und es ist möglich dass ich doch noch einmal nach Meran komme. Wenn so dann kannst du auch kommen aber nur wenn du keinen Posten hast. So bitte nimm dass [!] nicht für Gewissheit hin denn manches kann mir noch passieren.

Seitdem ich hier bin habe ich noch andere der Karten gemacht und die Namen eingeschrieben, den ersten Teil des Manuskriptes habe ich der Roy. Hort Soc.[2] abgegeben. Nun bin ich müde.
Mit den herzlichsten Grüssen verbleibe ich
Dein alter Onkel
J. F. R.

3

Cosmos Club
Washington D.C.

Dec. 26/52

Lieber Robert.
Ich wollte Weihnachten in Honolulu verbringen aber konnte keinen Sitz bekommen auf irgend einem Flugzeug. Ich kam von New York hierher nach Washington D.C. wo ich viele Freunde habe und habe auch gestern ein schönes Weihnachtsfest gefeiert.

Hoffentlich hast Du das Geld dass [!] ich Dir telegraphisch zukommen liess erhalten bitte schreibe mir wie es euch geht. Hast Du auch das Paket bekommen?

Ich bin Gott sei Dank wieder gesund und man sagt mir ich sehe sehr gut aus.

Wir haben halt eine gute Konstitution. Vor mir liegt schwere Arbeit. Alle meine Freunde fordern mich auf eine Biographie zu schreiben, und jetzt werde ich es auch thun sobald ich nach Honolulu komme, ich nehme mir keine andere Arbeit mit. Ich wünschte Du wärest hier. Aber jetzt ist es doppelt schwer ein Visum zu bekommen. Ich werde mich aber bemühen sobald ich in Seattle ansässig bin, aber im Falle es mir dort nicht gefällt verkaufe ich alles und komme nach Europa.

---

2 Die Phytogeographie Nordwestchinas. Der erste Band wurde für die Royal Horticultural Society fertiggestellt; deren Präsident, Lord Aberconway starb jedoch, und sein Nachfolger war an dem Projekt nicht interessiert. Daher wurde das Werk nicht veröffentlicht.

Wir werden sehen. Du weisst ich mache lange im voraus keine Beschlüsse.

Unter separatem Couvert sende ich Dir den Lieberoman [!] eines Nakhi Mädchens den Du haben wolltest.[3]

Die Totengesänge, oder die Lieder die beide singen ehe sie sich das Leben nehmen ist im Anhang.

Wenn es Dich interessiert ich kann Dir noch andere meiner Arbeiten schicken.

Grüsse mir Hans recht freundlich. Dir alles Gute wünschend verbleibe ich

Dein Onkel J. F. Rock

Eingeschlossen 9 Marken

4

Hotel Continental
Cambridge, Massachusetts

Jan. 3/53

Lieber Robert:

Wie Du siehst bin ich noch immer in New England. Ich erwarte Montag den 5ten einen Käufer für meine tibetische Kunstsammlung. Hoffentlich wird etwas darauf [!] werden.

Ich war einige Tage in New York dann gieng ich nach Washington wo ich Weihnachten verbrachte von wo aus ich Dir schrieb. Nachher fuhr ich mit dem Zug nach Connecticut wo ich zwei Tage mit einem alten Freund verbrachte und dann bei Auto nach Boston. Soeben fahre ich zurück von Needham wo ich Freunde von Nervi besuchte.

Und so geht es. Gestern hatte ich Lunch mit einem Vater Morse[4] der in Lichiang war und den ich mit meinem Flugzeug aus Lichiang nach Kunming brachte. Wir beide hatten Heimweh nach Yünnan.

---

3 The romance of ^{2}Ka-^{2}mä-^{1}gyu-^{3}mi-^{2}gky. A Na-khi tribal love story translated from Na-khi pictographic manuscripts, transcribed and annotated. By J. F. Rock F. R. G. S., Research Professor in Oriental Studies, University of Hawaii, Membre correspondant de l'Ecole Française d'Extrême-Orient. *Bulletin de l'Ecole française d'Extrême-Orient.*39.1939,1-155, Taf. 1-32.

4 Vielleicht William Reginald Morse (1874-), M.D.; er war seit 1917 mit der American Baptist Foreign Mission Society in China gewesen, seit 1932 als Dean, Medical College, West China Union University.

Ich bin mir noch nicht gewiss ob ich zuerst nach Seattle oder San Francisco reise, alles hängt ab, an dem Verkauf oder Nichtverkauf meiner Kunstsammlung. Es ist auch sehr möglich dass ich, wenn alles gut geht, und ich meine Bücher in Seattle in Ordnung bringe, nach Italien zurückkehren werde und Du kannst dann zu mir kommen und wir können weiter arbeiten. Ich glaube kaum dass ich in Amerika bleiben werde. Ich fühle es in meinen Beinen. Gesundheitlich geht es mir gut, keine Anfälle mehr. Hier gibt es sehr gutes Essen und sehr reichlich. Wie immer, Abends esse ich nichts.

Sobald ich selbst weiss wo ich stehe, werde ich Dich benachrichtigen. Ich werde dann wahrscheinlich nicht nach Honolulu fliegen, sondern von Seattle nach Canada reisen und von dort nach Europa fliegen. So wundere Dich nicht wenn Du von Canada hörst dass ich am Flug nach Europa bin.

Ich lege dem Brief einige Marken bei für Hans. Ich grüße euch beide recht herzlich, Euer alter Onkel

5

Hotel Edmond Meany
University District
Seattle 5

January 11/53

Lieber Robert:

Endlich nach 3tägiger und 3nächtiger Reise mit Empire Express kahm [!] ich heute morgens in Seattle von Boston an. Ich bin Tod müde. Ich muss aufrichtig sagen, es gefällt mir nicht. Oder bin ich rastlos?, weiter, immer wieder weiter, Zerstreuung ist das Einzige dass [!] mich am Leben hält. Aber wohin? Ich möchte am liebsten morgen nach Zürich fliegen und von dort nach Meran gehen um in Frieden mit Dir zu sein und zu schreiben.

Man hat mich sehr freundlich empfangen hier, aber mir gefällt es nicht. Ich bin allein und traurig. Soll ich nach Hawaii fliegen oder zurück nach Europa?

Morgen gehe ich zur Universität um meine Bibliothek zu sehen und die vielen Koffer auspacken. Ich nahm drei Koffer mit Tibetischen Bildern etc., mit hierher von Boston. 2 Kisten sind noch dort, alles andere habe ich verkauft, all das schwere, wie Buddhas und tibetische Gottheiten. Viel haben sie nicht eingebracht, aber genug um ein Jahr in Europa schön leben zu können. Ich will sehen ob ich alles andere hier verkaufen kann.

Ich glaube am besten wird es sein ich widme mich nicht mehr wissenschaftlichen Arbeiten die nichts einbringen, sondern schreibe für das Publikum. Ich habe manches erzählt von meinen Erfahrungen und alles sagt „ja warum schreiben Sie das nicht wieder, man würde sich reissen um das alles zu veröffentlichen, sie könnten damit viel Geld verdienen“. Und nun will ich das thun, aber nicht hier in diesem Lande. Alles ist furchtbar teuer. Die 3 Koffer zur Bahn zu bringen kostete $8.24¢ oder 208 Shillinge, ich fuhr mit der Bahn, da das Wetter so schlecht war und keine Flugzeuge abflogen. Nichts wie Schneestürme und Eis. Hier regnet es schon seit einer Woche und der Wind heult.

Vor 47 Jahren konnte man um 20 Dollars von Texas nach San Francisco fahren, heute kostet das $ 185.00 dazu das Essen, die Trinkgebühr, etc., es ist fürchterlich. Lange könnte ich das hier nicht aushalten. Darum zurück nach Europa. Wir können zusammen schön arbeiten und brauchen uns nicht überanstrengen. Ich bin zu alt dazu. Jetzt leichteres schreiben und Geld verdienen.
Mit herzlichen Grüssen verbleibe ich Dein alter, trauriger, einsamer Onkel J. F. R.

6 [Postscript]

Hotel Edmond Meany
University District
Seattle 5

2) Das Tempo hier ist fürchterlich! Es graut mir, Licht Reklame drehen sich wie im Ringelspiel automatische Menschen öffnen Thüre, laden dich ein! Alles ist künstlich, aber keine *Kunst.* Luxus überall, speziell in den Prachtzügen, herrliche Speisewagen, so etwas gibt es in Europa nicht, alles Diesel Züge, keine Kohlen mehr, alles ist rein und sauber. Kein Fenster braucht man öffnen, alles ist „air conditioned“. 3 Tage und Nächte war ich im Zug und mein Hemd war weiß wie Schnee. Aber ich fliege doch lieber, es ist so viel schneller und man spürt nichts. Ich glaube ich komme sehr bald aber ohne Bücher oder überflüssiges Zeug. Nachdem ich meine Erlebnisse geschrieben und Geld verdient habe ist Zeit genug um über wissenschaftliche Dinge zu schreiben. Wir werden sehen, bitte grüsse mir Hans. J. F. R.
16ten Jänner

Ich verzögerte den Brief aufzugeben und wollte Dir weiteres über meine Entschlüsse schreiben. Ich fliege weiter nach Honolulu und zwar am Donnerstag so viel ich weiss. Meine Bücher sind alle schön angekommen und sehr gut aufgehoben. Alle bewundern diese Bibliothek denn sie besteht aus Büchern die nirgends zu haben sind. Man hat mir ein schönes office eingeräumt wo ich arbeiten kann. Alle sind sehr liebenswürdig und thun alles was sie können um mich hier zu behalten.

Doch ich bin traurig und als ich alles auspackte und durchsah da kamen alle die guten Zeiten vor meinen müden Augen und ich wurde sehr traurig.

Nun genug. Ich hoffe ein Brief von Dir erwartet mich in Honolulu.

Bis dahin wie immer J. F. R.

Ich lege dem Brief Marken bei für Hans, darunter eine 10 Shilling oest., und auch einige 5 Shill. Das andere sind meistens Englische von den vielen Paketen die ich von London hierher sandte. Ich habe noch viele andere, sende sie später. J. F. R.

7

Honolulu, Hawaii, 12ten Februar/53

Lieber Robert:

Deine beiden Briefe an die hiesige Universität und an meine Adresse hier habe ich gestern nach meiner Rückkehr von der großen Insel Hawaii erhalten. Es waren schöne acht Tage. Man kann jetzt überall hinfliegen, in einer Stunde ist man auf Hawaii. Die Insel ist sehr gross, der grösste Berg ist Mauna Loa, der 18000 Fuss from Meeresgrund zum Meerespiegel reicht und dann nahezu 14000 Fuss über dem Meere steht, also über 31000 Fuss hoch ist. Er war gestern mit Schnee bedeckt. Wir fuhren mit Auto überall hin. Über das grosse Plateau zwischen den hohen Vulcanen und neuen so wie alten Lavaflüssen. Ich kaufte 400 Fuss Kodachrome Movies von den Eruptionen des Mauna Loa and Kilauea, die sind hochinteressant, sensationell. In 1950 flossen die Lavaströme von 9000 Fuss Höhe in den Ocean, es sind Tag wie Nachtaufnahmen, Lavaspringbrunne (fountains) 400 Fuss hoch spielten über den Bergabhängen, ich bringe die Movies mit nach Europa und Du wirst sie sehen.

In ein paar Tagen gehen wir nach Maui auf den grossen Berg Haleakala = Das Haus der Sonne; er ist über 10,000 Fuss hoch, 2000 Fuss ist die Tiefe des Kraters, und 21 Meilen im Umfang, mit 7 Aschen kegeln über 2000 Fuss Höhe. Dort wächst das Grosse

Silberschwert eine grosse Composite. Ich habe viele Samen gesammelt von den seltenen Bäumen hier die am Aussterben begriffen sind. Viele sende ich nach England. Ob Wien (Bot. Garten Rennweg) welche haben will? Ich sende Dir einige Bilder von mir aus meiner Jugendzeit in Hawaii und Japan. Ich lebe sowieso nicht mehr lange so sollst Du sie haben.

Wir gehen auch nach Molokai und Kauai, es wird mein letzter Besuch sein. Hier nichts als Einladungen man hat keine Ruhe, es sind alles alte, reiche Freunde. Sie wollen meine Bibliothek kaufen und sie von Seattle nach Honolulu bringen. Vorläufig habe ich nichts versprochen da ich die Bücher zu meiner Arbeit brauche.

Das Wetter hier ist herrlich, immer Sonnenschein und kühler Wind von den Bergen.

Ich habe um die Briefe die Du nach Washington schicktest geschrieben und werde sie nach meiner Rückkehr von Maui wohl finden. The National Geographic wusste meine jetzige Adresse nicht so konnten sie die Briefe nicht nachsenden.

Hier alle sagen ich sollte meine Erlebnisse niederschreiben und veröffentlichen. Nun vorläufig Schluss.

Mit herzlichen Grüssen an Dich und Hans, Dein alter Onkel J. F. Rock

Grüsse an Frau Ciani.

Beiliegend zehn alte Photographien und eine Ansichtskarte des Kilauea Vulkans.

8

Honolulu March 3/53

Mein lieber Robert:

Es tut mir unendlich leid dass Du noch nichts gefunden hast und das[!] es Dir nicht gut geht. Wie kann man Dir helfen. Wegen der Verhältnissen [!] respective Regeln kann man nicht Geld schicken eingeschlossen im Briefe. Cheques brauchen zu lange bis sie einkassiert sind, da bin ich ebenso schnell da. Ich fliege nächste Woche nach Vancouver British Columbia, dann nach Seattle und Europa wahrscheinlich nach Zürich. Da kannst mich dann dort antreffen und wir gehen zusammen nach Meran wo wir das Buch schreiben werden. Die Zeit hier ist schnell vergangen. Es war schön die alten Tage (Zeiten) wieder zu erleben. Es ist 33 Jahre her seit ich von Honolulu weg ging, war aber öfters da das vorletzte Mal in 1948.

Unter separatem Umschlag sende ich dir mein Buch über die einheimischen Bäume der Hawaii Inseln,[5] 1913 gedruckt. Es ist vergriffen und hat einen Preis von $ 50.00, wenn erhältlich. Ich habe drei Exemplare von dem Sohn des früheren Gouverneurs bekommen. Sein Vater hatte es damals veranlaßt das Buch herauszugeben. Die Lobeliaceen Monographie[6] und manches andere schicke ich dir auch behalte alles zum Andenken. Ich glaube lange werde ich nicht mehr leben. Für mich hat das Leben keinen Zweck mehr. Mein Heim in China ist weg. Honolulu hat sich sehr verändert und ist nicht mehr was es war. Voll von Touristen welche halb nackt mit übergrossen Bäuchen am Strande spazieren gehen, in unterhosen [!]. Es ist abscheulich. Auf den andern Inseln ist es besser da ist man noch so zu Hause. Auf der Insel Maui war ich mit Freunden auf 4000 Fuß Höhe am erloschenen Vulkan Haleakala. Das Leben hier ist furchtbar teuer, ein gewöhnliches Haarschneiden 35 Schillinge, ein Mittagmahl 85 Schillinge, und s.w. Im Hotel 625 Schillinge pro Tag ohne Trinkgelder also 25 Dollar da kann man in Europa eine halbe Woche leben.

Wir werden zum Pittscheider in Meran gehen dort kostet es 3000 Lire pro Tag per person alles inbegriffen ungefähr 4 Dollar pro Tag. Ich freue mich schon auf ein Wiedersehen.

Du schreibst schöne Briefe nur Schade dass Du dein Talent nicht verwerten kannst.

Ich will nicht viel schreiben sondern nur Dir wissen lassen dass ich Deinen Brief erhalten habe. Es tut mir leid dass Hans krank ist er

---

5 *The indigenous trees of the Hawaiian Islands.* By Joseph F. Rock, Botanist of the College of Hawaii, Consulting Botanist, Board of Commissioners of Agriculture and Forestry, Territory of Hawaii. With two hundred and fifteen photo-engravings. Published under patronage. Issued June 26, 1913. Honolulu 1913. [V], 518 S. inkl. 218 Taf. 4° Aufgeklebt: Edition limited to 1000 copies.

6 *A monographic study of the Hawaiian species of the tribe Lobelioideae, family Campanulaceae.* By Joseph F. Rock, botanist, College of Hawaii; consulting botanist, Board of Commissioners of Agriculture and Forestry, Honolulu. Author of «The indigenous trees of the Hawaiian Islands» and «The ornamental trees of Hawaii». With frontispiece and two hundred and seventeen plates. Issued February 20, 1919. Honolulu: Bishop Museum Press 1919. XVI, 395 S., Frontispiz, 217 Taf., Errata (Publications of the Bernice Pauahi Bishop Museum of Polynesian Ethnology and Natural History.) (Bernice P. Bishop Museum Memoirs. 7.2.)

soll doch ein Tropenheim besuchen. Da er im Krieg war soll doch die Regierung dafür sorgen dass er wieder gesund wird. Hier sind herrliche Spitäler wie erkrankte Soldaten alles bekommen die beste Fürsorge alles frei mit Gehalt dazu.

Nun schliesse ich mit herzlichen Grüssen für euch Beide und hoffe auf ein sehr baldiges Wiedersehen. Von Meran kommen wir nach St. Anton wenn es im ersteren zu heiss wird.
Wie immer Dein alter Onkel J. F. Rock

9

Honolulu, März 5/53

Mein lieber Robert:
Gestern kam Dein lieber Brief vom 12ten December der zweite den Du nach erhalt [!] des Weihnachtsgeschenkes erhalten hattest wird wohl später kommen. Briefe finden mich immer so lange ich noch am leben bleibe.

Heute schrieb ich an Frau Schuler und legte dem Brief ein Geschenk bei für Dich, ich bat sie es Dir nachzuschicken, auch schrieb ich ihr dass wir diesen Sommer wieder kommen werden und wir dieselben Zimmer mieten wollen. Wir gehen dann von Meran direkt zu ihr.

Heute kam ein Brief von Hans der am 26ten Febr. aufgegeben wurde. Es tut mir sehr leid das [!] er krank ist, er soll zu einem Tropenheim oder Tropenarzt gehen der diese Fieberkrankheiten versteht. Ich werde ihm bald schreiben oder diesem Brief einen an ihn beilegen. Ich lege auch 23 Marken bei und auf dem Couvert einige neue oder alte amerikanische. Die Nato Marken gibt es nur in dem Werte von 3 ¢ und keine anderen. Es werden immer neue 3 ¢ Marken hergestellt, sonst keine.

Deine Briefe erfreuen mich sehr und ich hoffe dass wir bald wieder zusammen sein werden. Hoffentlich kommt wegen der jetzigen Weltlage und Vorfällen nichts dazwischen.

Die Bücher habe ich noch nicht abgeschickt ich bin noch nicht dazu gekommen. Habe doch viel zu tun, ich schreibe die Titel aller der Fotographien die ich vor 40 Jahren hier aufgenommen habe cirka 2000 negative [!]. Ich werde die besten davon drucken lassen.

Es ist wirklich schön hier und meine alten Freunde hier laden mich immer ein. Sonntag fliege ich nach der Insel Molokai wo ich mit alten Freunden bleiben werde auf ein paar Tage.

Ich lege dem Brief eine Karte bei so dass Du sehen kannst wie es mit dem fliegen hier steht. Es ist doch möglich dass ich auch nach Kauai fliege denn sie ist eine der schönsten Inseln. Ich bekam von dem Direktor des National Parkes von Hawaii 1200 Fuss lang lebende Bilder von den Lavaflüssen der grossen Vulkan Mauna Loa, diese werde ich mitbringen sie sind Kodachrome (farbig) und sind ausserordentlich interessant. Sie zeigen grossen [!] Feuerquellen, die schnellen Lavaflüsse bei Tag und Nacht wie sie ins Meer fliessen. Es ist wirklich grossartig. Sie sind nicht zu haben, nicht zu kaufen. Es wird euch gewiss interessieren. Nun Schluss hoffentlich hörst du bald von Frau Schuler. Mit herzlichem Gruss an Euch Beide. Euer alter Onkel J. F. R.

10

Honolulu 12ten März/53

Mein lieber Robert
Deine Karte vom 9ten Jänner, deine Briefe vom 2ten Jänner und 6ten März alle heute erhalten, auch Frau Cianis Brief vom 6ten März worin sie mir den Erhalt des CARE Paketes ankündigte. Es freut mich so sehr von Euch allen zu hören. Ich habe wohl viele Freunde überall auf der ganzen Welt, aber ihr seit [!] doch etwas ganz anderes, wir gehören zusammen.

Ich kam gestern von der Insel Molokai zurück wo ich vor 44 Jahren forschte und Pflanzen sammelte. Es hat sich auch dort vieles verändert. Ich fühle ich gehöre nicht in die jetzige Zeit, ich lebe viel in der Vergangenheit. Nach China sehne ich mich nicht mehr, nur hie und da bekomme ich Heimweh. Ich bin ja ganz allein und so sehne ich mich nach Dir. Wir verstehen uns jetzt. Es tut mir leid dass der arme Hans solche Malaria Anfälle bekommt. Die muss man kurieren können.

Ich wollte gestern noch nach der Kauai Insel fliegen, aber ich habe viel zu viel zu tun, denn ich fliege am Dienstag den 17ten März nach San Francisco und am Sonntag den 22ten nach British Columbia. Von dort nach Seattle dann zurück nach Vancouver und dann direct nach Zürich – Meran. Hoffentlich wird nichts dazwischen kommen. Die politische Lage scheint sich sehr verschlimmert zu haben. Ich will deswegen schnell von den Inseln fort man weiss nie was noch werden wird. Hoffentlich wird Vernunft siegen und wir in Frieden leben können.

Alle meine Freunde wollen dass ich nach Hawaii zurückkomme und hier meine letzten Tage verlebe. Aber nicht in Honolulu es sind zuviele Touristen, es ist nicht wie vor 46 Jahren wo man jeden den [!] man auf der Strasse begegnete, kannte. Auf der Insel Maui ja, dort lebt man abgeschlossen und doch ist man in schneller Verbindung. Dort ist es kühl, denn Kula wo ich 8 Tage verbrachte ist auf 3300 Fuss Höhe, man hat eine herrliche Aussicht auf drei andere Inseln und man überblickt das unendliche Meer. Das Silberschwert[7] wächst auf 10,000 Fuss Höhe, often im Schnee und Frost. Es blüt [!] nur einmal im Leben und braucht 25 Jahre bis es zur Blüte kommt. Aber es wächst in grossen mächtigen Rosetten wie aus Silber gemacht. Ich bin traurig und doch fühle ich dass ich fort muss. Es ist ein Drang wie schon aus alter Zeit. In San Francisco erwarten mich Freunde auch in British Columbia, überall wo ich hingehe. Es ist komisch, ich komme mir wirklich wie ein Weltbürger vor und bin es auch. Alles sagt schreiben sie nur ihre Erlebnisse es würde ein grosser Erfolg werden. Und so hoffe ich dass wir zusammen arbeiten können. Ich nehme viele Bilder mit. Ich habe über 2000 negative [!] bearbeitet seit ich hier bin die Tittel [!] für sie geschrieben und die von den Palmyra Inseln drucken lassen. Wir können schöne Aufsätze schreiben und auch Geld dabei verdienen. Deinen Überrock will ich herrichten lassen etc. etc. etc. Wenn nur Friede bleiben würde, dann wäre alles gut. Ich möchte mich gerne mit Dir ausplaudern aber wir wollen schreiben und schöne Geschichten erzählen, mir kommen sie als etwas alltägliches vor aber wo immer ich zum erzählen [!] anfange die Leute staunen nur, so glaube ich doch dass etwas gemacht werden kann, aber wir müssen uns beeilen.

Ich lege diesem Brief noch einige Bilder aus meiner Jugendzeit hier in Hawaii bei. Von Tibet und China habe ich viele aber die sind teilweise in Seattle und teilweise in London.

Die Bücher die ich Dir versprochen habe gehen morgen an Dich ab. Bitte behalte sie zum Andenken an mich. Dass [!] grosse Buch von den einheimischen Bäumen ist nicht mehr zu haben, und kostet $ 50.00 Dollar wenn es zu haben wäre. Ich hatte hier noch drei

---

7 *Argyroxiphium*; Rock hat später darüber publiziert: A new variety of silversword [Argyroxiphium]. By J. F. Rock, Research Associate in Botany, Bishop Museum, and Marie C. Neal, Botanist, Bishop Museum. *Occasional Papers of Bernice P. Bishop Museum* 22.1957:4, 31-33, 1 Ill.

Exemplare. Die Lobelien Monographie und vieles andere werde ich Dir auch schicken.

Hast du die Broschüre über die Palmyra Inseln[8] erhalten? Es sind schöne Bilder darinnen, sie sind 1000 Meilen südlich von den Hawaii Inseln. Ich erzählte Dir in St. Anton Geschichten von dieser Reise wir werden sie nochmals veröffentlichen aber mit all den amüsanten Geschichten die mit dieser Reise verknüpft waren.

Es freut mich dass Frau Ciani das CARE packet erhalten hat und dass ihr auch etwas davon genossen habt.

Und nun schliesse ich denn es ist schon spät und ich muss noch über 1000 Photographien einpacken und morgen abschicken. Eine grosse Kiste mit Büchern die ich hier hatte geht auch ab, alles nach Seattle wo jetzt meine Bibliothek ist.

Seid mir alle herzlich gegrüsst und auf ein baldiges Wiedersehen hoffend verbleibe ich Euer alter Onkel

J. F. R.

Beiliegend auch ein Zweiglein von *Geranium tridens* = Dreizackiges *Geranium* vom Haleakala.

11

St. Francis Hotel
Union Square, San Francisco

18ten März 1953

Mein lieber Robert:

Wie Du siehst bin ich wieder am wandern. Gestern abends um 8 Uhr flog ich mit schwerem Herzen von Honolulu ab und kam heute morgen um 6.30 in San Francisco an. Was für ein Unterschied. In Honolulu schöner blauer Himmel, blühende Bäume, Palmen, Berge die in Pastellfarben wie verschleiert aussahen, das saftige Grün der Wälder, die bunten Vögel die jeden morgen [!] vor meinem Fenster auf *Cassia* Bäumen in voller Blüte sangen, und hier ein grauer

8 *Palmyra Island, with a description of its flora.* By J. F. Rock, with the cooperation of O. Beccari, A. Zahlbruckner, U. Martelli, H. L. Lyon, and M. A. Howe. Honolulu: College 1916. 53 S., inkl. 20 Taf.; 1 Kte. (College of Hawaii Publications, Bulletin. 4.) Auch: *Honolulu Star Bulletin.* 1916 (Ohne die Serienzählung) – The voyage of the Luka to Palmyra Island. By Joseph F. Rock. *Atlantic Monthly* 144.1929, 360-366

Himmel, keine Bäume, nur mit Wiesen bedeckte Hügel, Häuser die wie Bienenkisten aussehen so nahe aneinander gereiht. Mir wurde buchstäblich übel als ich das heute morgens sah! Bin vom Paradies auf die Erde gekommen! Meine alten Freunde in Honolulu wollten alle dass ich dableibe, mir war das Scheiden sehr schwer und ich musste schlucken um das Weh zu erdrucken.

Jedoch hier in diesem schönen Hotel fand ich zu meiner Ankunft schon Telefonanrufe und Einladungen, und so geht es überall wo ich hinkomme.

Am Sonntag fliege ich nach Vancouver und von dort nach Victoria auf der Vancouver Insel wo ich Freunde aus China besuchen werde, dann geht es nach Seattle und dann – nach Zürich-Meran. Nur soll nichts dazwischen kommen. Ein [!] Monat muss ich vielleicht doch in Seattle verbleiben denn ich muss doch die Begräbnis Zeremonien der Na-khi beschreiben und die dazugehörige Literatur übersetzen. Ich werde das Manuskript mitbringen und Du kannst es dann tipieren. Fribourg Anthropos warten schon darauf. Ich habe Dir eine ganze Menge meiner Veröffentlichungen geschickt, aber das ist nicht alles, manches ist schon vergriffen, und ich hatte Mühe diese zu finden die ich Dir schickte. Behalte sie zum Andenken.

Gesundheitlich geht es mir jetzt sehr gut. Alles hat gestaunt über mein Aussehen, 70 Jahre und keine Falten, rosiges Gesicht, aber meine Haare und einen leichten, schwingenden Gang. Das macht die fortwährende Luftveränderung. Bitte schreibe mir nach Seattle c/o Far Eastern Institute, University of Washington, Seattle 5, Wash. U.S.A. Mit herzlichem Gruss an Euch Beide, wie immer Euer alter Onkel
J. F. R.

12

Seattle, May 11/53

Mein lieber Robert:
Während der Nacht vom 8ten zum 9ten an welchem Tage ich über Canada nach London fliegen sollte, wurde ich plötzlich krank, Schmerzen und Fieber zeigten an dass ich eine grössere innere Blutung hatte die mich so schwächte dass ich auf ein abreisen [!] nicht denken konnte, speciell da ich direct, ohne irgend einen Aufenthalt nach London fliegen wollte und dafür Plätze reserviert hatte, Ich müsste [!] sie aufgeben, und anstatt heute morgen in

London zu sein, sitze ich hier und schreibe Dir diesen Brief der Dich wohl nicht erfreuen wird. Gott sei Dank kam der Anfall nicht während der Reise, und in vorhergehenden Nacht. Es geht mir wieder besser, aber ich habe mir vorgenommen ehe ich etwas anderes anfange, die Leichen Zeremonien der Na-khi, auf die Anthropos in der Schweiz seit letztem Herbst wartet, druckfertig zu machen, und das geht am besten hier wo ich alles Material habe. Ich werde fleissig arbeiten und am Ende des Sommers, oder im August nach Europa kommen wir können uns dann in Meran niederlassen und dort schreiben nach Herzenslust. Es tut mir sehr leid aber ich kann es nicht ändern.

Die Sommer sind hier sehr schön, nicht heiss, und es gibt grosse Schneeberge mit Gletschern nur 50 miles von dieser Stadt. The Olympic Gebirge erstreckt sich von hier nach Alaska, im Norden und nach Mexico im Süden. Es ist ein mächtiges Gebirge mit endlosen Tannenwäldern, mit ewigem Schnee gekrönten Bergen von welchen Mt. McKinley und Mt. Rainier die höchsten Berge sind, der Erstere 20,000 Fuss und der letztere über 14000 Fuss, also keine Minya Konka, aber sie erhöhen sich nahezu am Meeresufer, weil der Minya Konka seine Basis auf 14000 Fuss hat. So erscheinen diese Berge weil sie eben von der Ebene sich erheben viel höher. Dazu kommt noch dass die Schneegrenze hier auf 8000 Fuss anfängt und nicht auf 18000 wie am Minya Konka, denn die nördliche Breite ist zweimal die des Minya Konka, nämlich 54° anstatt 28°, und das macht den Unterschied.

Ich werde fleissig arbeiten und sobald ich die Arbeit fertig habe komme ich gewiss! So bitte gedulde Dich noch ein wenig. Hoffentlich hast Du meinen Cheque erhalten der Dich für eine Zeit aushelfen wird. Bitte schreibe mir nach Seattle an meine alte Adresse.

Mit herzlichem Gruss
Dein alter Onkel J. F. R.

13

Seattle May 24/53

My dear Robert:

Thanks for your letter. I am again O.K. as usual after a spell and shall remain so I presume for another 6 weeks or so. Glad you received the cheque. The same amount came in to me again the next day, unexpectedly. So I had better send you some more! Well I am working on these wretched funeral ceremonies and shall soon see the light of day, that is I will be through with it. Then the mss. needs to be typed. To send it to you with all this censorship business is too risky for I cant [!] have it get lost and do the work all over again. I will bring it with me to Switzerland for I still hope to come in June to Zürich.

Yes Seattle, i.e. the region about it is beautiful lots of Rhododendrons, & flowers, lovely forests and only 1/2 hour's drive from the town.

It is cool here just like in Li-chiang, and I have now many friends here.

Will think it all over and will let you know as soon as I am finished what I will do. The Vatican has just published a 153 page paper of mine (Museo Laterano)[9] will send you a copy; it came yesterday. With kindest regards to Hans and your good self, as ever J. F. R.

14

Merano, Febr. 13/55

Dear Robert:
Today I received a copy of the letter which Dr. Bornemann[10] wrote to the Publisher of my Funeral ceremonies, on your behalf.

After thinking the matter over I have come to the conclusion that it would be a very good idea to publish a comprehensive work on the Na-khi and their literature religion etc. A kleines Büchlein does not seem to satisfy me, „wenn schon denn schon". Why not get up a historic account from my *Na-khi Kingdom*, a description of the country (a geographic account) types of the people, their religion

---

9 The ^{2}Muan ^{1}Bpö ceremony or the sacrifice to heaven as practiced by the ^{1}Na-^{2}khi. By Joseph F. Rock, Research Fellow, Harvard-Yenching Institute of Cambridge, Mass., U.S.A. *Monumenta Serica.*13.1948, 1-160, Taf. 1-4; *Annali Lateranensi.*16.1952, 9-158.

10 Fritz Bornemann (1905-), Herausgeber des *Anthropos*; Ethnologe und Historiker der SVD: *Die Urkultur in der kunsthistorischen Ethnologie.* Mödling: St. Gabriel 1938. 148 S. (St. Gabrieler Studien.6); *Geschichte unserer Gesellschaft.* Roma: Collegium SVD 1981. 445 S. (Analecta SVD. 54); *P. Wilhelm Schmidt SVD, 1868-1954.* Romae: Collegium SVD 1982. 374 S. (Analecta SVD. 59)

with special to their poetic comparisons with nature, etc. I shall be very glad to so supply you with fine photographs, unpublished ones, original Na-khi texts to show their methods of writing, their way of combining symbols, their ideographic and phonetic uses.

This would make a book about at least 350 pages with say 50 Plates, and a map of the country. The Li-chiang map from my Nakhi kingdom could be copied. I think Ullstein might be the proper *Verleger*. I would write a preface and all in all a very decent and instructive book could be the result. If you want I could write to Ullstein when the time comes or to some other *Verleger* you may have in mind, but I don't believe in the *Büchlein* business. What do you say? You might also enquire how much it would cost to print such a book, with plates 5 x 7 inches and a textspiegel of the same size as my books have.

Drop me a line. I will return the two books shortly. You could start with the geographic part first.

Cordially yours,

J. F. Rock

15

Merano 20th Febr. 1955

Lieber Robert:

Dein Schreiben vom 16ten erhalten. Ich sehe daraus dass Du nicht praktischer geworden bist. An ein Drucken der Bilder in America wäre nicht zu denken, dass [!] könnte nur in Österreich geschehen. Die Tafeln in Amerika machen zu lassen wuerden [!] mindestens $6000.00 kosten und der Verleger in Österreich müsste noch eine ganze Menge Zoll bei der Einführung in Österreich zahlen. Das ist ganz ausgeschlossen. Die Bilder in *The ancient Na-khi kingdom* kosteten $ 10.000.00 und das war in 1947, oder richtiger 1946, und heute würde das das Doppelte kosten.

Ich glaube die St. Gabriel Missions Druckerei würde die Clichés machen und auch drucken. Papier in Österreich ist viel billiger wie in Italien oder America. Ich habe einen Durchschlag des Briefes den Dr. Bornemann in Deinem Interesse an die Redaktion schrieb erhalten. Du hast wohl denselben auch erhalten.

Ich bin seit einigen Tagen wieder krank mit inneren Blutungen und septischem Fieber, letzte Nacht war gräulich. Ich lebe nur noch von Milk und Grieskoch. Es wird halt bald zu einer Kriese [!]

kommen, aber hier möchte ich nicht sein wenn man einen operativen Eingriff nehmen müsste.
Mit herzlichem Gruss Dein Onkel Rock

P.S. Das Wetter hier ist furchtbar Alles im tiefsten, mehr wie meterhohen Schnee bedeckt.

16

Seattle, April 16/56

Dear Robert:
I am so sorry that in the pressure of work I have completely passed over the 14th your birthday, so belated congratulations. I remembered around Easter that it was your birthday and wrote to you saying so, but afterwards completely lost track of the days. This is a sign that I am getting old and forgetful!

Your good letter of recent day with enclosure (Pensions Preise) arrived and although the weather here is getting nicer every day, sunshine and bracing cold air, I do not want to sit here between 4 walls and contemplate the brick walls of houses, although I can see over most of them; so have decided to leave Victoria B.C. on May 1st for Montreal and on May 2nd at 4 p.m. for London, arriving there on May 3d at 8.30 a.m.

Shall be in London a week and then will fly to Vienna and Rome. In Vienna will decide where we will go for the summer. I am bringing along a manuscript, for the time being it is only a crude conglomeration of facts, I put together in 1954 while on Maui, in Kula, Mt. Haleakala. There are still some chapters missing and these will be supplied from either memory or diaries which I also intend to bring or send through the mails.

So we shall meet soon and we all shall go to Krantz for a good lunch.
Enclosed a few Dollars, as a belated birthday present.

Will write again before my departure, with kindest regards to your goodself, Hans, and Frau Ciani, always cordially your uncle Rock
The Ms. is The Funny Book as I call it and all my friends wish me to write.

17

Seattle, 27th April/56

Lieber Robert:
Hier einige Zeilen um Dir für Deinen Brief vom 21ten April zu danken, auch für die Einlagen. Wie gesagt ich fliege am 6ten von Vancouver nach Montreal, von dort 4.30 p.m. am 7ten nach London direct wo ich um 9 Uhr morgens am 8ten Mai ankomme. Am 16ten Mai bin ich in Wien. Ich überlasse es Dir mich irgendwo unterzubringen: Einzeln Zimmer [!] mit Bad, Frühstück eine Mahlzeit. Lange kann ich aber nicht bleiben denn ich muss nach Rome fliegen. Die Karten sind noch nicht gemacht für das Amnye Machhen Buch. Werde Dir von London schreiben genaue Zeit, Flugnummer, etc. Ich komme mit Swiss Air von London über Zürich werde mich aber dort nicht aufhalten.
Bis auf das Wiedersehen verbleibe ich Euer
Onkel Rock

Herzliche Grüsse an Hans, die Teufel Marie und sonst wohl niemand mehr!
P. S. Vielleicht wird ein Hotel auf kurze Zeit besser sein, aber ich überlasse es Dir, Du kennst meinen Geschmack! J. F. R.

18

Eden Hotel-Roma

Sunday, May 27/56

Dear Robert:
I suppose you have received my card by now. It rained hard on my arrival at 2 a.m. I found a nice room and bath awaiting me. The food is excellent. Met Tucci[11] and we came to the conclusion that Schubert[12] must revise those parts wrongly translated; Tucci offered

---

11 Tucci, Giuseppe (1894–1984), Indologe und Tibetologe an der Universität Rom, Gründer des Istituto di Medio ed Estremo Oriente. Vgl. Raniero Gnoli: *Ricordo di Giuseppe Tucci;* con contributi di Luciano Petech, Fabio Scalpi, Giovanna Galluppo Vallauri. Roma: IsMEO 1985. 79 S., Porträt (Serie Orientale Roma 55.)

12 Schubert, Johannes (1896 – 1976), Professor für Tibetisch und Mongolisch an der Universität Leipzig; Direktor des Ostasiatischen

his help if he should have difficulty, also that the second part of the book, that is Schubert's part, will be published later. In a way I am glad for my part & the book will thus not be connected with the work of a man behind the Iron curtain, and will be a separate publication.

Tucci also is determined to publish my Na-khi dictionary, in fact he insists that I let him publish it, for the major part of my Na-khi publications came out in his Serie Orientale Roma. So the publication of that work is assured. There will be 1500 pages and some plates figuring gods, spirits, Nâgas etc.

Had lunch yesterday with an old friend from Hongkong who lives there, he is a well known architect and has built many palaces for Rajas in India. He used to live in Bombay.

He urged me to write the *Funny Book.* We must get that done this summer. But not in Palermo, not in Meran, for it is too hot there in the summer. I fly to Zürich on June 3d and my address will be Hotel Tiefenau, Steinwies Strasse 8-10. I telephoned them while in Zürich Thursday evening, reserving a room for the 3d.

I hope all is well with you. Will send you $ 50 Dollars from Zürich, a safer place than any in Italy, where they pilfer letters which they suspect to have money in it.

Here it is quite warm but the nights are cool. My room is quiet, and I shall wait here till the maps arrive which I hope will be before the 3d of June. I will also make the last corrections in the last proof and the index to the book in the press.

There is so far no more news to write so will close. Hope you both are O.K.

Kindest regards, to you and Hans, Ciani & Anton.

Cordially J. F. Rock

Sent cards to Anton and Ciani.

Write me to this hotel what has arrived in detail.

Seminars. *Asienwissenschaftliche Beiträge.* Johannes Schubert in memoriam. Herausgegeben von Eberhardt Richter und Manfred Taube. Berlin: Akademie-Verlag 1978. 201 S., 28 Abb. (Veröffentlichungen des Museums für Völkerkunde zu Leipzig 32.); Eberhardt Richter: Johannes Schubert (1896-1976). *Namhafte Hochschullehrer der Karl-Marx-Universität Leipzig* 5.1984, 66-74; Verzeichnis der Arbeiten Johannes Schuberts. *Studia Asiae. Festschrift für Johannes Schubert.* Pt.1. Hrsg. von Heinz Mode u.a. Halle: Buddhist Centre Halle 1969, 329-331.

19

Sept. 4/56

Lieber Robert:
Herzlichen Dank für Dein Schreiben. Ich kann leider nicht nach Wien kommen da ich keinen Flugsitz nach Amerika bekommen würde bis ende [!] October und dass [!] ist zu spät.

Was das neue Buch anbelangt bitte lasse es nicht binden den [!] ich habe Rom gebeten sie sollen mir eins in Italien schön in Leder binden lassen.

Das Wetter hier ist kalt und regnerisch, scheusslich in fact. Such a rotten summer I have not seen for a long time.

Had a letter from Hawaii saying that the water was running 3-4 ft in the streets after a heavy rain last Friday. And yet a Seattle letter says hot and dry, the same for Vancouver.

So I say farewell, be good to yourself, glad we had the few weeks tiogether in St. Anton and Switzerland.
Cordially, your uncle Rock

20

Seattle Oct. 31/56

Lieber Robert:
Heute kamen die ersten zwei Pakete, Flora Hawaiiensis und Lobelioideae, Ich muss sagen dass das Gold welches zur Betitelung gebraucht wurde das billigste ist das es gibt, es muss dunkleres richtiges Goldgelb nicht messing-gelb sein. I hätte gern glattes Rexin nicht das farbige gewebe [!] gehabt, das Staub ansammelt, und schwer zum Anwischen ist. Auch sollten sie vorsichtiger sein mit dem Leim und ihn nicht auf das Titelblatt schmieren wie sie gethan haben, dann versucht es auszuradieren, und die Namen verwischt.

Bei den nächsten Büchern müssen sie sehr vorsichtig sein, und wie angemerkt nicht beschneiden, dass [!] heist [!] die Ränder, der Photostatierten Seiten, denn wenn etwas abgeschnitten ist, ist es wertlos. Nicht beschneiden!

Bitte sage ihnen, sie sollen sehr vorsichtig sein; sollten die Bücher noch nicht angekommen sein dann bitte Rexin glattes, nicht faseriges material [!] zum einbinden [!] benützen!! Und anderes dunkleres Gold, heiss eingepresst, man nimmt dazu Messing Buchstaben nicht Zink. Ich habe tausende Bücher binden lassen und verstehe etwas davon. Ein paar Schillinge mehr oder weniger spielt da keine grosse

Rolle. Na, aber ich bin 73, es wird nicht mehr lange dauern, aber ich habe immer gesagt: „What is worth doing is worth doing well!“ no half measures!!!

Besten Dank für Deine Bemühungen, lass Dir die Lobeliaceen Monographie auch binden auf meine Kosten.

Hat Hans das Bild schon bekommen?

Herzlichen Gruss

Rock

21

Seattle, Nov. 5/56

Lieber Robert.

Vielen Dank für Deinen Brief. Du fragst über die Har-la-llü Zeremonie, nein die ist noch nicht veröffentlicht und der Nâga Cult hat damit gar nichts zu tun. Es sind einige Anmerkungen drinnen und in der Legende der Har la llü geb ich Noten nummern an wo sie zu finden sind.

Das Buch über die Ka mä gyumigkyi ist ja der Kern der Ha [!] la-llü Zeremonie, ich wollte nur alle die anderen Bücher die dazugehören über 100, herausgeben, übersetzt und annotiert. Die Übersetzungen waren alle fertig aber sie wurden von den Japanern versenkt mit allen anderen, so werden sie nie mehr erscheinen. Die übersetzten Manuskripte in dem Kamägyümigkyi Buch sind die Kernbücher der Zeremonie. Wenn Du glaubst dass das Nâga Cult Buch damit etwas zu tun hat so kann ich sehen dass Du die Sache nicht verstehst, und keine Idee von Nâgas hast. Ich bin auch verliebt mit dem Kamägyumigkyi Buch und vielleicht wird mein Leben auch so enden wie das des Nze-bö-yü-lu-p‘er. Oft denke ich an diese hohen Alpenwiesen und Schneegipfeln des Jadedrachen Berges wo so viele ihr Leben beendeten. Auch Ho-chih-hui sein erster Sohn liegt dort irgendwo, man hat seinen Körper nie gefunden. Die Yuvu Sänge sind rührend, ich werde Dir gerne helfen mit der Übersetzung. Manchmal sind wörtliche Anspielungen die sich mit vorhergehenden Zeilen reimen. Ich hätte das sollen besser erklären. Schreibe mir Deine Übersetzung und ich werde Notizen machen wo Du vielleicht irrst oder falsch verstehst.

Morgen gehen das Schwert und Scheide in zwei Paketen ab, auch die fehlenden Nummern von den Geographic Artikeln. NB. Minyak

Konka[13] und die Sungma.[14] The account of my 1922–24 Expedition wurde von Grosvenor[15] geschrieben und in Nat. Geogr. veröffentlicht und sind darum nicht mit meinen gebunden. Schade dass die anderen schon so verschmiert sind. Vielleicht kann ich noch neue finden und Du kannst sie dann alle nochmal binden lassen. Soweit sind nur 4 Bände angekommen, es fehlen noch die anderen hast Du sie aufgegeben.
Mit herzlichem Gruss Dein Onkel Rock

22

Honolulu, Nov. 21st/56

Lieber Robert:
Soeben kam Dein lieber Brief vom 17ten November, zu gleicher Zeit kam einer von 25 West Park Road, Kew Gardens telling me that Miss Eagle, the 81 year old lady who always used to cook puddings for me in the evening, died suddenly of heart failure on Nov. 13th. She was a nice good old soul, worked hard all her life up to the last. She loved flowers and when she was through cooking, instead of lying down and resting, went on her knees and worked in the garden with her plants and flowers. It makes me feel very sad indeed. It shows how ephemeral human life is, here today, gone tomorrow for good. I wonder where my last day will find me? Time is getting short and really I often think, if I only could go soon, so that I could leave you both some money.

Today all the papers, those handwritten ones, and those you typed

---

13 Glories of the Minya Konka: magnificent snow peaks of the China-Tibetan border are photographed at close range by a National Geographic Society expedition. By Joseph F. Rock, leader of the National Geographic Society Yünnan-Szechwan Expedition, 1927-1930; author of «Seeking the mountains of mystery», «Life among the lamas of Choni», «Through the great river trenches of Asia», etc., in the National Geographic Magazine. *NGM* 58.1930, 385-437, 35 Taf., 24 Farbtaf., 1 Kte.

14 Sungmas, the living oracles of the Tibetan church. By Joseph F. Rock, author of «The glories of the Minya Konka»,«Konka Risumgongba, holy mountain of the outlaws»,«Life among the lamas of Choni» etc. in the National Geographic Magazine. *NGM* 68.1935, 475-486, 1 Taf., 8 Farbtaf.

15 The National Geographic Society's Yünnan Province Expedition. By Gilbert Grosvenor. *NGM* 47.1925, 493-498.

and we had in Kandersteg, arrived today in Honolulu. Will the Funny Book really ever see the light of day? If I write it and *if* I ever make some money with it, I will not forget you. My friend here will help me, and so we will see. All my diaries came too, so I have plenty to fall back on.

I find the price for binding the 2 volumes rather high, 600 shillings is a good deal. I think the best way to bind them (ich werde auf deutsch weiterschreiben) ist so wie ich es gesagt habe in meinem letzten Brief. Es ist viel besser wenn man nicht auf die andere Seite sehen muss, und dass jede nummer [!] vor einem ist und nicht auf der linken Seite. So bitte lieber nicht so binden lassen wie er und Du vorschlagst [!]. Wenn es auch etwas wellig ist so macht das nichts aus. Es ist viel bequemer man hat die Seite vor sich und muss nicht umdrehen. So bitte soll er es so machen wie ich es in meinem Brief vorschlug, es wird auch billiger sein. Es hat damit keine Eile, ich brauche die Bücher hier nicht, aber im Frühjahr in Seattle wenn er es bei 10ten Jänner fertig hat so genügt dass [!].
Mit herzlichem Gruss Dein Onkel
Rock

23

Honolulu Jan. 10/57

Lieber Robert:
Deine beiden Briefe vom 28ten Dec. und 4ten Jänner kamen pünktlich der letztere in 4 Tagen von Wien das ist wohl ein record. Habe auch von Frau Ciani gehört sie schrieb mir einen Brief für Weihnachten und eine Karte zu meinem 73ten Geburtstag.

Es tut mir leid dass das Schwert noch nicht gekommen ist, ich hätte es sollen als Luftpacket schicken, es war versichert habe aber keinen Einschreiben [---][16] hast Du das Datum an welchem ich es abschickte? [---] so bitte schreibe es mir. Das Wetter ist sehr [---] geworden und windig es regnet wieder mehr [---] nicht so viel wie letztes Jahr,

Gesundheitlich geht es mir ganz gut, kann mich nicht beklagen.

Von Anton[17] habe ich nichts gehört am 2ten [---] ist sein Geburtstag, da ist er 92 Jahre alt. [---] möchte ich nicht leben und noch dazu unter solchen Umständen.

---

16 Textverlust.

17 Anton, Rocks Stiefbruder.

[----] kommt mein Freund Franz Michael[18] er wird bis zum 13ten hier sein dann fliegt er weiter nach Japan. Vielleicht komme ich doch noch nach Europa nächsten Sommer. Im Frühling wird es nicht recht gehen, da das Wörterbuch noch nicht fertig ist. Ich habe kein Sitzfleisch mehr und möchte bald wieder weg oder dann in einen Ort wie Wien oder nicht weit davon bleiben.
Mit lieben Grüssen und bestem Dank für das Bindenlassen der Bücher, verbleibe ich dein Onkel Rock
Herzliche Grüsse an Hans habe schon lange nicht mehr von ihm gehört.

24

Honolulu, Jan 19/57

Lieber Robert:
Vielen Dank für Deine Karte und Brief vom 14ten dieses auch für die beiden photographien [!], sie sind sehr gut. Auch freut es mich dass das Amnye Machhen Buch schon angekommen ist. Dr. Soreto schrieb mir so dass ich schon fürchtete es sei sehr schlecht gelungen, aber *he wanted to be polite*: although I had written him twice, he again asked me where I wanted the books sent to, so I have not yet seen it. I cabled him to send them to Seattle except one which should come here by express. I also heard from Edinburgh and that was also complementary [!]. Are the pictures well reproduced? It is in a stiff cover.

I can't imagine what has happened to the sword. Could they have stolen it? It was insured for $25.00, perhaps it is still in the Customs (Zollamt)?

Franz Michael brought me several thousand typed cards of the dictionary and I am busy writing in the symbols, we had rubber stamp[s] made for the boxes in which the symbols are written so I don't have to draw them and so it goes quickly, like this:
I shall be writing to the post office in Seattle about the sword and

18 Michael, Franz (1907–1992); Professor für chinesische Geschichte an der University of Washington, Seattle. Marie-Luise Näth in *The China Quarterly* 138.1994, 513-516; *International Biographical Dictionary of Central European émigrés, 1933-1945.* München: Saur 1983, II, 815-816.

I do hope it will turn up. I should have sent it by air, then it would have been there by 1st of December. Neues gibt es nichts; heute regnete es; gestern war es warm, ja heiss. So die Bücher sind abgeschickt, es freut mich. Viele Arbeit habe ich, es ist jetzt Mitternacht und Zeit zum Schlafengehen. Ich arbeite noch, und wollte Dir aber doch noch schreiben. Mit dem fliegen [!] ist es nichts, ich muss das Wörterbuch jetzt zu ende bringen. Tucci ist wieder in Rom. Beiliegend einige Marken für Hans. Die auf dem Couvert sind neu, eben herausgekommen.
Sei vorsichtig und gib ihm das ganze Couvert. Herzlichen Gruss an Euch Beide Rock

25

Honolulu Jan. 24/57

Mein lieber Robert,
Deine Karte vom 18ten kam heute abend. Es freut mich dass das Schwert doch nicht verloren ging. Ja es ist das einzige im Westen mit Ausnahme eines das ich noch besitze in Seattle. Es ist dasselbe wie Du sagst dass Chan Chung-tien trägt mit Panzer bekleidet auf der Alpenwiese Ndaza gko am östlichen Abhang des Lichianger Schneeberges, Yü-lung Shan 玉龍山 Jade Drachen Berg. Nda-za gko war einer meiner herrlichsten Camping places. Dort habe ich auch die meisten manuscripte[!] des Nâga Buches übersetzt. Ich habe dort viel schönes [!] erlebt und genossen. Wir gingen auf Jagd, botanisieren, und photographierten zu unserer Herzens Zufriedenheit. Das waren wunderschöne Zeiten, ich erlebe sie nur noch in der Erinnerung. Ich habe viele Bilder aufgenommen von Chan Chung-tien mit Panzer und Schwert mit herrlichem Hintergrund im Föhrenwald. Ich werde Dir ein Negative senden und Du kannst es vergrössern lassen, oder ich werde es in Seattle vergrössern lassen und Dir in einer Rolle schicken. Die Zeit vergeht, ich wünschte Du wärest hier und könntest mir helfen, es gibt viele Karten die fehlerhaft typiert wurden und die müssen nochmals typiert werden, und alles muss noch einmal eingezeichnet werden.

Seit dem 14ten regnet es hier in strömen [!], in einer Woche hat es mehr geregnet als man in einem Jahr hier verbraucht. Ich hoffe dass der Südwind sich bald zum Nordwestwind ändern wird dann ist es wieder kühl und trocken. Viel neues gibt es nicht, mein Buch ist noch nicht angekommen, aber ich habe schon von vielen gehört darüber, die Tafeln sollen schön sein ist das wahr?

Nun ist es Abendbrot Zeit und ich muss schliessen. Ich hoffe Du wirst bald über die Romance of Ka-mä-gyü-mi-gkyi etwas senden.
Mit herzlichem Gruss Dein Onkel Rock
P.S. Ich hatte gestern an das Seattle Postamt geschrieben wegen des Schwertpacketes.
Beiliegend ein Dollar für Postspesen.

26

Honolulu, Febr. 4/57

My dear Robert,
Thanks for the card and letter of Jan. 24th I am really happy to learn that the sword did arrive. It is a very ancient one and has been in my possession since 1931 or even earlier. The old Tsong Kuan of Yung-ning brought it to me with the armor and helmet while I was camping on Nda-za gko where I photographed Chan Chung-tien dressed in it. (See Na-khi Kingdom); I think you spotted it as the same. It dates from the days of Kublai Khan. Das Gedicht möchte ich ändern, wenigstens die erste Zeile: Aus Asiens einsamen Bergen oder doch besser Tälern, denn du sprichst im zweiten Satz vom Drachengebirge. Es sind eigentlich nicht Täler sondern Alpen. Das Gedicht ist gut. You are improving! Ich hoffe Du sendest mir bald die Übersetzung des Yuvu (Gedichtes) Gesanges.

Ja das Wörterbuch schreitet vorwärts, ich habe nahezu alle die Karten die mein Freund brachte, ausgefüllt, und erwarte neues Material.

Wie ich Dir schon schrieb, ich habe aufgegeben nach Tasmanien zu fliegen. Es ist zu zeitraubend und ich komme lieber nach Europa. Mein neues Buch hat guten Empfang bekommen, auch der Direktor vom Far Eastern Institute schreibt: „It was a great thrill the other day to receive the copies of the Amnye Machhen Book. It is an exciting book and of course the maps are terriffic [!]. I am sure it will enjoy a very well deserved success.“ A Seattle newspaper gives a somewhat distorted account of it, see clipping! [nicht dabei.]

Just had news from India and Burmah. An airline has been established by the Chinese Reds from India to Lhasa (Bagdogra-Lhasa). Bagdogra is at the foothills of the Himalayas. I flew there from Calcutta in 1910.
There is little news. I am well, here it rains in torrents, but it is warm. Seattle has a dry winter, lots of snow, but sunshine.

I have neglected my correspondence, so will have to close as I have about 20 letters to write.
As ever your Onkel Rock
Herzliche Grüsse an Dich und Hans. Beiliegend einige Marken.
Gerne kannst Du Dein nächstes Gedicht Buch mir widmen wenn Du willst. Wenn ich nach Seattle komme werde ich Dir einen Saville Row Anzug schicken. J. F. R.

27

Honolulu, Febr. 5th/57

Mein lieber Robert:
Heute gab ich einen Brief an Dich auf, und heute Abend kam Deiner vom 29ten Jänner mit 5 Blättern des Na-khi Gedichtes. Ich habe es sofort gelesen und es hat mich sehr gerührt ich muss gestehen es klingt sehr schön nur hie und da müssen Erläuterungen hinzugefügt werden in Fussnoten um alles klar zu machen. Als prologue ist es sehr richtig die sieben Mundarten der Na-khi anzuführen und die Hehrkunft [!] der Mundharfe. Ich werde Dir ein Bild eines schönen Na-khi Mädchen's geben und auch eines Knaben der die Mundharfe spielt, das wird alles schön das Gedicht erläutern. Dazu noch ein Bild der herrlichsten Schneewiese auf dem Lichianger Schneeberg, und ein Bild von Lichiang selbst. Ich habe auch ein Bild einer Ahu mit grossem Stein wo sie sich anlehnen und wo ich einmal Überreste zweier gefunden habe als sie von den Krähen aufgefressen wurden.

Es tut mir leid dass ich The Romance of Ka-mä-gyu-mi-gkyi nicht hier habe so kann ich es nicht vergleichen, aber es ist in der Universitäts Bibliothek und ich werde es borgen.

Ich vermuhte [!] dass Zeile 3 Am Faden Sonne und Mond spielt sich zuruck [!] auf zwei runde Scheiben die die Mädchen auf dem Rücken tragen, die linke repräsentiert die Sonne, die rechte den Mond, der Faden vereinigt beide, den[n] beide sind mit demselben Seidenfaden gestickt (siehe Na-khi kingdom) Bild 76. Bild 84 zeigt die schöne Alpenwiese.
Febr. 7th Yesterday came your 2nd part:

Der letzte Teil des ganzen Gesanges scheint auf deutsch nicht so tragisch zu lauten wie auf english.

Die Na-khi have a romantic nature and [are] capable of poetic expressions and allegories, like when they say, and therebei [!] intimate *union* that when the paper comes in contact with water it melts. It must be realized that Chinese paper is very thin like Seidenpapier aber noch mehr faserig: the minute a drop of water touches it is dissolves.

The white silk handkerchief is carried by girls in the hand when they are dressed up for special occasions, see Plate 76, and is part of their finery. When a sentence appears which does not seem to have any relation to either the foregoing or following sentence, you will see by comparing the Na-khi text that is is a play on words, and yet has a meaning.

Zeile 46 has reference to the pleasures they will enjoy befor[e] suicide, so lets fill the measure with Gold to the full. Zeile 15 hast Du wohl nicht richtig erfasst, ich kann mich auf einen Wärmuthacker [!] nicht erinnern, ist es das Artemisia Kraut, wenn so dann ist es richtig aber Wärmuth kennt man nicht recht, ich glaube Artemisia *Na-khi Bbue* klingt besser. Es gibt dort zwei Artemisien, eine mit weisser, eine mit schwarzer Wurzel; beide wachsen überall auf fahlen Feldern und Gruben (watercourses) and along houses.

Well I will not continue now, but as soon as I have the Romance of Kamägyumigkyi I will compare each line and will give a complete résumé of your version.

Herzlichst, Rock

Beiliegend $ 1.00 für Portospesen.

28

Honolulu, February

Lieber Robert:

Eine geraume Zeit ist vergangen seit dem ich von Dir gehört habe. Was ist los?

Leider bin ich noch nicht dazu gekommen Deine Übersetzung zu vergleichen.

Ich erhielt wieder Karten und habe streng daran gearbeitet, heute abend, nahezu Mitternacht fertig gemacht. Jetzt haben wir herrliches Wetter, bin noch nicht nach Hawaii geflogen, werde aber in zwei Tagen wenn nichts dazwischen kommt.

Wie geht es Dir bist Du noch angestellt. Ich schrieb Hans von der großen party die meine Freunde machten und nun schicke ich Dir einen Zeitungsausschnitt davon. Es war einzig, wohl wird es so etwas nicht wieder vorkommen.

Neues gibt es sonst nichts, ich schreibe so dass Ihr ein Lebens-

zeichen von mir hat [!].
Ich komme bestimmt nach Wien, diesesmal wirklich das letztemal.
Meine Freunde werden mir eine Einsiedelei am Bergabhang bauen und da will ich dann wohnen.
Hoffentlich kommt ein Brief von Euch.
Herzlich grüsst Euch Euer Onkel Rock.

Einige Marken für Hans.

29

28th February/57

Lieber Robert
Heute kam Dein Brief vom 14ten, es freut mich dass Du die Dollar, ich sandte in je einem Brief einen Dollar Du hast den ersten nicht erwähnt. Ich möchte nur wissen dass Du sie bekommen hast, und das [!] die Briefe nicht gestohlen werden.
Nichts Neues, heute war ein fürchterlicher Wind.
Das Packet mit den Büchern ist angekommen, so schrieb man mir.
herzlichst [!] Rock

30

Honolulu 3d April/57

Lieber Robert:
Vielen Dank für Deinen Brief. Ich schrieb Dir von Puu Waawaa und legte dem Brief die Kritik bei über Deine Übersetzung, auch $ 5.00. Ich kam hier vor einer Woche von Hawaii zurück, und bin im Begriffe alles einzupacken und abzuschicken nach Seattle. Die Karten des Nakhi Wörterbuches gehen by Amer. Express in einer Kiste, so schwer sind sie. Auch meine Kleider, Schuhe etc. gehen by Packet post um nicht Excess baggage im Flugzeug zu bezahlen. So ist diese schöne Zeit auch vorbei.

Ich muss noch fleissig an dem Wörterbuch in Seattle arbeiten und dann kann ich nach Wien kommen; die Karten für den Atlas müssen auch noch gemacht werden. Ich habe noch viel Arbeit vor mir.

Dein Geburtstag kommt bald, ich glaube er ist am 14ten April, um sicher zu sein lege ich hier $ 5.00 bei, drink [!] ein Gläschen auf mein Wohl.

Morgen gehe ich noch einmal in die Berge dann Schluss. Sonntag

am 7ten April fliege ich nach Seattle.
Ich lege dem Brief einige Marken bei für Hans, er hat mir einen traurigen Brief geschrieben.
Nun herzliche Grüsse und alles Gute zu Deinem 43ten Geburtstag.
Wie immer Dein alter (junger) Onkel Rock

31

Seattle April 14th/57

Lieber Robert
Deinen lieben Brief vom 6ten erhielt ich vor zwei Tagen, besten Dank. Die Korectur [!] etc. war in Waimea aufgegeben auf der großen Ranch (Parker Ranch 3500 ft). Du schreibst ich antworte nicht. Ich war über einen Monat auf der grossen Insel Hawaii und liess mir die Post nicht nachschicken da ich auf einem Ort nicht länger wie 3-4 Tage blieb. Sobald ich geschrieben habe kommt am nächsten Tag Dein Brief, so kann ich nicht gleich wieder schreiben unsere Briefe etc. kreuzen sich.
Ich schreibe diesen Brief im Bett, so wie ich in diesem rauhen Klima ankam, bekam ich wieder Bronchitis. Ich huste und meine Lungen sind gestopft. Ich hatte sollen von Hawaii nicht fort, aber dieses Jahr war es später wie letztes, 15 Tage später. Auch das Paradies wird einem langweilig, aber ich arbeitete doch und beschrieb eine neue Lobelia nahezu während der letzten Minuten meines Aufenthaltes. Ich muss sagen ich bin nicht „happy" hier – Sie haben den Zins auf 175 Dollar im Monat erhöht das macht 4275 Schillinge im Monat, dazu kommt noch Wäsche und das Essen, etc. für das kann man in Wien ganz schön leben, hier muss ich mir selber kochen, das Geschirr waschen, etc.

Meine Freunde (von Honolulu) sind jetzt in Boston und sie schreiben ich soll sie im Mai in England erwarten. Hoffentlich geht die Bronchitis bald vorüber und wird nicht zur Lungenentzündung.

Bitte schreibe mir was ich Dir alles versprochen habe, Bilder zu Deiner Übersetzung, Anzug, etc. ich habe es nicht aufnotiert, wenn Du solche Wünsche hast, sag es mir, aber bitte schreibe sofort. Heute ist wohl Dein Geburtstag, und hast Du ein Gläschen getrunken auf aller Wohl? Hans schreibt traurige Briefe, es tut mir leid, die verdammten Russen die machen alle Menschen unglücklich und unsicher, hier wird gerüstet wie nie. Nur weg auf eine Alm wo man ruhig und in Frieden leben kann. Ich mochte am liebsten einpacken und alles mitnehmen weil ich noch kann, wer weiss was später werden wird. Da musste ich aber mit Schiff [!] fahren denn ich habe zuviele Koffer, 10 Koffer und 7 Stück Handgepäck, und noch die vielen Bücher die ich noch habe brauchen mehrere Kisten.

Ich lege für Hans einige Marken bei. Die gebundenen Bücher sind angekommen, aber schön gebunden sind sie nicht. Ich werde sie in England binden lassen, oder Du bist zu einem Schuster gegangen. Es muss doch in Wien eine schöne Buchbinderei geben? Ich werde selbst suchen wenn ich nach Wien komme. Inzwischen herzliche Grüsse an Euch Beide. Dein Onkel Rock

32

Seattle, Ostersonntag, April 21/57

Lieber guter Robert:
In meinem Leben ist eine grosse Wendung eingetreten; es ist wahr ich habe es erwartet, aber ich konnte mich nicht recht entscheiden. Ich bin noch immer krank mit der Bronchitis und habe Zeit gehabt nachzudenken. Meine Freunde in Honolulu haben mich gebeten ich sollte doch nach Hawaii zurückkehren und dort meine letzten Tage verbringen. Sie bauen mir eine Hermitage im schönen Nuuanu Thal gleich neben ihnen und kann dort ruhig arbeiten alle meine Sachen beisammen haben. Dieses ewige Hin und her, Bücher, Tagebücher, Bilder, Karten, Notizzen [!] etc,. etc. alles bei Post herumschicken um die ganze Welt, muss jetzt ein Ende nehmen.

Ich komme noch einmal nach Europa und zwar Ende Mai, vielleicht noch früher, ich werde nach Rom fliegen und dann den Sommer in Österreich verbringen, dann kehre ich nach Seattle zurück, werde alle meine Koffer etc. von dorthin, dass [!] heisst nach Honolulu schicken, mich endlich hinsetzen und richtig arbeiten und ohne Sorgen auf die Zukunft weiter leben so lange es geht. So ist jetzt meine Zukunft und Arbeit endlich gesichert. Ich bin wirklich froh dass es so gekommen ist.

Ich habe meinen Freunden in Boston geschrieben und ihnen definitive gesagt dass ich ihr Angebot bei ihnen zu wohnen angenommen habe und es mich freuen wird meine letzten Tage in Frieden und Arbeit verbringen zu können.

Heute Nachmittag telephonierten sie mir von New York. Wir haben volle 20 Minuten gesprochen. Sie sagten mir dass es sie riesig gefreut hat dass ich mich entschlossen habe nach Hawaii zurückzukehren und mit ihnen leben zu wollen. Sie kommen vielleicht auch nach Europa und wir werden uns in England treffen.

So siehst Du, ich habe ein schönes Ostergeschenk bekommen. Es freut mich Euch noch einmal sehen zu können, es wird aber jetzt gewiss das letztemal sein; ja immer kann man nicht über die Ozeane fliegen. Es ist die beste Lösung und ich kann mich dann hinsetzen und mit Freuden arbeiten. Es wird herrlich sein. Ich werde meine Verbindung mit der hiesigen Universität nicht aufgeben und wir werden weiter zusammen arbeiten.

Deine Briefe vom 6ten und 12ten April und Osterkarte habe ich erhalten besten Dank, leider kein Wort von Hans, es kränkt mich. Sobald ich in Wien bin können wir den Sommer hin zusammen an allen Deinen Übersetzungen etc. arbeiten. Bis dahin herzlichen Gruss Dein Onkel Rock

33

Zürich 11ten Aug./57

Lieber Robert,
Heute sind es 67 Jahre dass meine Mutter gestorben ist im Rudolfiner Krankenhaus in Wien. So will ich Dir jetzt einen Abschiedsbrief schreiben, der letzte von Europa. Ich fliege am 15ten Aug. nach London werde dort eine Woche verbleiben und dann nach Seattle über Canada zurückfliegen. October bin ich wieder in Honolulu, D.V. Ich lege dem Brief die 100 Schillinge bei die ich noch übrig habe von Salzburg. Die letzten Tage habe ich ein sehr interessantes Buch gelesen es heisst *Carneval der Einsamen* von Joachim von Kürenberg (Richard Wagners Tod in Venedig) Verlag Robert Mölich, Hamburg. Wenn ich es hier bekommen kann werde ich es Dir schicken. Es ist sehr lehrreich, spricht von Liszt, Eleonora Duse, d'Annunzio, und viele andere wie Verdi etc. Es wird Dir gewiss gefallen. Nun lebe wohl. Europa wird mich wohl nicht mehr wiedersehen, auch wir nicht. So scheiden wir freundlich wie am Züricher Bahnhof zu später Stunde.

Tucci schrieb mir noch ein paar Stunden vor seiner Abreise und stimmte ein das [!] der Index überflüssig sei. Ich habe noch bis gestern daran gearbeitet aber es ist zu viel ich kam zu Seite 101, und habe über 1500 Zettel geschrieben nur den Nâga Cult. Nun herzlichen Gruss wie immer Dein [Onkel Rock]

34

University of Washington
Department of Far Eastern and Slavic Languages and Literature
Seattle 5, Washington

Sept. 5th/57

Dear Robert
Your very poetic letter arrived today, two days after my arrival here. Had a nice flight across Canada and a delightful two days' rest in Victoria B.C. My friend Franz whom you met in Vienna awaited me at the airport and sped me home to my apartment. The cooking has started again but I don't mind; I get up at 5.30 every morning have breakfast and then go to bed again. I have a lot of work ahead of me so that I do not know where to begin.

Shall probably go to Hawaii in October but the day has not yet been decided upon.

I enclose a colored print of the first page of Lv-mber-lv-zaw ssaw (K'a-mä-gyu-mi-gkyi's classic) perhaps you can use it as a frontispiece in your paper on the Yu-vu.

There is little news except that I am well. It is hot here and humid a regular Indian summer. The peaches are still ripe and cheap, 10 ¢ a pound, also melons, plums, nectarins etc. Will write again later.
With all the best wishes
cordially your uncle J. F. Rock

35

Seattle Oct. 21/57

Lieber Robert
Many thanks for your good letter and whatever it is you so kindly sent me. Your good card was also received. I have been negligent in writing letters. Have been very busy trying to finish the dictionary but there is still a hell of a lot to do. Never write a dictionary. I have been well excepting one black-out of which I wrote to you. Had a letter from Hans in answer to mine written shortly after my arrival here [....] He is looking for 20000-25000 Schillinge, and [...] about selling the pictures and the stamps. He wrote [...] he had not seen you for 11 days, he thinks you [....] out of his way. He seems difficult to please. [...]ing to get married he has his divorce or must have [...]

for he said he started negotiation when he wrote me. [...] He is spending his holiday at his future wife's home. I wrote to him and told him he owed it to you to tell you [...] he will be leaving the Wohnung, he will move out to [...] his fiancee lives with her mother and will come [....] everyday till they find a place to live, to buy [...] Wohnung and furniture will cost the above sum.[...] can see by your letter he has not told you. He ought to for he cannot let you sit on the street by simply saying I go to be married. He should as I wrote to him, give you plenty time for you to adjust yourself. But please do not tell him that I told you in this letter. He has not followed my advice and has not told you that he will be married shortly and will give up the Wohnung. He said he told Frau Ciani, but he certainly should tell you for you are more concerned than anyone. As he has not told you I thought I ought to let you know so you can adjust yourself. But if you tell him that I said so then you will get into trouble with me.

I have acted in your interest. I told him that you should know, apparently he does not think so. So I am telling you now. I have forgotten what rent you are paying there where you are now a month. If he sells the pictures I shall be really angry but of course I have given them to him, but not to be sold. Stamps one can always get again but not such paintings.

Let me know right away what your reaction is and if you know that he is going to get married soon. Apparently you did not for otherwise you would not ask what he is doing in Gänserndorf. They have a furnished house, but it is too far from Vienna. I am sorry I can't give him the $1000.00 but at my time of life that is a lot of money. I shall be flying to Honolulu the end of this month or early November. I have a lot to pack. Maybe I can sell some things in Honolulu. Write me here to Seattle as quickly as possible.
Herzliche Grüsse Dein Onkel Rock

36

Honolulu, 10ten December 1957

Dear Robert:
Had a letter from your brother in which he thanks me for the stamps. He also told me that on the fifth of this month he will be divorced from his old wife in the courts of Vienna. He will then move out to his wife-to-be and will come daily to his work by train. He will pay

the rent of the old apartment till he has found a new place to rent. In January he will tell you. But I thought you ought to know now, so that you can arrange matters. He said nothing more about money. So you will be left alone till he finds a new home and he will pay the rent. I suppose this so that he has a place to come back to in case of necessity, or in case nothing should become of the marriage.

Under no circumstances must you tell him that I had written you all this.

Keep on to look out for a place in case he should not keep his word about paying the rent.

There is little news. Yes in one of his last letters he wrote that [he] would not take anything out of the old home, except the pictures belonging to him.

I returned from the Volcano but wished I had stayed on the big island. It is impossible for me to do any work as nothing except my books have arrived from Seattle. My dictionary and notes, etc. etc. are apparently still in Seattle. I enclose five Dollars, kaufe Dir ein gutes Weihnachtsmahl oder Spanferkel Lunch at New Years.
Mit all the best wishes for a Merry Christmas and Happy New Year, I remain your old Onkel Rock

37

Kaahaaina
Honolulu January 8/58

Mein lieber Robert.
Heute gedenke ich Euch [!] und dem lieben Wien aus alter Zeit und freue mich auf ein Wiedersehen in diesem Jahre. Die Zeit fliegt dahin so schnell dass ich oft denke „Nun ist der Abschied von dieser Welt nicht mehr ferne.“ Das Leben ist doch schön, wenn man nur halbwegs gesund ist und sich seiner Arbeit hingeben kann.

Ich arbeite fleissig an dem Wörterbuch wohl die letzte Arbeit über die Na-khi, denn das Wörterbuch trägt in sich alles was ich über die Na-khi, deren Literatur, Geographie, etc. etc. gelernt habe. Dann habe ich Zeit mich anderen Sachen zu widmen. Es ist meine Absicht im Herbst nach Europa zu kommen und zwar über Wien nach Rom und dann mich für eine Zeit in Meran niederzulassen um dort die Korrectur zu lesen. Es ist leichter dort denn es ist in Italien, man braucht keine Luftpost etc., und in einem sprung [!] ist man in Rome wenn es nötig sein sollte. Wenn Du frei bist und nach Meran kommen willst und mir mit dem Druck des Wörterbuches helfen willst dann können wir beide in Meran beim Pittscheider wohnen, da gibt es beides, österreichische und italienische Küche, und gute Küche. Ich glaube die Corras kommen auch nach Meran und so werden wir da alle beisammen sein.

Sollte Hans verheiratet sein, so werden wir schon den Zins des Hauses auftreiben können.

Ich werde meine Na-khi manuscripte [!] verkaufen, an Tucci seinem Institute, und somit zu den Druckkosten beisteuern. Meine Freunde wollen auch nach Europa kommen und ich werde so wie die Zeit erlaubt Cicerone spielen. Sie waren nie in Europa und sprechen nur Englisch. Wir wollen ein letztesmal fröhlich beisammen sein und arbeiten, denn die Korrectur wird viel Arbeit machen. Das sind meine Pläne und ich schreibe Tucci und werde ihm den Vorschlag machen.

Wie geht es Dir lieber Robert? Es tut mir leid dass wir nicht länger in der Schweiz verbleiben konnten, aber ohne Arbeit ist das für mich nicht möglich und Du hast das auch eingesehen. Wie geht es mit der Kamägyumigkyi, arbeitest Du daran? Ich glaube ich werde die Einleitung zum Wörterbuch in Meran schreiben.

Bitte schreibe öfters, ist das Packet aus Hawaii angekommen? Beiliegend 1 Dollar für Briefspesen.
Mit herzlichem Gruss verbleibe ich Dein alter Onkel Rock.
Viele Grüsse an Frau Ciani auch an Hans, bitte gib ihm die Briefmarken auf dem Couvert, sie sind eben erschienen.

38

McCandless Ranch
Kai Malino
South Kona, Hawaii

May 20/58

Lieber Robert
Vielen Dank für Karte & Brief. Bin wieder auf der grossen Insel Hawaii. Here ist es ruhig und man kann schön arbeiten.

Dein Englischer Brief lässt, was idiom [!] und Grammatik anbelangt noch manches übrig.
Ich sollte Dir auf englisch schreiben würde besser sein.
Die Geographic Magazine wird Dir geschickt.
Herzlichst Dein Onkel Rock

39

Kaahaaina
3860 Old Pali Road
Honolulu, Hawaii

Mein lieber Robert:
Many thanks for your welcome letter which reached me in Kona on Hawaii. I am glad you liked the little booklet. Alas dear Robert I am not as well as I need to be. I am suffering from spinal arthritis which makes it difficult for me to walk or get up. I can't lift anything nor can I bend over, with one word I have grown old. I suppose at 75 one cannot expect to do things one did when young. I could not run a race now like I did to the Villa Resi where you could not keep up with me. There is little news, the French are raising hell which is just like them. I have not yet had a chance to mail the Geographics but will do so tomorrow. I am receiving favorable reviews of the Amnye Machhen book and yesterday the Editor of the *Encyclopedia Britannica* asked me to write the articles about Tibet and West China at 2 ¢ a word, but I declined as I have too much to do. The dictionary is far from finished and it wont [!] be finished till the end of August. Will come to Vienna in mid September.

Yes I want to spend some time in Salzburg but not in the town itself but in the suburbs, a small villa like the Villa Resi.

I had forgotten the name of the Hotel where we stayed and consequently had a very bad time finding a good restaurant. Had a rotten meal there last year. I shall not forget the name.

Although I am sending stamps right along to Hans I receive no letter of acknowledgement, so will stop sending them. As I had been away in Kona I have as yet not sent him the suit and I suppose he is mad at me. I am living out in the country and it is not so easy for me to take a package to town.

As regards your Ka-mä-gyu-mi-gkyi story why not ask Anthropos in Fribourg, Posieux, Father R. Rahmann,[19] to publish it? They don't pay anything but you can get it published.

Ich werde Dir zwei paar Schuhe und Hemden senden. Leider kann

---

19 Rudolf Rahmann, 1902-1985, Herausgeber der *Monumenta Serica* von 36 bis 1948, hat sich selbst für die Na-khi interessiert; vgl. seine Remarks on the sacrifice to heaven of the ^{1}Na-^{2}khi and other tribes in South-west China. *MS* 13.1948, 395-408.

die alten typus [!] nicht mehr bekommen seit 38 Jahren trage ich diesen Typus Schuh und jetzt machen sie sie nicht mehr. Ich habe andere gekauft aber sie sind zu lang.

Ich bin letztes Jahr den steilen (senkrechten) Lift auf den Berg hinauf in Salzburg, es regnete in Strömen. Blieb nur über Tags dort und war tod krank als ich in Neunecken ankam. Nun Schluss.
Herzliche Grüsse, Dein Onkel Rock
Soeben kommt Deine Karte vom 12. Juni, vielen Dank.

40

Kaahaaina
3860 Old Pali Road
Honolulu, Hawaii

Lieber Robert.
Gestern kam Dein Brief vom 2ten Juni; vielen Dank. Von Hans habe ich seit April nichts gehört, er ist wohl böse auf mich dass ich ihm den Anzug noch nicht geschickt habe. Er wusste doch dass ich krank war und mich nicht anstrengen durfte wegen dem Herzanfall. Aber keine Zeile, auch schickte ich ihm manche Marke. Nun gut. Ich war in der Zwischenzeit wieder in ärztlicher Behandlung, hatte 4 Gedärm oder Magenblutungen. Seit drei Tagen geht es wieder besser. Und heute endlich kam ich dazu den Anzug einzupacken, er ist sehr schön, ganz neu, habe ihn nur drei oder viermal getragen. Im packet drinnen sind für Dich 4 Hemden, 2 paar Schuhe, ein paar Socken. Ich schicke das ganze Paket an Dich, man erspaart die Postspesen und dann ist Hans auch wenig zu Hause und wird dann niemand dasein das Packet zu empfangen. Er soll Dir die paar Schillinge bezahlen, ich glaube es sind 8 wie Du sagtest delivery charges. Auch schicke ich 6 oder 5, ich habe vergessen, Nat. Geogr. Magazines bis Juni 1958 (einschließlich). Mein Wörterbuch geht langsam vorwärts, leider nicht schnell genug für mich, denn ich will es jetzt wirklich loswerden. Nie hätte ich gedacht dass ein Wörterbuch-Lexicon so viel Arbeit in Anspruch nehmen kann.

Was den Na-khi Soldaten betraf so war dass [!] so. Er war in der Kaserne, eine kleine Baracke auf einem Hügel von dem man Li-chiang überblicken konnte. Seine Geliebte hörte (sie wohnte in Dü gkv, einem Dorf am Fuss des Schneeberges) dass die Soldaten den folgenden Tag abziehen müssten. Um ihn nicht zu verlieren ging sie nach Li-chiang und holte ihn aus der Baracke. Sie hatte ein altes

Schwert mitgebracht das aber in ein Tuch eingewickelt war. Sie holte ihn heraus sprach liebevoll zu ihm und bewegte ihn mit ihr zu gehen nach ihrem Dorf. Aussen angelangt zog sie das Schwert und ging hinter ihm her, schweigend. Ich kam eben vom Dorf und war am Wege nach Lichiang, zu Pferd und einige meiner Leute. Die kannten sie, auch den Soldaten; Lichiang Na-khi kennen sich alle; niemand fragte die beiden die stumm einhergingen er voraus, sie mit gezogenem Schwert hinterher. Meine Leute wussten was es bedeutete, sagten aber nichts zu mir. Am nächsten Tage hörte ich dass sie ihren Geliebten zwang mit ihr in den Tod zu gehen. Am Südost Abhang des Schneeberges ist ein sehr hoher Wasserfall der im Winter ganz trocken ist. Es ist eine steile schwarze Felswand mit einer tiefen senkrechten Rinne, circa 300 meter [!] hoch. Sie gingen beide auf dem Berge zur Stelle wo der Wasserfall über die Wand stürzte, dann stiess sie ihn hinunter drohend mit dem Schwert dann stürzte sie nach. Die Felsenwand heisst 'A-ngyü-t'an see Na-khi Kingdom p.213. Leider hab ich kein papier [!] mehr.
Herzliche Grüsse. Auf Wiedersehen im Herbst
Dein Onkel Rock

41

Kaahaaina
3860 Old Pali Road
Honolulu, Hawaii

Aug. 10/58

Lieber Robert:
Vielen Dank für Deinen lustigen Brief. Es freut mich dass es Dir gut geht. Du warst also in Graz, bin nur einmal durchgefahren. War wieder krank, Fieber 102°, ging in Hospital für XRays. Werde heute hören was los ist. Fühle mich jetzt sehr wohl. So können mich nicht wiederherstellen. Das Wörterbuch ist noch nicht fertig. Dein English ist besser aber noch nicht perfect. Das Gedicht ist schlecht, entschuldige, die Idiome sind falsch solche gebraucht man nicht.
Werde bald wieder schreiben. Hoffentlich ist das Packet schon angekommen.
Herzlichen Gruss
Dein Onkel Rock

42

Honolulu Aug. 16/58

Mein lieber Robert:

Dein lieber Brief vom 26ten Aug. von St. Anton heute erhalten; es hat mich wirklich gefreut von Dir so schnell zu hören. Die Geographic Magazines, es sind zwei, July & Aug., werden wohl bald ankommen. Ich kam gestern von Kona, Hawaii, zurück, hier ist es viel kühler. Von Hans kein Wort dass er den Anzug erhalten hat; es war das letzte Mal dass ich ihm etwas geschickt habe, wenn man nicht einmal eine Karte schreiben kann *der Anzug ist angekommen* dann zeigt dass [!] dass er kein Benehmen hat. Schluss; in Wien werde ich ihn auch nicht einladen; werde nur Dir schreiben, Du brauchst ihm nicht zu sagen dass ich, und wann ich komme. Er ist wohl böse dass ich ihm nicht sofort die 25,000 Schillinge schickte für die Scheidung, Hochzeit etc. Ich werde ihn nicht mehr erwähnen und will ihn auch nicht wieder sehen er ist zu grob. Wir wollen allein unsere Wege gehen und uns unterhalten. Ich habe nur einen Überrock und den brauche ich im Winter in Europe. Den Regenmantel braucht man in Hawaii. Andere Sachen kann ich Dir schicken, Winterleibchen, Socken, Hemden, Kravatten, etc.

Ich freue mich schon auf das Wiedersehen und werde länger in Wien verbleiben so dass wir länger beisammen sein können, denn es wird wirklich das letztemal sein dass ich nach Europa komme denn mit 75 Jahren soll man nicht so viel herumreisen, so grosse Strecken überfliegen. Dann habe ich auch noch viel Arbeit vor mir; ob ich alles fertig bringen werde? Ich fühle mich wohl aber ich glaube das Ende ist nicht weit weg. Ich träumte (war nicht ganz im Schlaf, halbwach) da händigte mir jemand eine Note auf der Hand geschrieben †26ten Februar -- 76. also gestorben am 26ten Februar im 76ten Lebensjahr. Also nur etwas über ein Jahr. Ich glaube aber nicht daran. Ich will meine Arbeiten fertig machen dann bin ich bereit!

Nun lieber Robert bleibe gesund wir wollen uns gut unterhalten in Wien. So lebe vorläufig wohl bis wir uns sehen Ende dieses Monats.

Wie immer Dein Onkel Rock

Beiliegend eine Broschüre die Dir eine Idee gibt von den Inseln und ihrer Lage.

43

Kaahaaina
3860 Old Pali Road
Honolulu, Hawaii

18th August 1958

Mein lieber Robert:
Vielen Dank für Deinen Brief vom 13 Aug. der heute ankam; 5 Tage von Wien sind wirklich schnell früher hätte es wohl einen Monat gebraucht. Es freut mich dass das Paket endlich angekommen ist ich fürchtete schon es sei verlorengegangen. Hoffentlich ist Hans mit dem Anzug zufrieden er ist von Saville-Row the best Taylor in England. Unter separat post the August number of the Nat. Geogr. Ich habe sie nie aufgehoben. Es tut mir wirklich gut zu hören dass Dir die Hemden etc. gefallen, wenigstens sind sie nützlich. Voriges Jahr habe ich die Sahara Schuhe (ein paar) ganz gute weggeworfen da ich dachte so was würde Dir gewiss nicht gefallen. Leider kann ich Schuhe die ich ungefahr 38-40 Jahre lang getragen nicht mehr bekommen, sie haben den Leisten weggeworfen und machen solche Schuhe nicht mehr. Ich bin ganz verzagt da mir andere nicht passen. Sie sind zu lang und nicht so bequem. Ich schrieb Hanan Shoe Store New York aber sie sagten die Schuhe sind unmöglich mehr zu bekommen. Nirgendwo kann ich solche Schuhe bekommen, aus Kalbhaut wohl aber nicht Zickenfell wie die letzten die ich Dir schickte. Vielleicht kann ich in England ähnliche bekommen.

Leider ist das Wörterbuch-Lexicon noch nicht fertig und ich bin nahezu verzweifelt die Zeit ruckt schnell vorwärts. Fertig muss es werden ich arbeite bis nach Mitternacht. War wieder leidig und gieng in die Straub Klinik um mich mit Xray untersuchen zu lassen wieder $ 100 Dollars weg. Ich leide an einem Magenbruch. Durch das ewige Sitzen hat sich der Magen durch die Diaphragme, oder ein Teil des Magens durchgearbeitet dadurch entstand ein Magenbruch. Die Sache ist graphisch so:

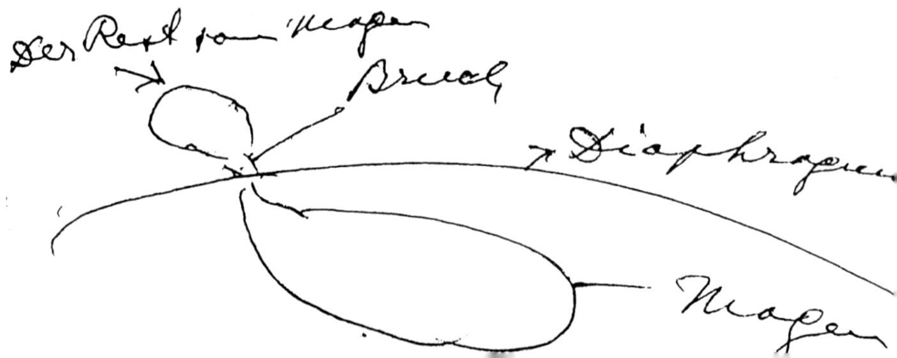

So kann die Nahrung nicht gleich durchkommen und bleibt oben stecken dass [!] dann Übelkeiten verursacht, eructionen, Erbrechen, etc. Im Duodenum ist ein grosses Diverticulum (it has even a fluid level) anstatt gerade zu sein. Ja wenn man alt wird haufen [!] sich viele Unannehmlichkeiten an. Aber sonst fühle ich mich sehr wohl und ich kann noch viel vertragen. Wie der Anton sind wir von gutem Teig gemacht! Der Mensch kann viel aushalten wenn er eine gute Körperbeschaffenheit hat. Komme was will ich komme nach Wien und bleibe dort bis Weihnachten. Wir werden uns herzlich ergötzen an allen den guten Sachen die Wien bietet, denn ich glaube es wird das letztemal sein dass ich nach Europa komme. Im Frühling nächsten Jahres gehen wir (ich und meine Freunde von Honolulu) durch Europa bei Auto dann fliegen wir nach SudAfrika und Australien, Tasmanien, New Zealand und Fiji nach Hawaii zurück. Dann so Gott will setze ich mich nieder und schreibe die Historical Geography of the three great principalities of West China. Ich werde auch diese Bände noch erleben. So siehst Du lieber Robert Dein Onkel hat noch viel vor sich und wird es auch durchsetzen. Es braucht Willenskraft und die, Gott sei Dank habe ich. Bitte schreibe ob Du noch irgendwelche Wünsche hast und ich werde sie erfüllen wenn sie mir möglich sind. Ich weiss Deine Wünsche sind nicht extravagant. Ich lege dem Brief ein Bild bei dass [!] auf dem Segelschiff *Luka* in 1913, ehe Du geboren warst, aufgenommen wurde; wir waren damals auf der Fahrt nach den Palmyra Inseln, eine unbewohnte Inselgruppe 6° über dem Äquator und 1000 Meilen südlich von Honolulu. Hast Du mein Büchlein über diese Reise?[20]

Das Wetter ist gut aber feucht und es regnet fast jeden Tag. Seattle hat trockenes Klima diesen Sommer, eine Dürre und über 100° Fahr. Hitze. So etwas gibt es hier nicht aber 95% Feuchtigkeit vor zwei Wochen.

Nun so schliesse ich mit der Hoffnung dass wir uns bald wiedersehen werden. Ich see [!] Euch schon vor dem Bureau auf der Ringstrasse warten auf den Autobus vom Flugplatz. Wie steht es mit der Kamägyumigkyi? ist sie nahezu druckfertig? Kann ich Dir helfen?

Mit herzlichem Gruss Dein alter junger Onkel Rock

Hoffentlich wird mir Hans schreiben. R.

---

20 s. o.

44

„Kalahiki“
McCandless Ranch
South Kona, Hawaii

Aug. 28/58

Mein lieber Robert:
Ich glaube dass ich Dir geschrieben habe was mir eigentlich fehlt. Jetzt geht es mir wieder ganz gut. Ich hatte auch immer Sciatica Schmerzen in den Knieen, und am ganzen rechten Bein. Ich nahm nur einige Cortizon Pillen und alle Schmerzen sind weg. Die Gelenke sind fügig und geschmeidig wie mit 25 Jahren statt 75. Frau Ciani kann sicher mit diesen Pillen ganz geheilt werden. Ich werde ihr sie mitbringen. Nicht einmal 1/2 Dutzend hatte ich genommen und alle Schmerzen waren weg.
Man sagt mir ich habe nie so wohl ausgesehen.

Hans hat mir noch nicht geschrieben sowohl ich auch gestern Post bekam. Er hat doch den Anzug schon angeschaut, er ist 5000 Schilling wert. Man kann ihn in Amerika nur um 200 Dollar bekommen. Du brauchst dass [!] aber ihm nicht zu sagen. es wird wohl erst Ende September werden bis ich kommen kann. Ich kam hier nach Kona um mich zu erholen und kehre Montag den 1sten September nach Honolulu zurück um meine Arbeit wieder aufzunehmen.

So Neues habe ich nichts. Meine Freunde gehen einer nach dem Anderen. Dr. Günther Köhler[21] den ich in Peking kannte als ich Bosshard[22] besuchte und der ein Bild von uns aufnahm ist am 11ten Juli gestorben in Dresden. Von Schubert höhre [!] ich nichts mehr, er scheniert sich wahrscheinlich denn Tucci wollte seine Übersetzungen nicht veröffentlichen. Tucci ist wieder in Swat. Letz[t]es Jahr hat er

---

21 Günther Köhler; Geograph, vgl. Walter Fuchs: Günther Köhler in memoriam, 1901-1958. *OE* 5.1958, 246-251, 1 Porträt.

22 Walter Bosshard (1892-1975), vgl. Peter Pfrunder, Verena Münzer, Annemarie Hürlimann: *Fernsicht: Walter Bosshard – ein Pionier des modernen Photojournalismus.* Bern: Benteli 1997. 238 S. (Reihe Schweizer Photographie 13) Seine bekanntesten Bücher im vorliegenden Kontext sind: *Durch Tibet und Turkistan.* Stuttgart: Strecker & Schröder 1930. XIV, 245 S.; *Kühles Grasland Mongolei. Zauber und Schönheit der Steppe.* Berlin: Deutscher Verlag 1938. 216 S.

die erste griechische Inschrift (Steinplatte) gefunden, die von der Zeit Alexanders des Grossen herrührt. Man hat immer nach solchen gesucht. Jetzt hat man wirklich Beweis dass die Griechen dort vorbeikamen.
Nun schliesse ich Grüsse mir Frau Ciani ich werde ihr Cortizon mitbringen es heilt alle Sciatica und Arthritis.
Herzlichen Gruss Dein Onkel Rock

45

Honolulu, Oct. 19/58

Dear Robert:
The reason why I have not as yet replied to your kind card and most welcome letter is due to my absence from Honolulu. I flew to Hilo in the big island and went *mountaineering*. Went to the summit bogs (swamps) of West Hawaii, in the Kohala mountains to 4500-5000 ft. To hunt plants. I had to get away from that damn dictionary which is nearly driving me crazy. A lot of editing will have to be done when everything is finished which will be soon. I will let them do the editing, I mean Tucci in Rome.

Not a word from Hans, well and good, I will remember it when I come to Vienna, I will not see him, that proud and arrogant individual. Wrote a large picture postcard to Anton, hope he got it. Before flying to Europe will have to go to Maui (a large island west of the big island) and go to the northern slopes of the Haleakala, a large extinct volcano. May be we can go to Rome or Palermo together and work there on the dictionary, introduction, etc. I suppose you read all about the pope's death etc. No I have not seen or heard any reviews (American) of the Amnye Machhen book. But I don't get to hear about American reviews.
There is little news. I feel well and hope to see you soon.
With kindest regards, herzlich, Dein Onkel Rock
Lege dem Brief einige Marken bei, tue was Du willst mit ihnen.
Deine Karte mit Johann Strauss kam auch an. Wie geht es Frau Ciani? Wir werden gewiss eine gute Zeit ohne Hans in Wien verbringen. R.

46

Honolulu Nov. 5/58

Mein lieber Robert:
Deine guten Briefe und Karten alle mit herzlichem Dank erhalten. Auch gestern kam ein trauriger Brief von Hans, er spricht von lebensmüde und allem möglichen. Jetzt hat er doch eine „Frau". Er schreibt jetzt hat der Robert Ruhe, als ob du Dich immer beschwert hattest, es war er der immer sagte ich kann mit Robert nicht leben, mit ihm in einem Raum sein. Er ist die ganze Lina, sie hatte immer etwas und war nie zufrieden. Ich werde nie vergessen als [ich] ihren Brief bekam nachdem ich ihr 1000 Dollar schickte; kein Wort des Dankes, sonder[n] „Ja wie lange soll ich denn damit auskommen." Er ist ein Bild (physisch) von ihr und hat auch ihr Gemüt. Solchen Brief kann man nicht, und ich werde ihn auch nicht beantworten.

Ich [bin] noch nicht fertig mit dem Wörterbuch. Wenn man das noch editieren müsste, verginge noch ein Jahr, am liebsten wäre es mir ich hätte damit nie angefangen. Man sollte die meisten Karten noch einmal typieren. In Seattle hatten so viele Typisten daran geschrieben und keine hielt sich an das Ms. und so ist ein Gemisch herausgekommen. Vieles habe ich wieder typiert aber es ist endlos. Ich gehe eben durch den Na-khi Naga Cult um alle Namen etc. einzuschalten das ist wie einen Index machen.

Aber zu Weihnachten bin ich bestimmt da. Wir wollen aber allein feiern; ich kann dass [!] nicht ertragen wenn einer den grossen Herrn spielen will, sich voll Wichtigkeit macht, und doch nichts dahinter steckt. Das Unteroffiziersein hat ihn beeinträchtigt und er sieht alle Menschen so an als wären sie Soldaten unter ihm. Es wäre zum lachen wenn es nicht so traurig wäre.

Ich bin wohlauf und gesund und hoffe dass Du dasselbe bist. Jetzt ist es immer noch heiss hier aber die Nächte sind sehr kühl.

Die arme Ciani, es tut mir sehr leid um sie, aber mit 80 Jahren muss man halt einsehen dass man nicht verlangen kann frisch und munter zu sein wie mit 40. Wenn sie nicht ein karges Leben hätte führen müssen dann ging es ihr jetzt auch besser. Heredität hat auch viel damit zu thun. Schau Anton an! Er hat kein reiches Leben gehabt und ist 93 Jahre geworden. Von unserer Vaters Seite waren alle langlebig. Ich war wieder in den Bergen aber auf dieser Insel und es tut mir gut vom Schreibtisch wegzukommen. Ich sehne mich schon mit den West China Fürstentümer[n] anfangen zu können aber zuerst das Wörterbuch aus dem Weg, dann etwas Wiener Luft, gutes Leben, Essen, Theater, Oper, etc.etc. mit dir zusammen, es wird ein Genuss ein.

Ich verkaufe wieder tibetische Sachen, drei Bilder und einen Gold Brokat Mantel für 3000 Dollars, mit dem Geld werden wir gut leben. Ich habe noch viele andere Sachen, die man hier schön verkaufen kann, wenn es notwendig ist. Ich brauche nicht viel ungefähr 50-60 Dollar im Monat, denn ich brauche mich ja um Lebenskosten nicht zu kümmern.
Beiliegend einige Marken.
Schreibe bald aber auf deutsch, das English klingt etwas fremd, denn ganz richtig ist es nicht, es ist etwas gespreitzt (*stilted* auf english), als wie es ein Fremder geschrieben hätte.
Herzlichen Gruss von Deinem alten Onkel Rock
Herzl. Grüsse an Frau Ciani und Anton.
Unter den Marken ist eine noch brauchbare 4 Sh. Marke.
P. S. Die Corras waren in Darjeeling und Kalimpong sind aber wieder in Hongkong.

47

Honolulu 31sten December 1958

Mein lieber Robert.
Wie Du siehst bin ich immer noch hier. Deine Karte und letzten Brief in dem Du mir die Ankunft des Paketes anzeigtest habe ich mit Dank erhalten. Ich schrieb Dir am selben Tag da ich das Paket abschickte und konnte dann glauben dass das Paket noch vor dem Brief ankam. Hoffentlich hast Du den Brief erhalten. Ich schrieb nicht bis ich nicht bestimmt wusste dass es unmöglich war für mich von hier weg zu fliegen. Es wäre eine Hast gewesen und die Arbeit hätte erst in Rom fertig gemacht werden müssen. Endlich ist alles fertig, nur noch die Titel der Divination manuskripte [!] müssen noch eingetragen werden. Ich wollte sie weglassen aber um alles vollkommen zu haben, habe ich sie dazugezogen. Ich will noch vor meiner Abreise die Einleitung dazu schreiben was meine Abreise noch verzögern wird. Sei mir nicht böse dass ich nicht, wie versprochen, kommen konnte.

Nun ich habe etwas was einen Philatelist interessieren würde, nämlich eine Menge Co[u]verts (Umschläge) mit eingeprägten Marken von 30-50 oder mehr Jahren zurück, manche von 1870 mit den alten eingeprägten Marken (Amerikanische). Wenn sie Dein Neffe haben will oder wenn Du sie mit Hans teilen willst im Falle er Interesse daran hat, so werde ich sie schicken. Ich lege dem Brief Marken bei auch so ein obengenanntes Couvert. Schreibe mit sofort ob Du sie haben willst oder nicht. Ich möchte sie gerne noch vor meiner Abreise absenden. Ich bin gesund, was ich auch von Dir hoffe.

48

Honolulu Febr. 18/59

Lieber Robert:
Vielen Dank für Deine Karte mit dem Heldenplatz im Schnee; es sieht alles sehr kalt aus, und ich glaube gut gewählt zu haben hier geblieben zu sein denn hier scheint die Sonne, es ist warm und Sträucher blühen.
Hans hat mir geschrieben und den Artikel beigelegt, er handelt über das alte Buch *The marching wind*[23] über das ich in meinem Buch im Anhang schrieb, sie gaben mein Bild von dem Amnye Machhen Berg das erste das jeh [!] aufgenommen wurde und auch das letzte.

The Na-khi Dictionary ging am 9ten Februar by Railway express nach Rome, in 3 Wochen soll es dort sein. Ich schreibe jetzt die Einleitung und den Aufsatz für Anthropos und dann geht es weg nach Wien. Ich bin gesund und munter. Vor zwei Tagen kam ich von Kona Hawaii zurück, eine Erholung nach der langen Schreibarbeit am Wörterbuch. Ich ging mit jeep, oder besser gesagt 4 wheel drive Auto ganz um den Berg Mauna Kea auf 7000 Höhe herum, es brauchte von 7 h morgens bis 6.30 h abends. Es war nicht das erstemal. Schnee lag am Gipfel, wolkenloser Himmel, die Sonne schien heiss. Es war herrlich. Heute regnet es vom grauen Himmel. Ich fühle komisch ohne wirkliche Arbeit zu haben und bin rastlos; so wird es bald fortgehen am Ende dieses Monats.
Mit herzlichen Grüssen und auf ein baldiges Wiedersehen
Dein Onkel Rock

---

23 Leonard Clark: *The marching wind.* London: Hutchinson 1955. 347 S. Es handelt sich um den Bericht über eine Reise zum Amnye Machhen im Jahre 1949.

49

Honolulu Febr. 19/59

Lieber Robert:
Nur eine Zeile um Dir anzuzeigen dass ich in Wien am 27ten März ankomme von Zürich, das ist ein Freitag Nachmittag um 5h glaube ich.
Hotel reservation already made from here, so will have nothing to worry.

I fly via Seattle-Victoria Vancouver, to Amsterdam direct, then Zürich-Wien-Rome. From there on everything is left open. I want to be in London in May for the Chelsea Flower Show.
So auf baldiges Wiedersehen.
Dein Onkel Rock
Samstag-Sonntag können wir dann ins Grüne gehen. R.
Vorläufig sage Hans nichts.

50

Honolulu Febr. 25th/59

Mein lieber Robert

Wie ich schon geschrieben fliege ich von hier am 14ten März nach Seattle bleibe dort bis 19th, von Vancouver B.C. geht es am 20ten nach Amsterdam wo ich am 21st um 1.00 p.m. ankomme. Am 23ten fliege ich von Amsterdam nach Zürich wo ich am selben Tag um 1.05 ankomme. Am 27ten verlasse ich Zürich und komme in Wien am Freitag den 27ten um 5.35 p.m. an.

Wenn Du frei bist lade ich Dich zum Abendessen ein. Ich bin wohlauf und freue mich des Lebens. Die Einleitung zum Wörterbuch macht guten Fortschritt.
Auf baldiges Wiedersehen und herzliche Grüsse
Dein alter Onkel Rock
Herzliche Grüsse an Frau Ciani.
Bleibe vorläufig 10 Tage in Wien. Hotel schon reserviert für diese Zeit. Ich komme mit Larin [?] Air Flug #240. R.

51

Honolulu den 27ten März/59

Lieber Robert:
Heute hätte ich in Wien ankommen sollen. Deine liebe Karte kam heute an vielen Dank für die Ostergrüsse, es tut mir leid dass ich zu spät bin mit den Ostergrüssen, ich bin vor zwei Tagen von der grossen Insel Hawaii zurückgekommen wo ich für 4 Tage auf dem grossen Vulkan Mauna Loa verblieb. Bin in die Wälder und habe botanisiert. Auch zu Deinem Geburtstag komme ich zu spät. Wir werden ihn doch feiern wenn auch verspätet.

Ich komme bestimmt nach Wien fliege am 12ten April von Amsterdam mit KLM ab. Ich schrieb Hans dass es möglich sein wird dass ich vielleicht doch ein paar Tage in Amsterdam verbleiben werde. Wir (ich und er) werden uns später treffen. Bleibe nur eine Woche in Wien hoffe aber später noch einmal zu kommen. Oder wir können irgendwo in der Schweiz auf Urlaub gehen wenn Du einen hast. Meine Freunde von Honolulu kommen im Mai und wir werden uns in Kopenhagen treffen. Alles andere mündlich.
Es grüsst Dich herzlichst Dein Onkel Rock
Recht herzliche Grüsse zu Deinem 45ten Geburtstag!!
Time marches on.

52

Zürich 3ten Juli/59

Lieber Robert:
heute kam das eingeschriebene Buch: Na-khi Nâga Cult etc. Vielen Dank.

Seit meinem letzten Brief war ich in ärztlicher Behandlung, aber Gott sei Dank, es war nichts Schlimmes. Ich fuhr auf der No 8 Strassenbahn vom See nach der Steinwiesstrasse.[24] Wie immer gab es viel Leute beim Conducteur und ich musste warten bis er all die Karten durchzwickte, eine dumme altmodische Sache. Ich konnte mich nicht anhalten, und auf einmal lief der Wagen schnell um eine Kurve und ich flog mit aller Wucht gegen die Stange und meine linken Rippen. Es schmerzte sehr und ich fürchtete sie wären gebrochen und die Lunge beschädigt denn beim Atmen hatte ich

---

24 Ins Hotel.

grosse Schmerzen. Es stellte sich heraus dass ich Quetschungen erlitten hatte, aber Umschläge mit Essigsaurer Tonerde linderten die Schmerzen und ich bin wieder ganz wohl. Röntgen Untersuchungen zeigten keine gebrochene Rippen. Auch liess ich dass [!] Herz untersuchen und der Arzt sagte ich hätte (was ich schon wusste) eine linkseitige Erweiterung des Herzens, brauchte mir aber keine Sorgen zu machen. Er gab mir etwas für mögliche Anfälle. Ich sagte immer Unkraut verdirbt nicht.
Mit herzlichem Gruss Dein Onkel Rock

53

Zürich, 4ten Juli/59

Lieber Robert:
Dein lieber Brief vom 29ten erwartete mich hier zu meiner Ankunft aus Bonn gestern nachmittag. Ich hatte in Bonn Dr. Meisezahl[25] besucht, er ist Regierungsrath, befasst sich aber in seiner freien Zeit mit Tibetanischer Literatur. Wir besuchten zusammen das Guttenbergische Museum [!] wo sie die erste Bibel die Guttenberg [!] druckte aufbewahrt ist. Dort ist auch das Institute, Forschungsstelle Papiergeschichte. Der Vorstand ist eine Dame. Dieses Institut interessiert sich sehr über die Na-khi mss. Ich schickte ihnen eines welches sie jetzt (das Papier) mikroskopisch untersuchen. Das Papier ist nämlich von ganz anderen Pflanzenrinden hergestellt wie irgend andere asiatische mss. Wie ich Dir aus Bonn schrieb wird das ms. das ich ihnen gab in faksimile herausgegeben, transcribiert übersetzt und dann das Resultat über die Papieruntersuchung veröffentlicht.

Nun habe ich folgenden Vorschlag: Ich erzählte Dr. Meisezahl von Deiner deutschen Übersetzung des K'a-mä-gyu-mi-gkyi mss, und er war darüber begeistert und sagte, es wäre am besten man publiziere es in Mainz. Es würde dann als separated Buch erscheinen mit allen Na-khi textseiten !] in facsimile reproduciert. Die Forschungsstelle Papiergeschichte veröffentlicht jedes Jahr ein Geschenkbuch für die Freunde des Vereins der Zellstoff und Papiergeschichte und Dein Buch würde als solch ein Geschenkbuch erscheinen mit farbigem Titelblatt.

Ich verstehe ganz gut dass Du es nicht im Anthropos gedruckt

25 Richard Othon Meisezahl; vgl. Helmut Eimer: R. O. Meisezahl (1906-1992). *Oriens*.34.1996, 1-9; *ZAS* 24.1994, 234-237, Porträt.

sehen willst und Dein Argument dass es nur wieder Wissenschaftler sehen werden und nicht Poesie Liebhaber ist gut genommen. Aber die Geschenkbücher sind etwas ganz anderes, ich habe einige davon gesehen in Mainz und sie sind sehr schön, elegant und kunstvoll. Ich würde Dir raten es dort zu drucken und es wurde als viel grössere Sache erscheinen mit allen Seiten in facsimilie (Lichtdruck) wiedergegeben, dazu noch die Farbige als Titelbild. Alle wissenschaftlichen Bemerkungen können ausbleiben, das Ganze kann als literarisches Werk erscheinen. Ich würde auch ein Vorwort schreiben und die Sache wie Du selbst angibst erklären, nähmlich [!] dass es als wissenschaftliche Arbeit schon herauskam auf English, und dass Du das Ganze damit poetisch vorführen willst um die poetische Seele und tiefen Gefühle des sonst primitiven Na-khi Stammes zur Kenntnis zu bringen.

Hoffentlich bist du damit einverstanden. So schön würde es Niemand veröffentlichen.

Bitte schreibe mir was Du darüber denkst, es wäre wirklich närrisch einen solchen Vorschlag abzuschlagen. Ja Samstag auf der Fahrt nach Koblenz-Mainz hatte ich einen Herzanfall, es sind keine rheumatischen Schmerzen sondern Beklemmungen die einem die Kehle einschnüren und das [!] ganze Unterkiefer zerdrücken; es dauerte eine halbe Stunde, aber die hatte mich sehr geschwächt.
Deine Karte kam soeben an, vielen Dank.
Deinen Brief erwartend verbleibe ich Dein alter Onkel Rock

54

Honolulu Febr. 5/60

Lieber Robert
Dein lieber Brief vom 11ten Jan. soeben erhalten, er wurde mir nach Kona Hawaii nachgeschickt und kam jetzt zurück da ich schon weg war.

Was den Koffer anbetrifft so wäre es besser ich kaufe Dir einen, oder bringe einen mit, lasse ihn bei Dir und schicke meine alten Sachen per post [!] nach Honolulu zurück.

Die Schuhe – sie sind schön gesolt [!] (das Solen kostete $6.oo oder 125 Schillinge) sie sind aber schön.

Von den sogenannten Sahara Schuhe[n] habe ich keine, ich hasse sie, aber wenn Du willst werde ich Dir ein paar kaufen sie kosten $ 11.00. Die neuen Hanan Schuhe wie ich sie Dir jetzt schicke kosten $ 27.50.

Ich arbeite an dem Buch für das Gutenberg Museum und sobald es fertig ist, komme ich dann. Ich schrieb über die ganze Kultur der Na-khi, auch Geschichte, Geographie des Landes etc. Kurz gebunden.

Wenn ich immer nach Europa reise dann komme ich nicht dazu die drei Fürstenthümer zu schreiben.

Washington D.C. State Department schickte jemanden herum zu sehen was ich noch an Material habe. Sie fürchten, da ich schon 76 Jahre alt bin, dass ich mein Material nicht verwerten werde, und sie wollen finanziell behilflich sein, alles herauszubringen.

Sie interessieren sich sehr über West China und Tibet. Sie sagten was sie in meinen Büchern finden ist nirgends anderswo zu finden. Jetzt erst finden sie das heraus. Auch von Foundations habe ich Briefe bekommen. Sie wollen meine Sachen publizieren. Wenn man dem Tode nahe ist dann kommen sie.

Wir werden die Kamaegyumigkyi herausgeben. Gib mir eine Abschrift von dem was ich Dir in der Schweiz vor zwei Jahren diktierte über die Nakhi, um vielleicht mögliche Fehler zu verbessern.

Es ist schon spät abends und so lebe wohl.
Es grusst [!] Dich Dein alter Onkel Rock
Herzliche Grüsse an Frau Ciani und Hans.

55

Febr. 18/60

Mein lieber Robert
Vielen Dank für Deine beiden Briefe die am selben Tag ankamen. Ja ich arbeite Hals über Kopf noch an der Einleitung für das Gutenberg Museum wo das Papier der Na-khi mss. untersucht wird.

Mit dem Diktieren geht es nicht denn da muss viel nachgesucht werden in den Tagebüchern und Stoßen von Notizzen [!] über Namen etc. Man kann einen Roman dictieren aber nicht eine wissenschaftliche Arbeit. Ich [!] habe eben Anton geschrieben am 24ten dieses ist sein Geburtstag 95 Jahre alt. So lange will ich nicht leben. Am 16ten dieses wäre mein Vater 127 Jahre alt gewesen. Ja wo sind die Zeiten.

Es ist mir unmöglich das Datum anzugeben wann ich in Wien ankomme; zuerst muss ich die Antwort von Marburg erhalten dass die Manuskripte angekommen sind den[n] vorher wird nicht bezahlt.

Ich lasse mir den Cheque nach Honolulu schicken. So wie auch meine Arbeit fertig ist dann kann ich Dir wissen lassen. Ich werde am besten durch das Reisebureau hier in Honolulu telegraphisch Zimmer bestellen vielleicht im Grand oder Bristol. Bis dahin alles Gute. Dein Onkel in Hast J. F. Rock

56

Honolulu Febr. 29th/60

Mein lieber Robert:
Heute sende ich Dir ein Buch dass [!] ich Dir zu Weihnachten schicken wollte aber es war schon vergriffen. Erst gestern kam eine neue Auflage heraus. Es ist voll von farbigen Bildern und handelt über die National Parks of the U.S.A. Hawaii ist auch inbegriffen. Hoffentlich bekommst Du es.

Das Jahr schreitet vorwärts und man wird älter. Ich habe viel Arbeit vor mir. Ich werde benachrichtigt dass das State Department jemanden von Washington nach Honolulu schickt um über das Zusammenfassen alles meines Materials zu unterhandeln. Sie sind speziell in mein Kartenmaterial interessiert. Ich sprach schon einmal davon, wir wollten einen Atlas von West China herausgeben, 50 Karten jede Karte mit 50 Bildern illustrieren. Wenn das zustande komt, so bin ich sehr erfreut.

Die Barth Foundation schickt ihren General Secretär nach Honolulu, er kommt am 11ten März um mit mir die drei grossen Fürstenthümer West Chinas zu verhandeln. Jetzt erst wachen sie auf. Sie sagen solches Material kann man nirgends finden.

Es wäre sehr schade wenn alles verlorenginge. Vielleicht glauben sie ich [sei] schon mit einem Fuss im Grab.

Es wird eine Heidenarbeit sein, aber ich lasse mich nicht binden. Ich sagte ihnen Anfangs Mai fliege ich ab. Meine ganzen Fotographien, sie müssen alle mit Etiketten versehen werden und in 60 Albums angebracht werden. Die grossen 60 Albums, es waren 62, zwei sind gestohlen worden in Seattle, hatte ich vor Jahren in Hanoi, Indochina machen lassen. Meine ganzen Tagebücher sollen typiert werden. Das alles wird viel Zeit verbrauchen.

Ich werde Dir schreiben nachdem die Herren angekommen sind was aus allem werden wird.

Ich hörte von Deutschland, Bonn, dass sie meine Arbeiten in Asiatische Forschungen herausgeben wollen. Nun wir werden sehen.

Am 24. März ist Frau Ciani ihr Geburtstag; ich werde ihr etwas schicken.
Nun herzliche Grüsse von Deinem noch rüstigen Onkel Rock

P.S. Ich habe die Schöpfungsgeschichte schon einmal vollkommen übersetzt und zwar im Journal of the West China Border Research Society, Chengtu. Es ist in dem Zhi-mä Zeremonie-Buch citiert im Anfang, wo alle meine Arbeiten citiert sind.

Auch im Na-khi Nâga Cult ist das meiste darin übersetzt, es sind verschiedene Versionen da. Sehe nach im Na-khi Nâga Cult unter: ^{1}Ts'o ^{2}mber ^{3}saw, da ist auch von der Schöpfung die Rede, aber mehr noch im ^{2}Muan ^{1}bpö Ts'o mber t'u.

Das tibetische Schwert ist so wie abgebildet in the National Geographic Farbenbild, ich glaube im Artikel über The mountain god of the outlaws (Konkaling) da sind drei oder vier Muli Soldaten alle mit Schwertern, oder the Yung-ning Abbot mit seinen Soldaten. Es ist so ein Schwert mit Silber beschlagen, und Coralen [!] oder Türkisensteine besetzt; vielleicht kann ich Dir doch eins schicken.

Wenn wir können zusammen arbeiten ich werde keine Arbeit mitnehmen, nur vielleicht das Büchlein dass [!] Du nicht hast wo die Schöpfung beschrieben ist, das Töten der Seele, der Ursprung der Na-khi Bücher oder Tso-la Bücher. Das ist aber übersetzt im Nakhi Nâga Cult unter dem Titel ^{1}Bpö ^{1}p'a ^{3}gko ^{1}shu, es ist ein komisches Buch. J. F. R.

57

Honolulu April 26/60

Mein lieber Robert:
In continuation of my last letter I would like to say that the Nakhi Mss. etc. have all arrived safely in Marburg and I feel much relieved. The Mss I sent to Tucci 500 in number also have arrived. I am asking Marburg to send me 10000 Deutsche Marks to Honolulu and to keep the rest in Marburg till my arrival. However I do not know when that will be.

Am 11. März kommt der Research Director der Barth Foundation um mit mir über die Bücher die ich die Absicht habe herauszugeben, zu sprechen. So wird endlich daraus etwas werden und ich brauche nicht nebst der vielen Arbeit noch mein eigenes Geld zu steuern [!]. Erst jetzt werden sie alle wach!

Ich werde durch das Reisebureau in dem ich meine Karte kaufen werde, Hotelzimmer in Wien, Rom, etc. reservieren lassen, dass [!] ist viel besser als wenn man es in Wien macht. Es wird wohl Ende April sein bis ich abfliegen kann, diesmal mit Jet, keine Bummel-Avions mehr.

Die Maschine kaufe ich Dir in Europa. Ich schickte Dir einige National Geographic Magazines; auch beauftragte ich Anthropos Dir meine letzte Arbeit zu senden. Werde Dir bald wieder schreiben.

Ich schrieb Hans und schickte ihm Marken.

Das März Nat. Geogr. Mag. hat einen guten Artikel über den Vulkan Ausbruch in Hawaii mit schönen farbigen Bildern.[26] Ich werde ihn Dir schicken; die Schuhe wirst Du wohl schon bekommen haben. Die sogenannten Sahara Schuhe ich kaufe ich bald [!] und sende Dir sie.

Ich arbeite immer noch an dem Buch für Mainz Geschichte Papier Forschungs Guttenberg Museum. Da es aber für diese Zeitschrift zuviel sein wird so soll es als ein Band *Asiatische Forschungen* Universität Bonn erscheinen. Ich werde mich also eine geraume Zeit in Deutschland aufhalten.

Mit herzlichen Grüssen Dein Onkel Rock

58

Bad Nauheim 9th May/60

Mein lieber Robert:

Deine Karte habe ich gestern erhalten, vielen Dank dass Du meine Briefe aufgegeben hast. Gestern schickte ich Dir einen eingeschriebenen Flugbrief mit 100 Shillingen die ich noch übrig hatte. Ich hoffe Du hast sie bekommen.

Nauheim ist schön, das Hotel ist schön gelegen, aber das Essen ist schlecht, schwer und geschmacklos wie die Deutschen selbst. Die Kellner sind unfreundlich und ich mochte nicht länger bleiben.

Vorgestern war ich in Wiesbaden und habe bei Harrassowitz einige Bücher gekauft. Gestern waren wir in Darmstadt bei Prof. Jayme Direktor des Institutes für Cellulosechemie wo man das Papier der Na-khi Handschriften untersuchte. Die Arbeit war fertig und sie

---

26 Vermutlich: Simpich, Frederick Jr.: Fountain of fire in Hawaii (Kilauea Iki Crater). Photos by Robert B. Goodman and Robert Wenkam. *NGM* 1960:März, 303-327.

sind zu interessanten Resultaten gekommen. Die Arbeit wird in Rom mit meiner herauskommen so dass Alles was man über die Na-khi weiß zusammen erscheinen wird.

Hier sieht man hässliche sehr fette dicke Frauen die Körper wie Fässer haben und auf entweder sehr dicken oder dünnen Stelzen gehen. Die Männer sehen grob und dick aus mit furchtbaren Bäuchen vom Saufen und Überessen. Sie machen hier die Kur durch. Es würde mir auf die Dauer unmöglich sein hier zu leben und ich möchte nur raus und sobald wie möglich.

Morgen fahre ich nach Marburg zur Westdeutschen Bibliothek und hole mir das Geld. Sie wollen ich soll einen Katalog über alle Na-khi Handschriften schreiben für welchen sie bezahlen werden denn dafür haben sie viel Geld und Übersetzungen gewisser Handschriften können da eingeschoben werden.

Morgen Abend komme ich wieder zurück nach Nauheim und Sonntag fahre ich nach Bonn wo ich bis zum 22ten verbleibe.

Ich hoffe nach Wien zurückzukommen, erwarte einen Brief von Frau Girond in Bonn, hoffentlich kommt sie nicht. Dann komme ich bestimmt nach Wien.

Auf baldiges Wiedersehen und herzliche Grüsse

Dein Onkel Rock

59

Hotel Königshof

Bonn, Koblenzer Str. 9-11

15ten Mai 1960

Mein lieber Robert:

Vor einer Stunde kam ich hier an und fand Deinen lieben Brief vor vom 12ten Mai. Glad you received the money. Bad Nauheim hat eine schöne Umgebung aber das Essen ist fürchterlich speziell wenn man von Wien kommt. Die DMark kostet oder kommt 4.17 auf den Dollar und der Schweizer Franken 4.31 auf den Dollar, der Schilling 25.93 auf den Dollar. Ja ich habe mich etwas geirrt. Das Hotel war sehr modern, gross und schön. Die Taxi waren billig. Wir sind von 9 Uhr morgens bis 6 Uhr Abends gefahren und gestanden und haben nur 50 Mark pro Tag bezahlt, ungefähr 12 Dollar mit Trinkgeld. Wir fuhren nach Darmstadt, Marburg, Wiesbaden und Frankfurt nur zwei Rasttage.

In Marburg hat alles geklappt. Wir haben verabredet dass wir einen Katalog machen werden über alle Na-khi Handschriften, für diese Arbeit bezahlen sie (sie haben viel Geld und bekommen es von der Forschungsgemeinde).

Hier in Bonn habe ich ein schönes Zimmer Aussicht auf den Rhein, sehr ruhig und sehr angenehm, aber hier bezahle ich 35 Deutsche Mark pro Tag, equal to 219.36 Shillinge, also doch billiger wie letztes Jahr im Krantz. Hier werde ich mich etwas ausruhen. Es freut mich sehr dass man Dir die Anerkennung gibt die Dir gebührt, wieviel gibt man Dir denn? Arbeite an dem K'a-ma-gyu-mi-gkyi, wir werden es herausgeben! Ja ich komme nach Wien und dann geht es in die Schweiz. Ich will nur sehen was mein Freund aus [—] London vorhat. Wenn ich im September noch in Europa [—] wir nach Palermo. Herzlichst Dein Onkel Rock

60

Bonn 18th Mai/60

Mein lieber Robert:
Die letzten drei Tage habe ich am Radio zugehört über die Verhandlungen in Paris. In meinem ganzen Leben habe ich so eine wilde, ausgelassene, freche Kundgebung wie sie Kroushew brachte nicht gehört. Es war natürlich dumm dass die Amerikaner gerade vor der Konferenz Spionage Flugzeuge noch durch Russland schickten, dann erzählten sie noch Lügen und hatten die Affrond zu sagen dass diese Spionageflüge wiederholt würden. Dass die so dumm sein würden hätte ich nicht gedacht.

Die Lage ist sehr kritisch, die deutsche Mark ist etwas gefallen aber noch nicht viel.

Ich glaube unter den Umständen es wird besser sein ich fliege nach Honolulu zurück, aber es ist noch nicht sicher, ich werde in England abwarten was geschehen wird. Heute hatte ganz England Flug Alarm und Prüfung der Flugstreitkräfte, kein gutes Zeichen.

Ich habe heute Kruchew gehört, er schrie wie ein Wilder, und die Reporter sagten er schüttelte beide Fäuste. Es ist Propaganda für die Sateliten [!] und China denen es nicht gefiel dass er zu freundlich wurde mit Amerika. Am Son[n]tag fliege ich nach London und erwarte einen Brief von Dir.
Herzliche Grüße von Deinem Onkel Rock

P.S. Erhielt einen Brief von M. Géraerd[27], sie kommt nicht nach Wien.

61

Park Lane Hotel
Piccadilly
London W.1.

26ten Mai/60

Mein lieber Robert.
Deinen sehr willkommenen Brief vom 22ten habe ich gestern erhalten. Es freut mich unendlich dass Dir diese schöne Anerkennung zuteil wurde. Dein Brief ist von Freude und appreciation erfüllt und ich bin stolz auf Dich. Gern will ich Deinem Wunsch Folge leisten und Dir den Shakespear [!] besorgen als Geschenk zur Anerkennung Deiner dichterischen Arbeiten. Ich hoffe dass es Dir gelingen wird die Poesie, wie sie die Na-khi selbst fühlen, auf deutsch wiederzubringen. Jedesmal, wenn grosse starke Männer diese Legende (?) sangen kamen ihnen die Thränen in die Augen. Mich selbst hat es immer sehr berührt. Hoffentlich gelingt es Dir.

Ich bekam ein Schreiben vom Hotel Prinz Eugen dass mir bestätigt dass mein Zimmer vom 31ten Mai bis 7ten Juni reserviert ist. So fliege ich Montag am 30ten nach der Schweiz (Zürich), bleibe dort bei 31ten und komme am selben Tag nach Wien, mein Billet ist schon gekauft. Am 7ten Juni fliege ich zurück nach Zürich und werde dann irgendwo ein paar Wochen in den Bergen verbleiben. Ich komme nicht mehr nach London, sondern fliege von Zürich nach Amsterdam – über den Nordpol nach Vancouver Brit. Columbia und Honolulu, dass[!] ist der kühlste Weg.
Auf baldiges Wiedersehen am 31ten Mai in Wien. Telephon 651741.
Herzlicher Gruss von Deinem alten Onkel Rock

27 Die Schreibung ist undeutlich, möglicherweise lautet der Name: Girond oder Giroud.

62

Park Lane Hotel
Piccadilly
London W.1.

May 26/60

Lieber Robert
Ich ging sofort in die Buchhandlung und bestellte ein 3-bändiges Werk von Shakespear [!] auf Indiapapier gedruckt (viel teurer wie gewöhnliches Papier 50% teurer). Es ist eben erschienen or wird im Juni erscheinen kostet ungefähr $20.00. Die Oxford Press gibt es heraus. Es wird Dir im Juni zugeschickt.
Bis nächsten Montag auf Wiedersehen Rock

63

Zürich, 26ten Juni/60

Mein lieber Robert
Oft denke ich an Dich und wünschte dass Du kommen könntest. Das Wetter war sehr warm und jetzt regnet es schon zwei Tage und es ist kühl. Vorgestern war ich in der Oper hier und habe *Fidelio* gesehen. Ausgezeichnete Sänger und Sängerinnen vom New York Metropolitan (Gastsänger). Es ist aber eine traurige Sache und nichts Fideles als der Name ansagt oder ansagen soll.

Gestern erhielt ich einen Brief von dem großen Verlag Steiner in Wiesbaden, die wollen meine Tagebücher veröffentlichen oder auch Na-khi Literature. Wenn ich wieder in Honolulu bin werde ich – wie wir in Wien besprochen haben – die Sintflut der Na-khi, obwohl schon einmal veröffentlicht, nochmals und besser herausbringen, denn die story ist wirklich interessant, auch das „Töten der Seele".

Wie geht es mit der K'a-mä-gyu-mi-gkyi, wenn du das richtig fertig bringst, will ich zur Veröffentlichung die nötige Summe – wie schon einmal vorgeschlagen – beisteuern.

Von Honolulu höre ich wenig und nicht viel Interessantes. Mein Freund Paul der in Wien war, hat mir von Bangalore Süd Indien geschrieben. Calcutta wäre fürchterlich gewesen, und er erkrankte so dass er nicht aus dem Bett steigen konnte.
Schreibe bald. Kaufe Dir Rotwein (aber guten) und drinke [!] ihn auf unser Wohl. – Wie immer Dein alter Onkel Rock
Beiliegend 20 Franken Swiss. R.

64

Nikko Kanaya Hotel

October 23d/60

Mein lieber Robert:

Kam mit Auto von Tokyo hier an. Ich war in Nikko vor 42 Jahren, unglaublich wie die Zeit dahinflog. Ich bin müde und gedachte Deiner: Ich soll in meinem Alter nicht herumfliegen; Du hast recht. Meine Freunde die 20 Jahre und 10 Jahre respective jünger sind, sind beide stark erkältet; mir fehlt nichts und bin wohlauf. Wir bleiben 5 Tage hier dann mit Auto nach Myanoshita. Morgen fahren wir mit Auto nach Kegan Wasserfall, dann zum See Chusendji. Dann nach Hakone-Kyoto. Meine Freunde wollen nach Kyu-shu ich gehe aber nicht und fliege am 18ten nach Hongkong. Ich freue mich auf Darjeeling and the Eternal Snows The Himalayas. In Kalimpong habe ich viele Freunde und auch sind dort Na-khi und es wird möglich sein zu erfahren was mit meinen Leuten in Likiang geschehen ist.

Ich hoffe dass Du gesund bist, erkälte Dich nicht. Ich schlafe bei offenem Fenster, hier ist es kalt, ein Bach rauscht bei meiner Terasse [!], der Wald von Ahorn ist rot gefärbt, die Sugi (Cryptomeria) Bäume über 300 Jahre alt streben gegen den Himmel. Die Tempel sind alt und schön, das moderne Japan ist hässlich, keine Kimono mehr; die Mädchen tragen Hosen oder andere Kleider des Westens.

Oct. 24th/60

Ein schöner Morgen, die Berge von der Sonne geküsst leuchteten rot, der Frost rosa, die Nadelhölzer erinnern an St. Anton. Während der Nacht träumte ich von Lina, Deiner Mutter. Ich war in Wien und suchte sie auf, aber sie war nicht zu Hause, zwei Mädchen sagten sie wäre in einem Kaffeehaus und ich ging mit ihnen sie aufzusuchen. Ich sah sie in einer Gesellschaft von Männern und Frauen an einem Tisch sitzen, die Männer lachten. Ich beobachtete alles durch ein grosses Fenster weil die Mädchen hineingingen und ihr sagten ich sei draussen und möchte sie sehen. Sie kamen heraus und sagten sie wollte mit mir nicht sprechen da ich ihren Brief nicht beantwortete. So ging ich weg ohne sie zu sehen. Ich hatte aber keinen Brief von ihr bekommen. Im nächsten Moment war ich vor dem Hause am Franzenring 22 und zeigte auf das Thor welches sich ungemein vergrösserte und sagte zu den Mädchen: „In diesem Haus war ich geboren“. Dann erwachte ich. Warum solches in Nikko träumen?

Kannst du den Traum erklären? Wenig Leute sind auf der Strasse die auf die Berge führt, alles ist ruhig und man freut sich des Lebens.

Bitte schreibe mir c/o Imperial Hotel Tokyo wo ich am 18ten November wieder eintreffe; ich sorge um Deine Gesundheit denn Hans schrieb dass Du krank warst.
Herzliche Grüsse von Deinem Onkel der wünscht Du wärest mit ihm.
Wie immer Rock

65

The Fujita Hotel
Mikyanoshita, Hakone, Japan

Oct. 28/60

Mein lieber Robert:
Heute hier angekommen nach einer schönen Auto Fahrt von Tokyo über Yokohama und Kamakura. Der 600 Jahre alte Buddha steht noch. Sein Tempel wurde vor 300 Jahren von einer Meerflutwelle hinweggerissen, aber er hielt stand. Das Erdbeben von 1923 hat seinen steinernen Sockel zerstört. Jetzt hat sein Kopf zum Zerspringen angefangen und er wird jetzt repariert. Vor 28 Jahren stand ich vor dem grossen Dai Butsu oder wehmütigen Buddha. Und vor 42 Jahren war ich in diesem sehr schönen Hotel. Sie haben einen Annex zugebaut der wirklich herrlich ist. Leider habe ich mich sehr erkältet und huste viel darum gehe ich sofort zu Bett. Morgen werde ich Dir weiter schreiben .
Oct. 29th 8 a.m.
Ein schöner Morgen. Wegen meiner Erkältung bleibe ich im Bett und werde 5 Tage in diesem ruhigen bewaldeten Ort verbringen. Herzliche Grüsse Dein Onkel Rock

66

Oberoi Mount Everest
Darjeeling

Dec. 8/60

Mein lieber Robert.
Ich sitze allein im Zimmer bei einem Kohlenfeuer in diesem alten Hotel. Die Aussicht über das Himalaya Gebirge ist unaussprechlich herrlich, es ist kein Wunder dass Tibeter die Berge vergöttern und

anbeten. Vor 47 Jahren war ich hier, da existierte dieses Hotel noch nicht, aber die Villa in der ich wohnte steht noch. Vor ein paar Minuten läuteten die Tempelglocken unaufhörlich, sie erinnerten mich an die Zeit als ich selbst in einer Lamaserie (in Choni) nahezu drei Jahre verbrachte.

Heute morgen glühten die Schneeberge rosa von den Strahlen der aufgehenden Sonne. Ich verfiel in tiefe Gedanken über das Wesshalb [!] und Warum und Woher! Alles und wir selbst sind ein Rätsel dass [!] Niemand wird lösen können.

Ich besuchte einige Freunde die der Tod noch verschont hat, andere sind schon ins Jenseits gewandert, unter ihnen der Sohn des letzten Königs Thibaud[28] von Birma, der in diesen Bergen im Exile lebte. Vor 10 Jahren hat uns jemand zusammen photographiert, Nebeski war auch dabei und eine kluge deutsche Frau die ein paar Monate nachher auch starb. Ich bin noch immer übrig. Ich kam hierher um die heiligen Berge noch einmal zu sehen ehe ich meine Augen für immer schliesse, das war mein Wunsch und jetzt ist er erfüllt.

Über 4000 tibetische Flüchtlinge sind hier und in Kalimpong, viel Armut und Schmutz.

Meine Freunde kommen am 10ten oder 11ten, dann werde ich nicht so einsam sein. Aber ich liebe das einsame Leben das Zeit zum denken lässt.

Meine Freunde die vor Geld strotzen, thun mir doch leid, die Frau lässt nichts wie Kleider machen, sie denkt nichts, 20 paar Schuhe machen zu lassen. Wenn ich die Armut dieser Flüchtlinge sehe, kleine Kinder in Fetzen gekleidet, nichts zu essen, und wenn ich dann die Frau ansehe, die sich im Spiegel schaut ob jener Stoff besser steht wie ein Anderer, da werde ich traurig. Nun Schluss mit diesen Betrachtungen, das Leben ist halt so.

Hoffentlich ist alles angekommen. Ich werde Dir bald wieder schreiben. Wenn möglich fliege ich nach Nepal, denn nach Indien komme ich bestimmt nicht mehr.

Die herzlichsten Grüsse sendet Dir Dein noch immer rüstiger Onkel

Rock

Die Marken die auf dem Couvert sind, sind neue indische, bitte gib sie Hans.

---

28 Thibaw, 1858-1916.

67

An Bord des Lufthansa-Senator-Dienstes
In flight from Montreal Canada to Chicago

Febr. 4th/61

Wie meine Karte anzeigte flog ich doch nach U.S.A. mit der Lufthansa. Dein letzter Brief erreichte mich noch in Frankfurt. Frau Giraud[29] telephonierte mir nach Frankfurt und sagte die Krapfen wären angekommen und die besten die sie je gegessen hatte. Ich danke Dir herzlichst dass Du sie geschickt hast.

Der Flug soweit war schön. Wir flogen von Frankfurt in einem Schneesturm ab. In Paris war es wieder heiter. Und dann den ganzen Weg nach Montreal und jetzt nach Chicago fliegen wir im Sonnenschein. Da wir mit ihr fliegen dann wird es immer früher anstatt später. In Frankfurt ist es schon 10 Uhr nachts oder mehr, und hier ist es 6 Uhr Abends oder weniger. Wir flogen um 11.30 von Frankfurt ab und kamen in Goosebay Labrador um 1.30 mittags an.

Aber in Frankfurt war es 7.30 p.m. Alles ist erfroren und mit Schnee bedeckt aber im Flugzeug ist es warm. Das Essen sehr gut, Plenty of caviar, Gansleberpastete, etc. etc.

Ich freue mich schon auf Honolulu, überall in Europa ausser Wien, hat es geregnet. Es war nicht schön in Frankfurt, eine hässliche Stadt.

Werde dies in Chicago aufgeben und Dir von San Francisco auch noch schreiben. Vielen Dank für alles.

Mit herzlichen Grüssen an Dich, Hans, Frau Ciani und Anton bin ich wie immer Dein alter Onkel Rock

68

St. Francis Hotel
Union Square, San Francisco, 19

Febr. 5th/61

Mein lieber Robert:

Auf dem Lufthansa Flugschiff schrieb ich Dir einen Brief der in Chicago aufgegeben wurde, hast Du ihn bekommen? Ich dankte Dir für die Krapfensendung an Frau Giroud auch für Deinen letzten Brief den Du mir nach Frankfurt schicktest.

---

29 Die Namensschreibung ist undeutlich, s. o.

Hier ist doch ein ganz anderes Tempo, eine Grossartigkeit „Grandeur“ die es in Europa nicht gibt. Als wir von dem schrecklichen geschmacklosen Frankfurt abflogen herrschte ein Schneesturm, ich dachte kein guter Anfang. Aber schnell waren wir über den schmutzigen Wolken und flogen im blauen Himmel. In Paris war es ruhig, kein Schnee aber doch düster. Um 1.30 p.m. flogen wir schon über Goose Bay, Labrador, 7.30 p.m. Frankfurter Zeit. Alles gefroren. In Montreal herrschte eine sibierische [!] Kälte. In Chicago tiefer Schnee, und als wir über die Rocky Mountains hinweg waren tief dunkelblauer Himmel und die Gestirne funkelten. Um 9 p.m. kamen wir glücklich in San Francisco, müde aber wohlbehalten an. Alte Freunde erwarteten mich und ich fuhr nach der Stadt. Es ist 10 a.m. Gut geschlafen, gut gefrühstückt, ein neuer Mann. Heute geht es in die Berge, es ist Sonntag. Ich glaube am Donnerstag entweder nach Seattle oder direct nach Honolulu zu fliegen. Werde Dir bald wieder schreiben.
Mit herzlichen Grüssen an Dich, Hans, Frau Ciani und Anton, verbleibe ich wie immer Dein alter Onkel Rock

69

Zürich, 3ten Juni/61

Lieber Robert:
Vielen Dank für Deinen Brief vom 2ten den ich eben jetzt 2h nachmittags erhielt. Ich schrieb sofort einen Eilbrief nach Wien zu Anton. Hoffentlich kommt er dem Tod zuvor. Ich glaube immer noch dass er 100 Jahre alt werden wird.

Ich bleibe bis 14ten hier und mache dann zur Erholung eine Mittelmeerfahrt nach Griechenland und Turkei. Am 29ten Juni bin ich wieder in Zurich. Anfang Juli fliege ich über Frankfurt-Amsterdam-Vancouver nach Honolulu zurück. Ich verliess Honolulu am 14ten Mai und nach in 14 Stunden in London. Meine Freunde überraschte ich, ich verliess sie in New Dehli [!] letzte Weihnachten. Sie wollten dass ich mit ihnen mit Auto nach Schottland fahre es ist aber noch zu kalt dort, es hat geschneit vor ein paar Tagen. Hier ist es auch kalt.
Herzliche Grüsse, wie immer,
Dein Onkel Rock

70

Zürich 7ten Juni/61

Lieber Robert:
Deinen Brief der die Nachricht vom Tode Antons brachte habe ich heute Morgen erhalten. Ich muss sagen dass ich diese Nachricht heute erwartete. Schade dass er nicht bis 100. lebte. Es ist das letzte Kapitel dass [!] in meiner Kinderzeit begonnen hatte und jetzt sein Ende nahm. Der Anfang war nicht gut. Ich erinnere mich auf das erste Begegnen mit ihm. Es war knapp nach dem Tode oder war es noch vor dem Tode meiner Mutter, ja es war vor dem Tode der Mutter denn ich war 5 Jahre alt. Mein Vater und ich fuhren nach Brunn a/S und besuchten einen grossen jungen Mann der damals 24 Jahre alt war. Wir waren in einem Garten wo er als Gärtner angestellt war und ich ging zwischen ihm und Vater den Garten auf und ab. Wer er war oder was mein Vater mit ihm sprach wusste ich nicht. Nur nach einigen Jahren nachher erfuhr ich dass er mein Halbbruder war. Er kam einmal am 19ten März nach Wien und dekorierte den kleinen Altar mit weissen Kamellien zu meinem Namenstag. Dann später kamen andere Zeiten die besser vergessen sind und von denen ich Dir einmal erzählte. So ist dieses Kapitel geschlossen, es hat lange gedauert. Für Euch ist es nicht das Letzte und jetzt bin ich an der Reihe und hoffentlich wird es nicht solange dauern.

Was die Kerze anbetrifft so mag sie dafür gedient haben wie die Lampe die man den todten Na-khi in die Hand giebt während man ihre Seele ins Jenseits befördert. Du kennst das Hä-zhi p'i.

Heute morgen erfuhr ich dass Pater Wimmer[30] den ich noch im Jänner dieses Jahres besuchte auch seinem Krebsleiden erlegen ist. Zwei Schläge auf einmal – der Letztere hat auch gelitten hat aber etwas von Wert hinterlassen. Er schrieb noch dass [!] Inhaltsverzeichniss des Supplements der Lobeliaceen.

Ich benannte gerade noch eine neue Art die auf den Koolan Bergen auf Oahu gefunden wurde nach ihm: *Trematolobelia Wimmeriana* Rock. Sie konnte aber nicht mehr in sein

30 Franz Elfried Wimmer (1881-1961), aus Niederösterreich stammender Verfasser des Standardwerks: *Campanulaceae-Lobeliodeae*. Leipzig: Engelmann 1943. 260 S. (Engler: Das Pflanzenreich.106.); Teil 2. 1953. VIII, 262-813 S. (Das Pflanzenreich 107.); Supplement. 1968. X, S. 815-1024 (Das Pflanzenreich 108.) Vgl. *Taxon* 10.1961, 239-240, Porträt.

Supplementum aufgenommen werden. So werde ich sie in Honolulu veröffentlichen.

Meine Freunde sind in London geblieben und gehen erst in zehn Tagen nach Schottland. Es ist möglich dass sie später in die Schweiz kommen. Wir sollen zusammen im August nach Skandinavien fahren und zwar in ihrem Auto. Ob ich gehe weiss ich noch nicht. Wenn ja, dann kommen wir im September nach Wien, sonst, wenn ich nicht mittue, fliege ich anfangs Juli nach Honolulu zurück.
Dies ist jetzt wirklich mein letzter Europa Besuch.
Herzliche Grüsse Dein Onkel Rock
Bleibe bis 15ten hier komme am 29ten nach Zürich zurück. Schreibe mir nach Athen, Griechenland c/o American Express hierher. R.

71

Honolulu 24ten Febr./62

Lieber Robert:
Es ist lange her dass ich von Dir gehört habe und ich wundere mich wie es Dir geht. Bist Du noch in der Wittelsbachstrasse oder bist Du umgezogen. Ich vernehme durch das Radio dass in Europa eine schreckliche Kälte herrscht, speziell in Deutschland. In Amerika ist alles verschneit. Hier, eben heute scheint die Sonne herrlich und es ist schön warm, alles ist grün, aber vor ein paar Tagen war es auch hier kalt. Der Thermometer fiel auf 56° fahrenheit ungefähr 14°C dass [!] ist für Honolulu sehr kalt. Auf Mt. Haleakala auf der Insel Maui ging der Schnee 1000 m den Berg hinab bis auf 7000 Fuss. Mauna Kea und Mauna Loa waren verschneit und auf dem Ersteren sind Leute Schi gelaufen.

Seit einem Monat leide ich an derselben Sache an der ich 1948 leidete [!] trigeminale Neuralgie wessen ich nach Europa und Boston kam um mich operieren zu lassen. Jetzt sind die Nerven wieder zusammengewachsen und die unaushaltbaren Gesichtsschmerzen sind wieder da. Medikamente nützen wenig, so werde ich mich wieder einer Operation unterziehen müssen. Das wird meine EuropaReise verzögern oder gar zunichte machen. Und so geht es wenn man alt wird.

Was machst du es würde mich sehr interessieren was Du jetzt tuhst [!]? Das Wörterbuch ist noch nicht heraus aber der erste Band wird jetzt gedruckt. Der zweite ist nahezu gesetzt nur die letzten 150

Seiten erwarte ich um Korrektur zu lesen. Das Wörterbuch hat mehr Arbeit gemacht wie irgend jede andere Arbeit. Ich schickte Tucci 500 Na-khi mss. um die Kosten teilweise zu bestreiten. Er verkaufte sie um $10,000 Dollar. Wenn ich das gewusst hätte, hätte ich sie selbst verkauft und ihm einen Teil davon gegeben. Ich besitze noch 229 Na-khi mss. sie sind die Besten, schön illuminiert. So geht es im Leben.

Nun bitte schreibe mir bald denn ich will wissen wie es Dir geht. Von Hans höre ich öfters aber er scheint von Dir nicht viel zu wissen. Mit herzlichen Grüssen, verbleibe ich Dein älter werdender Onkel Rock

72

[Febr. 1962?]

Mein lieber Robert.
Soeben erhielt ich Deine freundlichen Wünsche zu meinem 78ten Geburtstag am 13ten für welche ich Dir herzlich danke. Mit meinem Herzen geht es besser aber die trigeminale Neuralgie wegen der ich 1948 nach Europa und Boston kam von China hat jetzt wieder angefangen; ich bin in ärztlicher Behandlung aber medicamente helfen da wenig, sie wollen einen operativen Eingriff machen, dass [!] heist [!] ein Loch in den Kopf machen auf der rechten Schläfe und dort die drei nerven abschneiden. Das will ich aber nicht. Es gibt halt immer etwas. Ich weiss nicht gewiss ob ich unter den Umständen nach Europa komme. Auch das Geld wird knapp.

Die Korrektur des zweiten Bandes des Wörterbuches ist 50% gemacht, die letzte Seite war 250, es wird ungefähr noch 250 oder 260 Seiten ausmachen. Der erste Band ist druckfertig, Gott sei Dank. Der zweite ist weniger schwer, sie machen weniger Fehler und so geht es schneller. 60 Tafeln kommen dazu die sind fertig, auch eine Landkarte von der Präfektur Lichiang.

Ich ediere jetzt den 8ten Band meiner Tagebücher welche eine Sekretärin (auf Kosten der Regierung) tipiert. Sie werden in fünf Kopien tipiert, drei davon für mich, mit elektrischer Schreibmaschine getiept [!]. Es ist noch viel zu thun. Ich muss mich an die Fürstentümer daran machen sonst werden sie nie herauskommen. Die Karten werden in Seattle gemacht.

Das Kulturleben der Na-khi soll auch noch geschrieben werden, über die Hälfte habe ich fertig. Dazu kommt noch die chemische Papieruntersuchung der Na-khi Handschriften von Prof. Jayme von Darmstadt.

Ich glaube sollte ich so lange leben wie Anton ich würde nicht fertig werden.

Die Katalogisierung der Na-khi Handschriften über 1000, wartet auf mich in Marburg. Ich habe keine Zeit krank zu sein und noch weniger zum Sterben.

Wenn ich nach Europa komme dann mache ich Wochenende Abstecher nach Wien um Leberknödelsuppe, Kalbschnitzel etc., etc. zu geniessen.

Es freut mich dass Frau Ciani das Paketchen erhalten hat. Ja es ist von mir, ich habe es aber nur ein oder zweimal gebraucht. Danke ihr für Ihre Wünsche und grüsse sie mir recht herzlich.
Auch herzliche Grüsse an Dich. Wie immer Dein alter Onkel Rock

73

Honolulu 23ten März/62

Lieber Robert:
Herzlichen Dank für Deine Karte und Glückwünsche zum Namenstag. Here [!] in Amerika feiert man keine Namenstage und ich vergesse es ganz.

Du schreibst ich antwortete Deinen ausführlichen Brief nicht. Dein Brief war die Antwort auf meinen Brief den ich Dir sandte, so bin ich Dir keinen schuldig. Zweitens war ich so überhäuft mit Arbeit dass ich jeden Tag bis 1 Uhr morgens arbeitete. Heute habe ich die letzte Korrektur des ersten Bandes 637 Seiten nach Rome geschickt das ist sehr ermüdende Arbeit. Jetzt ist wieder neuer Proof da 369 Seiten des zweiten Bandes. In Rome haben sie eine Menge Karten verloren und ich musste vieles mehrmals schreiben und die Symbole zeichnen. Das alles nimmt viel Zeit in Anspruch. Jetzt muss ich noch die schon lange angefangene Arbeit über *Na-khi life and culture* fertig schreiben dass [!] als eine separate Arbeit mit einem Aufsatz von Prof. Jayme (Darmstadt/Deutschland Institut für Cellulose Chemie der technischen Hochschule) über das Na-khi Papier, zusammen in Rome veröffentlicht wird.

So ist jede Minute kostbar. Ich habe nicht die Zeit zur Verfügung die Du hast. Dazu leide ich noch an dieser schrecklichen trigeminalen Neuralgie. Ich soll mich einer Operation unterziehen will aber warten bis meine Arbeit fertig ist. Ich schreibe diesen Brief um Mitternacht.

Ende April hoffe ich in Wien zu sein. Die Deutsche Forschungsgemeinde Godesberg Deutschland bezahlt alle Reisekosten,

Aufenthalt in Marburg etc. etc. Das ist das erstemal seit 1945 dass die Na-khi Arbeit etwas eintragen wird.

Die Historische Geographie der beiden Fürstentümer Choni und Muli müssen etwas warten, so auch die Landkarten von denen ich fünf inzwischen fertig machte, fünf sind schon letztes Jahr gezeichnet von mir und werden in Seattle herausgegeben.

Ich sandte Frau Ciani eine Kleinigkeit zu ihrem Geburtstag. Deiner ist um Ostern?

Herzliche Grüsse Dein Onkel Rock

74

Honolulu 6ten April/62

Mein lieber Robert:

Herzlichen Dank für Deinen lieben Brief. Ja ich weiss ganz genau wann Dein Geburtstag ist. Ich erinnere mich als wäre es gestern gewesen als Du zu den Füssen Deiner Mutter lagst. Nun ich wünsche Dir alles Gute, vielen Erfolg in Deinem Dichten, Gesundheit und alles was du Dir wünscht. Beiliegend 10 Dollar für ein Geburtstagsessen und ein Gläschen Wein beim Krantz.

Wenn Du Schuhe brauchst, schwarz und braun, so sage es ich werde Dir sie schicken. Ich selbst gebrauche jetzt sehr wenig da ich nicht viel ausgehe. Kleider brauchst Du wohl jetzt keine. Meine alten sind zu schlecht um sie jemandem zu senden.

Die Nervenschmerzen halten an und ich glaub ich muss doch in das Hospital um mich operieren zu lassen, es wird nicht besser. Man hat mir telephoniert, der Präsident der hiesigen Universität, dass sie mir am 12ten April den Ehrendoktor verleihen wollen. Ich wollte absagen aber man hat mir gerathen ich soll ihn annehmen. Mein Wörterbuch macht mir Kummer. Es ist eine Heidenarbeit. Ich hätte mir nicht vorgestellt dass es so schwierig sein würde. Ich komme nicht dazu eine andere Arbeit anzufangen. Vielleicht ist es meine letzte.

Nun sei mir recht herzlich gegrüsst und nochmals alles Gute von Deinem alten Onkel Rock

75

Honolulu June 3/62

Mein lieber Robert:
Deinen lieben Brief, in dem Du für die zehn Dollar die ich Dir schickte, danktest, habe ich nicht beantwortet, da er ja eine Antwort auf meinen Brief war. Deine schöne Oster Karte hätte ich bestätigen sollen und Dir für Deine guten Wünsche danken sollen. Karten habe ich keine hier und ich gehe nicht oft in die Stadt wo es sehr heiss und ich habe auch nicht immer die Verbindung.

Ich habe viel Arbeit gehabt denn der zweite Band ist immer noch nicht vollständig gesetzt. Der erste ist im Druck Gott sei Dank. Inzwischen habe ich ein kleines Buch geschrieben über *The Life and Culture of the Na-khi of the China-Tibet Border* mit 12 Bildern. Ein zweiter Teil auf Deutsch stammt von Prof. Dr.-Ing. Jayme von dem Institut für Cellulose Chemie von der Universität Darmstadt. Er und eine Frau Dr. Harder haben das Papier sehr alter Na-khi Handschriften auch mikroskopisch untersucht und ihren Befund in einem Aufsatz kundgemacht. Tucci druckt ihn auch mit meiner obigen Arbeit.

Jetzt habe ich momentan Atemzeit und so schreibe ich Dir, vergib mir daß ich nicht gleich geantwortet habe.

Meine Freunde sind nicht hier und ich bin allein mit den Angestellten, die aber nicht im Hause schlafen. Frau Marks soll am 11ten dieses zurückkommen, sie sind in San Francisco. Ich liess ihnen wissen, dass im Falle sie nicht am 11ten kommen können sondern am Ende des Monats, ich am 19ten nach Amsterdam fliegen werde und von dort direkt nach Wien wo ich am 13ten ankommen werde. Kommen sie aber am 11ten – und sie baten mich nicht vor Ihrer Ankunft abzureisen – dann fliege ich am 15ten nach Vancouver Canada und verbleibe dort zwei Tage und fliege am 17ten nach Amsterdam komme aber dann nicht nach Wien, denn ich soll am 24ten in Marburg die schwere Arbeit dort beginnen über 1000 Na-khi manuscripte [!] zu katalogisieren und beschreiben.

Ich komme dann im Herbst. Es mag möglich sein dass ich dann eine letzte Reise mache, ehe ich zu alt werde, nach Süd Afrika, Madagascar, Mauritius, Australien, New Zealand, Tahiti-Honolulu. Das ist natürlich noch ein Traum, der sich aber verwirklichen kann.

Ich werde noch vor meiner Abreise einige Sachen wie Schuhe etc. senden.

Bis dahin recht herzliche Grüsse und lasst uns hoffen auf ein letztes fröhliches Wiedersehen.
Dein alter Onkel Rock

76

Marburg/L. Aug. 19th/62

Lieber Robert:
Thank you for your cards from le Grazie. I am very glad that you are having a profitable time from a cultural Standpunkt and also that you had a trip to Firenze, but two days is hardly enough for a town like Florence. There are the two famous art galleries, Galleria Pitti and the other the name of which I have forgotten. It must be rather hot at this time of the year.

Marburg is a beautiful town old – very old. The University was founded in 1270; there is a castle on a hill and the old town is on the slopes surrounding the castle wie Küchlein ihre Henne. Die Leute hier sind sehr lieb zu mir und erfüllen alle meine Wünsche – sie sind aber bescheidene. Ich bekomme viele Besuche von Professoren, Indologen, Sinologen, Tibetologen etc. und all[e] interessieren sich für die Na-khi Sprache und Schrift. Gestern war ein Indologe zu[m] Mittagessen. Ich zeigte ihm die alten Na-khi Handschriften und den ersten Band meines Wörterbuches, da sagte er – „ja was ich weiss wissen auch andere, aber das was sie [d.i. Sie] bringen ist alles neu, das weiss Niemand anderer“. Auf das hatte ich nicht so gedacht. Ein Sinologe und ein Sanskritist wollen meine Schüler werden um sie in der Na-khi Sprache zu unterrichten, aber ich habe leider keine Zeit für Unterrichten. Die Katalogisierung der Na-khi Handschriften macht Fortschritte, ich habe schon über 260 durchgearbeitet. Von denen habe ich 90 übersetzt. Da kommt noch vieles neue heraus, Götter und Dämonen etc., die nicht im Wörterbuch verzeichnet.

Der erste Band des Wörterbuchs ist schön geworden, er hat über 558 Seiten und 29 Tafeln, er ist schön gedruckt auf schönem Papier, auch die Tafeln sind schön. Der zweite Band ist viel dicker geworden, das macht aber nichts; es werden ungefähr 600 Seiten sein. Sie werden gebunden.

Die Arbeit hier wird in zwei Bänden herauskommen und mit schönen farbigen Tafeln die schon jetzt gemacht werden, es wird eine schöne Arbeit sein, wahrscheinlich meine letzte was Na-khi anbetrifft. Ich habe als Assistenten einen Dr. Janert[31] er ist Indologe, spricht English und kann sehr gut typpen. Er bleibt bis October. Er wohnt hier im Hotel alles auf Kosten der Deutschen Forschungsgemeinschaft und der Westdeutschen Bibliothek.

Leider hatte ich vor ein paar Tagen einen sehr bösen Anfall der trigemischen Neuralgie, er hat für eine Stunde angehalten und ich wurde nahezu verrückt. Es ist so furchtbar dass man sich leicht das Leben nehmen könnte. Es gibt hier gute Kliniken, eine grosse für Neurologie, überhaupt sind hier viele und alle möglichen Kliniken so ist man medizinisch gut versorgt. Gestern und heute gab es keinen solchen Anfall mehr.

Schreibe mir bitte bald wieder, vielleicht kannst Du hier auf Besuch kommen, später, wenn Du willst.
Nun grüsse ich Dich recht herzlich und verbleibe wie immer
Dein alter Onkel Rock

77

Marburg Samstag den 15ten September / 62

Mein lieber Robert:
Deinen willkommenen Brief vom 12ten habe ich heute (soeben) erhalten. Vielen Dank. Die ersten Eindrücke die man erfahrt [!] wenn man von einem Niveau des Lebens in ein anderes geht bleiben wohl. Ich kann mich nicht an alles das erinnern von dem Du schreibst. Ich kann Dich noch sehen wie Du den Bahnsteig entlang gingst in Venedig. Ich hätte Dich damals nahezu vermisst denn da ich mit dem Dolmetscher der American Express plauderte als der Zug schon angekommen war. Ja das waren damals noch andere Tage! 1933, nahezu 30 Jahre, Du warst damals 19 Jahre alt und ich 49, heute bin ich nahezu 79! Wie lange wird der Sojourn auf dieser Erde noch dauern? Anton ist heimgegangen, Frau Ciani ist noch da und ich das letzte Blatt am Baum, wann wird es fallen?

Wir waren aber noch einmal im Royal Danieli als wir am Wege nach Cortina d'Ampezzo waren, wir kamen spät Abends an und assen aufgeschnittenen Schinken, der war sehr gut. Robert, Du und ich haben viel erlebt. Meine Jugend war nicht erfreulich wenn ich da

---

31 Klaus Ludwig Janert (1922-1994), 1963 Professor für Indologie an der Universität Köln.

zurückdenke graut es mich. Wenn man mit 5 Jahren keine Mutter mehr hat ist das schlecht bestellt mit der Erziehung eines Kindes. Ich könnte Dir viel erzählen von dem Du nichts weisst.

Es freut mich dass Du gutes Wetter hattest. Hier war der ganze Sommer nicht schön nur ungefähr drei sonnige Tage. Der Herbst scheint schöner zu werden wenigstens hat er gut angefangen. Wie wir das erstemal zusammen in Venedig waren hat es geschneit, wir fuhren damals nach Zürich und Du sagtest mir Good bye am Bahnsteig denn Du fuhrst nach Wien und ich erwartete meinen Freund Ceresole, der ist auch schon lange tod [!].

Ja ich arbeite um zu vergessen. Alles ist nicht so schön jetzt in Honolulu wie ich höre. Mein guter Freund Paul [R. Weissich] schreibt mir „sie können froh sein dass sie nach Deutschland fuhren denn everything is very strange". Ich weiss nicht ob es sich ändern wird. Wir werden sehen. Jedenfalls bleibe ich dieses Jahr bis nach Weihnachten in Europa. Ich hoffe wir werden uns mehrmals ehen. Denn diesesmal ist es wirklich das letztemal dass ich in Europa bin.

Was machst Du in Venedig? und wo gehst Du von dort hin? Du schriebst Du würdest am 3ten September wieder in Wien sein. Ich selbst mache eine Pause und komme Ende dieses Monats nach Wien! Dann werden wir beim Krantz wieder essen. Es wird wohl nicht so schwer sein dann dort ein Zimmer zu bekommen.

Jetzt gibt es ein neues Hotel in Venedig habe ich gehört dass das Royal Danieli in den Schatten stellt. Erkundige Dich wie es heisst und wo es ist. Vielleicht gehen wir noch einmal zusammen nach Venedig „ein Abschluss einer Geschichte die vor 29 Jahren angefangen hatte".

Herzliche Grüsse von Deinem alten Onkel Rock

78

Marburg / Sept. 23/62

Mein lieber Robert:

Ich schrieb Dir nach Venedig hoffentlich hast Du meinen Brief erhalten.

Ich schriebe [!] Dir auch kürzlich dass ich sehr krank war und noch bin – mit meiner Trigeminal Neuralgie. Die Schmerzen sind fürchterlich, alle Drogen helfen nichts nur für eine Stunde dann sind sie wieder da. Ich bin ganz desperate geworden und habe mich entschlossen mich einer Operation zu unterziehen. Am Montag den 24ten fahre ich nach Giessen in eine chirurgische Nervenklinik um dort operiert zu werden, ich habe schon einmal eine solche gehabt.

Ich weiss nichts von dem Hospital oder wo es ist. Schreibe an meine Adresse im Kurhotel Ortenberg, Marburg/Lahn wo mich Dein Brief erwarten wird. Ich werde nicht lange in Giessen bleiben hoffentlich.

Mit herzlichen Grüssen Dein alter Onkel Rock

79

Marburg Sept. 27/62

Lieber Robert:

Danke für Deine Karte aus Venedig die mir nach Giessen in das [!] Neurologisch-chirurgische Klinik nachgeschickt wurde. Niemand kann wissen was ich an Schmerzen durchgemacht habe. Ich wurde am Montag sofort nach meiner Ankunft operiert. Der Trigeminus nerve wurde mit elektrischem Strom verbrannt oder verkocht. Da die Klinik eine alte war brachte man mich in dieses schöne Kurhotel zurück. Ich muss sagen alle waren sehr lieb zu mir. Der Nerven Spezialist Chirurg Prof. Pia war sehr nett, ein feiner Mann. Als ich ihn fragte wie viel ich ihm schuldete, sagte er: Es war gern gethan. Er kommt morgen von Giessen mich zu besuchen mit seiner Frau. Sie thun hier alles was sie thun können. Auch meine Honolulu Freunde haben mir von Honolulu nach Giessen telephoniert, das Gespräch hat über 150 Dollar gekostet. Ich bin sehr schwach aber die Schmerzen sind weg. Diese Methode des Nervenverbrennens soll für immer dauern und die Schmerzen werden nie zurückkommen. Ich bin zu schlapp zum Reisen. Ich muss mich erst erholen ehe ich reisen kann und allein kann ich nicht. Wenn ich nicht nach Wien kommen kann dann kannst Du vielleicht hierher kommen und wir gehen nach Wien zusammen. Aber es wird noch eine Zeit dauern.

So kommt halt manches was man nicht erwartet. Ich hatte vom Schmerz einen Nervenzusammenbruch.

Herzliche Grüße Dein alter Onkel Rock

80

Marburg, 1sten Oct. /62

Mein lieber Robert:
Deinen Brief vom 29ten eben erhalten. Ich wunderte mich wo Du warst da ich von Dir nichts hörte.

Ja es geht mir besser, keine Schmerzen aber die halbe Gesichtsseite fühlt sich taub an. Es gibt aber noch brennenden Schmerz und Zuckungen. Auch kann ich nicht richtig essen. Aber ich bin froh daß es vorüber ist. Ich schrieb in meinem Brief ob Du her kommen könntest Du erwähnst nichts davon. Warum?

Ja mein lieber Robert Du wirst immer Probleme haben wo das hinführen wird? Nun das ist Deine Sache. Wenn Du so ein freies Leben führen willst dann musst Du halt dass [!] in Kauf nehmen. Wenn man ein berühmter Schriftsteller ist und viel von seinen Büchern einnimmt dann kann man so leben. Robert Du bist alt genug und weist [!] was Du willst. Aber ich fürchte dass Dich diese Probleme verfolgen werden.

Nun ich sende Dir hier $ 50.oo dass [!] sind 1300 Schillinge also genug für den Zins und für ein Glas Wein und ein Wiener Schnitzel. Ob ich nach Wien kommen kann ist eine Frage. Es kommt drauf an wie ich mich fühle, ich bin noch schwach. Schreibe mir über Deine Pläne.
Herzliche Grüße Dein Onkel Rock

81

Marburg Oct. 5/62

Lieber Robert.
Sowie Dein Brief ankam in dem Du mir schriebst Du brauchtest Geld für den Zins habe ich mich hingesetzt und Dir $ 50.oo geschickt. Fünf Tage sind vergangen und keine Antwort. Ich glaubte doch Du könntest eine Karte schreiben und mir sagen daß Du das Geld erhalten hast und mich nicht sorgen lassen.
Herzliche Grüße
Dein Onkel Rock
Ich bin sehr erschöpft und nicht wohl. R.

82

Marburg/L. Oct.20/62

Mein lieber Robert.
Herzlichen Dank für Deinen lieben Brief und auch für die Franziskerln die mich hier erwarteten.

Das Wetter ist sehr schlecht hier, grau, kalt und traurig. Der Flug war gut und Dr. Voigt erwartete mich in Frankfurt und wir fuhren mit Taxi nach Marburg.

Hier kann ich nicht schlafen, mein Herz pocht und hüpft herum und wirft mich nahezu aus dem Bett. Das macht mich sehr nervöse. So habe ich mich entschlossen ende [!] nächster Woche nach San Francisco zu fliegen mit der Lufthansa direct mit nur einem kleinen Aufenthalt in Montreal. In Frisco bleibe ich über Nacht und fliege am folgenden Tag nach Honolulu zurück. Ich werde froh sein wieder in der Sonne sitzen zu können und die grünen Berge Blumen und alle meine Freunde wieder um mich zu haben.

Hier sind alle gut zu mir, sie wollen ich soll nächstes Jahr im Frühling wieder kommen. Wir werden sehen.

Es war schön ein paar Tage Wien zu geniessen. Ich hätte länger bleiben sollen.
Sei mir herzlich gegrüsst wie immer Dein Onkel Rock

83

Marburg/L. Oct. 26/62

Lieber Robert.
Vielen Dank für Deine Karte vom Opernhaus.
Ja ich fliege am 30ten von Frankfurt um 10.30 with [!] der Lufthansa direct nach S. Francisco und komme dort am selben Nachmittag um 5.15 p.m. an. Am 1st Nov. fliege ich nach Honolulu weiter.

Die Weltlage ist sehr kritisch, aber in Österreich wird nichts geschehen. Die Sachertorten sind angekommen vielleicht kannst Du von Zeit zu Zeit mir eine nach Honolulu schicken, ich werde Dir das Geld dazu schicken, vielleicht einen Apfelstrudelteig der ist jetzt zu haben. Dr. Hain, Corra ihr Schwager weiß wo man in [!] bekommen kann.

Nun wenn nichts passiert komme ich nächsten Frühling nochmals.
Sie wollen unbedingt dass ich wieder komme. Wir werden sehen.
Es geht mir von Tag zu Tag besser. Schreibe mir nach Honolulu. Wie

immer Dein Onkel Rock
Die Bücher bekommst Du sobald sie erscheinen, Du bist auf der Liste.

84

Hotel Frankfurter Hof
Frankfurt am Main

Oct. 30 / 62

Lieber Robert.
Deinen lieben letzten Brief habe ich noch gestern in Marburg erhalten. Es war alles sehr lieb von Dir mir so zu schreiben. Dr. Voigt kam mit mir nach Frankfurt, es war eine schöne Fahrt durch die herbstliche Landschaft, den traurigen Bäumen mit ihren gelben, roten aber fallenden Blättern. Es mahnte mich dass der Abend meines Lebens auch da ist. Ich glaube ich bin dieses Jahr sehr gealtert.

Ich sprach zu Dr. Voigt über den Tonband und es wird eine Kopie machen lassen. Aber noch besser ich werde mir in Honolulu einen Apparat kaufen und dort in aller Ruhe selber eine Aufnahme machen, dass [!] wird viel besser sein. In ein paar Stunden fliege ich ab.
Sei mir herzlich gegrüsst, Dein Onkel Rock

85

Honolulu Oct [d.i.Nov.!] 4 / 62

Mein lieber Robert:
Jetzt bin ich wieder in Honolulu, aber was für ein schrecklicher Flug von Frankfurt nach San Francisco.

Wir flogen zuerst nach London da war es warm, alles war grün und die Sonne schien herrlich. Dann hiess es dass wegen Gegenwind und der tanks die nicht soviel *fuel* aufnehmen können in Island gelandet werden muss und zwar auf Keflavik; ich war einmal dort zu Gast bei einem amerik. General vor 6 Jahren. Ein eisiger Wind wehte in diesem öden Lande. Wir flogen dann über Grönland nach Montreal in Canada dort herrschte eine sibirische Kälte, wir landeten in einem Schneesturm. Wir hatten 3 Stunden Verspätung.

Wir flogen um 12.15 p.m. Frankfurter Zeit ab und landeten in San Francisco (Frankfurter Zeit) 6.30 a.m. den nächsten Tag. Ich war 25 Stunden ohne ein bisschen Schlaf; das ist für junge Leute anstrengend.

In San Francisco war schönes Wetter, alles war grün und war

warm 77°F. Ich blieb 1 1/2 Tage um mich auszuruhen und flog dann nach Honolulu weiter, 4 Stunden, 2300 Meilen. Ich war froh wieder in Honolulu zu sein es ist so schön warm. Meine Freunde waren in San Francisco und wir waren zusammen, sie *motored* nach Carmel und ich sollte mit, wollte aber nicht denn ich habe viel zu tun und auch wollte ich langen Motorfahrten ausweichen.
Gesundheitlich geht es mir gut, hoffe Du bist gesund.
Mit herzlichen Grüssen Dein Onkel Rock

86

Honolulu Dec. 2nd / 62

Lieber Robert:
Deine Karte vom 26ten Nov. came [!] vor zwei Tagen und ich danke Dir für dieselbe und die Grüsse die sie brachte. Ich würde gerne jetzt in Wien sein trotz der Kälte. Das Lied: *Tauet Himmel den Gerechten* hatte ich längst vergessen, ich kannte es jedoch sehr gut.

Hawaii ist nicht weit jetzt doch denjenigen die nie da waren scheint es weit entfernt zu sein. Aber wenn eine Gedankenbrücke besteht, ist man gleich da.

Von Marburg schrieb man mir dass dort alles verschneit ist und man mit Pelzschuhen gehen muss. Hier ist Sommerwetter und doch beklagen sich die hiesigen Leute über die Kälte +15° C. und in der Stadt +20–22°C. Es regnet viel und ein sehr starker Wind weht von N.O.

Eben bin ich fertig meine Weihnachtskarten über alle Welt zu senden, nahezu 100 Karten. Meiner Ansicht nach ist es eine Geldverschwenderei.

Unter separater Post (Packetpost Luftpost) sende ich Dir ein paar schwarze Schuhe und die seidenen Pyjamas die wir zusammen damals kauften. Ich habe sie nur ein oder zweimal getragen. Bei mitte [!] Dezember schicke ich Dir ein paar Dollar für ein Weihnachts-NeuJahr Festessen. Wie steht es sonst mit Dir? Wie geht es der Frau Ciani? Hans ist krank schon seit ein paar Wochen nicht mehr im Geschäft. Ich habe erst jetzt von ihm gehört. Suche ihn auf, er ist doch Dein Bruder!
Neues weiss ich nichts. Die Vögel singen, viele der Bäume blühen und alles ist schön grün, gar nicht wie Weihnachten.
Herzliche Grüsse von Deinem Onkel Rock

87

Vienna January 31, 1958

My dear Uncle:
thank you very much for your Aloha-Christmas-present which arrived yesterday; it gave much fun to taste all these different Hawaiian flavours which here are unknown.

I gave one to Anton-Onkel and he was very much delighted. the Red-Pepper Jelly I gave to Hans – he likes hot things and he is in a bad mood again, no one knows why. Perhaps his divorce does not work, yet he does not speak about. Now he is on leave for a week or so, I don't know, he borrowed my suitcase and left.

Well, I am glad to hear you had a good time at Kona with your friends. I am awfully sorry that I have troubled you by asking for bishops; never mind, forget it. There is after all no such thing as security in life; „it is in giving that we receive –" St. Francis of Assisi says.

Winter is very cold here. I bought myself a new wintercoat and feel fine. Well, there is nothing interesting to tell for this time. I am glad you will really come this year and am happily looking forward.

With deep affection and Aloha yours

88

Vienna, April 17, 58

My dear Uncle:
thank you so much for your dear letter and the cheque of $ 10.– about which I sent you immediately a card. I was touched to see that you thought on my birthday giving such a nice present and all good wishes. I am getting on very well in my job and so have hopes to continue.

You are very right what you did say about Hans. I have about the same experience. I sent him a small present for his birthday but never received an answer nor did he thank nor reply. I do not understand that. I am paying my rent for the apartment and do not depend on his charity what I, as a fact, never did, for the only help was, that he paid the rent for the apartment before when I had no job what he had to do anyhow living there; even though I often paid but not regularly, yet he liked to put it like he would support me, what he never did.

Well, I am very glad to have this job at the NCWC and will carry on. (of course I have to pay back debts I made before I got this job and so are burdened with that but am glad to be able to do anyhow). Therefore your dear gift to my birthday was a great help to me and I am very grateful dear Uncle.

I hope you find my English knowledge has improved and I am making no mistakes. It is very important in my job now to know good English and to be able to express myself in all ways and matters. I again have to thank you for this, you have troubled a lot to correct me while being with you – I rather say you were troubled correcting me, or by correcting me.

Waiting for news from you I hope you are enjoying a vacation (still taking some work with you what you should not) in the hills of Hawaii and that you will recover
Aloha nui loa herzlich Dein

89

Wien, 27. Juli [1958]

Lieber guter Onkel:
herzlich Danke [!] ich Dir für Deinen lieben Brief und die fürsorgliche Unterrichtung mit Hans seinem Brief. Ich hatte keine Ahnung, denn wie Du ja richtig vermutest, er sagt mir darüber kein Wort. Well I do my best to keep my job. I am always the very first in the office in the morning and never stayed at home a day when I am sick, and I had a bad bronchitis in Winter and later on a *Bauch-Grippe* and had to take Entero-vioform ... so I really take it seriously, and therefore I have to ask you for a good word from an American bishop because that would save my place and job and help immensely, though you said you cannot do anything in this matter, that is a pity, for one never can tell how things go – and a good recommendation is most valuable.

As long as I have this job I have no fear to hold the Wohnung in the Wittelsbacherstrasse and I enjoy it indeed to be alone and in peace ... every morning while having breakfast I am reading loud English poems and prayers celebrating and practising, starting so my day in good spirit and mind. And I make a bow in front of the Buddha and touch my sword before I am leaving.

The other day I tried again to make a poem in English: (twenty

years ago I tried in Hong Kong)
Dark ancient silk
from forbidden-town Lha-sa
walls up my sole camp
with the image of Buddha
his hand blessing the ground;
cross-barred by the sword
hanging overgrown
with the skin of brown leather
wherein glooms the iron
since Kublai Khans [!] historic days,
the smoky jurts of his Hordes
in solitude valleys,
the Dragon-Mountains white sleep.

I hope you like it; I shall be very grateful by correcting me.
I feel there will be no war and I am very certain you will come to Vienna and am gladly looking forward!
With all my best wishes and much love
Alles Liebe

Ich bat Dich im letzten Brief um Reviews über Dein Buch vom Amnye Ma-chhen? Du erwähntest es gar nicht, bitte sende mir eine oder die andere Buchbesprechung, ich bin sehr neugierig und würde mich sehr freuen!

90

Vienna August 1st 1958

My dear Uncle:
returning from Graz, Steiermark, I found your dear letter, and I do understand your feelings, your being fed up with even paradise conditions and with a labouring work of gigantic volumen – yes ELIEN! you need a change, a real good soft Wienerschnitzel mit Kartoffelschmarrn and Salzburgernockerl! Round up your Dictionary and bring it to an end, pack and come! In the meantime I will have finished my visits to the different offices of the voluntary agencies. I just have been in Graz and in a week or so start through then other parts of the country but have to be back with the first of September and hope you will be on the way to come by that, and stay for a really long time enjoying all what Vienna has to offer.

You always have been staying too short a time when you have been in Vienna one must take time for everything and make it gemütlich! You never have been a „trockener Wissenschaftler“ you were full of life and vitality every time! And you are right to enjoy and embrace life with all their treasures and values not so much these in the museums but – as you said – which Sacher and Kranz-Ambassador has to offer! When you come it will be just the time for a good ausgezogenen Apfelstrudel und auch noch Zwetschkenknödel. The season starts with the best operas and the famous Vienna Philharmoniker-Konzerts, the best in the world there is no doubt. The heat will be over by then I think, now it is just tropical and I am suffering a lot? it is terribly hot. I am hoping for a cool Autumn and am happily looking forward to your coming1 I am glad you feel well and active, and I will do my best to manage you a comfortable place in a good hotel, but it will be very difficult; do you know a way? Vienna is just overcrowded with tourists until end of September, but I can get you a good room with bath in the Hotel de France am Schottenring, I know a Hungarian waiter there who promised to manage that. So please be kind enough and tell me in time the date of your arrival long ahead to be able to arrange things.
Well, I hurry to post this letter; it is about ten in the night and I go to the Hauptpost to mail it right now.
I wonder how your plans will be –
much love and all the best
Yours, Robert

91

Vienna October 25th 1958

My dear Uncle:
this morning I received your dear letter from Oct. 19, and I am very glad to learn that your silence was due to your stay in Hilo and you are in the best health mountaineering and hunting plants! So I do not need to worry about your health indeed. Well I understand that botanizing is a good change after doing etymology, and your good old system is in the best way balanced.

I am sorry that Hans has not yet written to you; as a fact he is now ill, has a kind of flue, I told you in my last letter, but of course there was all the time before he could have sent you a letter and his gratitude for the excellent suit and all the stamps you sent to him. I have the impression his mind is somehow blocked and he cannot overcome it, and after all, he has a heavy hand for writing and has difficulties to express his thoughts. When he still lived at Wittelsbacherstrasse I had plenty of opportunity to watch him: and I saw, that it took him often a fortnight to compose a letter to you, dear Uncle. The longer time – I rather say: the more time passed, the more grew his helplessness to find excuses and words for, and especially hard it is for him to thank, so strange that is – but I watched that, herein he has great difficulties. And you are right, dear Uncle, to say he is proud and arrogant, but he clings to it in his weakness, his over-repression of primitive instincts.

Thank you for continuing sending, enclosing stamps, I have a grown up nephew who is collecting and very happy if he can get such very rare ones as you always giving us. This nephew Heini (a son of Lisi, the latter perhaps you will remember) collects especially stamps from Germany and is very keen for stamps from Israel, and I sometimes have occasion to help him by saving a stamp for him, as you do for Hans.

Today I passed by at the renewed Hotel Imperial, near the American Express Co., and asked about the price of a room with bath, single: 250.-- Schilling without breakfast, and they can make reservations ordered 8 days ahead. You see, dear Uncle, the prices have gone up here; the Dollar is today 25.65 S, but the *Imperial* is the best Hotel we have in Vienna now. In the Grand Hotel have moved in the Atomic Commission, Sacher is full most of the time and the Kranz-Ambassador will not be much cheaper I am sure.

I would like to come to Rome with you, this grandest town of all historic towns, yet I only may get not more than about 8 days vacations from my office, and that, I am afraid you will not find worth while for such short a time with all the travel expenses. But in Vienna, there of course, I am, after Office time, always to your disposition, dear uncle, any may certainly be able to help you finishing your Dictionary and starting your *Three Principalities of Western China* !!!

Happily looking ahead to your coming
bleibe ich mit lieben Wünschen Dein Robert

92

Istituto Italiano per il Medio ed Estremo Oriente
Via Merulana 248
Il Presidente

Rome 22 Gen. 1964

Dr. Robert J. Koc
Wittelsbacherstrasse 4
Vienna-2, Austria

Dear Dr. Koc,
I have returned home only a short while ago, from an absence of several months spent on our archeological sites in Asia, and I have here found your card of August 25, from Lerici, which has unfortunately gone without an answer, having reached here after my departure in July last. I regret this occurrence all the more as you have been unable to come to Rome in the course of the summer, and to meet my collaborators, especially Prof. Gargano who is in charge of publications.

I assume you know that the Second volume of the Na-khi Dictionary is now at the stage of the second proofs; the first proofs of the book had been revised by the regretted Prof. Rock himself, but his last very severe illness has prevented him from carrying the work much further, and there seems to be a great number of corrections and alterations to be carried out. The proofs are now in the hands of Mrs. Lester Marks, who is earnestly devoted to doing the best of this heavy task. To our regret, this involves much delay in publication, and high added costs, for the re-making of several sections. Although I feel hesitant about it, I would like to ask you wether you think that for your part, or in association with other heirs of the regretted Professor, you might in some way contribute to these heavy and unexpected outlays, so as to facilitate publication of his book. I assure you that we are doing all we can to meet the circumstances, and to speed up work as far as possible.
With my best regards, and wishing you a happy New Year,
Sincerely, Giuseppe Tucci

93

Robert J. Koc
Serra di Lerici, La Spezia

June 12th 1967

Excellenz Dr. Giuseppe Tucci
Roma, Piazza Vescovia 1

Dear Sir,
may I kindly aske [!] you about the forthcoming of my uncles [!], DDr. J. F. Rock's Na-khi Encyclopedic Dictionary about which I have no news after I received the first volume?

I remember a letter of yours two years or so back, that you have trouble to carry on, although you had received money (or the value in Nakhi original manuscripts you sold in Marburg) from my uncle of 12.000.-- U.S. Dollars which was expected to cover the whole expenses.

I suppose Mr. Paul R. Weissich has as executor helped with some funds to that effect.

Would you please be kind enough and let me know to the about address, where I stay until October, about the matter. I feel my duty to care as principal heir and nephew of the late DDr. Joseph Rock.
With my best wishes,
Yours very sincerely
Robert J. Koc

94

Istituto Italiano per il Medio ed Estremo Oriente
Il Presidente

Rome, 27 Giu. 1967

Mr. Robert C. [!] Koc
Serra di Lerici, La Spezia

Dear Mr. Koc,
I have received your letter of June 12, to which I have not replied earlier, owing to my absence from Rome, and the fact that I had to take some time, so as to give you a complete view of the situation, regarding the publication of your regretted Uncle's work, the Na-khi Dictionary, part II.

As we have repeatedly advised you, the work proceeds slowly at the printers', due to many difficulties, as after the demise of the very regretted Prof. Rock, the main trouble lies with the revision of the proofs, an extremely heavy and complex task by itself. Such a task had been entrusted, as you know, to Mrs. Lester Marks, who for several reasons, has kept them for a very long time. When the proofs have finally reached us, fresh and tiresome checkings were still to be made and after new corrected proofs were made, it was clear that they required further careful revision. We now hope to return them to the printers, for the definitive printing order. Doubts and uncertainties are still cropping up, although we have received a visit from Mrs. Lester Marks herself, some months ago, that has enabled us to talk about this matter more at length. We have no specialists available, well-versed in the field of studies pursued by Prof. Rock, and it ensues that the revision is proceeding rather slowly. However, we hope to bring out the volume for the end of this year.

You therefore need not worry about your late Uncle's work, as it is in our own interests to have it published, both from a scientific point of view, as for a consideration of costs. In this connection, I would now like to settle straight the matter of contributions which stands in the following way: Prof. Rock had contributed One Thousand Dollars as a personal gift, in addition to One Thousand Dollars remitted us through the Far Eastern Institute of the University of Washington, D.C., and he had then donated us a Series of his Mss., with the authorization of having them purchased by the Westdeutsche Bibliothek of Marburg/Lahn, to the purpose of emplyoing the proceeds of the transaction towards publication of his Dictionary. As matters have turned out, the total of such contributions has not proved sufficient to cover publication costs, which have unfortunately run to a much higher amount. However, I have to make it clear that Mr. Paul R. Weissich, Executor of Prof. Rock's Estate, with whom we had exchanged some correspondence as far back as 1963, has not contributed any funds to this enterprise, in no way whatever.

As soon as the volume will be available, we shall have pleasure in sending you a copy of it, and in the meantime I beg you to accept the assurance of my consideration with all best wishes.

Yours truly,
Giuseppe Tucci

95

Robert J. Koc
Serra di Lerici, La Spezia

June 30th 1967

To Il Presidente
Mr. Giuseppe Tucci
Instituto [!] Italiano per il
Medio ed Estremo Oriente
Via Merulana 248
Roma

Dear Sir:
I am very glad about your full letter and thank you for giving me a picture as things are so far, it is very kind of you indeed. I can imagine that the costs are immense and the troubles great and I think Mrs. Lester Marks could help you a little in the matter.

As about this Mr. Paul R. Weissich[32], executor of the last will of my dear uncle, that is a chapter by its own and a very dark business he did. But to your knowledge, which may help you, dear Sir, it was Mrs. Marks who wrote me in her first letter after my dear uncle died, that they created a fund, a Dr. J. F. Rock fund (if that is correct, I have her letter not with me), and I expected, that money will have been sent to you, to the Instituto [!] Medio ed Estremo Oriente that is one point of view.

Another is, that the Deutsche Morgenländische Gesellschaft, Marburg/Lahn, West Germany inherited with item VII, of my uncles [!] last will: „All notes and manuscripts pertaining to geography, history and ethnology of China and Tibet, as well as such photographs of my collection which will illustrate Historical Geographies of the former Kingdom of Mu-li and of the Principality of Cho-ni in west China; also all my manuscript maps of West China.“ That did not include the original Na-khi manuscripts my uncle still had and which was coloured and especial [!] valuable, yet Mr. Weissich gave them too to Marburg and said there were only seventeen. Another thing is that the Deutsche Morgenländische Gesellschaft bought a lot of Mss of Na-khi origin just before from my uncle, though I do not know if they ever have sent it before he died,

---

32 Paul R. Weissich (1925-), Direktor des Foster Garden in Honolulu, Rocks Testamentsvollstrecker.

his bank account does not show it, and Mr. Weissich as executor in power so afterwards may have received it but – used otherwise, he never told me so. That is the other point of view, and so the Deutsche Morgenländische Gesellschaft would have all reasons to help you, dear Sir, to cover the costs of printing the II volume of the Na-khi Dictionary. The sum I miss is nine thousand U.S. Dollars. I myself received only a very few thousand due to the strange manipulations of Mr. Weissich, who for instance kept extreamly [!] long the objects of art till he urged several time sent it: insured it for 50.000.00 U.S. Dollars, which gave me enormous troubles in London and expenses as dock charge and I received not more than something about 2,000.-- (two thousand!) U.S. Dollar on the best market for asiatic [!] art in London! Is that not very strange? What happened with these object of arts [!] in the meantime, do you think my dear uncle collected worthless art? Certainly not. I thought Mr. Weissich is doing that out of idealistic reasons and will help to get the works of my dear uncle published and so, and I wouldn't mind, if the money of my dear uncle is used in such a way, but I am afraid, Mr. Weissich did not so.

But never mind about my own troubles, what I am going to say is, that the Deutsche Morgenländische Gesellschaft has all interest that the II. volume of the Na-khi Dictionary will come forth and shall give money for it! The [!] got the lot of it from my dear uncle, and have now most of the Na-khi Mss, and need therefore especially this work of my dear uncle, so I think it would be adviceable [!] to try. But there was some misunderstanding between me and director Dr. Voight [!], who liked to buy the object of arts [!] from me, which of course I was interested to do and told him to get in touch with Mr. Weissich, for I believed it can be bought directly from Hawaii place and that the objects of art need not going round first to Vienna with all the expenses of custom etc. – but see that Mr. Weissich did some monkey business, for I wrote later about three times to Dr. Voight in Marburg but he never replied. So he may have handled it out with Mr. Weissich, and worthless things were sent to me to London, where a friend, a British Ambassador, sold it for me by Christies [!]. You can imagine my troubles, dear Sir, through about five year [!], so long Mr. Weissich draged [!] things out! And I am the principal heir by my dear uncles [!] last will, getting almost nothing.

So to be successful at Marburg, you better do not mention my name please, not that I mean, but in your interest, in the interest of my dear uncles [!] work.

Please forgive me that I mentioned my own troubles, but all is mixed up, and I think you, as a dear friend of my dear uncle, should have a picture about the things as they are.

I am glad that the II. volume of the Na-khi Dictionary shall be published this year and am happily looking forward to receive it, for which I do thank you by this.

I stay here untill [!] the beginning of October; in Vienna my address is: 1020 Vienna, Wittelsbacherstrasse 4/19 Austria.

I remain with all good wishes
yours very sincerely
Robert J. Koc

BRIEFWECHSEL MIT C. S. SARGENT

Rock an Sargent

1.	15. 7. 1924
2.	16. 7. 1924
3.	19. 7. 1924
4.	24. 7. 1924
5.	1. 8. 1924
6.	9. 8. 1924
7.	10. 8. 1924
8.	19. 8. 1924
9.	22. 8. 1924
10.	28. 8. 1924
11.	1. 9. 1924
12.	9. 9. 1924
13.	30. 9. 1924
14.	4. 10. 1924
15.	o. D.
16.	5. 11. 1924

Sargent an Rock

1.	18. 7. 1924
2.	22. 7. 1924
3.	20. 8. 1924
4.	25. 8. 1924
5.	14. 10. 1924
6.	11. 12. 1924
7.	4. 4. 1925
8.	6. 5. 1925
9.	18. 8. 1925
10.	18. 8. 1925 an Endicott
11.	28. 8. 1925
12.	28. 8. 1925 an Endicott
13.	15. 9. 1925
14.	8. 10. 1925
15.	29. 10. 1925
16.	9. 11. 1925
17.	18. 11. 1925
18.	25. 11. 1925
19.	31. 12. 1925
20.	8. 1. 1926
21.	2. 2. 1926
22.	1. 3. 1926
23.	22. 3. 1926
24.	26. 3. 1926
25.	27. 4. 1926
26.	10. 5. 1926
27.	7. 5. 1926
28.	7. 5. 1926 Kellerman an Sargent
29.	11. 5. 1926
30.	18. 5. 1926
31.	21. 7. 1926
32.	10. 8. 1926
33.	9. 9. 1926
34.	14. 9. 1926
35.	27. 9. 1926
36.	27. 10. 1926
37.	13. 11. 1926
38.	15. 11. 1926
39.	18. 11. 1926
40.	23. 11. 1926
41.	2. 12. 1926
42.	16. 12. 1926
43.	12. 1. 1927
44.	21. 1. 1927
45.	3. 2. 1927

Rock an Sargent

1

COSMOS CLUB
WASHINGTON, D.. C.

July 15, 1924

Prof. C. S. Sargent
Arnold Arboretum

My dear Prof. Sargent,
Your very kind letter and package of plants arrived safely. Allow me to thank you most sincerely for this gift for Hawaii. I can assure you that it will be appreciated as much as I appreciate it. If there is anything that I can do for you especially, while in the Orient, please let me know. My San Francisco address till the 26th of this month is Chancellor Hotel. After that in care of Dr. T. Burkill, Director Botanical Garden, Singapore, will reach me.
Thanking you again for the valuable gift and the courtesies extended to me at the Arboretum, I remain
Very faithfully yours,
J. F. Rock

Please give my best regards to Dr. Rehder. [1]J. F. R.

2

ST. BOTOLPH CLUB
4 NEWBURY STREET, BOSTON

July 16, 1924

Prof. Charles Sprague Sargent
Director Arnold Arboretum
Jamaica Plain

---

1 Alfred Rehder. 1863-1949, Botaniker am Arnold Arboretum. Vgl. Clarence E. Kobuski: Alfred Rehder. *Journal of the Arnold Arboretum* 31.1950, 1-38.

Dear Professor Sargent,
In compliance with your request I am writing you today regarding the proposed expedition into Asia. I have carefully considered, in the light of my past experiences in China all the necessary equipment for such an expedition especially into a region about which we know next to nothing and have come to the conclusion that the initial expense will naturally be heavier than for the successive years in the field. I would propose engaging again my seed & plant collectors who are thoroughly familiar with the drying of plants, packing & collecting of seeds etc. They have a fair knowledge of Tibetan, speak Chinese and Nashi. My cook will also be available, I found him a very valuable adjunct as he is half Tibetan and speaks Tibetan fluently, also Nashi and Mandarin, this makes it unnecessary employing a Tibetan interpreter. I could arrange to have these men meet me in Yünnanfu which town can be reached by train from Haiphong.

The equipment which would have to be bought here would consist of two good large strong tents, rifle and colt automatics for protection against wild animals. Photographic supplies, camera, lenses, films, developers, developing tent etc. I would propose buying light, wind proof airmen's coats lined with fleece for the high windswept regions north of the Kokonor, both for self and my men.

I would advise that the proposed expedition buy its own transport animals either yaks or camels as necessity requires. All other material can be purchased in Shanghai. The amount necessary for the first year including cost of equipment and salary of the leader would be about 14000 Dollars, the remaining years 12000 Dollars per annum. I would ask for salary the sum of 500 Dollars per month out of which I would buy my subsistence in the actual field but not while in transit to China or in Shanghai or Peking where Hotels will have to be frequented.

I propose to explore the Amne Machin range, the Mountains northwest of Hsining fu which place would serve as the basis of the expedition until such a time when it will be advisable to push on into Chinese Turkestan, the Altai Mountains etc. If it is desired I shall be willing to push on into Persia. The period required to carry out the proposed journeys and carry on the work satisfactorily would at least be three to four years.

I sincerely hope that I shall have the priviledge of leading such an expedition for the Arnold Arboretum into these unknown regions of Asia. I will do my utmost to bring such an expedition to a successful completion. Hoping sincerely that you will be able to soon inform me that I can start out next October, I remain very sincerely yours
Joseph F. Rock

up to 26th of this month
Samoset Hotel Rockland Maine
after that National Geographic Society, Washington

3

THE SAMOSET
ROCKLAND MAINE

July 19, 1924

Prof. Ch. S. Sargent
Arnold Arboretum
Jamaica Plain, Mass.

Dear Prof. Sargent,
In response to your kind letter of June 18th I sent you a night letter which is self-explanatory. I just want to make sure that there is no further misunderstanding. I am sorry that I did not express myself more clearly. The salary of the leader which will be myself is included in the 14,000 of the first year and $ 12,000 the second. It may be very likely that the second year's expense may be only $ 10,000, and if so I will naturally carry that amount or any left over to the third year. I am not familiar with the cost of things in Kansu and beyond. In Yunnan it was a different proposition. The Yunnan currency was lower than Shanghai currency and therefore things were correspondingly cheaper. For example, I received the last year 115.00 dollars silver Yunnan for 100 Shanghai, but 153 paper in Yunnan for 100 Shanghai. But

paper is useless in the interior. The same will probably hold good in Kansu. I shall keep the cost as low as possible.

If you wish me to explore the Kansu range the first year I shall be glad to follow your wishes, much will depend on the condition of the country, and that will have to be decided on the spot. Whenever possible I shall follow your instructions.

Regarding the return journey from Kansu little can be said now and what you suggest may be the most feasible, but conditions change constantly in an unsettled country, like China.

I shall be stopping over in Boston on my way to Washington, which will be about the end of this month or possibly earlier. I hope that by that time you will be able to tell me whether our plans can be carried out or not. I hope sincerely that we will be able to arrange this expedition, as I am very keen to get back into the field. I love the work and shall be happy to do my best, for I am exceedingly enthusiastic in the carrying out of such work. The more difficult it is the better I like it.

If you wish me to come to Boston sooner I shall be glad to come at any time, just drop me a postal card in care of this hotel, and I will come at once.

With kindest regards and thanking you for your letter, I am, faithfully yours,
Joseph F. Rock

4

THE SAMOSET
ROCKLAND MAINE

July 24, 1924

Prof. Ch. S. Sargent
Director Arnold Arboretum
Jamaica Plain, Mass.

Dear Professor Sargent,
Your kind letter of July 22 just at hand. I am leaving Rockland for Boston by the Monday night boat on the 28th of this month which will get me in Boston on the 29th on Tuesday. I shall spend

a few days in Boston, say up to the morning of the second of August, to allow plenty of time to discuss matters.

I hope we will be able to come to some definite arrangements by that time. I would naturally like to know more or less definitely, as I have a good deal to do as yet which I shall have to finish ere leaving. It will also be necessary to make definite bookings as boats across the Pacific are very crowded.

With kindest regards and thanking you for your kind letter, I remain, faithfully yours,

J. F. Rock

5

HOTEL MCALPIN
NEW YORK

August 1, 1924

Prof. Ch. S. Sargent
Director Arnold Arboretum
Jamaica Plain

Dear Prof. Sargent,

I am leaving today or tomorrow for Washington, D. C., after a successful stay in New York. I booked on the *President Wilson* sailing from San Francisco September 30th. The next boat leaves October 15th which I thought perhaps a bit too late.

I also went to Abercrombie & Fitch, the greatest sporting goods store in New York and looked over tents and other camping outfits. It appears that the tents will have to be made to order, as all their stock-tents are too light a material. As my stay is such a short one I have wired you asking you to kindly send definite answer to care National Geographic.

With kindest regards,
Faithfully yours,
J. F. Rock

6

Washington, D.C.
Aug. 9th, 1924

Prof. Ch. S. Sargent
Arnold Arboretum
Jamaica Plain, Mass.

Dear Professor Sargent,
Your letter of August 6th is at hand.
In regard to the expenditures to be incurred for the next fiscal year, and the allowance for the same as stipulated „Fourteen thousand dollars“, I would like to make the following statement.

In my previous dealings, as with the Government, the whole sum allotted was given to me before leaving this country. I became automatically temporary special disbursing agent for the Government.

I would rather that the whole sum for the first year be turned over to me and I shall render accounts for expenditures etc. through subvouchers. As distributing agent I was authorized to pay my salary every month from the fund allotted.

The only way possible to carry out the work properly is to turn over to me the sum for the first year. If you require a bond I can furnish one. This sending of moneys to China is not a good way of arranging matters. When in Peking I shall have to make arrangements with the postal authorities to buy my checks at stated intervals on Shanghai, the amount being paid in Kansu currency. If I have not the proper amount ready at hand it will be impossible to make such an arrangement. I must be able to satisfy the postal authorities that I have funds to say 5000 or 6000 dollars on hand in a bank before I can make such an arrangement.

I can only say from previous experience that the only way to arrange this without getting into trouble or handicaps is to advance me the whole sum, I shall act as disbursing agent and will be responsible for the money as far as possible, if you require a bond I will furnish one but if you think it cannot be advanced then I do not know how we can arrange this.

I am carrying at the present three accounts, one in Rangoon for Yunnan payments, and two in Shanghai, one in American dollars and one in Shanghai dollars. This is to avoid having all money converted into Shanghai dollars. The latter fluctuates considerably and hence would be a loss should all the money be converted into Shanghai dollars. I shift from one account to the other as far as I require Chinese currency.

To avoid mixups I would suggest that the whole sum be turned over to me in the form of draft or cheque on New York or Washington in my favor, with a statement for what purpose the sum is to be used, such as expenditures, salaries, etc. This I will sign, and at the end of six months, or at the end of the year will render you an account or an itemized statement as to expenditures.

This is the best way and has proved satisfactory to all concerned. It relieves you of the trouble to send part of it here and part of it there and it relieves me of worrying whether it has been sent to the proper place and if on receipt in China or Rangoon it was handled properly. Bankers in China exchange foreign currency first into taels and then into Shanghai dollars, so as to get a double exchange charge.

So please if you will allow me, to make this the least complicated I would ask you to mail me a cheque or draft on Washington or New York for the whole amount, say 14000 dollars. This is the best way of doing it, and practically the only way I would care to do it. This is for the safeguarding of the work. I shall be responsible for all expenditures and will furnish receipt for all items for which receipts can be obtained. I will give you at the close of the year an account for expenditures as I have done for the National Geographic.

I have wired to China re my assistants and have sent 300 rupees to Likiang, and 100 dollars to the American Consul in Yunnanfu. I have also sent a cable to Hongkong to my boy, an English-speaking assistant who has travelled with me for 2 1/2 years. He is in Hongkong and I asked him to await my arrival in October. I have booked for September 30th on the *President Wilson*. On my way back to Washington I stopped in New York and made arrangements as to tents and other necessary camping outfit. I have received yesterday from Calcutta detailed maps of the region I am to explore. These I had ordered from Peking last May by letter. I shall bring these maps with me to Boston.

My connection with the National Geographic ends on the 15th of this month. I am not connected with the Department of Agriculture save in an honorary way. I would like to inquire what my title will be and when I will be considered as being in the employ of the Arnold Arboretum.

There is no hurry about sending the draft or check but I would like to have it sometime before leaving for San Francisco.

Hoping this to be satisfactory, I remain

Faithfully yours,

J. F. Rock

* If attacked by brigands and robbed of cash carried on caravan will you hold me responsible?

7

HOTEL SCHUYLER
NEW YORK

Aug. 10, 1924

Prof. Ch. S. Sargent
Arnold Arboretum
Jamaica Plain, Mass.

Dear Prof. Sargent,

I have come to New York today to make the necessary purchases which I will order to be sent in your care Arnold Arboretum. I will repack them on arrival in Boston into army trunks which I shall bring with me.

I thought it wise not to delay as some things, like tents, will have to be made to order, and the time is getting short. I will return to Washington probably on Tuesday when I hope to have finished with all my purchases. I shall temporarily advance the

money for these purchases for which I will submit bills in my accounts.
With kindest regards,
Very truly yours,
Joseph F. Rock

8

NATIONAL GEOGRAPHIC SOCIETY
WASHINGTON, D. C.

August 19, 1924

Prof. Charles S. Sargent
Director, Arnold Arboretum
Jamaica Plain, Mass.

Dear Prof. Sargent,
I have just returned from New York where I had gone to make the necessary purchases for the coming Arnold Arboretum Expedition. I thought it wise not to wait too long as some of the articles, like tents, had to be made to order. The tents in stock were all of light duck and not strong enough for the remote regions where I expect to work. I ordered two small travelling tents for daily use while on the march and one larger tent while camping for the season in the Amne Machin and Richthofen range, and one tent for my assistents and equipment. I bought all such other articles here – that is, in New York – as cannot be procured in the Orient. I also ordered all my photographic supplies, as camera, etc. directly from Eastman.

My personal equipment chargeable to myself I bought at the Army stores in Washington. A number of things I can better buy in Shanghai and have therefore not bought them here in the U. S.

I conferred with the Manager of the Far East Division of the American Express Co. in New York regarding the transportation of equipment directly to Shanghai and Hongkong. It will be much cheaper to send it through the Express Co. than by taking it along personally and paying the excess baggage rate. On the steamer the express charges and freight charges are practically the same. The former way (express) is more convenient and I will not have to run about to handle the unloading, etc., of the trunks which otherwise would have to be done by Cooke & Son who are expensive. I have bought cheap but yet strong, serviceable army trunks I am sending to New York to Abercrombie & Fitch. Most of the equipment will be packed in these trunks, which will save buying wooden boxes and paying packing charges, besides they can easily be opened for customs inspection. I will personally supervise the packing in New York the first or second of September, the Express Company will then take charge and deliver them to me both in Shanghai and Hongkong; that is, I have divided the material which is to go to Shanghai and Lanchowfu and which is to go to Hongkong. This will not necessitate bringing all to Boston. I shall bring 350 lbs. baggage such as I am allowed free on train and ships as per transpacific ticket.

I have written Mr. Davis, the American Consul in Shanghai, re permits for landing a gun and Colt automatic, as well as regards courtesies from the Chinese Customs authorities to facilitate the quick landing of the material.

I have also written to Mr. Myers, the American Consul in Yunnanfu, to get certain permits there and to ask the French Consul, Mr. Bodard[2], to inform the Customs authorities at Hongkong to give me every facility and to extend to me such courtesies as they did when I passed through Indo-China last time.

I have also written the Standard Oil Company, Mr. Kirkman in charge, to be good enough and again help me with my baggage through the customs as he did last time without a hitch. There is neither an American Consul nor an American Express Co.'s agency in Hongkong or Hanoi.

So you can see, Prof. Sargent, I have taken all the necessary steps to assure a good beginning. I have wired to Likiang through Mr. Myers the American Consul at Yunnanfu, asking the Mission in Likiang to inform my men (8) to meet me in Yunnanfu the last part of October. I have sent a 300 rupee check to the Mission

---

2 Césaire Auguste Albert Bodard, 1883-1969, war 1922-1925 Konsul 1. Kl. in Yünnanfu. Ab 1926 wirkte er als Generalkonsul in Batavia.

which can be sold to the Postmaster in Likiang for local currency silver. To the Consul I sent a check for 100 dollars Shanghai to be converted into Yunnan currency to pay for expenses for the boys and incidentals, should they arrive before myself.

Yesterday I had a letter from my head collector in response to one I wrote him from Peking last May, wherein he assures me that he will come either to Yunnanfu or Tengyueh or Bhamo with as many of my collectors as I should wish and when I wanted them. This is all very encouraging.

I am certain that the expedition will be very successful and beg to assure you Professor Sargent that I will do all in my power to make it a great success.

I hope that you have found time to write either to Dr. Taylor or Mr. Fairchild[3] regarding my collaboratorship and special passport for China, Japan, Indochina and Hongkong. With such a passport I can prevail upon the authorities for assistance while with an ordinary passport I cannot do so to a great extent; besides it will facilitate matters when dealing with foreign customs, etc.

Will you kindly give me the name and address of the Belgian gentleman in charge of the Peking-Hankow Railway. I would also appreciate a letter of introduction to him and the necessary credentials „to whom it may concern“.

The financial end of the expedition we can arrange on my arrival in Boston the early part of September.

I am at present engaged in labelling the large collection of plants I brought back with me from my last expedition. It will, however, be impossible to bring the woody plants with me, as Mr. Maxon[4] wants to check up the number of specimina etc. It will require many boxes to ship the woody collection to the Arnold

---

3 David F. Fairchild, 1869-1954, Botaniker, war eine Zeitlang am Bureau of Plant Industry tätig. Er ist durch seine Weltreisen, seine Bücher und den Fairchild Tropical Garden bei Miami bekannt.Vgl. *Plants & Gardens* NS 23.1967:3, S. 68-70.

4 William Ralph Maxon, 1877-1948, Spezialist für Pteridophyta. Assistent, später Kustos am U.S. National Museum. Vgl. Rocks Korrespondenz mit Maxon in: H. Walravens: *Joseph Franz Rock (1884-1962): Berichte, Briefe und Dokumente des Botanikers, Sinologen und Nakhi-Forschers.* Stuttgart: Steiner 2002, 153 ff.

Arboretum. The lilies are being laid aside and I shall bring them with me for Mr. Wilson.[5]
Hoping that you are well, I beg to remain
Faithfully yours,
Joseph F. Rock

9

Washington, D. C,
August 22, 1924

Professor Charles S. Sargent
Director Arnold Arboretum
Jamaica Plain, Mass.

Dear Professor Sargent,
Many thanks for your kind letter of August 20th.
If you desire, the date of the expedition shall commence on the day of my sailing from either Boston or New York. Re the check for 14000 dollars, I would be grateful to you if it could be sent to me the first week in September or on the first of the month to New York City, care of Mr. R. A. Foulks, General Manager Far East Development Dept. The amount will best be mailed in the form of a draft. I am asking you to send this to New York at that particular time because I shall have to stop in New York on my way up to Boston to check over all the material brought and see to it that it is promptly dispatched by the American Express Co. I will then not need to return to New York from Boston but will go directly from Boston to San Francisco. My last letter to you will have explained all this.

I shall make out a receipt in duplicate for the amount sent, one of which I shall send to Mr. Adams as you request and the other to you personally.

I am sorry to learn of the transfer of Mr. Hers[6] which will

---

5 Ernest H. Wilson, 1876-1930, Botaniker und Forschungsreisender; er hatte durch drei Reisen auch Ostasienerfahrung. Seit 1919 war er als Stellvertreter Sargents Vizedirektor des Arnold Arboretum. Vgl. *Dictionary of American Biography* 20.1936, 321-322

6 Wohl Joseph Hers, 1884-1965, belgischer Eisenbahningenieur und

necessitate his absence from Peking. However, I have made many pleasant contacts with a number of people in Peking holding official positions and it will not be difficult for me to arrange matters. I will need no one in Peking or elsewhere to attend to botanical work, as all my shipments of plants and seeds will be in such shape that they will need no handling by anyone. I shall make arrangements with the American Express Co. in Peking and the Postal authorities to ship my herbar. collections directly from Peking or Shanghai to the Arboretum. I have always travelled alone and rather do so this time than engage European assistants, and as I stated I have many friends in Shanghai, Peking and Taiyuanfu who will be very glad to help me. I am certain that I can make satisfactory arrangements.

Both Dr. Marlatt[7] and Dr. Taylor are out of town on vacations. I have spoken to Dr. Kellermann the associate chief and he will look into the matter. Dr. Fairchild has returned to Washington two days ago. Please write a note to Dr. Kellermann Associate Chief, Bureau of Plant Industry, and to Dr. Fairchild asking them to speedily arrange the necessary formalities for a special passport which I think is imperative not so much for China proper as for the countries around China.

I am still obliged to write an article for the National Geographic to which I am devoting my time in the forenoon, while in the afternoon I am working on the collection of plants. We have labelled nearly all of the first lot, several thousand numbers, and will be beginning on the next lot next week.

I expect to be in Boston not later than the tenth of September, when I hope all my work here will have been brought to a satisfactory close.

Again thanking you for your letter, I remain,

Faithfully yours, Joseph F. Rock

P.S. Kindly inform me when draft has been mailed. If you desire it can be mailed to me to Washington care of the National

---

Amateur-Dendrologe, der 1919-1924 in Nordchina botanische Sammlungen anlegte.

7 Charles Lester Marlatt, 1863-1954, Entomologe, seit 1912 Leiter des Federal Horticultural Board.

Geographic 16th & M Street, Washington, D.C. and I can take it up to New York with me. J. F. Rock

10

COSMOS CLUB
WASHINGTON, D.C.

Aug. 28th, 1924

Dear Professor Sargent,
Many thanks for your good letter. I saw Dr. Kellermann, the Associate Chief of the Bureau. Everything is O.K. I received today a letter from the Secretary of Agriculture enclosing a letter to Mr. Hughes, Secretary of State, in which Mr. Wallace requests that I be given a special passport. The latter I will receive next Saturday, August 30th.

I shall be leaving here about the fifth or sixth for New York. This will bring me to Boston about the tenth.

I have sent today to the Arnold Arboretum four trunks containing books, photos, supplies, etc. One is empty. Please hold same until my arrival. I will have to repack them there.
Everything has gone very smoothly. I am certain the trip will be a great success.
As ever faithfully yours,
J. F. Rock

I gave the State Department my title as Botanical Explorer.

11

THE RACQUET CLUB
WASHINGTON, D.C.

Sept. 1, 1924

Dear Prof. Sargent,
I have received your kind letter announcing the fact that the treasurer of Harvard College has forwarded the check for 14000 dollars to the American Express Co. New York. Mr. Foulks of the above Company wrote me yesterday stating that the check had

been received and is being held for me.

I am waiting for these holidays to pass and then will at once get busy securing the various visas and my passport. Owing to the visit of the Prince of Wales, the State Department, or Mr. Hughes did not find time to sign the passport. It was supposed to have been ready but he has not yet signed it. Special passports must be personally signed. The Department also gave me an engraved letter to various authorities giving this trip official status. This will facilitate matters greatly, especially with the French in Indochina. I am translating all this into Chinese, which documents I shall have certified by the Chinese Legation here in Washington. English written papers or documents, are of not much value in the interior of China.

I still have several thousand sheets of plants to label and I hope to get finished with that by the end of next week. There is an endless variety of things to be done ere starting on such a long trip.

Hoping to see you about the tenth of this month I remain, with kindest greetings,
Faithfully yours,
Joseph F. Rock

I shall send a receipt for the check (draft) as soon as I shall have received it from Mr. Foulk[s]. J. F. Rock

12

New York
Sept. 9, 1924

Prof. Charles S. Sargent
Arnold Arboretum
Jamaica Plain, Mass.

Dear Professor Sargent,
I acknowledge herewith the receipt of a cheque on a local Boston Bank for the sum of Fourteen Thousand Dollars $ 14,000.00 from

the Treasurer of Harvard College to be used for the Arnold Arboretum expedition.
Sincerely yours,
Joseph F. Rock

13

San Francisco, Calif.
Sept. 30, 1924

Prof. Charles S. Sargent
Arnold Arboretum
Jamaica Plain, Mass.

My dear Prof. Sargent,
I had an uneventful trip across country and a very pleasant stay with Mr. Morton.[8] His botanist, Mr. Teuscher[9], seems to be of the opinion that I should collect seeds in large quantities, as they want seeds for distribution as an exchange. He wrote out a memorandum which I enclose herewith. I shall only take orders from the Arnold Arboretum and not from a young fellow like this man Teuscher. However, he is an enthusiast and means well. I hope you will let me know your opinion on this, but I suppose that I can shape my own plans and shall carry out the plans as I outlined them to you before I left. Too many cooks spoil the broth. I shall be guided by our arrangements as outlined. I shall do the best I can, but I can only serve one institution.

I had not known of the demands proposed by Mr. Teuscher, of whom I had never heard ere this. They shall get all the willows and poplars and all I can possibly secure and more. But if seeds are to be collected in large quantities for the planting of hundreds of trees of each species and still have seeds for distribution wherewith to

---

8 Joy Morton, 1855-1934, Besitzer der Morton Salt Company, gründete 1922 das Morton Arboretum in Lisle, Illinois. Er wurde anfangs durch C. S. Sargent beraten.

9 Heinrich (Henry) Teuscher, 1891-1984, Landschaftsgärtner, Gartenbauspezialist, Botaniker (Spermatophyta). Er plante 1936 den botanischen Garten in Montreal, dessen erster Kustos er wurde.

establish an exchange, that will be another matter. As parcel post does not exist, the sending of seeds by letter post registered in large quantities will materially increase the cost. However, I suppose Mr. Teuscher is an enthusiast and thinks it all very easy. Of course he has no experience, being just fresh from Germany.

Please do not censure him or write him about this but simply write to me please and let me known what my status is in regard to the Morton Arboretum. So far I recognize only the Arnold Arboretum, under whose direction I am going out to China, and as I understood it the seed and plant collections will be divided between the Arnold Arboretum and the Morton Arboretum. Mr. Morton was very nice indeed. He certainly has a wonderful place there and a splendid opportunity of building up a fine institution, and I shall do all I can to help him by bringing in all that it will be humanly possibly to send and bring.

I managed to get everything off. I had to make additional purchases here in San Francisco and had to buy one more trunk. Everything has arrived with the exception of the Aneroid barometer which I had repaired by the Taylor Instrument Co. I suppose it could have reached me if sent by air mail.

One more point. I had a lot of difficulties regarding the shipment of firearms and ammunition owing to the war raging in China. The Pacific Mail had instructions not to take contraband, but I managed to persuade them to take them on board by making L. A. P. official business and having them addressed c/o American consul whom I have already notified. So everything is O.K. and I anticipate no trouble in China. I sent the Shanghai freight as unloading optionally either at Shanghai or Hongkong.

I secured from the Golden Gate Park a copy of Schneider's *Handbuch der Laubholz-Kunde* which will help me a great deal.

I wish to thank you dear Prof. Sargent for the splendid opportunity you gave me to make this exploratory trip to the Amne Machin and Richthofen range possible. I will do my best to make it a success.

Mr. Bangs[10] need not fear that I shall return without birds.

---

10 Outram Bangs, 1863-1932, Kustos für Säugetiere am Museum of

Again expressing my sincere appreciation I remain as ever,
Faithfully yours,
Joseph F. Rock

Kindest greetings to Mrs. Potter[11], Mr. Wilson and Rehder and the ladies. J. F. R.

P. S. I have always taken complete notes on the components of a given forest and have at all times made notes on the phytogeographic aspect of a range. When travelling I am taking notes as to the plants encountered en route. These notes are of course augmented by herbar material.
What I said about seeds in this letter Mr. Teuscher told me personally and is not found in the memorandum. J. F. R.

My address will be
J. F. Rock c/o American Consul Yunnanfu, Yunnan, China (up to about November)
Then c/o Y.M.C.A., Chengtu, Szechwan, China

I enclose another letter from Mr. Maxon of the Smithsonian which is self explanatory. I do not think that this should be looked upon as an exchange, but we may say that we will send them a set of the plants I shall collect in China on this trip. J. F. R.

Comparative Zoology der Harvard Universität. Rock sammelte Vögel im Auftrage des Museums. Vgl. Birds collected by Dr. Joseph F. Rock in Western Kansu and Eastern Tibet. By Outram Bangs and James L. Peters. *Bulletin of the Museum of Comparative Zoology at Harvard College* 68.1928,311-381.

11 Nathalie Potter war Sargents verwitwete Tochter.

14

PACIFIC MAIL S.S. CO.
S.S. PRESIDENT WILSON

October 4th, 1924

Dear Prof. Sargent,
We are to arrive in Honolulu on Monday the 6th. The weather has been fine and everything went well. All my trunks and boxes including guns and ammunition are on board. I hope everything will be alright in regard to landing the same in Shanghai. Please inform Miss Sawyer that the American has arrived before the departure of the boat so that I have everything that was sent to me to San Francisco. I have sent the remainder of the amount due to the Eastman Hockhouse Co. in case Miss Sawyer should receive a bill from them she can refer to my letter stating that I have paid the seven dollars due them by check on Washington. I have also sent a check to the Taylor Instrument Co., New York, for the repair of the Aneroid Barometer.

I have received a very kind letter from Mrs. Potter which I appreciate very much indeed. I have answered it and I hope Mrs. Potter will have received my letter promptly.

With kindest greetings and best wishes I remain, as ever, faithfully yours,
J. F. Rock

Plase give my best regards to Mr. Wilson, Rehder and the ladies. J. F. R.

15

S. S. EMPRESS OF CANADA

[no date]

Prof. C. S. Sargent
Arnold Arboretum
Jamaica Plain, Mass.

Dear Prof. Sargent,
Just a line to advise you that I had no trouble in landing my freight, ammunition etc. in Shanghai. A permit had been obtained from Peking, the Customs had already been informed and all was passed without examination. I only had to pay duty on the arms and ammunition. Arrangements were made immediately for the transshipping of the material to Chungking, Suifu and Chengtu through the American Express in Shanghai and the Canadian Methodist Business Agency in Chungking. Arrangements were made with the Mission Treasurer in Chengtu to honor my checks. A permit had been secured from the Hongkong police through the Pacific Mail Agent in Hongkong to land my other arms and ammunition (for bird collecting) in Hongkong for transshipment to Haiphong. A wire was waiting me in Hongkong from Likiang informing me that 8 men had left for Yunnanfu on Oct. 1st. A letter from the American Consul at Yunnanfu stated that the French authorities at Haiphong had been informed through the French Consul at Yunnanfu regarding my arms etc. and a permit had been obtained from the Chinese authorities at Yunnanfu to import the same. My boy and Tibetan cook are at Hongkong and are awaiting me there. I remained six days in Shanghai to arrange for transshipping goods, made personal contacts with various people and institutions which will be of assistance to me in the interior. I also arranged all of my financial matters.

The day I arrived in Shanghai the war came to an end, only Peking was still in trouble. The situation in Peking did not clear until the day of my departure for Hongkong yesterday. All wires were cut, no trains were running and no mail could be sent, hence I could not get into communication with Mr. Hers.

Everything was so far gone without a hitch and according to schedule as I planned it in Boston and Washington. It was very wise that Mrs. Potter did not go East for there was or still is grave danger. However, the situation may clear up now since the Christian general[12] has taken charge of Peking. Tsao Kun[13] the

---

12 Feng Yü-hsiang 馮御香, 1882-1948, einflußreicher Militärführer in Nordchina. Vgl. Howard L. Boorman: *Biographic dictionary of Republican China.* New York, London: Columbia Univ. Pr. 1967-1971. II,

President has abdicated or resigned and Wu pei fu[14] has been degraded and cut off from his army, the latter of which is surrounded. Word was sent from Yunnanfu that Tang Ki Yao[15] the Governor had been assassinated by his own men but was later stated to be only a rumor. Sun Yat Sen has declared a Soviet republic in Canton and looted and burned the city. Damage is estimated at 50 million dollars and a thousand people were killed. What will happen there since the Christian general is in command in Peking is problematical. There is however a bright ray appearing on the Chinese horizon and things may become settled. There is the danger however that Szechwan will once more declare its independence since Wu peifu [!] its subjugator is out of office and power.

It seems an endless and interminable struggle. The surrounding country of Shanghai is in ruins and thousands of people are homeless and have fled to the International settlement. The International settlement was barricaded and surrounded by barbed-wire entanglements, all wooden bridges connecting the settlement with Chinese territory had been demolished to prevent Chinese troops from rushing the city. Shanghai is indeed a city of paradise and contrast, while war was raging outside and machine guns and tanks were guarding the boundary of the settlement symphony concerts were going on, theatre, dancing, etc. Shanghai itself was quiet. The Red Soviet flag is now flying on the Russian consulate opposite the Astor House. The present Soviet Consul general, who has been twice sentenced to jail by the Mixed Court (by an American judge) is now up again before Court and will in all

---

37-43.

13 Ts'ao K'un 曹錕 (1862-1938) war Präsident von Oktober 1923 bis November 1924. Vgl. Howard L. Boorman: *Biographical dictionary of Republican China*. New York, London: Columbia Univ. Pr. 1967-1971. III, 302-305.

14 Wu P'ei-fu 吳佩孚 1874-1939, Militärführer in Nordchina. Vgl. Howard L. Boorman: *Biographic dictionary of Republican China*. New York, London: Columbia Univ. Pr. 1967-1971. III, 444-450.

15 T'ang Chi-yao 唐繼堯, 1881-1929, Militärgouverneur von Yünnan. Vgl. Howard L. Boorman: *Biographic dictionary of Republican China*. New York, London: Columbia Univ. Pr. 1967-1971. III, 223-225.

probability pay another visit to the Shanghai jail; then Shanghai will be without a Russian Soviet Consul general. Consuls do not enjoy immunity and since Russia has given up exterritoriality rights he comes before the Mixed Court whose word is final and there is no appeal. Pleasant prospects for the representative of a great (in territory) country.

I will keep you informed regarding m successional movements from time to time.

With kindest regards to yourself, Mrs. Potter, Wilson, Rehder and all friends, I am

Faithfully yours,

J. F. Rock

16

Haiphong, Nov. 5, 1924

Prof. Ch. S. Sargent
Arnold Arboretum
Jamaica Plain, Mass.

My dear Prof. Sargent,

Just a line from this place to tell you of my safe arrival with all my belongings. I had no trouble with the French Customs and they passed everything without examination and free of duty, even the guns and ammunition. I wrote a letter to the Commissioner of Customs in Hanoi, and he granted me free entry of everything. I am asking the same of the Chinese Customs at Yunnanfu. In Shanghai I only had to pay duty on the rifle and cartridges, all else went free.

I am leaving Saturday for Yunnanfu and thence my long voyage will commence. My boys are all up there waiting for me. I did not take my Siamese boy along as he was very impertinent to me in Hongkong and so I am going to make one of the Nashi boys number one boy. It will also be much cheaper.

Will write you again soon. I have not forgotten about the Pine from Indochina and will inquire in Hanoi.

With kindest regards and good wishes for your continued good health, I remain
Faithfully yours,
Joseph F. Rock

Kindest greetings to Mrs. Potter, Mr. Wilson, Rehder and all the ladies at the Arboretum. J. F. Rock
I made good time from S. Francisco to this place – one month and four days.

Angkur Gomba, a monastery 60 Li from Taochow old city. May 26, 1925

My dear Prof. Sargent.
As I wrote to you I went once more to the present headquarter of the Living Buddha of Labrang at Angkur gomba or Djrakur gomba as the Tibetans call it. I am exceedingly sorry to say that the Regent of Labrang the father of the Living Buddha declared war on the Mohammedan general of Sining. The camp I had intended visiting had broken up and the Tibetan chiefs have all returned. War has been the decision and this on the strength of word that came from the Lanchow Tuchun that the Chinese will be on their side and will wipe out all the Mohammedans. The fight is expected to commence in six days. 70,000 Tibetans have assembled with their weapons. The Chinese general of Pingliang is also joining the Tibetans who have been bled white by Machi. What I am going to do I am unable to say. I came mainly to find out what they had decided and I am of course exceedingly anxious; it is difficult to know what to do. We have such a lot of things and my men, mules cannot be had and with 22 yaks we cannot move very fast. I had hoped that Kansu was to be a peaceful place but I have been sadly mistaken. It was providential that I did not go to Sining and further north as that will be the center of the fight. The Mohammedans of Liangchow are about ready to come down and

attack Lanchow. But I have been told that now the Mohammedans are all very much afraid as the Chinese are joining the Tibetans and this will mean a general massacred all the Mohammedans. It is most difficult to decide what to do. The Labrang Regent gave me a feast today and we discussed the situation from every angle. He said if I once get to Radja on the Yellow River I would be all right. He said if I would take the Mohammedans along they would surely be killed on the road as the Tibetans would spot them immediately. He said he would procure me 5 Tibetans who can handle yaks and would send 2 Lamas with me who would explain my presence etc. to Tibetan troops we would be bound to meet en route. He said I would have nothing to fear from the Tibetans but the Mohammedans now dress in Tibetan clothes and do most of the robberies and now with the war on he fears for my life should I go in. He said that two lamas would go with me to Radja gomba on the Yellow River and would thence return. He is also giving me three letters to the three tribes of the Ngoloks which live at the Amne Machin, one is the Rimong, the others are known as Konser and Konkai, they are now very friendly to Labrang as Labrang is fighting Machi, and they are joining in the fight for Machi has made them pay a water and grass tax.

In five days I will know what we will do, go in to the Amne Machin and risk it or if it is too dangerous. I have to think of my 12 men who have families. There will be no alternative but to retreat as fast as possible towards Sungpan. My money has not yet come from Lanchow and I fear that it may be too late. I am assure you my dear Dr. Sargent, that I am exceedingly grieved at this awful tragedy that is hanging over us or rather over Kansu. I can only hope that we will escape safely somewhere, but where that somewhere will be no one knows. I will write again as soon as I have decided what we *must* do.

With kindest regards and best wishes I remain faithfully yours
Joseph F. Rock

Let us hope for the best.
Kindest regards to Mrs. Potter, Wilson, Rehder and all the friends at the Arboretum.

P. S.
Even here at Angkur gomba elev. 10,500 ft. not a shrub has a single leaf on. We had since I left home, continuous snowstorms and we had an awful trip to this place. But I am glad I came, travelling was so uncomfortable in those blizzards over passes that I almost wanted to return, but we pushed on and I am glad we did for we learned much.
J. F. R.

There may be one other alternative and that is go into the Tebbu country, west of Choni. From here (Angkur gomba) I can overlook the mighty snowrange which lies southwest of Choni, it is an enormous rocky range about 17,000 ft. in height and is said to contain much forest, no one has worked there. At the present the range is covered with snow and I do not think that the plants found there will be too tender for Massachusetts. May be I will have to go in there to escape the bloodshed and murders that will soon take place in this region. At any rate I will do the best I can under the circumstances which I can assure you are more serious than any I have yet had to face.
J. F. R.

10 p.m.
We had another consultation this evening, word has come that the missionary at Labrang has left the place also the postmaster. They now asked me to wait ten days longer, by that time things will have come to pass. They said Choni will be safe and if it comes to the worse I could cross the Tao River into Tebbu country. They have agreed however to escort me to Radja they fear however that Mohammedans dressed in Tibetan clothes may attack us, and they wish me to give them a letter stating that I am going in on my own responsibility and in case of my being murdered or looted they will not be held responsible. I am writing such a letter in Chinese and English and am placing my seal (Chinese seal) upon it.

We have spent so much money by this time that it would be a shame not to go in. I realize the responsibility to enter a country which is in the most peaceful times, a nest of nomad robbers, and in times of war it is more than hazardous. I had left money with the people in Labrang to do some buying for me, sheepskin coats etc. for my men. They have now fled and I have neither coats nor money. In Hetso I was only able to get four coats made. This is not the time for buying sheep skins, as the grass is still to short in fact it has just sprouted and sheep are not killed at this time as they are too thin. And yet we cannot wait as by the time we shall come out of Radja it will be bitterly cold and these garments will be needed.

Such trips take more thought and time as is realized. A poor community like this where one cannot even buy the cheapest kind of grass paper and where things are sold on the instalment plan so to say it is mighty difficult to fit out a dozen men with proper clothing and buy the necessary things for the road. Everything has to be talked over and discussed for days until one can get anything done at all. I will continue my letter later.
J. F. R.

Taochow Old City
May 29th, 1925

We have returned to Taochow yesterday.

Sargent to Rock

1

July 18, 1924

Mr. Joseph F. Rock
Samoset Hotel
Rockland, Maine

My dear Mr. Rock,
I have read carefully your letter of July 16th. You speak of expenses for the first year as $ 14,000, and the other years as $ 12,000. This is not quite clear to me because you say „the amount necessary for the first year including cost of equipment and salary

of the leader would be about $ 14,000." Later you say, „I should expect a salary of $ 500 a month." Is this $ 500 a month or $ 6000 a year in addition to the $ 14,000, making the whole cost of the first year $ 20,0000, and if so, who is the leader who is to get the $ 6,000 a year? I take it for granted that you are the leader.

If the cost of the whole expedition, including your salary, is $ 14,000 for the first year and $ 12,000 for the two following years, I think we could manage a three year expedition, but if it is to be $ 20,000, and $ 18,000 it is more than we could afford.

My understanding is that you would devote the first year to the Amne Machin Range in Thibet, the range in which you are especially interested; the second year, I should like to have you devote to the Richthofen range in Kansu, or possibly it might be more convenient to devote the first year to the Kansu range, in which I am most interested in; the third year would be devoted to the Southern slope of the Altai Range, and you might perhaps much more easily come out by the Trans-Siberian Railway, either coming East or West, than undertaking the long and difficult route out by Persia, which perhaps had better be left to another expedition.

I think on your way from Rockland to Washington you had better stop for a further discussion of this matter. I am anxious that an arrangement can be made by which a plan can be carried out.

Yours faithfully,
Director

2

July 22, 1924

Mr. Joseph F. Rock
Hotel Samoset
Rockland, Maine

Dear Mr. Rock:

I was very glad to get your night letter and your letter of July 19th.

I think there is no doubt but what we can make all the necessary arrangements for this expedition.

The Wilsons are going away on the first of August for a few days, and I should like to have him here when you come. Would't it be possible for you to come, say on Thursday of next week, as I suppose we shall have a good many things to talk over.

Yours faithfully,
Director

3

August 20, 1924

Dear Mr. Rock:

I have your letter of the 17th of August. Don't you think that the date of the beginning of your expedition should be the date of your sailing from New York? The check for $ 14,000 could be dated a few days before and sent you in time to reach New York on the day before you sail, and then you could pay from it the amounts you have already expended for material, time, etc.

I wrote several days ago to Dr. Taylor about the passport. If I do not get a favorable reply from him I will write to Fairchild on the subject. I have also written to Professor Marlatt about authority to get things in without examination at Washington but I daresay we shan't accomplish much with him.

If you will let me know if the check, which I think should be sent in the form of a draft on New York, shall be sent to you two days say before you sail I will attend to it. You will, of course, make out a receipt for this amount and send to the Treasurer of Harvard College, Mr. Charles F. Adams, 50 State Street, Boston, and also one to me here.

I had a short note a couple of days ago from Joseph Hers, our Peking friend, saying that he had left the military service in China and joined what he called the *financial services*, whatever that may be, and that this would take him away from Peking into another part of China. His address is in care of Société Belge d'Entreprises en Chine, Peking. Whether he will be able to attend to botanical work I do not know. Probably he would not have as many men in his employ to use for this work. I shall of course give you a line of introduction to him. Possibly you may not think it worth while to make that long journey unless you have other friends in Peking whom you can depend on for help. We can talk about this and many other things when you get here.

I will write you another letter tomorrow telling you just what we want to obtain. Let me know as soon as possible when you are likely to be here.

Yours faithfully,

Mr. J. F. Rock
Nat. Geographic Society
Washington, D.C.

4

ARNOLD ARBORETUM HARVARD UNIVERSITY
JAMAICA PLAIN, MASS.

August 25, 1924

Dear Mr. Rock:

I have your letter of August 22nd and have asked the Treasurer of Harvard College to send you the draft for $ 14,000 on September 1st in care of the American Express Company, 65 Broadway, New York City, attention of Mr. R. A. Foulks, General Manager Far East Development Department.

Yours very truly,

Mr. J. F. Rock
National Geographic Society
Washington, D.C.

5

October 14, 1924

Dear Mr. Rock:

I am much obliged to you for sending me the correspondence from Morton. Of course it is perfectly impossible to send seeds in large quantities, and foolish to send seeds of Chinese plants already established in the United States and Europe, except in cases where seeds can be obtained from species growing much further north than they have before been collected, as there is a chance of such seeds producing hardier plants.

I have told Morton that our plan was to send everything here, herbaria, seeds and cuttings, in care of the Arboretum to the Department of Agriculture in Washington; and that on the whole I think he had best withdraw from the plan of paying half the cost of the expedition and allow it to be paid by the Arboretum. If this is done I should send him, as I should Kew and the Garden in Edinburgh, one of the herbaria and as many seeds as ought to be sent him. Of course it is desirable that the distribution of these should be as large as possible both in this country and Europe, especially those which have not been collected before. I hope he will agree with this plan.

In thinking the matter over more carefully, I am convinced that it was a foolish idea to think that an expedition of the difficulty and delicacy of yours could be generally managed by more than one person under the immediate supervision of one leader. If you had to receive orders or even suggestions from Morton as well as from me you would, I am afraid, soon get into hopeless confusion. Wilson and I have had so much experience in Chinese matters, about which nothing is known at the Morton Arboretum, that I hope we shall get on without trouble if he accepts my suggestion.

I hope you are having a pleasant voyage and will reach China safely. With kind regards,

Faithfully yours,

Mr. J. F. Rock

Yunnan, China

6

December 11, 1924

Dear Mr. Rock:

I am glad to get your letter without date written from the steamship *Empress of Canada* and to hear that you have had no trouble at Shanghai, also that everything was progressing favorably with you. I do not feel very sure that this letter will reach you, and fortunately I have nothing important to say.

I suppose that we shall not get any material from you before the middle of the winter of 1925/26 for it will be too late to send out cutting of Willows and Poplars before the autumn of 1925. I hope you will number in one series everything you collect, including cuttings, herbarium specimens and seeds, and in the case of the seeds if you can add the genus it will be an assistance to us in distributing them. The number, the generic name when possible and the locality are all the information we need. The species can be worked up later from the herbarium specimens which I suppose will not come until you return yourself.

There is no news to send you from here. Every one is well, and so far we have had little winter and practically no snow.

With kind regards and all good wishes for your health and success, I am

Faithfully yours,

Mr. J. F. Rock
Chengtu, Szechwan, China
(c/o Y.M.C.A.)

7

April 4, 1925

Dear Mr. Rock:

I was very glad to receive yesterday your letter of February 15th from Chengtu and to learn from it that you were still well and in good spirits in spite of your hard journey and many trials and tribulations.

In any case you need not give the Morton business another thought. He understands fully the situation and agrees entirely with me in the wisdom of giving up my first suggestion that your expedition should be a joint affair between this Arboretum and his. Of course we shall send him a portion of the seeds and cuttings we receive from you and do everything else we can to help him.

Wilson seems to think that your troubles in reaching your destination are nearly over and I hope he is right in this. You will of course be late for early spring flowers in Tibet but that cannot be helped. The seeds and fruiting specimens are more important, and the cuttings of Willows and Poplars would of course be made in the autumn.

Rehder has just called my attention to the fact that Komaroff[16] has described a *Picea crassifolia*, a new species found in the Kukunor region, latitude 37°98'40". This he says grows in the forests of *Picea Schrenkiana*. The latter is a tree which we know little about and are not at all sure if the species cultivated under that name in Europe and the United States is the real thing. Of course these two species are good plants for you to look out for.

We are having what promises to be a very early spring here, and as yet there is no indication that much damage has been done by the winter which, with the exception of a few extremely cold days, has been a mild one. There was no snow unfortunately, but lately we have had a good deal of rain and if we get through the next ten days all right I do not think we shall have any trouble.

It is delightful to have dealings with a man as intelligent, brave and cheerful as you are, and I shall be greatly mistaken if your journey does not turn out to be the most important botanical expedition which has ever gone out from the United States.

Let me hear from you as often as you can, and with all good wishes, believe me,

Faithfully yours,

---

16 V. L. Komarov, vgl. *Bulletin de la Société des Naturalistes de Moscou* NS 45.1936, 67-75, mit Porträt; B. V. Skvortsov: Vladimir Leont'evič Komarov, 1864-1945. *Proceedings of the Harbin Society of Natural History* 5.1946, 5-8, 1 Taf.

Mr. J. F. Rock
Si-ning Fu, Kansu
via Peking, China

8

May 6, 1925

Dear Mr. Rock:
I have received your letter of the 21st of March from Mienchow. You are certainly having a hard time and I am surprised at your courage. Wilson and I both agree that you are going to get through all right. A man as energetic, intelligent and courageous as you are is sure to overcome the kind of difficulties you encounter, but it is a slow and troublesome business.

I communicated at once with the Museum about the $ 2000 they were to pay in July and enclose a letter written by Barbour which explains itself. I have therefore asked the Treasurer to make the draft $ 12,000 instead of $ 10,000 as you certainly ought not to be short of money. This he has done and I have no doubt that the draft is already in Shanghai. I gave Mr. Adams your direction as follows:

„Drafts are to be sent in U.S. Currency to the credit of Joseph F. Rock with the American Express Company of Shanghai, China. This draft must be on New York, American Express Company, 65 Broadway, and in transmitting the same the Shanghai Express Company's office must be directed to credit the amount of the drafts to the U.S.Dollar Account of Joseph F. Rock and not the Shanghai Dollar account.“

Every one is well here, and we have a wonderful spring with more flowers than I have ever seen in the Arboretum before after what has proved a mild winter for the plants. Mrs. Potter is now somewhere between Honolulu and San Francisco, and I only hope that she has not heard of the death of her cousin John Singer Sargent, the great artist, who was a particular friend of hers. If she heard of it, as she might have in Peking or Japan, the news I fear has greatly interfered with the pleasure of the last part of her journey. She leaves the ship at San Francisco and is coming home by rail via Santa Barbara.

Let me hear from you as often as you can for we are naturally worried. If there is any trouble about the money sent by Mr. Adams better send me a cable.
With all good wishes for better luck in the future, I am
Faithfully yours,

Mr. Joseph F. Rock
c/o The Reverend F. D. Learner
Sinningfu [!], Kansu

9

August 18, 1925

Dear Mr. Rock:
I received a few days ago by the same mail your letters of May 20th and 26th, and yesterday there came (I do not understand why) from the Department of Agriculture your letter of the 24th of June which is the latest word we have had of you. They fill Mrs. Potter and me with the greatest admiration for your courage, patience, energy and wise management, and it begins to look now as if you would pull through all right and make a great success of the expedition.

The cable announced in your letter of May 20th reached me. When I received your letter of May 20th I sent a cable to Dr. John Ferguson[17] in Peking telling him the situation you were in and asking him to do anything he could to help you. On the 8th of August I received this reply from him which was in cipher: „Telegraph immediately through foreign office to Governor who has replied ‹Luetoan›, Magistrate has been instructed to investigate and report by wire. will inform you.“ Ferguson will I know do everything he can to help you.

I am having § 540.30 mentioned in your letter of June 29th deposited in your account in the American Security & Trust Company of Washington, D.C., this being the amount of your salary which you had to expend for the expenses of the expedition.

---

17 John Calvin Ferguson, 1866-1945, Berater der chinesischen Regierung. Vgl. R. H van Gulik: Dr. John C. Ferguson's 75th anniversary. *MS* 6.1941, 340-356.

I suppose it is a little too early to talk about getting seeds out of Kansu. Do you think it would be advisable to send them to our Embassy in Peking to forward? I have no doubt I could manage this if you think well of the plan, or I daresay that Dr. Ferguson would attend to it for us. We are on very friendly relations with the Southeastern University at Nanking. Three persons in the botanical departmnt there have studied here in the Arboretum, and Professor H. H. Hu[18] has just left here after two years of study. He will be the head of the botanical department of the University, which I hope to be able to help become a great institution as there is a chance of obtaining money for the purpose and they already have twelve hundred acres for an Arboretum and Botanic Garden. I do not know if Professor Hu could be of any assistance to you in any way but he knows about our expedition and some of your troubles, and as he is very devoted to the Arboretum I am sure he would do anything to help you if his assistance would be of any value. I believe him to be a man with a great botanical future.

I hope I shall hear from you again soon and that the news will continue to be as good as when you last wrote or even better. I cannot give your message to Mr. Wilson as he is away on a vacation until the 1st of September, but I am sure he will write you on his return. Every one else is well and the Arboretum has never looked as well at this season as we have had unusually heavy rains during the summer.

With kind regards and good wishes, I am,

Always faithfully yours,

10

August 18, 1925

Dear Mr. Endicott:

Will you please have deposited for the Arboretum $ 540.30 in the American Security & Trust Company of Washington, D.C., to be

---

18 Hu Hsen-hsu [su] 胡先驌 1894-1968, Botaniker, war 1923-1925 an der Harvard Universität tätig, 1925-1928 als Professor an der Tung-nan-Universität.

placed to the account of Joseph F. Rock, and charged to our expedition in China.
Yours very truly,

A. L. Endicott, Esq.
Cambridge, Mass.

11

August 28, 1925

Dear Mr. Rock:
I received yesterday your letter of July 10th from Choni describing your troubles on your trip into the Tebbu country and saying that you had decided to pick up and leave for the Kokonor with the expectation of reaching the northern mountains of the Kokonor the end of this month. You certainly are having a horrid time. Something I am afraid has interfered with this plan for I received on August 19th a cable from Ferguson in Peking stating „Magistrate reports 13th of August Rock well, working, located Choni", which loooks as if you had returned to that city. I do not understand it but I have no doubt that you will overcome all your difficulties.

I shall ask the Bursar of the University to add to his deposit in Washington the sum of $ 35.14 which I understand is still due.
Let me hear from you as often as possible. With all good wishes, believe me
Always faithfully yours,

Mr. Joseph F. Rock
Choni, Prov. of Kansu
China

12

August 28, 1925

Dear Mr. Endicott:
The other day I asked you to deposit in Washington for the account of Joseph F. Rock $ 530.30. Will you kindly add to that

deposit the sum of $ 35.15 and charge to the expenses of our Chinese exposition [expedition].
Yours very truly,

A. L. Endicott, Esq.
Cambridge, Massachusetts

13

September 15, 1925

Dear Mr. Rock:
You wrote me on June 29th that you had packed up everything and was just starting for the Kokonor region, but I have a cable from Mr. Ferguson stating that he had heard from the Chinese magistrate that on the 13th of August you were well, working and located in Choni. I can only account for this that you failed to reach Kokonor and returned to Choni. I suppose that I shall get a letter from you in the course of a few days now. Naturally I am anxious about you and about what happens to make you change your plans.

You speak of sending your material via Chinese Turkestan. This may be all right and you no doubt know more about the safety of this route than we do here. In any case I believe you had best retain herbarium specimens and bird skins until next year when conditions may be more favorable, and send this year only seeds. These should be numbered of course, with the same number which will be on the herbarium specimens. I suggest that you separate the seeds into three different sendings, one lot to go via Chinese Turkestan if you believe in that route, one by mail in small packages directed to the Arboretum, and the third also in small packages adressed to John C. Ferguson, 3 Hsi-Chao Hutung, Peking, China.

I am writing today to Mr. Ferguson asking him to attend to reforwarding these packages to the Arboretum direct. I feel quite sure that he will do it but if he is unwilling to undertake this I will suggest by cable another method of sending them. I do not think that any of the seeds which will be in such small packages need be sent to Washington as we at one time proposed. I think we can get

them through all right without this formality.
Trusting that all is at last going well with you, and with my kindest regards, I am
Faithfully yours,

Mr. J. F. Rock
Choni, Kansu, China

14

October 8, 1925

Dear Mr. Rock,
We have been immensely interested and pleased at the receipt of the five packages of seeds addressed to the Arboretum which came via Shanghai and were delivered to us here without any examination or trouble. They came apparently in about two months and arrived a little more than a year after you sailed from San Francisco. I feel very encouraged.

A short time ago I wrote to the Federal Horticultural Board and have now received from Dr. Marlatt, the Chairman, the following:

„In the case of seeds collected by Mr. Rock, we will furnish the addressed tags (12 enclosed herewith) which you will kindly mail to him for use in connection with shipments that he may wish to make from time to time. He can combine as large a package of different kinds of seeds under each tag as he finds convenient. The address on these tags has been changed to the Board's Boston office, 408 Atlantic Avenue, and Mr. Smith, in charge of that office, has been instructed, promptly on the receipt of any such parcels, to deliver them to the Arnold Arboretum and make the examinations at the Arboretum so that your or your assistants can see the goods on delivery and as they are unpacked, and can if necessary assist in the disinfection of the seeds. I feel perfectly sure that you have no objection to treatment of the seeds if such treatment should be necessitated by any conditions of infestation or probability thereof, knowing of course that your interest is just as much as our own to exclude any new foreign pests.“

This is certainly very satisfactory to me and I do not believe that the disinfection under these conditions will amount to anything, while if they went to Washington there would be long delays. I enclose the tags as Dr. Marlatt requests and I think in future that instead of sending the seeds via Peking, as suggested previously, or via Turkestan, that you send them all addressed to the Arnold Arboretum, Boston, Massachusetts, with one of the enclosed tags placed on the package. The difficulty about sending to Washington is shown by the fact that the seeds sent directly to the Arboretum arrived nearly a week ago but we have heard nothing yet of those which you sent to Washington.

Our plan is to distribute your seeds to Kew, the Royal Botanic Gardens at Edinburgh, the Jardin des Plantes in Paris, the German Botanical Garden near Berlin, the Botanic Garden in Sapporo, northern China [!], and the Southeastern University at Nanking where I hope eventually an important botanical establishment will be built up. I think also that if there are seeds enough to around whe should send a few to Dr. Morton who at my suggestion agreed to bear half the expenses of the expedition. This arrangement I canceled after your visit to him as I was afraid it might make complications which would be disagreeable for you. I should like to know if this disposition of your seeds is satisfactory to you for of course it can be changed.

On September 5th of last year, Mr. Kellerman, Acting Chief of the Bureau of Plant Industry, wrote me as follows:

„We have discussed with Mr. Rock the matter of obtaining plant material for us incident to his explorations for the Arboretum. Mr. Rock is familiar with our work and in general with our plans because of his past experience with us. I would suggest that instead of specifying the plant material we desire, we leave it to his judgment to send us such plants as he may be able to obtan incident to his explorations for the Arboretum, forwarding the plant material with his shipments to the Arboretum. The plants can then be forwarded to us from the Arboretum. If you prefer, Mr. Rock can make the shipments direct to the Department.“

I do not know that this discussion entails anything on you but I hope not. Of course the sending of plants will be out of the question, and I haven't a very high opinion of the work done by the Bureau of Plant Industry in Washington. If you feel committed in any way, or would like to have us do so, we can send a few of your seeds to this Bureau as they arrive. As you know, I want to make your expedition as useful as possible, and to carry out any of your ideas which may relate to the accomplishment of this.

I have read, of course, your article in the National Geographic Magazine relating to your present journey. I am not sure that it is a good plan to write too much for publication at present about this expedition, and in any case I think such articles should be sent via the Arboretum in order that we can read them before they are published. Of course we shall want to preserve in the Arboretum a set of all your photographs, and your films will be the property of the Arboretum just as those taken by Mr. Wilson on his various journeys for us are preserved here.

I am very anxious to get your next letter and to learn if you have gone to Tibet or are still at Choni. It is evident that you are discovering a lot of new and probbaly extremely valuable conifers and we hope that we may receive seeds and cones of these this year.

I have been writing for the Journal of the Arboretum a short note about your expedition and tried to join to it a biographical note about yourself. This, however, I was not able for want of information to make very satisfactory. We ought to have all the information we can get about you on file here, and if you can find a spare half-hour some day I hope you will jot down answers to the following questions: The date of your birth, the nationality of your father and mother, your age when you made the journey to Egypt with your father, and if you made any other journeys as a boy. We have no information at all about you between your journey to Egypt with your father and your appearance in Hawaii in 1919 as we make it out. Will you tell me what you did before you went to Hawaii and where you obtained your botanical education and training? I have the dates in the preface of one of your books of your journey around the world when you were working on the Hawaiian flora and I am not sure that you left Hawaii late in 1919 or early in 1920. All I know about this is that

I received from you in August 1920 a note written in Washington. When did you leave Washington that year for your employment by the Department of Agriculture, and when did you transfer to the Geographic Society? And was all your work in Yunnan done for this Society or partly for the Department of Agriculture? For such a distinguished man, it is astonishing that so little has been published, or if more about you has been published that it has been so completely overlooked in our library, where I am happy to say we have what I believe is a complete set of all your published works.

I send you hearty congratulations on your safe arrival in the north after certainly what was a most terrible journey, so terrible that I think it would be wise for you to remain in north central Asia until you have completed the work in Kansu, Tibet and on the southern slope of the Altai mountains. You will never want to go back there again and now that you are there it is wise to finish as much as you can, always provided your good health holds out.

With my sincere thanks for all you are doing for the Arboretum, and with all good wishes for the New Year, believe me

Always faithfully yours,

Mr. J. F. Rock
Choni, Kansu, China

15

October 29, 1925

Dear Mr. Rock:

I have recently received from you letters from Hsiningfu dated September 4th and 12th. That of the 12th arrived two or three days before the letter dated September 4th.

I regret all the difficulties you are having and especially the sickness of your cook, which must be a serious interference with your work. I hope he will recover. I feel so uncertain about your plans that I am sending this letter in duplicate, one copy to Lanchow and one to Choni.

As I have written you, the five packages of seeds which you

sent direct to the Arboretum arrived quickly and in excellent condition, but I have heard nothing as yet from those which you sent for the Arboretum to the Department of Agriculture.

I enclose copies of my correspondence with C. C. Thomas of the Department of Agriculture. When I wrote him on October 16th I did not know that you had sent seeds to the Department directly and supposed that this seed had been taken by him from one of the packages which you had sent to the Arboretum in Washington. I do not know what your obligations are to the Department but if you wish any of your seed to go there it should be sent though the Arboretum and avoid future confusion as you will see by Thomas' first letter. They are giving your seed an entirely different number from the one we have used in planting it here and distributing it to our correspondents. Personally I should never think of sending valuable seeds to the Department of Agriculture as there are many more important and better conducted horticultural and botanical institutions in the world able to make better use of the extremely rare seed we hope to get from you. Of course we shall be governed by your wishes in this matter, but I think I can properly insist that everything you send from China should pass through the hands of the Arboretum.

This letter, if it ever reaches you, as I feel sure it will in time, ought to be in your hands not long before Christmas. I cannot venture to wish you a Merry Christmas and a happy New Year, but I can hope that your health will remain good and that you will be able to accomplish all that you expect to accomplish in 1926.
With kind regards from us all, and all good wishes, I am,
Faithfully yours,

Mr. J. F. Rock
Choni, Kansu, China

A copy of this letter to Mr. Rock was sent to Lanchow, Kansu, China.
P.S. The people at the Museum of Comparative Zoology in Cambridge would be glad to have the bird skins sent by parcel post at any time which may be convenient to you. The packages should

be simply addressed *Museum of Comparative Zoology, Cambridge, Massachusetts.*

16

November 9, 1925

Dear Mr. Rock:

I have just received your letter of September 16th from Tankar announcing the sending to Washington on that day for the Arboretum of seven packets of seeds and I am writing there today asking them to hurry them forward as rapidly as possible.

The first lot of seeds you sent came here direct so quickly that I am sorry the lot sent by you on September 16th did not come in the same way as I haven't yet received the seeds of the first lot which I understand you sent for the Arboretum to Washington.

You say you haven't heard from me for several months, I wrote you on October 14, 1924, to Yunnan-fu, on December 11, 1924, to Chengtu, on April 4, 1925, to Sinning-fu, on September 15th to Choni, on October 8th to Choni and Lanchow. I suppose the only letter of mine which you could have received previous to your writing on September 16th was the one written on May 6th.

My letter of October 8th was the most important one for it included labels to be put on the packages of seeds to insure their direct delivery to the Arboretum. Lest this letter may not have reached you I am enclosing a copy and am asking them to send me from Washington another supply of the labels. With good luck my original letter of the 8th of October and the one I am writing to you today ought to reach you early in December, and I think it would be wise to delay until then sending any more seeds so that you may use these special labels. If they leave Kansu by the 10th of December they ought to be here early in February, early enough of course to sow the seeds both here and in Europe.

We have had rather a poor autumn with less beautiful autumn foliage and fruit than usual, and I see by the papers that a hard winter is predicted.

I hope you are well and that all goes well with you.

With kind regards, I am

Faithfully yours,

Mr. J. F. Rock
Tankar, Kokonor Border, Kansu, China

17

November 18, 1925

Dear Mr. Rock,
I am very glad to be able to report that the six packages of seeds which you sent via Washington have arrived safely and in good condition. I received from Washington fourteen more of the tags to put on your packages of seeds and as I do not feel very sure about your winter address I am sending half of them to Lanchow, Kansu, and half to Choni, Kansu, by registered mail.

I enclose a letter from Mr. Ferguson which I received this morning. I am sure we can count on his help if it is needed.
With kind regards and sincere good wishes for the New Year, I am
Faithfully yours,

Mr. J. F. Rock
Tankar, Kokonor Border, Kansu, China

18

November 25, 1925

Dear Mr. Rock:
the following is from a letter received from Smith of the Edinburgh Botanic Garden which I think may interest you:

„If I can be of any service in identification of some of the herbaceous material of Rock's Kansu collection I trust you will afford me the opportunity. I am particularly interested in *Primula* and would very gladly examine and report on the Primulaceae of his expedition if you wish me to do so."

Probably there are species of *Primula* in the regions you are going to visit which are still undescribed and may prove valuable garden plants. It might well be remember his interest in them if the opportunity offers.
I hope all is going well with you and that we shall soon hear from you again.

With kind regards,
Always faithfully yours,

Mr. J. F. Rock
Tankar, Kokonor Border, Kansu, China

19

December 31, 1925

Dear Mr. Rock,
I have your letters of October 9, 25th and 28th which all came by the same mail. As this is the last letter I shall write in 1925 I will begin it by wishing you a happier time, or at least a more comfortable one in 1926 than you had in 1925, which I am afraid you will look back on as the most disagreeable year of your life. There is one thing to be said, at any rate, you cannot possibly have a worse time in the future than you have had in the past.

I am sorry to say that the package of Willow cuttings which you sent directly to the Arboretum and the seeds which you sent via Washington have not yet appeared. I am sorry that you sent seeds to Washington rather than direct to the Arboretum where we have facilities for getting them now, and if they go to Washington the authorities there always help themselves liberally to your seeds. As I have written you, Washington is not the best place to handle such valuable seeds, and we have numerous correspondents in the extreme northern part of the world where they will be appreciated and I hope will be well taken care of. In the future I hope you will send everything directly to the Arboretum.

I agree with you that is is a surprising thing to find Rhododendrons growing so far north and in such an inhospitable region. The discovery seems to me very important for if these species do not have value as ornamental flowering plants they may prove of immense value when crossed with more southern or handsomer species in producing a new race which will prove hardy in regions like this where we can grow none of the southern Chinese species.

I am addressing this letter to Kansu, although I am not sure that you will make this your winter headquarters, but probably if you do not you will leave directions there to have letters forwarded to you. We have sent you, as requested, a quantity of oiled paper by letter post, as I agree with you it is absolutely necessary that you should have it. I should suppose that material packed in sphagnum and enclosed in oiled paper will require an outer covering, but I don't know whether burlap is the best material or not, for this. I should think that wrapping paper and cotton cloth for the outer wrapping would be sufficient.

There is not much news to write you. We have had as yet no snow but a few days of excessive cold, that is the thermometer has been down during two or three nights to 7° Fahrenheit. This is bad as the plants need a covering of snow in such cold weather.

Everybody at the Arboretum seems well and I think we have been busier here than ever before, so many demands are made on us for material and information, and we are preparing for a great deal of spring planting. Mrs. Potter is well and passing the holidays with her daughter in Brookline. She has taken an apartment in New York for the winter but in February is going to Italy for a couple of months, and next summer expects to go to Paris with her daughter. She is certainly a restless person. She is much interested in your trip and very often asks what I have heard from you.

Good-bye for the moment, and with all good wishes believe me
Faithfully yours,

Mr. J. F. Rock
Kanchow, Kansu, China

20

January 8, 1926

Dear Mr. Rock:

I received several days ago your letters of November 10th and 18th. They were followed by two packets of Willow cuttings and the seeds of the Juniper and *Picea* sent direct to the Arboretum. The Willow cuttings were dried up and dead on arrival I am sorry to say; the seeds were all right. I am particularly glad of this Juniper seed which is either a new species or possibly *Juniperus saltuaria* Rehder & Wilson which appears to be a widely distributed species at high altitudes in western central Asia, although we have never succeeded in getting seeds and it is not in cultivation. This tree was collected by Purdom on the Peling-shan in Kansu, by Wilson on the mountains of northwestern Szechwan, and it is probably the Juniper collected by Przewalski in the Kokonor region and on the upper sources of the Hoang-Ho. Your seed seems to be in condition to germinate I am glad to say.

The seeds which you sent to Washington arrived only today. These are Nos. 13, 284-12,304, inclusive, 13,307, 13,308, 13,309, 13,310, 13,311.

When you wrote on November 10th and 18th my letter of September 15th was the last you had received, and the letter of October 8th, the important one in which I told you why I had made an arrangement with Washington to have all your mail sent here direct and examined, enclosing special tags for your packages had not been received. You have, I hope, received it before this. In case of an accident I enclose an extract from that letter of October 8th which tells you about my arrangement with the Federal Horticultural Board. I sincerely hope that you will send nothing more to Washington. They appear to help themselves liberally to the Arboretum seeds and to give them new numbers. These seeds are too vaulable for the Washington people to handle and I want to save them all for the botanical gardens and forest stations in the colder Parts of the northern hemisphere.

I note that you are going to send everything except your mail packages to the American Consul-General at Shanghai. Please charge him to put everything on one of the steamers which sails from Shanghai direct to Boston, coming of course through the Panama Canal. There are two of these lines and each of them run good steamers. I think it will save trouble if the bird skins as well as the Arboretum material are shipped in this way to the Arboretum, and we can arrange to have the skins sent to Cambridge. The Consul-General at Shanghai should be asked to have a bill of lading for the freight shipped from Shanghai to the Arboretum sent to me here by mail.

You have certainly had a bad time and I hope you are now

somewhere where you can get the much needed rest and have comfortable quarters. You don't say where you expect to pass the winter but I take it for granted it will be in Choni, at any rate I am sending this letter to you there and shall continue to write you there until I get instructions to the contrary.

There is no news from the Arboretum to write you. Everyone is well here, and we have so far had no snow and only a little cold weather but a great deal of rain and fog, and on the whole an unsatisfactory autumn. We badly need snow and a great deal of it to protect the plants which are not entirely hardy.

I hope that I shall hear from you again, and with all good wishes, I am

Faithfully yours,

Mr. J. F. Rock
Choni, Kansu, China

21

February 2, 1926

Dear Mr. Rock:

I am very glad to report that I have received your letter of November 29th from Labrang and finished December 8th at Choni, also your two letters from Choni of December 9th and 13th, and to learn from them that after all your hardships and hard work your were still cheerful and full of hope for future achievement.

We have now received the numbers of the seeds sent by you on the list enclosed. By this you will see that only a part of the seeds of the last sending are here but I expect the remainder by the next mail. The following is a list of the seeds which had germinated before the end of January:

Caprifoliaceae #13275, Fraxinus #13243, Berberis #13256, Berberis #13263, Berberis #13272, Berberis #13273, Berberis #13295, Cotoneaster #13269, Cotoneaster #13270, Nitraria Schoberi #13240. Rhododendron #13278, Rhododendron #13279, Rhododendron #13302, Rhododendron #13303.

I shall share your seeds, as you suggest, with Dr. Lyon[19] at Honolulu, the Department of Agriculture at Washington, the Golden Gate Park at San Francisco, and Frank L. Morgan at Hoquiam.

It is needless to say that the Arboretum Journal will be more than glad to publish anything you may care to write about the features of vegetation you have seen. I hope you will do this for us.

I hope you will not be shy about sending me a biographical note about yourself. It is not for publication but we should have on file some account of a man who is doing so much for the Arboretum. I shall write today to Dr. Ferguson and to the American Consul at Shanghai in the hope of securing authority for the latter to ship your birds and herbarium specimens without the necessity of opening the boxes or paying a duty.

I can well understand the cost of traveling has been much higher than was contemplated before you started on this journey when all the difficulties you have encountered and the attendant expenses were not expected. Do not let this worry you. The Arboretum will be only too glad to meet all your financial requirements, and I feel confident that you will use such economy as may be necessary.

I see by papers that William E. Brigham[20], who was a classmate of mine at Harvard, has died in Honolulu. I suppose you knew him well. He was a man of energy and a great deal of ability but unfortunate sexual peculiarities made him unpopular in this country at least, and his real ability I do not think has been appreciated here. I have heard a great deal about the Bishop Museum of which he was Director. I suppose you know it well. There has been much said in its praise.

We have been having some very cold weather here lately and

---

19 Harold Lyon, 1879-1957, Gartenbauspezialist für die Hawaiian Sugar Planters Asociation.

20 William Tufts Brigham, 1841-1926, A.M. der Harvard Universität 1865, 1888-1918 Direktor des Pernice M. Bishop Museum in Honolulu. Er starb am 29. Januar 1926. Vgl. *Who was who in America*, 1897-1942. (1943), 139.

only a very little snow, and I am afraid a good many plants in the Arboretum will suffer between now and the 1st of May.
Hoping that this will find you well and somewhat rested, and with all good wishes, I am
Faithfully yours

Mr. J. F. Rock
Choni, Kansu, China

22

March 1, 1926

Dear Mr. Rock:
Since I last wrote you I received your letters of January 1st, 3rd, 4th, and 8th and your accounts; also your five packets of seeds which came here direct (Nos. 13,744, 13,202, *Betula* species, number 12131 and 12799) and your photographs and cones. In all we have received 397 packets of seeds and 27 packets of cuttings, all in good condition and a good many of the seeds have already germinated. I suppose that all your collections of last year, with the exception of the cases containing the herbarium specimens and the birds, have now been received. I will write today to our Minister in Peking and see if we cannot get some help from him.
We have sent you $ 2000. to Shanghai. This is the additional amount, I understand from your letter, which you need for this year; and I have a memorandum to send you in May $ 15,000 for your next year.

We have sent you all the stationery and material you have asked for with the exception of the brown wrapping paper. If this is sent in quantity by registered mail it will cost so much that the Arboretum would be put out of business. The opinion here is that China is a much better country than the United States in which to obtain wrapping paper. We have had printed and sent to you a set of your photographs.

Your first year, in view of all the difficulties, dangers and hardships, has certainly been a wonderful success and I have nothing but admiration for what you have accomplished.

Your photographs are wonderfully taken and most interesting

and valuable. I don't know how you got the idea that we didn't want to get photographs of the scenery. These are always important and interesting additions to our collection, and you may be sure you cannot send us too many of them. Photographs of the bark of large trees close to the tree are valuable for they give a better idea of the bark than any description.

It is impossible to give any advice at this distance about your future movements, and all I can say in a general way is that now you are in that part of the world, which may never be reached by a botanist again, or certainly not for many years, we ought to get as much information and material as is possible in the time you are willing to devote to it. I shall be disappointed if you are not able to visit the Amne Machin for I have an idea that this will prove a very interesting region. I think that this would occupy the rest of 1926, that you would not get to the Altai before the spring of 1927, and that if you come home via Chinese Turkestan, where there may be much to collect, you might not be back here until the late autumn of 1928, that is four years from the time you left.

You understand, I hope, that having entire confidence in your judgment I leave all plans to you with the understanding that you will be supplied with money needed for a longer or shorter stay.

We have had a very hard February with more snow and worse roads than we have had for years, and today it looks like anything but the first day of March. Every one is well, and Mrs. Potter called Saturday for Spain with the idea of being back the first of May and returning to Europe again next summer. I shall hope to hear from you again soon.

Always faithfully yours,

P.S. The Lily bulbs came at the same time as the cones and, as you request, some of them have gone to Washington.

23

March 22, 1926

Dear Mr. Rock:

Your letters of January 18th and 19th have been duly received. Like all your letters they are most interesting and give me the greatest pleasure. Do not be afraid of writing too many of them. We have also received in excellent condition seven packages of *Salix* and *Populus* specimens and the new consignment of your photographs, a set of which will be returned to you next week. We have also the specimens of herbaceous plants which I think will enable us to determine or have determined the plants raised from your herbaceous seeds which we have distributed with those of the woody plants.

It may be interesting to you to know that we have distributed among the best cultivators in the northern part of the three hemispheres three thousand seven hundred and eighteen packages of your seeds, a good result for your first season's work I think. Some of the seeds here which Judd has have germinated well and our gardner Sander, who is the best man so far as I know anywhere for seedlings, Rhododendrons and Azaleas, has already picked out from the first pan of *Rhododendron* seeds which he planted fifteen hundred plants which are growing well. I am looking forward, as you know, with a great deal of interest to the Rhododendrons which you have discovered for it is possible that they may not have large of handsome flowers but they will certainly be of immense value to cross with more tender species which are not hardy in this part of the world.

I find that in my recent letters to you I did not send Marlatt's latest instruction about forwarding your seeds. The following is a quotation from his letter of February 17th:

„With respect to your second communication of the same date the Board is perfectly willing to allow Dr. Rock to send the plant material to which you refer by letter mail in registered sealed packages, in view of the restrictions which are placed on the Post Office Department to permit entry and delivery of such material when so tagged that it will come via the inspection service of the Federal Horticultural Board of this Department, it would seem to be better to have Mr. Rock instructed to address this material to the U.S. Department of Agriculture, Federal Horticultural Board, Boston, Mass. Mr. Rock's name on the package will be a clear indication that it is intended for you, and any additional information can be placed inside the package. Such action will, in the first pace, be a compliance with the requirements, and, in the second place, will avoid any risk of the package being diverted by some postal official en route. Furthermore, if any packages not properly marked reach Mr. Smith he will be instructed to see that they are promptly turned over to you. As hitherto, the opening, inspection, and any necessary treatment can be made at the Arboretum after they are freed from Customs, unless some of the material is so infested that it would seem desirable to have it given more technical inspection."

This means that the tags which we have sent you will not be needed on the seed packages but they will be addressed as Marlatt directs in this letter. We are in close touch with him and I am sure that he will not subject it to any injurious treatment. Of course things like photographs, herbarium specimens and anything not requiring action by the Federal Horticultural Board should be addressed to the Arboretum.

I am afraid it is still cold where you are. Here it is a very late spring, and this is really the first day that feels like it.

Don't forget to write me as often as you find time to do so, for your letters give me great pleasure.

With kind regards,

Always faithfully,

J. F. Rock, Esq.

24

March 26, 1926

Dear Mr. Rock:

Since I wrote you the other day I have had a visit from Bangs who came over to read some of your recent letters and talk over the bird situation. He agrees with me that this bird collecting business must not be given up even if you have expended the 200 pounds contributed by the Cambridge Museum for the purpose. Barber, the man who furnished the money, is away and will not be back until next month. I think there is no doubt, if the 200 pounds is already expended, that an additional sum can be obtained from friends of the Museum, so I hope that you will be able to continue this work as the region is of course a most important one and the birds which are there are entirely unrepresented in any museum. Perhaps you will let me know how much of the 200 pounds you have already expended when you are writing.

It is astonishing how successful your seeds are germinating, and there are already up here and at my place in Brookline thousands of your plants. It doesn't look as if a single seed has failed. We shall certainly have our hands full in taking care of the plants in a few years. Glad to see, too that some of your cuttings have already started to grow, and at this end of the line the situation is most satisfactory.

I hope everything is going well with you, and with kind regards and all good wishes, believe me,

Always faithfully yours,

Mr. J. F. Rock
c/o Postal Commissioner
Lanchow, Kansu, China

25

April 27, 1926

Dear Mr. Rock:

I have received your letters of January 26, February 12 and 26. According to the request in your letter of the 26th I have written to Taylor in Washington and urged him as strongly as I possibly could to have your passport hurried up and sent to you. I think he will probably attend to it.

I still hope that you will visit the southern slope of the Altai mountains, and I do not expect you home before the autumn of 1928 as it will be of course be possible to do some collecting in Chinese Turkestan and other places on your way home.

I enclose a copy of a letter which originated in the United

States Legation at Peking and I suppose has been sent to me from Shanghai. We certainly cannot furnish information here they are looking for but possibly you will be able to do so.

I am glad to say that your seeds are germinating in the most satisfactory manner and that from your first year's collection there should be more than a million plants distributed through the northern United States and northern Europe. I hope you will be equally successful this year in Tibet.

With kind regards and all good wishes for your health, believe me,

Faithfully and gratefully yours,

Mr. J. F. Rock
Lanchow, Kansu, China

26

May 10, 1926

Mr. J. F. Rock
c/o Postal Commissioner
Lanchow
Kansu, China

Dear Mr. Rock:

Since writing you on Saturday, I have received this letter from Mr. Kellerman, a copy of which I am enclosing along with photostat copy of your passport, in the hope that it may reach you.

I have not heard yet about the best way to send money to you but I am doing my best to find out. I will be able, I hope, to write again on this subject in the course of two or three days.

Faithfully yours,

Director

27

United States Department of Agriculture
Bureau of Plant Industry
Washington
Office of the Associate Chief of Bureau

May 7, 1926

Mr. C. S. Sargent
Arnold Arboretum, Harvard University
Jamaica Plain, Massachusetts

Dear Dr. Sargent,
In a letter just received from Mr. Rock, it would appear that he has not yet received the passport which was forwarded to him by registered mail on February 10 to the address given in his letter requesting the passport, namely, c/o Postal Commissioner, Lanchowfu, Kansu, W. China. Mr. Allanson also wrote to Mr. Rock regarding the passport on March 6 and again on March 9; apparently none of these letters have reached Mr. Rock up to the time of mailing the letter, just received, on March 8 from Choni, Kansu.

I am enclosing herewith a photostat copy which we had made from the original passport before it was mailed. If there is any address that occurs to you that is likely to be reasonably prompt in reaching Mr. Rock, I would be very glad if you would forward a copy of this letter together with the passport direct to Mr. Rock.
Very truly yours,
K. F. Kellerman
Associate Chief of Bureau

28

May 7, 1926

Mr. J. F. Rock
c/o Postal Commissioner
Lanchow, Kansu, China

Dear Mr. Rock:
I have received your letters of March 6th and 9th from Kansu. You have evidently failed to receive several of my letters, and I am sending, by registered mail to the Postal Commissioner at Lanchow, copies of my letters of November 13 and 25, December 31, January 8, February 2, March 1 and 26, and April 27. If you ever get these, you will see that on March 2 we sent you $ 2,000.

to Shanghai as you requested, and we have tried to do everything you have asked us to do.

I am trying to arrange to cable the $ 2,000. you failed to receive to the Postal Commissioner at Lanchow but it seems doubtful if this can be accomplished. Then there is $ 15,000. which will be sent to you in May, and I have not found out yet how this is going to reach you. It is quite evident that things are in a very poor condition in eastern China. I have every faith, however, in your tact, ability, and courage, and somehow or other you are going to pull through.

Faithfully yours,

29

May 11, 1926

Mr. J. F. Rock
Choni, Kansu, China

Dear Mr. Rock,

I received this morning letters from you from Choni dated March 13, 18 and 23, the last in answer to the receipt of my letters of December 23, 1925. On the 7th of May, I sent you copies of my letters of November 13 and 25, December 31, January 8, February 2, March 1 and 26, and April 26. c/o Postal Commissioner at Lanchow, and on May 10 sent you, in his care, a copy of letter from F. K. Kellerman with photostat copy of your passport.

We know that the check for $ 2,000. which we sent you reached Shanghai but have been unable so far to trace it any further. We are doing our best to discover the best way to get money to you, and I shall hope to write you again on the subject both to Lanchow and Choni in the course of two or three days. I am sending this letter to Choni although in your letter dated December 29 you wrote „Address all letters for the next ten months or so in care of the Postal Commissioner at Lanchow.“ Although you say you are not going to Lanchow, and that there may be difficulty there, you have given me no new address, and I think we should continue to write you there. This I am sending to Choni as there is nothing more to write to Lanchow.

Of course I want you to carry out the plan of visiting the southern slope of the Altai Mountains but I fully realize that without sufficient money this cannot be done. If there is any way of getting the money to you, you will certainly have it.
Faithfully yours,

Director

30

May 18, 1926

Dear Mr. Rock:
I have this morning your letter of March 26th and 29th. I am glad to hear that at least the wrapping paper has reached you and I hope you will get the check before long. I am worried, of course, lest you should be short of money but I don't know what more we can do about it. You haven't changed your address from Lanchow but from what you write of conditions there I will send this letter to Choni in the hope that it may sooner or later reach you. What surprises me is that you are so cheerful and apparently little discouraged by all the hardships and anxieties which you have met with on this journey.

I enclose a copy of a communication which I received from Peking this morning and which looks encouraging, but whether it means that your cases of specimens can be shipped from there at once or not I do not know. Judging from your telegram, you won't be back in Choni before November but possibly you have made or can make some arrangements by which letters can be forwarded to you.

We have had a cold and late spring but the flowers are beginning to appear slowly. So far nothing has been injured in the Arboretum and everyone here is well.
With kind regards and good wishes, believe me,
Always faithfully yours,

Mr. J. F. Rock
Choni, Kansu, China

31

July 21, 1926

Dear Mr. Rock:

My last word from you was your cable dated at Shanghai May 13 and received here the same day, and my last letter was received July 13, dated at Labrang and was written consecutively from April 29 to May 4, the day before you left Labrang for Tibet.

Your courage and cheerfulness and certainty of future success fill me with the greatest admiration, in view of all you have had to endure. The results of your first year are remarkable and most satisfactory to us here.

After thinking carefully over the situation and the condition of affairs in northern and central Asia, I have reached the conclusion that it would be wise to abandon the journey to the southern slope of the Altai mountains with the long journey home through Chinese Turkestan and India, and after your return to Kansu to remain there until the early spring of 1927, devoting the winter to shipping your seeds and other material by mail and packing the remainder, then during the season of 1927 further exploring the Tebbu region and returning as soon as the work connected with this is finished. This is one plan. Another is that you consider your exploration finished when you get back to Kansu from Tibet. By adopting the first plan you would, if all goes well, be back here during the winter of 1928, and by the second a year earlier.

At this distance I do not feel that I can decide which of these plans is the wiser and I will leave it to you to decide. There is money enough for either, and you need not worry about money. The difficulty is to get it to you. It is my hope and intention that you will find at Lanchow when you get there $ 16,000. unless I hear that you will need a larger sum.

December 29th last you gave as your address for ten months c/o Postal Commissioner, Lanchow, Kansu, and I shall consider this your address until I hear to the contrary. Choni I now understand is only a small suburb of Lanchow which therefore would seem to be a safer place to address money and letters.

Did I write you that we had received from you in good condition four hundred and two packages of seeds and forty-two lots of cuttings, and that we have distributed three thousand three hundred and eighteen packages of your seeds, sending them to every important botanical and horticultural establishment in the northern United States, northern Europe and eastern Siberia. Instead of giving Judd any of your *Rhododendron* seeds I gave them to Sander who must have raised many hundred thousand plants which are doing well and will certainly more than supply this part of the country. Your photographs, too, are very interesting and I think well worth the cost of your expedition up to the present time.

I hope you will find this letter when you get back, but I hardly expect to hear from you again before January.

With every good wish for your safe and successful journey this year, I am

Faithfully and gratefully yours,

Mr. J. F. Rock
c/o Postal Commissioner
Lanchow, Kansu, China

32

August 10, 1926

Dear Mr. Rock:

I was delighted to receive yesterday your letter of the 8th of June from Radja Gomba on the Yellow River. Its receipt surprised me for I had not expected to hear from you for some time to come. You certainly have had a difficult but I should judge a most interesting trip, and so far are getting on with your usual courage and cheerfulness for which I heartily congratulate you. The account of the big mountain you have seen is most interesting, as is the prospect of new Spruces from regions never before visited by a white man.

There is nothing new in regard to your specimens or nay acknowledgment of the draft for $ 2000. which I suppose is still in Shanghai; nor have I been able to make an arrangement to send you additional money which certainly should leave here no later than September.

We have had some very hot weather, but not a great deal of it. July started with what promised to be a serious drought but abundant rains have done away with this danger to the Arboretum which I think looks remarkably well, better perhaps than it ever has before.

I do not suppose you will get a letter from me now until you get back to Choni as the only address I have for you is in care of the Postal Commissioner, Lanchow, Kansu.

I should be very glad to see you and I take this opportunity to congratulate you again on what you have already accomplished, and on your skill in dealing with the troublesome natives.

With all good wishes, I am as ever
Faithfully yours,

Mr. J. F. Rock
c/o Postal Commissioner
Lanchow, Kansu, China

33

September 9, 1926

J. F. Rock
Choni, Kansu
China

Dear Mr. Rock:

I was delighted and surprised to receive your letter of the fifteenth of June as I had not expected to hear from you during the summer. This came two or three days ago and this morning your package of seeds arrived having come here direct and escaped the government abuse.

We are going to send your money now through the Hongkong-Shanghai Banking Corporation, being assured that this institution could deliver it to you if anyone could do so. Kidder Peabody & Company of Boston will cable six thousand dollars to you at once through the Hongkong-Shanghai Banking Corp. and the balance of six thousand dollars by mail. I hope that this will reach you by November 1st or earlier and the Hongkong bank has promised to let us know if the money was delivered. Apparently this Hongkong bank seems to be a more enterprising institution than any at Shanghai or Pekin[g].

I feel now very strongly that you should return to the United States next winter, bringing with you probably all your seeds rather than sending them by mail and all your other material. You have been in that rough country long enough and borne more hardships with greater cheerfulness than anyone I have ever heard of. It won't do to overdue [!] it, and if you can get your herbarium back here you will have work enough at the Arboretum for I should think some years. If you want to travel again I don't think you should go to China, and we can find an easier region for you to explore. I should like to feel, however, that your connection with the Arnold Arboretum was to be a permanent one if possible.

I shall address this letter to Choni, which I hope you will reach in November. Ths means that you ought to be here by the early spring with your material. I want of course very much to see you again, but although I am pretty well I am getting older every day and may have to drop out before long. I certainly don't want to do this until you are back and arrangements are made for your working here.

Mrs. Potter has been in Marimbar taking the cure which I don't think she needed. I suppose she is now in Switzerland and returns to Paris with the expectation of being here late in October. I have given her your message and she said she was going to write to you at once.

With all good wishes for your safe return to Kansu and later to the United States, I am
Always faithfully yours,

34

September 14, 1926

Dear Mr. Rock:

I have received through the Secretary of Agriculture copies of correspondence which came to him through the Department of State in Washington, including a 10 g letter written to you by E. F.

Stanton, American Vice Consul at Kalgan, China, urging upon you the danger of trying to visit the southern slope of the Altai Mountains, and saying that it would be very problematical if you would be allowed to proceed direct from Kansu to these mountains as it is usually necessary for a foreigner to proceed to Urga and after being looked over by the Secret service either to be allowed to go on or turned back. I do not know whether you have ever received this letter. As I hope you have given up the Altai Mountains it makes no difference. If you are still anxious to go and have not received the letter I can send you a copy of it.

I am more than convinced that the wise thing for you to do is to come home as soon as possible after you return to Choni from your trip to Tibet.

With kindest regards and all good wishes, believe me,
Always faithfully yours,

Mr. J. F. Rock
Lanchow, Kansu, China

35

September 27, 1926

Dear Mr. Rock:

I have received this morning your letter of the 10th of August written at Choni. I am very sorry that you did not have a more successful journey in Tibet, but I cannot tell you how glad I am that you are back in Kansu and that you will have time this autumn to finish up the Tebbu country and so be ready to return to America next winter or in the late autumn.

I realize all the troubles you have had and are having, and I congratulate you on being safely back in Choni, although judging from your letter you may have serious difficulty in getting away from there. As I have already written you, the money has been sent to the Hongkong bank and I shall hope to hear that you have received it. Of course it will be impossible for you to leave China until those funds reach you.

There is no news here. I am glad to say that your seedlings are doing well and that we shall have an immense crop of the Rhododendrons at Brookline, to say nothing of those which will be raised by persons wo whom we have sent the seeds.

I cannot tell you what a relief it will be to shake your hand again, and this I shall hope to do before 1927 has passed.
With all good wishes for your health and further success, I am,
Faithfully yours,

J. F. Rock
Choni, Kansu, China

36

October 27, 1926

Dear Mr. Rock:
I have received your cable of October 20th asking me to remit funds immediately to the American Express Company at Shanghai and no one else. We had already sent $ 12,000. for you to the Hongkong & Shanghai Banking Corporation and have now cabled them to send this money to the American Express Company at Shanghai. As this may be a slow proposition we shall cable you tomorrow $ 6000. to the American Express Company at Shanghai.

Your letter of the 22nd of August reports that you are downhearted for the first time. No one certainly has a better reason to be disheartened than you. Do not forget that it is as important to discover that no plants grow in a country as it is to find what grows in it, and for this reason I consider your Tibet journey a great success. Altogether we have distributed about five thousand packets of seeds collected by you among all the northern botanical and horticultural establishments in North America, northern Europe and extreme northern Asia. The photographs which we have already had from you are worth what the trip has cost and you will probably have a lot more to send us.

Mrs. Potter is now in Brookline and has been reading with the greatest interest all your letters received since she left America for Europe on July 1st. She agrees with me that it is time for you to get out of China and come back to America where you will have several years' occupation, I fancy, in working up your collection. I had hoped that you would be able to start home this winter as soon as you had put in order the material collected this autumn in the Tebbu country, but it has occurred to me that if you find there are any trees or important shrubs of which you have not already collected elsewhere flowering specimens it might be wise to make a spring campaign in the Tebbu for the purpose of collecting flowers of these plants. This means, of course, the delay of your return until next summer. This matter must be left to your judgment. I only want you to understand that I am perfectly satisfied with what you have already accomplished and that we want you back to the United States as soon as possible.

With kind regards and my highest appreciation for all that you have accomplished on this journey for the Arboretum and the world, believe me,
Always faithfully yours.

Mr. J. F. Rock
Choni, Kansu, China

1 copy of this letter sent to him at Shanghai, China, c/o American Express Co.

37

November 13, 1926

Dear Mr. Rock:
There is now or there should be in the hands of the American Express Company at Shanghai $ 20,000. for you. On October 21st Kidder Peabody Company cabled the Hongkong & Shanghai Banking Corporation to pay $ 12,0000. to the American Express Company in Shanghai, and on October 28th the Bursar was instructed to cable a draft for $ 5000. to the American Express Company in Shanghai for you. In addition to the $ 18,000. there are $ 2000, which ought long ago to have been in the hands of the American Express Company at Shanghai but which apparently has been lost if it has not turned up recently.

I hope you will cable me when you receive any money from the

American Express Co. at Shanghai, and make any other suggestions you can by which you can be helped. I do not know that there is anything which can be done from here until we hear from you.

I am glad to read in your letter to Wilson that you have returned from the Tebbu country and, as I have already written you, it is up to you to decide whether you will return to America this winter or make a spring trip to the Tebbu to get flowers of the plants of which you have collected fruit this autumn. It would be a great relief to me to have you safely back in this country.

Always faithfully yours,

Mr. J. F. Rock
Choni, Kansu, China

38

November 15, 1926

Dear Mr. Rock:

Since writing the enclosed I have received your letter from Choni dated September 22nd. I do not see that there is anything further I can say or do to relieve your financial difficulties. By this time you must have received money from Shanghai. I shall be very glad to write to the Postal Commissioner at Lanchow thanking him for his kindness and assistance to you.

In your letter from Choni of September 22nd was enclosed your letters of September 9th and 14th from the Tebbu country which, according to your accounts, must be a wonderful place. Of course it is up to you to decide whether you should stay and return there in the spring for flowers or return to America as soon as you can settle your affairs and get the money from Shanghai to pay what you owe. I am most anxious that you should get back here as soon as possible but of course it would be a mistake not to get as much complete material as you can from the Tebbu country.

I wish I had something better to write you about your money affairs but I hope you realize that we have done and are doing everything we can to get money to you.

Always faithfully yours,

Mr. J. F. Rock
Choni, Kansu, China

39

November 18, 1926

Dear Mr. Rock:
I have received this morning the enclosed which means, I hope, that you have at last received some money.

We received on the 16th 52 packages of seeds, a package of 11 cuttings of *Populus* (No. 14859) and another package of 16 cuttings of *Populus* (No. 14646). I haven't seen the seeds as yet, but I understand they are in good condition. They were inspected by the Government people in Boston and sent out here for inspection, and the inspector has allowed our man here to do the fumigating, so I think there is no question but that they will be all right. I am very anxious to see the list of seeds as I shall hope to now in the course of a couple of days.

I hope to hear from you again soon and that you are in less trouble financially. Let me know if there is anything more I can do.

With kind regards,
Faithfully yours,

Mr. J. F. Rock
Choni, Kansu, China

40

November 23, 1926

Dear Mr. Rock:
I have your letter of the 28th of September describing your plan of sending the collections of dried plants and birds by registered mail directly to Boston. This under all the circumstances, although expensive, seems the thing to do and I hope you will carry out your plan this winter. I suggest that the herbarium specimens be addressed to the Arnold Arboretum and that the packages containing the birds be sent to Dr. Outram Bangs, Museum of Comparative Zoology, Cambridge, Masachusetts. I will write him

that I have written you this and will write you if he does not approve of the plan. We shall be very glad indeed to get the herbarium specimens for if anything should happen to them we could not determine your plants until they flower and fruit here, which in the case of trees might be twenty-five or thirty years.
I believe I have already written you that we have received this autumn in excellent condition fifty-three packets of seeds and two packets of Poplar cuttings.

I agree with you that it would be wise to remain another year in China so as to explore as thoroughly as possible the Tebbu country and get as many specimens of flowers in the spring and fruits in the autumn as possible. This I hope will enable you to get home early in 1928 when you may be sure of a warm welcome. We are very anxious to have you safely back in the United States.
With kind regards and all good wishes,
Faithfully yours,

Mr. J. F. Rock
Choni, Kansu, China

41

December 2, 1926

Dear Mr. Rock:
We have received your letters dated October 7th and 10th from Choni which are still unanswered, one hundred and seventy-one photographs, fifty-three packets of seeds in excellent condition, and two packets of Poplar cuttings.

I was strongly advised by a banking firm in Boston in which I have the greatest confidence to send the $ 12,000. for you to the Hongkong & Shanghai Banking Corporation, supposing that it was in Hongkong. The money was sent not to Hongkong but to Peking. On September 11th, Kidder Peabody & Company of Boston cabled $ 6000. and mailed $ 6000. to the Hongkong & Shanghai Banking Corporation, and this firm on October 21st cabled to the same Corporation to transfer this money to the American Express Company at Shanghai, requesting you to cable when you received it or any part of it. On October 29th, fearing there might be some mistake in this, the American Express Company cabled to you at Shanghai $ 6000. which you should have received long ago.

Political matters in China seem to be getting worse and worse, and I am getting seriously worried about you and cabled you today to the American Express Company at Shanghai to tell you that it would be all right to return home at once or as soon as possible. You have had a very long and hard journey, and I do not want you to be entirely broken down by it. You are too valuable for that and the sooner we can see you in this country the better we shall be pleased. Of course I realize it is an important matter to complete the Tebbu collection next year, but I do not think what you would gain by it would justify your running further risks and the danger of a breakdown. You have made already a magnificent collection, and I think you have been long enough in China for your own good and the general good of the world.

If you do decide to stay over another season I hope you will be able to send, as you have suggested, the herbarium specimens by mail. They are of course the valuable part of your collection for without them it will be impossible to determine any of your plants until those which are growing here have produced flowers and fruit; this might mean a period of forty or fifty years, which under all circumstances would be a most serious misfortune. If you come home this winter you probably could bring the specimens more safely than you can send them by mail, but this is a matter which you on the ground can decide better than any one else.

Your plan of sending the birds by mail in small packages is highly approved by Bangs and the other Museum authorities. Bangs says in his note to me: „Rock should keep account of what the packing of the birds costs and we will of course pay that also. I am very anxious to see his birds as he must have some very choice things. One thing, please tell Rock, and that is to ship his birds to us addressed simply to the Museum of Comparative Zoology, Cambridge, Massachusetts, and not to me personally, as the Customs people pass anything so addressed O.K., but I personally often have to go to some trouble when things come addressed to me.“

Of course you realize how highly I value the work which you have done for the Arboretum. I realize fully all the trouble, anxiety and danger you have been submitted to, and I will say again it is for this reason that I think it would be wise for you to come home just as soon as you can get there.
Faithfully yours,

P.S. We have received since this was written 132 negatives from you in two packages. I am arranging to cable at once $ 12,000. to the American Express Company at Shanghai for your use, fearing that you may never receive the money unfortunately sent to the Hongkong & Shanghai Banking Corporation at Peking. If this money reaches you, or any part of it, please cable me. If you haven't succeeded in obtaining any of the money from Peking I shall ask them to return it to Boston.

Mr. J. F. Rock
Choni, Kansu, China

42

December 16, 1926

Dear Mr. Rock:
The last letter received from you is dated October 12th. You will see by the enclosed that the American Express Company at Shanghai should now have received $ 24,000. for you. I hope this is true and that you are able to get what you want from Shanghai. The American Express Company have not cabled me here on the subject and of course I want to know that you have received the funds. I do not know what else I can do in this matter but if there is anything do not fail to let me know by cable.

I agree with you that it is desirable that the Tebbu forests should be fully explored, but Mrs. Potter and I have worried about you. You have had such a hard journey that we have dreaded a nervous breakdown for you, so I cabled you to use your own judgment about coming home and I hope you will do this. Your health is more important to the world than the exploration of the

Tebbu forests. On the ground you can judge of such matters much better than we can here.

Since October 1st, 1925, to date we have received 476 packets of seeds and after planting a full set we have made up 3790 packets and sent them to all the principal botanical and horticultural stations in the extreme northern part of North America, Europe and Asia. We have also received 521 negatives; those which came recently have not been developed. As soon as they are will be sent to you, also copies of your maps.

I am writing Bangs what you say in your last letter about ammunition and I am sure he will attend to it. I notice, too, what you say about your change of plan in sending the herbarium material to Boston and I only hope it will work well.

Conditions in Western China appear to be getting worse and it will be a great relief to see you safely in this country again.

I notice with much interest what you say about your diary and your proposition of writing a travel journal giving descriptions of the country, etc., also your suggestion of writing a preliminary article for some widely read journal. Just now I do not feel very enthusiastic about the *National Geographic Magazine*, for they have recently included in their list of explorations for 1926 your journey in China and Tibet, claiming it as their enterprise and saying nothing about the Arboretum. I wrote the President of the Society on the subject and he does not appear to have been very enthusiastic about my letter. Certainly it was a low down trick on their part. If you want to publish a preliminary article I think the Royal Geographic Society of London would be much better.

I understand fully that this expedition has added immensely to the geographical as well as to the natural history information of the world, and I think this information should be published after your return to the Arboretum, but this matter of course I am willing to leave with you. I do not think that with the present condition of things in western China another expedition from any country will visit the region which you have explored.

I notice what you say about the bad packing of envelopes and will see in the future if we cannot do better in this matter. Under the new arrangement with our Government your packages of seeds

come here now unopened and are examined here at the Arboretum, so I see what an excellent packer you are. I am most anxious for your herbarium material, for if anything should happen to it either in China or en route it will be impossible to determine your plants until those raised from your seeds flower and produce fruit here, and this may be in twenty-five to fifty years or more. This would mean that many persons to whom I have sent your seeds would get tired of caring for plants which they could not name and that many of them would be lost.
With the compliments of the season and all good wishes for the New Year, I am,
Faithfully yours,

Mr. J. F. Rock
Choni, Kansu, China

43

January 12, 1927

Dear Mr. Rock:
We have lately received from you letters dated October 30th (containing a wood specimen of a Juniper), October 31st, November 4th and 6th. We have also received from you in good condition one hundred and sixty lots of seeds, thirty-four packets of cones, and five bundles of Poplar and *Salix* cuttings. The five parcels announced in your letter of November 6th containing maps, diaries and other personal belongings have not yet arrived and will not be openend until your return.

I of course I am always anxious about you and most anxious to see you safe in the United States before I pass into the next world. For this reason, and as a matter of economy, I do not think we can consider your plan of making specimens of the wood of the trees from which you have collected or may collect seeds.

I understand that you are ready to devote the spring and autumn of this year to further exploration in the Tebbu country in order to make the collection of the seeds and herbarium specimens in that important region as complete as possible. If you do this I hope you can leave China in time to be back here in the spring of

1928. Since your letters were written, political affairs in western China are certainly much worse, but I believe the foreign powers will prevent the Cantonese force from attacking Shanghai.

I hope that the attack of flu is over by this time, but it is a bad thing and I fear that you haven't been able to take very good care of yourself.

I am very glad to find that you say nothing in recent letters about a shortage of funds, so I hope the money sent to the American Express Company at Shanghai reached them all right and that you are able to get what you need. Do no hesitate to cable here if anything is going wrong that I can help you about. You have certainly made a wonderful success and I am more than pleased with what you have accomplished.

I am showing your letter to Mrs. Potter and she will write to you. Cheer up (not that you need this advice), and with all good wishes, believe me,

Always faithfully yours,

Mr. J. F. Rock
Choni, Kansu, China

44

January 21, 1927

Dear Mr. Rock:

I received his morning your cable of January 20th and I am much relieved to hear from you again, as affairs in China seem to be in a worse condition than ever and I do not see much chance of improvement in that benighted country.

Your packages of personal belongings have not arrived yet but I think we have all your seeds and cuttings in good condition.

We are having a most disagreeable winter with very changeable weather and already have had thirty-seven inches of snow. Now it is as warm as summer and sickness in this country is unusually prevalent this year.

Keep well and do not take any unnecessary risk.

Always faithfully yours,

Mr. J. F. Rock
c/o American Consul
Tatsienlu, Szechwan, China

45

February 3, 1927

Dear Mr. Rock:

We have received recently letters dated Choni, October 24, 25, and 28, November 18, 22 and 28, December 14 and 16, 1926. we also received on December 31, 1926, thirty bundles containing one hundred and sixty packages of seeds and twelve packages of cones, one wood specimen and five bundles of cuttings; on January 15, 1927, eleven bundles containing forty-three packages of seeds, and on January 25 eleven bundles containing forty-three packages of seeds and two packages of negatives.

In your letter of October 25th you ask for my approval of your selling surplus or unneeded material. This I think would be a very wise thing to do.

The additional draft for $ 3000. was sent several days ago to the American Express Company at Shanghai for you, and I have today asked the Bursar to have deposited in the American Surety and Trust Company at Washington $ 1000. for you.

On November 22nd, 1926, you wrote that you had sent via Lanchow eighty-six boxes of herbarium specimens and twenty boxes of birds to the American Express Company in Shanghai by registered mail, but I haven't heard from Shanghai that any of these have reached the Express Company. I cabled them yesterday on the subject, asking them to wire me a report, but as yet have received no answer. I confess I feel a good deal of anxiety about these collections, still am hopeful for the best.

I am sending this letter in duplicate to the two addresses you sent me, and I shall hope to hear soon that you are progressing well on your journey toward Boston where you may be sure you will be very welcome.

No news here. The weather is mild for this season of the year and the snow has entirely disappeared.

With kind regards and all good wishes, I am,

Faithfully and gratefully yours,

Mr. J. F. Rock
c/o China Inland Mission
Tatsienlu, Szechwan, China

Note: One copy of this letter sent to Mr. Rock, c/o Mission, Likiang, Yunnan, China

BRIEFWECHSEL MIT JOHANNES SCHUBERT 1935-1961

Johannes Schubert (1896-1976) ist vergleichsweise wenig bekannt. Die Ursache mag sein, daß er als Bibliothekar im Deutschland der Hitler-Zeit und als Professor in der DDR weniger Außenkontakte und weniger Publikationsmöglichkeiten hatte als Gelehrte in anderen Ländern. Immerhin liegt eine Gedenkschrift für ihn vor, die zwar nur bescheidene biographische Angaben, dafür aber ein Schriftenverzeichnis liefert:[1]

Asienwissenschaftliche Beiträge. Johannes Schubert in memoriam. Herausgegeben von Eberhardt Richter und Manfred Taube. Berlin: Akademie-Verlag 1978. 204 S., 44 Taf. (Veröffentlichungen des Museums für Völkerkunde zu Leipzig 32).

Johannes Schubert stammte aus Annaberg, wo er am 7. 9. 1896 geboren wurde. Sein Vater, Guido Emil Schubert, war kaufmännischer Angestellter, seine Mutter Minna Ida Hausfrau. Er besuchte die Höhere Bürgerschule und das Realgymnasium in Plauen, legte das Abitur aber erst 1920 in Leipzig ab, bedingt durch den Weltkrieg und familiäre Umstände. Nach kurzen philosophischen und theologischen Studien wandte er sich der Orientalistik zu und studierte ab 1924 in Leipzig und ab 1926 in Berlin Chinesisch (bei August Conrady, Otto Franke und Erich Schmitt), Tibetisch (bei Friedrich Weller und August Hermann Francke), Türkisch (bei August Fischer) sowie Mongolisch und Mandjurisch (bei Erich Hauer und Erich Haenisch). Nach seiner Promotion mit einer

---

1 Darüber hinaus liegen vor: Eberhardt Richter Johannes Schubert (1896-1976). *Namhafte Hochschullehrer der Karl-Marx-Universität Leipzig* 5.1984,66-74. – E. Richter, M. Taube: In memoriam Johannes Schubert. 7.9.1896-2.8.1976. *Wissenschaftliche Zeitschrift der Karl-Marx-Universität Leipzig,* Gesellschafts- und sprachwiss. Reihe 1977, 169-171. H. Walravens: Briefwechsel Johannes Schuberts mit Ernst Schäfer und Bruno Beger. *NOAG* 175-176.2004, 165-224. – Herr Dr. Manfred Taube, Schüler von Johannes Schubert und bester Kenner seines Œuvres und seines Nachlasses, war so freundlich, das Manuskript durchzusehen; ich verdanke ihm zahlreiche Ergänzungen und Korrekturen.

Arbeit über Tibetische Nationalgrammatik trat er 1928 in den wissenschaftlichen Bibliotheksdienst ein und war bis 1954 als Bibliotheksrat und Fachreferent für Orientalistik an der Universitätsbibliothek Leipzig tätig. Als Interessenschwerpunkt zeigt sich schon früh Zentralasien, Lamaismus und Bibliographie. Seit 1952 war Schubert auch am Ostasiatischen Seminar der Universität tätig, dessen Leitung er von 1959 bis 1965 innehatte. 1960 erhielt er dort einen Lehrstuhl für Tibetologie. Er hatte Gelegenheit zu mehreren Reisen nach Ostasien, so nach China (März – August 1955), in die Mongolei (1957, 1959 und 1961) sowie nach Assam und in das Himalaya-Gebiet (Oktober 1958 – Januar 1959).

Es mögen Schriftprobleme gewesen sein, die Schuberts Interesse an den Na-hsi geweckt haben. 1930 hatte er einen Artikel *Tibetische Schriftprobleme*[2] publiziert und sich auch über *Lateinschrift oder Nationalschrift in China* geäußert.[3] Jedenfalls war er von Rocks Arbeiten und Publikationen sehr beeindruckt und hat sich, wie dem Briefwechsel zu entnehmen ist, selbst intensiv mit den Na-hsi-Texten befaßt, freilich nur auf der Grundlage von Rocks Publikationen, da ihm sonst kein Material zur Verfügung stand. Aus dieser Sicht ist sein Beitrag *Na-khi-Piktographie. Notizen über eine wenig beachtete Bilderschrift*[4] durchaus beachtenswert.

Die folgende Korrespondenz belegt die Zusammenarbeit zwischen beiden Gelehrten, gewissermaßen *by proxy*, denn nach dem Briefwechsel ist nur ein persönliches Treffen in Leipzig belegt. Darüber hinaus nimmt auch das Arbeitsumfeld Konturen an: Krieg in Deutschland und eine schwierige Nachkriegszeit, Krieg auch in Ostasien und erschwerte Arbeitsmöglichkeiten. Unter günstigeren Vorzeichen hätte die Kooperation der beiden Forscher gewiß gewichtigere und dauerhaftere Ergebnisse gezeitigt. Den Briefen sind zwei Rezensionen Schuberts beigefügt, die er Rock als Anlagen übersandt hatte. Auch späterhin hat Schubert Werke Rocks besprochen: The ^{2}Zhi ^{3}mä funeral ceremony. Wien 1955. *OLZ* 52.1957, 543-544; Bilderschrift-Literatur der Na-Hsi. *OLZ* 62.1967, 117-123 [zu: *Na-khi manuscripts* sowie *Na-khi-English dictionary.*]

---

2 *Buch und Schrift.*4.1930, 51-67.

3 *Archiv für Schreib- und Buchwesen.*4.1930,145-148.

4 *Buch und Papier. Buchkundliche und papiergeschichtliche Arbeiten, Hans H. Bockwitz zum 65.Geburtstag dargebracht.* Leipzig: Harrassowitz 1949, 114-142.

Übersicht[5]

1. 19. 9. 1935
2. 27. 9. 1935
3. 19. 11. 1936
4. 1. 4. 1937
5. 30. 7. 1937
6. 4. 11. 1937
7. 18. 12. 1937
8. 7. 6. 1939
9. 27. 7. 1947
10. 22. 9. 1947
11. 20. 1. 1948
12. 22. 2. 1948
13. 17. 3. 1948 Von Günther Köhler an Rock
14. 15. 5. 1948
15. 27. 6. 1948
16. 14. 10. 1948
17. 27. 10. 1948
18. 17. 12. 1958 Von Rock
19. 21. 12. 1948
20. 7. 1. 1949 Von Rock
21. 29. 4. 1952
22. 24. 8. 1953
23. 30. 8. 1953 Von Rock
24. 3. 9. 1953
25. 20. 9. 1953
26. 5. 10. 1953
27. 22. 10. 1953
28. 16. 9. 1954
29. 26. 11. 1955
30. 27. 2. 1956
31. 4. 6. 1956
32. 8. 5. 1958
33. 6. 2. 1961
34. 7. 5. 1961

Rezension von *The ancient Na-khi kingdom of Southwest-China.*[6]
Rezension von *The Na-khi Naga cult and related ceremonies.*[7]

---

5 Die Briefe 6-8 befinden sich im Besitz der Staatsbibliothek zu Berlin, Orientabteilung (Depositum der Deutschen Morgenländischen Gesellschaft), die übrigen gehören dem Archiv der Hunt Library, Carnegie-Mellon University Pittsburgh. Für die Möglichkeit des Abdrucks sei beiden Institutionen bestens gedankt.

6 *OLZ* 48.1953,465-466.

7 *OLZ* 50.1955,79-81.

1. Leipzig, den 19. Sept. 1935

Sehr geehrter Herr Dr. Rock!
Als ich am 4. Sept. aus meinem Ferienurlaub zurückkehrte, fand ich Ihren Brief vom 31. Juli vor. Etwa 8 Tage später kam dann Ihre eingeschriebene Sendung mit dem Amnyi-Machen-Text an, und heute ging Ihr Brief vom 19. Aug. ein, dem die Photographie des Minya-Konka-Kloster-Textes[8] beilag. Für Ihre lieben Zeilen und für die Übersendung der Texte danke ich Ihnen herzlichst. Ich hoffe Ihnen meinen Dank in der Form noch besonders abzustatten, daß ich Ihnen eine einigermaßen anständige Übersetzung der beiden Texte schicken kann. Freilich bis zum Januar / Februar 1936 wird die Sache wohl dauern. Sollten Sie jedoch die Übersetzung des Amnyi-Machen-Textes für Ihre Arbeit über dieses Gebirge früher benötigen, so teilen Sie mir das bitte mit und ich werde Sie nicht im Stiche lassen. Mit Freuden entnehme ich aus Ihrem Schreiben vom 31. Juli, daß Sie voraussichtlich im kommenden Frühjahr nach Europa und vielleicht auch nach Leipzig kommen. Ich darf Ihnen wohl heute schon mitteilen, daß Sie in der Zeit Ihres Leipziger Aufenthaltes mein Gast sind. Bei dieser Gelegenheit könnten wir manche Arbeit usw. natürlich viel besser besprechen, als das schriftlich zu machen ist. Herr Prof. Hadl[9] hat mir übrigens Ihre Grüße auch schon bestellt und mir mitgeteilt, daß Sie ihm einen Artikel für die Zeitschrift *Artibus Asiae* zugesagt haben.[10] Auch

---

8 Manfred Taube vermerkt: "Gemeint ist das Mahâtmya für das Kloster Mi-ñag Gaṅs-dkar (vgl. dazu J. F. Rock: The glories of the Minya Konka. *National Geographic Magazine* 58.1930, 385-437, und Arnold Heim: *Minya Gongkar. Forschungsreise ins Hochgebirge von Chinesisch Tibet.* Bern, Berlin: Hans Huber 1933, S.70 [mit Abb.]). In der in Anm. 11 genannten Aufstellung der 1950 abgeschlossenen Arbeiten steht der Titel Die Gründungsinschrift für das Kloster des Mi nyag Gaṅs dkar – ich vermute, daß dieser Text damit gemeint ist. Auch diese Übersetzung wurde meines Wissens nicht veröffentlicht. (Anfang der 50er Jahre erhielt ich von J. Schubert eine Kopie des tibet. Textes und möchte ihn demnächst publizieren.)"

9 Vgl. Robert Fazy: Richard Hadl [1876-1944]. *Artibus Asiae* 1940/45, I-III, Porträt. – Hadl übernahm Artibus Asiae nach Salmonys Emigration und führte sie schließlich von der Schweiz aus weiter. Nach Kriegsende wurde wurde sie wieder von Salmony übernommen.

10 Ein Beitrag von Rock ist tatsächlich in dieser Zeitschrift erschienen: J. F. Rock: The birth and origin of the Dto-mba Shi-lo, the founder of the

glaube ich, daß der Aufsatz, der ja – wie Sie mir eben schreiben - über den Ursprung der Moso Bücher (Handschriften) handelt, außerordentlich interessant wird und überaus wichtig ist. Es muß wirklich in Bezug auf osttibetische Dinge und damit auch über die Moso nun einmal mehr veröffentlicht werden als das bisher geschehen ist. Abzüge meiner Arbeiten schicke ich Ihnen, sobald wieder etwas im Druck erschienen ist. Bei der Übersetzung des Amnyi-Machen-Textes werde ich in gewissen sachlichen Dingen Sie ab und zu doch um eine Auskunft bitten müssen. Ich habe mir den Text schon einmal vollständig – wenn auch nur flüchtig – angesehen und kann sagen, daß er für das richtige Verständnis nicht so ganz ohne ist und auch manche falschen Schreibungen (phonetischer Natur) usw. enthält, so schön wie er an sich auch geschrieben ist. Nun Sie werden schon bald von mir wieder hören.
Mit den besten Grüßen
Ihr sehr ergebener Johannes Schubert

Die Photographie des Minya-Konka-Kloster Textes behalte ich sehr gern und danke Ihnen auch dafür noch besonders. Die Photographie des Amnyi-Machen-Textes[11] mit dem – echt Tibetisch ! – sehr langen und eigenartigen Titel ཨཱ་གཉན་ཆེན་པོ་མ་རྒྱལ་སྤོམ་རའི་མཚོན་རྟོགས་བསྐང་བཤགས་བསྟོད་བསྐུལ་དང་བཅས་པ་དབྱར་ཀྱི་དཔལ་མོའི་རྒྱུན་འཕྲིང་དངོས་གྲུབ་ཀྱི་ཉི་མ་བདུན་འབར་ཞེས་བྱ་བཞུགས་སོ

Mo-so shamanism, according to Mo-so manuscripts. *Artibus Asiae* 7.1937, 1-85, inkl. Taf. 1-16.

11 Manfred Taube vermerkt dazu: “Gemeint sind zwei Ritualtexte, die wie ein roter Faden fast den gesamten Briefwechsel durchziehen (letzte Erwähnung 8. 5. 58). Entsprechend einer im Nachlaß von J. Schubert befindlichen Aufstellung von 1950 handelt es sich um folgende Arbeiten: 1.‹Das Ritual für die Berggottheit des Amnye-matschen. Aus dem Tibetischen übersetzt, erläutert und herausgegeben. (Erste Übersetzung des Ritualtextes. Gliederung der Zeremonie. Deutung schwieriger Textstellen. Manuskript seit 1938 bei Dr. Rock; sollte in China gedruckt werden)›; 2.‹Ritual für die Berggottheit des Amnye-Matschen (Gsol-mchod). Text, Übersetzung, Erläuterungen.› Schuberts Bearbeitung beider Texte sollte im Rahmen von Rocks Buch *The Amnye Ma-chhen range and adjacent regions* (Roma 1956) erscheinen, wurde aber – vermutlich wegen Tuccis Bedenken (siehe Briefe vom 4.6.56 und 8.5.58) – nicht veröffentlicht. Verbleib des tibetischen Textes und des Schubertschen Manuskriptes unbekannt.”

[Lha.gñan. chen.po. rma.rgyal.spom.ra'i. mṅon.togs. bskaṅ.bśags. bsdod. bskul. daṅ. bcas.pa. dbyar.gyi. dpal.mo'i. rgyan.'phreṅ. dṅos.grub.kyi. ñi.ma. bdun.'bar. żes.bya. bżugso] erhalten Sie mit der Übersetzung in bestem Zustande zurück.
Sch.

2. Leipzig, den 27. Sept. 1935

Sehr geehrter Herr Dr. Rock!
Recht herzlichen Dank für Ihren Brief vom 28. Aug., der gestern hier ankam. Ihre Zeilen mit den Mitteilungen über die Nakhi werden ja immer interessanter! In der Meinung, daß das literarische Material, soweit Sie es übersetzt haben, unbedingt veröffentlicht werden muß, stimmen wir wohl beide überein. Dazu kann ich Ihnen versichern, daß ich mein möglichstes tun werde, Ihnen im Hinblick auf die Veröffentlichung zu helfen. Soweit ich aus Ihren Briefen einen Einblick in das in Frage stehende Material genommen habe, würde ich an Ihrer Stelle folgendes machen:

1) den ersten Band, Geographie und Geschichte des alten Nakhi-Reiches enthaltend, herausbringen.

2) die eigentliche Literatur, d.h. die Reihen von Zeremonien in Einzelpublikationen in Zeitschriften (wissenschaftlich !) oder *Beiheften* zu solchen Zeitschriften veröffentlichen.

In dieser Hinsicht habe ich schon gestern mit Prof. Hadl Fühlung genommen. Er ist gern bereit, die Einzelpublikationen zu drucken, und zwar als Beihefte zur Zeitschrift *Artibus Asiae*. Soweit ich in seinen Betrieb Einblick habe, wird er die Sachen ja ausgezeichnet herstellen und im Hinblick auf die Reproduktion der Original-Texte denkt er mehr an Nachzeichnungen als an irgendeine photomechanische Wiedergabe der Originale. Ich persönlich würde mich freilich – gerade in diesem Falle – nur für die genaueste photomechanische Reproduktion entscheiden können. Vielleicht könnten Sie ja – wenn Sie an Prof. Hadl schreiben – das gleich zur Hauptbedingung machen, daß nur beste photographische Wiedergabe der Originaltexte in Frage kommt! Eine andere Frage dürfte hinsichtlich der Finanzierung zu regeln sein, und was ich Ihnen da jetzt schreibe, bitte ich aber ganz vertraulich behandeln zu wollen: Prof. Hadl – er ist nicht Dr. und hat seinen Professortitel von der Kunstschule in Florenz oder da irgendwo – steht in seinem Betriebe eigentlich allein da. Ob und wieviel

Arbeiter er beschäftigt, weiß ich selbst nicht. Auf alle Fälle ist mir aber bekannt, daß er selbst mit am Setzkasten arbeitet und auch sonst – neben seiner Verlagstätigkeit viel technisch bei seinen Drucken selbst macht, so z. B. Zeichnen und Herstellen chinesischer Schrifttypen, weil ihm chinesischer Satz fehlt. Die in der Zeitschrift *Artibus Asiae* benutzten chinesischen Charaktere hat er selbst gezeichnet und geschnitten. Mir und auch soviel ich weiß anderen gegenüber hat er stets die finanziellen Schwierigkeiten betont, die er habe, und was die Zeitschrift betrifft, so glaube ich auch, daß das die Sache nicht trägt und daß viel am Bestehen nur dem ideellen Gedanken Hadl's, die Wissenschaft zu fördern, zuzuschreiben ist. Wieweit aber sonst sein Betrieb sich aufrecht erhält und wieweit nicht, das kann ich nicht wissen. Honorar zahlt er für die Arbeiten in der Zeitschrift nicht. – Nach diesen Zeilen nun werden Sie auch das folgende verstehen, was ich Ihnen im Auftrag von Prof. Hadl schreiben soll, daß er nämlich gern Ihre Arbeiten drucken will und daß er evtl. wenigstens um eine kleine finanzielle Unterstützung dabei bitten würde. Sie können sich ja mit Prof. Hadl deswegen nochmals direkt in Verbindung setzen. Er hat Ihnen übrigens schon geschrieben betr. des Artikels für *Artibus Asiae*, und zwar hat er den Brief durch Luftpost geschickt. Er bittet Sie, ihm mitzuteilen, ob dieser Brief durch Luftpost wirklich schneller zu Ihnen gelangt, damit sich evtl. weitere Korrespondenz viel rascher abwickeln läßt. – Sollten Sie wünschen, daß ich mich noch wegen Ihrer Sache mit anderen Verlegern in Verbindung setze, so will ich das gern tun und mich, soweit es sich um günstigste Bedingungen für Sie handelt, voll und ganz einzusetzen; denn das halte ich Ihnen gegenüber für meine Pflicht. Freilich wird mancher Verleger die Sache selbst erst in Augenschein nehmen wollen. Doch ist darüber vielleicht noch zu reden. Sollten Sie – wie Sie mir mitteilten, tatsächlich im kommenden Frühjahr nach Leipzig kommen, so wäre es am einfachsten, wir könnten diese Angelegenheit hier persönlich regeln. Doch ganz wie Sie denken, das sind nur meine Vorschläge! Wegen Übersetzung der Texte, die Sie mir schickten, hatte ich schon im letzten Briefe entsprechende Mitteilung gemacht. Was den Text im Kloster am Mi.nyag.Gaṅs.dkar anbelangt, so muß ich sagen – wie Sie schon in Ihrem Artikel in *Nat. Geogr. Mag.* andeuten –, daß es sich nicht um eine besondere Geschichte des Klosters (etwa wie um einen Dkar-chag) handelt, sondern mehr um die Gründung von Klöstern in buddhist. Sinne allgemein und um den Berggott im besonderen. Der Name Gaṅs-dkar und Gaṅs-chen kommt des öfteren vor. Auch eine Datierung ist am Schluß, die ich aber noch nicht eindeutig bestimmen kann.[12] Ferner enthält der in tibetischer Kursive geschriebene Text sehr viel Abkürzungen, die aufzulösen nicht ganz leicht ist. Unter diese fällt auch der Name des Berggottes, der nur einmal im Texte und zwar gleich im zweiten Absatz der fünften Zeile (wenn man die Lâñcâ-Überschrift mitzählt) vorkommt, indem von einer Meditation des Gottes die Rede ist. Sie schreiben, er heißt Rdo.rje lu tru, wobei die letzten beiden Silben phonetisch anzusehen sind.[13] Wenn das der Fall ist, so würde ich an einen schrifttibetischen Namen རྡོ་རྗེ་ཀླུ་ཕྲུག་ Rdo.rje. klu.phrug (also = Naga-Sohn) oder ཀླུ་གྲུབ་ klu-grub denken. In der Abkürzung des Namens im Texte, wenigstens kann ich die Stelle nur so verstehen, steht aber r/j/d/o/e – l/t/u, wobei letzteres doch deutlich auf eine hier schriftlich fixierte Form lu.tru geht. Sollten Sie Gelegenheit haben, von irgend einem Lamaisten (Tibeter) darüber Auskunft zu erhalten, so wäre ich Ihnen für eine Mitteilung sehr dankbar. – Übrigens danke ich Ihnen auch für die Notiz über den jetzigen Abt von Kumbum (im Briefe an Prof. Hadl), die ich mir in mein Handexemplar notiert habe und bei Herausgabe des ganzen Textes und Übersetzung des Kumbum-Führers mit berücksichtigen werde.[14] Ebenso danke ich Ihnen für die Übersendung der

---

12 Manfred Taube: "Das Kolophon enthält nur den Hinweis, daß der Text in der 1. Hälfte eines *chu-stod*-Monats verfaßt wurde."

13 Manfred Taube: "Die von Schubert zitierten Silben ‹rdo-rje ltu›[!] beinhalten keinen Götternamen, sondern sind Abkürzung für rdo-rje lta-bu, ‹vajra-gleich›. Der von Rock genannte Name, Dorjelutru, entspricht schrifttibetischem rDo-rje-blo-gros (auf Zeile 41 des genannten Textes erwähnt). – Die in Schuberts Briefen vom 26. 11. 55 (letzter Abschnitt) und 4. 6. 56 (2. Abschnitt) angedeuteten Bedenken zu seiner Übersetzung beziehen sich vermutlich nicht nur auf die zwei ‹Amnye Machen-Texte›, sondern auch auf dieses Blatt vom Kloster Mi-ñag Gaṅs-dkar."

14 Manfred Taube: "In der Schubert-Sammlung der Bibliothek des Museums für Völkerkunde Leipzig befindet sich ein Blockdruck mit dem Titel *gDan-sa chen-po sKu-'bum Byams-pa-gliṅ-gi gdan-rabs Mu-tig 'preṅ-ba-daṅ rTen-daṅ brten-par bcas-pa'i dkar-chag ched-du brjod-pa rNa-ba'i bcud-len źes bya-ba*; Vorarbeiten für eine Edition oder Übersetzung des Textes fanden sich nicht im Nachlaß. Vgl. den ca. 5 Jahre zuvor fertiggestellten Aufsatz ‹Die Äbte von Kumbum› (*Artibus*

Photographie des Wandgemäldes im Gaṅs-dkar-Kloster, das den Rdo.rje lu.tru darstellt. Die Kleinigkeiten sind darauf doch besser zu erkennen als in Ihrem Artikel im *Nat. Geogr. Mag.*

Für heute möchte ich damit meine Zeilen schließen und bleibe mit den besten Grüßen
Ihr sehr ergebener Johannes Schubert

Leipzig C 1, Mozartstraße 13

3. Leipzig, den 19. Nov. 36

Sehr geehrter Herr Dr. Rock!
Verbindlichsten Dank für Ihren Brief vom 1. November 36. Ich bin gleich nach Empfang desselben bei Prof. Hadl gewesen und dort haben wir Ihnen auch gleich die Postkarten sowohl nach Peiping als auch nach Yünnanfu geschickt. Hoffentlich haben Sie den entsprechenden Brief von Prof. Hadl, den ich zum Teil vorgelesen bekommen habe, erhalten. Aus ihm geht ja hervor, daß Prof. Hadl die Aufsätze sehr gern druckt. Was nun die gewünschte Übersetzung des Amnyi Machen-Textes betrifft, so werde ich Ihnen diese vollständig nun sehr bald senden. Die Analyse haben Sie wohl inzwischen bekommen. Heute kann ich Ihnen auch die Analyse des Dgra.lha-Textes[15] senden, dessen vollständige Übersetzung dann auch nicht lange mehr dauern wird. Aus diesem Text habe ich Ihnen einige Stellen notiert, in denen vom Rma.chen.spom.ra die Rede ist, im Falle Sie das für Ihre Arbeiten schon dringend brauchen. Darf ich Sie vielleicht bitten, mir immer gleich das mitzuteilen, was Ihnen etwa in der Analyse schon unrichtig oder nicht den Realien entsprechend vorkommt. Haben Sie ähnlichen Zeremonien beigewohnt? Wie sahen die besonderen

---

Asiae 4.1930/32,220-235)."

15 Manfred Taube: "In seinem Aufsatz Das Ritual für Rta Thugs.dkar.po (*MIO* 8.1961, 51-63) erwähnt Schubert auf der ersten Seite, daß er dabei sei, ein Ritual für den Dgra-lha mit dem Titel *Rgyal brṅan lha bsaṅs bžugs so* zu übersetzen; seine Absicht, seine Übersetzung an Rock zu schicken (siehe Briefe vom 15. 9. 53, 5. 10. 53, 16. 9. 54), wurde meines Wissens nicht realisiert. Über den Verbleib des tib. Ms. und der Übersetzung Schuberts ist mir nichts bekannt."

Gtor.ma's aus? Hier versagen ja die Wörterbücher und sonst alle Hilfsmittel völlig; so z.B. bei dem im Dgra.lha-Text genannten chab.gtor (Wasser-Gtor.ma) und Dmar.gtor (Roter-Gtor.ma). Die Texte sind sehr schwer zu verstehen; ich bin aber jetzt so ziemlich dahinter gekommen; es fehlt manchmal nur noch ein wirklich sachlich richtiger Ausdruck.

Besitzen Sie überhaupt – weil ich gerade daran denke – einheimische tibetische Kalender für die einzelnen Jahre? Da sind hier keine aufzutreiben, und ich brauche eigentlich dringend einige für eine Arbeit, die in *Artibus Asiae* erscheinen soll. [Es ist die Übersetzung eines tibetischen astrologischen Werkes, das lediglich die 八卦 *pa-kua* und ihre Beziehungen zum Menschen behandelt.[16]] Falls Sie also solche Kalender besitzen sollten, würden Sie mir leihweise dieselben zur Verfügung stellen?

Mit Freuden stelle ich aus Ihrem Briefe fest, daß Sie viele chinesische Bücher 地方志 der Grenzprovinzen Westchinas gekauft haben. Gerade diese müssen nämlich einmal systematisch durchforscht werden und erst kürzlich, am 13. Oktober, habe ich hier in Leipzig auf der Tagung der Gesellschaft für Völkerkunde zu Leipzig einen Vortrag gehalten, betitelt: *Einheimische Quellen zur völkerkundlichen Erforschung Tibets und des chinesisch-tibetischen Grenzgebietes*,[17] in dem ich auf die Durchforschung auch dieser Literatur hingewiesen habe. Allerdings hatte ich den Eindruck, daß unter den Zuhörern von Tibet niemand etwas gewußt noch verstanden hat! Nun, der Vortrag wird auch gedruckt, und ich werde Ihnen ja dann einen Abdruck senden.

Übrigens hatte mir seinerzeit Herr Baron von Staël-Holstein[18] einige Sonderdrucke geschickt, und ich hatte ihm im damaligen

---

16 Manfred Taube: "Es handelt sich um den im Brief vom 22. 2. 48 unter Nr 1.2 genannten Text (siehe dort)."

17 Manfred Taube: "Siehe dazu Brief vom 15. 5. 48 und die dazugehörige Anmerkung."

18 (1876-1937). Vgl. Ernst Schierlitz: In memory of Alexander Wilhelm Baron von Staël-Holstein. *Monumenta serica* 3.1938,286-291; Johannes Schubert: Baron A. von Staël-Holstein† *Artibus Asiae* 7.1937, 227-229. – M. Taube: "In einem Brief vom 9. 12. 1932 berichtet Rock an B. Laufer, daß er in Peking mit ‹Baron Stael de Holstein› zusammengetroffen sei (*Kleinere Schriften von Berthold Laufer*. Teil 3. Stuttgart: Steiner 1985,404). – Siehe dazu meine Bemerkung zum Brief vom 3. 9. 53."

Dankschreiben die Übersendung einiger Sonderdrucke meinerseits versprochen. Leider ist diese Sendung von mir bis heute nicht abgeschickt worden. Ich komme mir deshalb sehr schuldbewußt vor und getraue mir die Sache nicht mehr abzuschicken. Dürfte ich Sie bitten, falls Sie noch in Peking weilen, vielleicht Herrn Baron von Staël-Holstein einmal zu fragen, ob ich ihm die zurechtgelegten Sachen noch schicken darf?

Doch nun möchte ich schließen, da der Brief gleich morgen Freitag mit der China-Post noch weg soll.

Mit den besten Grüßen
Ihr stets ergebener
Johannes Schubert

Leipzig C 1, Mozartstraße 13
Anhang 5 Blätter: Analyse [liegt nicht bei]

4. Leipzig, den 1. April 1937

Lieber Herr Dr. Rock!
Gestern erhielt ich Ihren freundlichen Brief vom 11.März. Ich danke Ihnen herzlichst für die Mitteilungen, aus denen ich auch ersehe, daß es Ihnen wieder besser geht. Wenn Sie nach Europa kommen in diesem Sommer, so wäre das ja großartig, und wir könnten ja vieles besprechen, was sich schriftlich gar nicht so machen läßt. Jedenfalls sind Sie dann – wie ich Ihnen schon früher einmal schrieb – hier in Leipzig mein Gast. Meinen diesjährigen Sommerurlaub hoffe ich im August zu bekommen und will da nach Mittenwald (ins Karwendel) oder – wenn ich die Devisengenehmigung bekomme – ins Ötztal (nach Längenfeld), um dort mit meiner Frau einige Hochtouren und Klettertouren zu unternehmen. Das ist immer meine einzige Erholung im Jahr, sonst komme ich kaum in ein Restaurant oder sonstwo hin.

Was nun Ihre sehr wichtigen Fragen betrifft, so kann ich dazu folgendes schreiben:

1) Sie haben sehr recht, wenn Sie, falls ein Verlag die Mo-so-Arbeiten (Zeremonien usw.) übernimmt, darauf bedacht sind, daß das auch mit Sicherheit veröffentlicht wird, d.h. eben unter allen Umständen gedruckt wird.

2) Die Veröffentlichung durch die Ecole Française d'Extreme-

Orient stellt freilich die ganze Angelegenheit völlig sicher, wie mir scheint.

3) Wenn Prof. Hadl die Veröffentlichung übernimmt, wird der Druck bestimmt vorzüglich; auf alle Fälle besser als anderswo. Die völlige Sicherheit der ganzen Publikationen ist freilich vielleicht nicht so sehr garantiert: denn einmal ist Prof. Hadl so um die 60 Jahre alt, hat vermutlich niemand weiter zur Hand, der im Falle seines Todes die Sache zu drucken durchführen würde und könnte. Finanziell wird es ihm wohl schwer möglich sein, das allein durchzuhalten, und ich muß Ihnen hier im Vertrauen gestehen, daß er im Stillen auf einen Zuschuß von Ihrer Seite hofft. Ich habe ja ziemlich viel Einblick in seinen Verlag [– es sind nur wohl 3 Räume, von denen ich wiederum nur 2 kenne; dann ist noch ein Fräulein da, welches die Kontorarbeiten usw. erledigt und die übrigens außerordentlich viel tätig ist und das ganze Unternehmen durch ihre Arbeit, ihren Blick usw. mit hält und fördert; Prof. Hadl und dieses Fräulein unternehmen eigens den Satz (Handsatz) und nur ab und zu wird ein Drucker oder zwei angestellt.] Aber so alles kennt man natürlich doch nicht, finanzielle Dinge gleich gar nicht. Mir gegenüber – er ist da allerdings sehr offen – behauptet er jedenfalls stets, es sei an seinen Verlagswerken nichts verdient und die Zeitschrift (*Artibus Asiae*), die nach den geschilderten Verhältnissen allerdings bestimmt eine große Leistung darstellt, scheint sich vielleicht geradeso zu tragen. An dieser hängt er mit der größten Liebe. – Ihre Arbeiten müßten ja freilich extra, als besondere Werke erscheinen und – wie gesagt – so ganz allein kann Prof. Hadl das finanziell, meines Erachtens nach, nicht machen. Er würde sich bestimmt weitgehendst anstrengen, das auf alle Fälle! – Am besten wäre natürlich schon, wenn Sie einmal hier nach Leipzig kämen, könnten mit ihm selbst sprechen, in alles Einblick nehmen und dann entscheiden. Bitte behandeln Sie diese Mitteilungen von mir streng vertraulich und ich werde Herrn Prof. Hadl auch von Ihrem Brief vom 11. März lieber gar nichts sagen und die Grüße deshalb zunächst besser nicht ausrichten.

Hoffentlich hat Ihnen inzwischen der erste Teil des Amnyi-Machen-Textes gefallen und Freude gemacht. Der andere Teil ist in Arbeit (Schreibmaschine). Ich besitze leider selbst keine[19], so

---

[19] Die Schreibmaschine in der Bibliothek darf ich dazu nicht benutzen. [Anmerkung Schuberts.]

daß die Angelegenheit – wenn ich etwas mit der Maschine schreiben will, immer erst etwas umständlich ist. Mit den freundlichsten Grüßen Ihr [...]
Johannes Schubert

5. Leipzig, den 30.7.1937

Sehr geehrter Herr Dr. Rock!
Anbei unter Einschreiben wieder einen Teil des Amnyi-Machen-Textes. Bitte, nehmen Sie es mir nicht übel, wenn es noch nicht der Schluß ist. Aber es war beim besten Willen nicht möglich, bis jetzt die Arbeit ganz fertig zu schreiben. Im MS übersetzt liegt sie schon seit März vor! Auf alle Fälle freue ich mich aber sehr, daß die Arbeit noch in diesem Jahre gedruckt werden soll! Was den tib. Text anbetrifft, so würde ich vorschlagen, einmal den Originaltext (vielleicht etwas verkleinert!) abzubilden und dazu den richtig in Versen abgeteilten Text in Umschrift, den ich Ihnen zusammen mit dem Schluß der Übersetzung sende, abzudrucken.

Morgen trete ich auf vier Wochen meinen Urlaub an und werde mich bis Ende August im nördlichen – deutschen Teil – des Karwendel aufhalten, um dort einige Hoch- und Klettertouren zu unternehmen. Dann – im September – geht es an den Schluß des Textes. Es wird dieses Jahr übrigens auch noch eine andere Arbeit (tibetische Nationalgrammatik) von mir im Druck erscheinen.[20]

Was Ihren Vorschlag betr. einjährigen Aufenthalt in China betrifft, so wäre mir das natürlich äußerst erwünscht und angenehm. Doch hätte ich mich darüber gern einmal mündlich mit Ihnen ausgesprochen.

In der Hoffnung, daß Sie doch noch einmal hier nach Deutschland kommen werden, bleibe ich
Ihr stets ergebener
Johannes Schubert

---

20 *Das Sum-cu-pa und Rtags-kyi-'ajug-pa des Großlamas von Peking, Rol-pai-rdo-rje;* herausgegeben, übersetzt und erläutert. Leipzig: Offizin R. Hadl 1937. 104 S. (Artibus Asiae Supplementum 1.)

6. Leipzig, den 4. Nov. 1937

Sehr geehrter Herr Dr. Rock!
Indem ich hoffe, daß Sie indessen meine unter Einschreiben und mit Flugpost gesandte Postkarte erhalten haben (und hoffentlich auch inzwischen doch noch den zweiten Teil des Amnyi-Machen-Textes) sende ich Ihnen heute den versprochenen Brief mit der in Ihrem Brief vom 25. Sept. gewünschten Auskunft. Ich beantworte zunächst Ihre Fragen:

Bei dem mir skizzierten Bilde (Pantheon) handelt es sich um die Darstellung der berühmten Lamas der gelben Kirche mit den zugehörigen Schutzgottheiten.

བླ་མ་ཆོད་པའི་ཚོགས་ཤིང་ Bla-ma chod-pai tshogs-shiṅ

(Der Name des Gesamtbildes) heißt: Der Baum (shiṅ, Stammbaum) der gesamten Körperschaft der[jenigen] Lamas [der gelben Kirche], die [hauptsächlich] zu verehren sind.

Auf dieses Bild muß sich auch der von Ihnen genannte Text བླ་མ་མཆོད་པ་ དཔེ་ཆ་ *Bla-ma mchod-pa dpe-cha, Buch für die Verehrung der Lamas* beziehen.

[Das mir genannte zweite Bild (Pantheon), mit Buddha in der Mitte, ལམ་རིམ་ ཚོགས་ཤིང་ *lam-rim tshogs-shiṅ* ist *Der Baum (shiṅ, Stammbaum) derer, die den Stufenweg des Heils beschritten haben.*]

Die auf dem ersten Bilde oder für die Teile des ersten Bildes genannten Ausdrücke, richtig geschrieben wie folgt:

1) མཉམ་ལེན་བྱིན་རླབས་རྒྱུད་ mñam len byin rlabs rgyud

2) རྒྱ་ཆེན་སྤྱོད་རྒྱུད་ rgya-chen spyod rgyud

3) ཟབ་མོ་ལྟ་རྒྱུད་ zab-mo lta rgyud

sind Titel von Tantra's, die sich auf die beiden Richtungen beziehen, die im Lehrgebäude Tsoṅ-kha-pa's bestehen, nämlich

a) auf den praktischen Weg der Ausübung der Lehre im ausgedehnten Umfang (rgya chen spyod)

b) auf den theoretischen und mystischen Weg der tiefen oder inneren Anschauung (zab-mo lta-ba).

Vermutlich sind also auf dem Bild jeweils untereinander diejenigen Lamas dargestellt, die die betr. Wege besonders beschritten haben.

mñam len byin rlabs müßten die sein, denen beide Wege geläufig waren, nämlich:

Rdo-rje 'achaṅ (ist Vajradhara), 'Ajam dpal dbyaṅs (ist Mañjuśrî). grub thob ñi ma 'adzin ist der Zauberer (Mahasiddha) Ñi-ma 'adzin = *der die Sonne hält*, deshalb faßt er einen Sonnenstrahl. gsan-'adus (=Guhyakâlâ) sind dämonische Schutzgottheiten. Ebenso Phyag rdor 'abyuṅ bdud und der Yi-dam Skye-rdor. Zu der Umwelt gewisser Lamas gehören auch die 35 Stoṅ-śag. Der Ausdruck [śog-lṅa] ཤོག་ལྔ་ Steht für [so-lṅa] སོ་ལྔ་ d. ist 35.

Dus-gsum saṅs-rgyas sind Buddhas der drei Zeiten und Dpao-mkhâ 'agro sind Dpao's und Dâkinî's (Luftwandlerinnen). Für Dpao gibt es auch noch die weibliche Form Dpao-mo, die hier dem Ausdruck Mkhâ 'agro-ma entsprechen müßte.

Das ist alles, was ich Ihnen zunächst sagen kann, ohne das Bild gesehen zu haben. Bestimmt eine sehr interessante Sache!

Nun lieber Herr Dr. Rock! Daß Sie sich einsam fühlen, glaube ich bei Arbeiten auf unserem Gebiete sehr gern, das geht einem aber auch hier so. Können Sie denn nicht nach Europa kommen und die Sachen hier in Europa fertig bearbeiten? Da ginge natürlich alles viel schneller. Außerdem glaube ich, daß die Kriegsgefahr dort bei Ihnen ja viel größer ist wie hier. Bei uns denkt niemand an Krieg. – Jedenfalls kann ich Ihnen nochmals versichern, daß ich Ihnen gern auf alle Fälle für die Bearbeitung Ihres Materials mit helfen will. Es dauert eben nur manchmal etwas länger, da ich ja täglich volle 8 Stunden mit anderen Arbeiten (in der Bibliothek) beschäftigt bin. Sonst läßt sich aber alles machen. Auch die Texte in tibetischer Kursive werden entziffert!

Die Grüße an Prof. Hadl habe [ich] inzwischen auch bestellt. Er ist sehr gern bereit, noch mancherlei von Ihnen – in großartiger Ausführung! – zu drucken!

Mit den besten Grüßen
Ihr stets ergebener
Johannes Schubert
Leipzig C 1, Mozartstraße 13

Wie schon mitgeteilt, sende ich Ihnen in ca. 3-4 Wochen den Schlußteil des Amnyi-Machen-Textes. Sollte der andere Brief doch verloren gegangen sein, so müßte ich Ihnen den zweiten Teil auch noch einmal mitsenden. Sch.

7. Leipzig, den 18. Dezember 37

Lieber Herr Doktor Rock!
Indem ich Ihnen den Empfang Ihres Briefes vom 20. Okt. 37 sowie den Empfang der beiden tibetischen Texte (maṇḍala und Tshogs-shiṅ) bestätige, sende ich Ihnen gleich die Übersetzung des Maṇḍala-Textes mit. Bei einer Veröffentlichung müßte in diesem Fall doch der *Originaltext* mit publiziert werden, wegen seiner Eigenart. Wenn Sie über die Herkunft des Textes genau Auskunft geben können, müßte das alles noch mit gesagt werden. Auch wer der Schreiber war und in welcher Gegend er beheimatet war. Den Text selbst, den ich mir abgeschrieben habe, sende ich Ihnen mit dem anderen, den ich mir während der Weihnachtstage abschreiben werde, zusammen zurück, Anfang Januar. Dann schicke ich auch den Rest des Amnyi-Machen-Textes, *der fertig ist*, den ich aber noch nicht geschickt habe, weil ich glaubte, daß der Maṇḍala-Text zunächst Ihnen lieber ist.

In dem einen Briefe, in dem ich Ihnen die Fragen vom Tshogs-shiṅ-Bild beantwortete, muß die eine Gruppe གསང་འདུས་ gsaṅ 'adus [**nicht** gsaṅ-bdud] heißen; gesprochen Sang-dü. Ein ähnliches Bild ist auch in dem Buche von Alice Getty, *Gods of Northern Buddhism*[21] (am Schlusse), farbig veröffentlicht worden.

Doch heute soll das inzwischen genügen und ich sende Ihnen noch, wenn auch nachträglich eingehend, die besten Wünsche zum Neuen Jahre und zugleich für Ihre weiteren Arbeiten! Ich hoffe, im neuen Jahre Ihnen von mir aus schneller zur Hand gehen zu können.
Mit den besten Grüßen
Ihr stets ergebener
Johannes Schubert

8.
Herrn Dr. Jos. Franc. Rock
Dalat
Französ. Indo-China

Flugpost via Marseille - Hanoi.

---

21 Oxford: Clarendon Press 1928. LII, 220 S.

Absender:
Dr. Joh. Schubert
Universitäts-Bibliothekar
Leipzig C1
Mozartstr. 13

Leipzig, den 7 Juni 1939

Lieber Herr Dr. Rock!
Soeben Ihren Brief v. 8. Mai mit bestem Dank erhalten. Die goldene Inschrift am Haupteingang (Innenseite) des Klosters von Choni ist nach den tibetischen Lettern zu lesen: „oṃ padmo ṣṇî ṣa bi ma le hûṃ phaṭ» und entspricht in genauem, analysierten Sanskrit: „OṂ Padma uṣṇîṣa vimale HŪṂ PHAṬ“ und ist zu übersetzen: OṂ! DU [Vokativ auf e des Neutrums] reiner [Vimala] Lotus [Padma] – Scheitelschmuck [uṣṇîṣa]! HŪṂ PHAṬ! – Inzwischen sind Sie ja nun auch in den Besitz meines Briefes mit den Namensdeutungen (Amnye-machen) gelangt und ich hoffe, daß das Ihnen weiter hilft. Sie können mir – wie schon mitgeteilt – gern die anderen Namen aufschreiben, denn dies und jenes läßt sich bestimmt feststellen. Auch die tibetische Schrift über das Muli-Reich könnte nach dieser Richtung hin ausgewertet werden! – Ihre Notiz über das chinesische Labrang-Buch hat mich sehr interessiert, und ich will versuchen, hier ein Exemplar des chinesischen Druckes zu bekommen.
Inzwischen verbleibe ich mit den besten Grüßen
Ihr stets ergebener Johannes Schubert

8a.

II. In dem in den *Litterae Orientales* von Harrassowitz (1938, April Heft 74) Nr 264a angezeigten Buche von Walter Eugene Clark: *Two lamaistic pantheons*[22], kommt der Name der grünen Tara, auf dem Bilde, das Sie mir freundlicherweise schenkten, zu einem Teil (Rje.btsun.ma.kâ.pa.li-Târa རྗེ་བཙུན་མ་ཀཱ་པ་ལི་ཏཱ་ར་ <die Hauptfigur>) vor, aber auch nur erwähnt! Eine Abbildung oder Statue ist nicht davon im Buche enthalten. Wenn mir alle Dinge des Bildes völlig klar sind, werde ich Ihnen ausführliche Mitteilung machen und denke auch an eine Veröffentlichung, was wohl auch ganz in Ihrem Sinn liegt?

---

22 Cambridge, Mass: Harvard Univ. Press 1937. (Harvard-Yenching Institute monograph series 3.) 2 Bde.

9. Leipzig, den 27. Juli 1947

Lieber Herr Dr. Rock!
Mit diesen Zeilen versuche ich das erste Mal direkt wieder mit Ihnen in Briefwechsel zu kommen. Über Herrn Koc[23] in Wien habe ich es bereits einigemale versucht. Von Ihnen ist jedoch bis heute hier noch nichts eingegangen. In den Jahren, während denen wir nichts von einander hörten, hat sich ja so viel ereignet! Immerhin sind wir, meine Frau und ich, so leidlich durchgekommen, wenn auch zweimal ausgebombt. Meine Bücher konnte ich – durch vorherige Sicherung – retten und unsere wissenschaftliche Korrespondenz ist gleichfalls erhalten. Sobald unsere briefliche Verbindung wiederhergestellt ist und regelmäßig läuft, kann die wissenschaftliche Arbeit, die sehr notwendig und von internationalem Nutzen ist, wieder weiter gehen. Gesundheitlich geht es uns leidlich, ja – von den allgemeinen Zeitverhältnissen abgesehen – vielleicht sogar gut. Der Kampf ums Leben bleibt uns dabei freilich nicht erspart! Was das berufliche anbetrifft, so bin ich durchaus noch in meiner alten Stellung und Position. Schreiben Sie die Briefe an die Universitätsbibliothek, da geht es vielleicht noch am schnellsten. Leider kann ich von hier aus noch keine Flugpost senden und muß deshalb sehen, wie lange ein Brief an Sie und die Antwort zurück unterwegs sein wird.

Jedenfalls bin ich bestrebt, sobald als möglich meine Beziehungen zu Ihnen wieder aufzunehmen und ich würde mich außerordentlich freuen, bald von Ihnen zu hören.
Mit den herzlichsten Grüßen
Ihr Johannes Schubert

10. Leipzig, den 22. September 1947

Lieber Herr Dr. Rock!
Endlich ist unsere briefliche Verbindung wieder zustande gekommen! Vor einigen Tagen erhielt ich Ihren Brief vom 31. Juli ds. Js., dem auch das andere Schreiben vom 30. Januar beilag. Es ist mir ganz unerklärlich, wie dieser, an die Bibliothek adressierte Brief, als unbestellbar zurückgehen konnte. So danke ich Ihnen nun heute für Ihre Mitteilungen herzlichst und hoffe, daß Sie

---

23 Robert Koc, Rocks Neffe in Wien.

inzwischen auch meinen Brief – mit welchem ich gleichfalls versuchte, unseren Briefwechsel wieder anzubahnen – erhalten haben. Wollte ich Ihnen alles das, was wir seit dem Jahre 1939, d.h. seit Unterbrechung unserer Korrespondenz erlebt haben, in Einzelheiten schildern, so gebe das einen ganzen Roman. Die Hauptsache ist jedenfalls: wir leben alle noch, und es geht uns – von allgemeinen Zeiterscheinungen abgesehen – gut. Schwer bedauere ich den Verlust Ihrer Handschriften mit den Texten und Übersetzungen der Nakhi-Literatur; ich bewundere jedoch Ihren Mut, die ganze Arbeit von Neuem zu beginnen. Sie ist ja freilich für unsere Wissenschaft ungemein wichtig. Wir müssen auf diesem Gebiete tüchtig arbeiten, um vorwärts zu kommen. Da denke ich gerade an das Moso-Wörterbuch, von dem wir sprachen, als Sie uns seinerzeit mit Ihrem Besuch beehrten. Ich halte diese Arbeit für sehr dringend und wesentlich. Ihre neuerdings erschienene Publikation, über das *Ancient Na-khi kingdom of Yünnan....* wird mich äußerst interessieren, und ich möchte hoffen, daß es – wie Sie mir schreiben – durch Ihre Freundlichkeit und mit Ihrer Mühe gelingen wird, ein Exemplar hierher zu bekommen. Bisher sind ja wohl Büchersendungen aus dem Auslande noch nicht möglich gewesen. Der Wissenschaftler ist deshalb zur Zeit noch etwas abgeschlossen im Hinblick auf die Erscheinungen der Fachliteratur in den Ländern der Welt. So ist es mir sehr willkommen, in Ihrem Briefe von dem neuen Buche Bacot's (Tun huang-MSS)[24] zu lesen, von dem mir bis dato nichts bekannt war, und ich wäre Ihnen auch fernerhin sehr dankbar, wenn Sie mir immer einmal über solche Neuerscheinungen etwas mitteilten. Sollte ich nun dank Ihrer Hilfe ein Exemplar Ihres neuen Buches bekommen, so bin ich gern bereit, über das Buch etwas zu schreiben – vielleicht in Form einer Geschichte der Erforschung der Moso oder so ähnlich –, das dann hier und vielleicht auch bei Ihnen in irgendeiner wissenschaftlichen Zeitschrift veröffentlicht werden kann.

Sie wollten aber auch gern von mir und meiner Tätigkeit hören und ich will Ihnen das Wesentlichste schreiben: Zunächst versehe ich – wie schon früher – meinen Dienst als Bibliotheksrat an der hiesigen Universitäts-Bibliothek. Daneben fange ich nach und nach

---

24 *Documents de Touen-houang relatifs à l'histoire du Tibet* / J. Bacot, F. W. Thomas, Ch. Toussaint. Paris: Geuthner 1940 [=1946]. 204 S. (Annales du Musée Guimet. Bibliothèque d'études 51.)

an, meine Spezialstudien wieder aufzunehmen. Es dauert jetzt alles nur etwas länger und schreitet nur successive vorwärts, da der täglich erneute Kampf um den Lebensunterhalt viel Zeit für sich in Anspruch nimmt (wie lange ich das noch weiterhin gesundheitlich aushalte, ist nicht abzusehen). Ich habe verschiedene Arbeiten (so z.B. über das Reismandala[25], über ein Bild im Campo Santo zu Pisa, über das *Vaidûrya dkar.po*[26], über lamaistische Medizin u.ä.) zum Teil als Materialsammlung, zum Teil auch schon manuskriptmäßig (auch der Dgra-lha-Text ist teilweise übersetzt) vorliegen, aber noch nicht zum Abschluß gebracht. Das soll nun nach und nach werden. Wenn ich an Ihre geplante historische Geographie von Nordost-Tibet denke, so wäre eigentlich die Zusammenarbeit mit Ihnen an Ort und Stelle das gegebenste und würde für die internationale Fachwissenschaft ein glänzendes Ergebnis liefern: – aber es bleibt sicher nur ein Ideal, an dessen Verwirklichung doch nicht zu glauben ist. Auf jeden Fall muß unser Briefwechsel auf dem Laufenden bleiben und ich beantworte zunächst Ihre Fachfragen und bearbeite die entsprechenden Texte usw. so, wie Sie mir dieselben zustellen können und so gut ich in der Lage bin, Ihnen – hoffentlich werden auch bald die Postverkehrsverhältnisse mit dem Auslande besser – die Bearbeitungen wieder zukommen zu lassen.

Mit großer Spannung sehe ich nun Ihrer Antwort entgegen und freue mich schon ganz außerordentlich, daß wir uns nun doch wieder – wenn auch nur brieflich – austauschen können. Mit Ihrem Herrn Neffen stehe ich auch in Verbindung, so daß ich von ihm auch immer mal etwas höre. Es wird Sie vielleicht sehr interessieren, daß Herr Dr. Günther Köhler in Dresden ist[27]. Er ist Dozent an der Technischen Hochschule und gleichzeitig an der Bergakademie Freiberg/Sachsen. Er besucht mich manchmal in Leipzig, und wir wollen jetzt zusammen ein kleines Buch herausgeben.

---

25 Veröffentlicht später als Das Reis-Maṇḍala. Ein tibetischer Ritualtext. *Asiatica. Festschrift Friedrich Weller zum 65. Geburtstag gewidmet.* Leipzig: Otto Harrassowitz 1954, 584-609.

26 M. Taube: "Vermutlich ist der in den Bemerkungen zum Brief vom 22.2.48 genannte Aufsatz gemeint (‹Tibetische Äquivalente ...›)."

27 Vgl. Walter Fuchs: Günther Köhler in memoriam, 1901-1958. *Oriens extremus* 5.1958,246-251. – M. Taube: "Geograph, Prof. in Peking (1932-1938), Freiburg, Dresden, Berlin; bereiste 1933/1934 den Nordosten Tibets."

Mit den herzlichsten Grüßen – zugleich von meiner Frau, die Ihnen für die besonderen Grüße bestens dankt – bleibe ich
Ihr ergebener Johannes Schubert

Wie lange der Brief gehen wird, bis er Sie erreicht, weiß ich nicht. Flugpost haben wir hier leider keine.

Soll ich wegen Ihres anderen Buches einmal mit Leipziger Druckern und Verlegern Fühlung nehmen? Ich werde inzwischen einmal unverbindlich hören, wie es aussieht mit der evtl. Herstellung Ihres Werkes im Druck. Sch.

11. Leipzig, den 20.1.48

Lieber Herr Dr. Rock!
Vielleicht darf ich mich der Hoffnung hingeben, daß Sie meinen Brief vom 22.9.47 inzwischen erhalten haben. Heute versuche ich, Ihnen einige Notizen durch Vermittlung Ihres Herrn Neffen zukommen zu lassen, in der Meinung, daß dies ein Weg ist, auf dem die Nachricht Sie doch etwas schneller erreicht. Diese Nachricht ist jetzt aber nicht wissenschaftlicher Art, sondern entspringt ganz der Not und dem Zuge der Zeit. Ich möchte Sie nur bitten, soweit das möglich sein sollte, mir gelegentlich einmal etwas Tee zu senden. Das ist das Getränk, das ich hier am meisten entbehre und bei meinen Arbeiten des öfteren benötigte. Wenn Sie darüber hinaus noch in der Lage wären, meine Anschrift an ein evtl. amerikanisches Hilfskomitee weiterzuleiten, um evtl. einmal in den Genuß eines Liebesgabenpaketes zu kommen, so wäre ich Ihnen noch weit mehr zu Dank verpflichtet, als es jetzt schon der Fall ist.

Noch weit dankbarer bin ich Ihnen aber, wenn Sie mir diese Zeilen nicht übel nehmen, sondern sie auffassen als das, was sie sind, d.h. als ein frei und ehrlich ausgesprochenes Desiderat, wie es ganz in der Natur der Dinge liegt.
Mit den herzlichsten Grüßen
Ihr Johannes Schubert
Leipzig C 1, Lampestr. 6

[Beischrift von Koc:] Lieber guter Onkel, dieser Brief Dr. Schuberts ist begreiflich, man muß sich die große Not vorstellen. Recht liebe Grüße Dein Robert, 18. Feb. 48

12. Leipzig, den 22. Febr. 1948

Lieber Herr Dr. Rock!
Ihr Brief vom 23. November 1947 ist vor einigen Tagen hier eingetroffen. Es dauert nach hier also etwas über 8 Wochen. Für Ihre hochinteressanten Mitteilungen danke ich Ihnen außerordentlich. Was nun Ihre derzeitigen Arbeiten betrifft, so muß ich immer wieder sagen, daß ich Ihren Mut bewundere! Denn es gehört sehr viel dazu, umfangreiche wissenschaftliche Arbeit noch einmal von vorn anzufangen. Selbstverständlich haben Sie sehr recht, wenn Sie sich nur auf die wichtigsten Texte – und das sind ja noch reichlich viele – beschränken. Den Abschluß sollte aber doch ein gutes Wörterbuch bilden, denn dies ist das Notwendigste. Könnten Sie nicht vielleicht auch jedem Einzeltext ein Spezialwörterverzeichnis beigeben, das alle Wörter des Textes unter Hinweis auf die betr. Stellen enthält. Dadurch kann eine Arbeit nur gewinnen! Leider ist ja z. Z. mit Büchersendungen nicht viel zu machen hier nach Leipzig; aber ich wäre Ihnen doch sehr dankbar, wenn Sie versuchten, mir Ihre Veröffentlichungen – wie früher – zukommen zu lassen. Man ist sonst doch zu sehr abgeschnitten vom eigentlichen Forschungsgebiet und das ist an dieser isolierten Stelle, an der man bei solchen Studien in Europa eben sitzt, doch gewaltig spürbar. Mit besonderem Interesse habe ich gelesen, daß Sie einige Jahre in einer Karma-pa-Lamaserie verbrachten und dort auch die Sprache lernten. Vielleicht darf ich mich bei Enträtselung schwieriger Stellen in den Texten auch an Sie um evtl. Auskunft wenden, zumal Sie die Möglichkeit haben, auch Eingeborene zu Rate zu ziehen. Sobald der Dgra-lha-Text fertig übersetzt ist, erhalten Sie von mir Nachricht. Hoffentlich gelingt es doch, trotz aller derzeitigen Schwierigkeiten, die leider z. Z. gegeben sind. Ich weiß nicht, ob Ihnen bekannt ist, daß ich umfangreiches Material (Kesar) daliegen habe und nur noch auf die Gelegenheit, die Arbeit abzuschließen und zu veröffentlichen lauere.[28] Das würde sicherlich

---

28 M. Taube: "Erschienen sind lediglich ‹Einige Notizen zur Erzählung von Geser Xan› (*MIO* 7.1960, 382-394) und 6 Jahre später eine ausführliche Rezension zu R[olf] A. Stein: *Recherches sur l'épopée et le barde au Tibet.* Paris: Presses Universitaires de France 1959. XI,646 S. (Bibliothèque de l'Institut des Hautes Études Chinoises 13.) in *OLZ* 61.1966, 505-509. Darin werden (Sp. 357; vgl. auch *MIO* 7, S. 385 Fußnote 27) 3 unpublizierte Ge-sar-Versionen genannt: 2 Schreibhefte,

auch Herrn Dr. Stein[29] interessieren und ich möchte Sie bitten zu veranlassen, daß wir zusammen in Verbindung kommen. Vielleicht ist das schon nach der Richtung hin wichtig, daß nicht etwa jeder zugleich am selben Thema arbeitet. – Herrn Dr. Köhler habe ich übrigens Ihre Grüße bestellt. Er ist manchmal bei mir in Leipzig, sonst aber Leiter des geographischen Instituts der Techn. Hochschule Dresden und unterrichtet gleichzeitig auch an der Bergakademie in Freiberg Geographie. Sicherlich wird er Ihnen bald schreiben. Ich selbst aber hoffe bald wieder auf gute Nachricht von Ihnen und bleibe mit den besten Grüßen
Ihr sehr ergebener
Johannes Schubert

1) Wäre es Ihnen möglich über Charakter und Gebrauch der beiden Bücher 1) མཎ་ད་ལ་དཔེ་ཆ་ [maṇ.da.la.dpe.cha], 2)

---

18 und 26 Seiten, eines von Gau Phun-tshogs geschrieben und von Samuel 'Byor-ldan glossiert, das andere von einem bka'-blon von lCags-pa bei Leh kopiert (beide über A. H. Francke zu J. Schubert gelangt), und außerdem eine unvollständige Abschrift Schuberts ‹von dem leider verlorenen Exemplar im ehemaligen [!?] Museum für Völkerkunde zu Berlin›. In einer 1950 geschriebenen Liste eigener Arbeiten nennt Schubert als ‹im Manuskript teilweise fertig› *Das Problem der Gesarsage. Mit 2 bisher unveröffentlichten Fassungen (Fragmente).* Schon zu Lebzeiten hat J. Schubert tibetische Blockdrucke und Manuskripte an das Völkerkundemuseum Leipzig verkauft, Manuskripte des Geser-Epos sind in Schuberts zugehörigem Textverzeichnis jedoch nicht genannt. Auch über den Verbleib eines vielleicht von J. Schubert begonnenen Ms. ist nichts bekannt."

29 Rolf Stein (1911-1999), Tibetologe und Sinologe, später Professor am Collège de France. Rock stand mit Stein in Briefwechsel. Vgl. Kuo Liying: Rolf Alfred Stein (1911-1999). *Cahiers d'Extrême-Asie* 11.1999/2000,XI-XXXX; *Tantric and Tibetan studies in honour of R. A. Stein.* 1-3. Bruxelles 1981. (Mélanges chinois et bouddhiques.20.) Verf. u.a.: *La civilisation tibétaine.* Édition définitive. Paris: L'Asiathèque (1987). 252 S. *Les tribus anciennes des marches sino-tibétaines. Légendes, classifications et histoire.* Paris: Presses universitaires de France 1961. 105 S. (Bibliothèque de l'Institut des hautes études chinoises.15.) In Hinsicht auf das Gesar-Interesse ist zu nennen: *L'épopée tibétaine de Gésar dans sa version lamaïque de Ling.* Paris: PUF 1956. 399 S. (Annales du Musée Guimet. Bibliothèque d'études 61.)

བསོན་གྱི་སྤར་ཁ་དཔུངས་ཤིག [bson.gyi. spar.kha. dpuṅs.śig[30]] zu erfahren?

2) Ist die Abschrift oder Photokopie der Textstelle aus dem Vaidûrya-dkar.po zu bekommen, die die sogenannte *Schildkröten-Methode* (ru.sbal.tshul) schildert?

Meine Frau läßt für die Grüße bestens danken und wünscht Ihnen alles Gute weiterhin. Meinen anderen Brief, den ich Ihnen über Ihren Herrn Neffen sandte, haben Sie sicherlich inzwischen erhalten. Ich habe mir darüber doch noch sehr Gedanken gemacht und hoffe, daß Sie mir das Schreiben nicht übel nehmen, denn es ist aus der Not heraus entstanden.

13. Dresden, den 17. März 1948
Bayreuther Straße 21

Sehr verehrter, lieber Herr Dr. Rock!
Durch Dr. Schubert wurden mir Ihre freundlichen Grüße aus Likiang übermittelt, so daß ich nun beruhigt bin, Sie wieder in Ihrem alten Arbeitsfeld tätig zu wissen. Allerdings vermutete ich Sie in Yünnan, für das Sie sich von jeher besonders begeisterten. Es ist ja bedauerlich, daß auch Ihnen der Krieg da draußen so unangenehm mitgespielt hat. Um so bewundernswerter bleibt Ihre Energie, nun nochmals von Neuem, mit der Nashi-Literatur zu beginnen. Sie können versichert sein, daß ich Ihrer gern gedenke und nur bedauere, daß unser letztes Treffen in Peking so kurz war. Viele unserer alten Bekannten leben nicht mehr. Der Apotheker Göring[31] starb; Dr. Müller[32] scheint sich übel aufgeführt zu haben,

30 M. Taube: "Dieser Titel wird in Schuberts Aufsatz ‹Tibetische Äquivalente der I-ching-Hexagramme› (*MIO* 4.1956, 411-420) erwähnt; in Fußnote 1 (S. 411) ist von ‹der künftigen Publikation des in dem Aufsatz genannten tibetischen Büchleins *Bson gyi spar kha dpuṅs shig*› die Rede (auch dies ist meines Wissens nie erschienen, auch davon kenne ich keine Vorarbeit); die Stelle, auf die in dieser Anmerkung verwiesen wird, findet sich auf S.413 (kurze Beschreibung dieses Textes). – Auch die unter 2) genannte Textstelle aus dem *Vaiḍūrya dkar-po* sollte der Übersetzung dieses tibetischen Textes zugute kommen."

31 Alfred Göhring, Deutsche Apotheke, Peking.

32 Vermutlich Herbert Mueller, der nach dem Krieg wegen angeblicher

so daß eigentlich nur Dr. Boss[h]ard[33] noch großzügig in China reist. Es würde mich freuen, von Ihnen selbst Näheres zu erfahren, vor allem, ob Sie nun ganz nach Likiang übergesiedelt sind und was Sie trotz Ihres beachtlichen Alters noch an Forschungsreisen vorhaben.

Von de Terra[34] hatte ich gute Nachricht aus Mexiko, wo er wieder einen homo sapiens ausgegraben hat, den wievielten weiß ich nicht, aber anscheinend muß er mit der Smithsonian Institution gut zusammenarbeiten. Über seine Familienverhältnisse weiß ich nichts Näheres, wäre Ihnen aber dankbar für Mitteilung, ob Sie seine Frau kennen. Haben Sie de Terra einmal im Laufe der Jahre getroffen?

Ich selbst habe mit viel Arbeit an der hiesigen Technischen Hochschule zu tun, hoffe aber, daß ich im Laufe der Zeit wenigstens zur Zusammenfassung meiner chinesischen Erfahrungen kommen werde. Allerdings muß ich ganz offen gestehen, daß ich vorläufig den Lauf des Hwang ho[35] östlich von Kaifeng nicht weiß.

«Nachrichtenspionage» (news intelligence) in den sog. Shanghai-Prozeß verwickelt und zu mehreren Jahren Haft verurteilt wurde. Vgl. H. Walravens: *Herbert Mueller (1885-1966), Sinologe, Kunsthändler, Jurist und Journalist. Eine biobibliographische Skizze.* Berlin: Bell 1992 [1993]. 206 S. 4° (Han-pao tung-Ya shu-chi mu-lu 45.)

33 Walter Bosshard (1892-1975), Photojournalist, Teilnehmer der deutschen Zentralasienexpedition 1927/28 unter Emil Trinkler, deren zweibändige Wissenschaftliche Ergebnisse Günther Köhler 1932 herausgab (Berlin: Dietrich Reimer) Verfasser der Bücher: *Durch Tibet und Turkestan. Reisen im unberührten Asien.* Stuttgart: Strecker & Schröder 1930 (engl. Ausg.: Hazards of Asia's highlands and deserts. London: Figurehead 1932. [Pioneer series.]); (Hrsg.:) *Botanische Ergebnisse der deutschen Zentral-Asien-Expedition 1927-1928.* Zürich 1932. (Repertorium specierum novarum.31; Mitteilung aus dem Botanischen Museum der Universität Zürich.143.); *Kühles Grasland Mongolei.* Berlin 1938 (Neuaufl.: Zürich: Fretz & Wasmuth 1949, Frankfurt/M.: Büchergilde Gutenberg 1950). – Vgl. auch Peter Pfrunder, Verena Münzer: *Fernsicht: Walter Bosshard – ein Pionier des modernen Photojournalismus.* Bern: Benteli 1997. 238 S., 217 Abb.

34 Helmut de Terra, geb. 1900, Anthropologe, Paläologe; gleichfalls Teilnehmer der Trinklerschen Zentralasienexpedition. Später war er in Mittelamerika tätig.

35 Vgl. auch Günther Köhler: *Der Hwang-ho, eine Physiogeographie.* Gotha: J. Perthes 1929. 104 S. (Petermanns Mitteilungen. Ergänzungsheft 203.)

Ich wäre Ihnen für diesbezügliche Aufklärung außerordentlich dankbar. Sie sehen, es scheitert schon an den einfachsten Tatsachen!

Darf ich Sie heute bitten, alte Bekannte von mir bestens zu grüßen, wenn Sie ihnen begegnen. Ihnen selbst wünsche ich alles Gute und hoffe doch, wieder von Ihnen zu hören.
In alter Verbundenheit bin ich Ihr ergebener
Günther Köhler

14. Leipzig, den 15. Mai '48

Lieber Herr Dr. Rock!
Zunächst habe ich Ihnen herzlichst für Ihre beiden Briefe aus Zürich (v. 28.4.) zu danken, die ich vor wenigen Tagen erhielt. Ich beantworte vorerst den wissenschaftlichen Teil:

Von meinen Aufsätzen in der *Oriental. Literaturzeitung* 1933, Nr 8/9[36] und in den *Tagungsberichten der Ges. f. Völkerkunde*, 2. Tagung, 1936[37], besitze ich nur je einen Sonderabdruck. Meine Versuche, noch je ein Exemplar im Buchhandel zu erlangen, waren vergeblich – so erklärt es sich auch, wenn ich Ihnen erst heute antworte – und ich schicke Ihnen deshalb meine beiden Exemplare [für mich lasse ich die Sachen hier einmal abschreiben; photokopieren ist teils nicht möglich, teils zu kostspielig] und bitte Sie dieselben anzunehmen. Sollte inzwischen einmal wieder etwas von mir gedruckt werden, so sende ich Ihnen jeweils postwendend ein Stück. Auf die Notizen über die Sruṅ.ma's, die Sie mir freundlicherweise zu senden gedenken, bin ich sehr begierig; denn ich will eine zusammenfassende Arbeit, über diese gesamten Erscheinungen in Tibet, publizieren. Der Dgrâ.lha-Text ist noch in Bearbeitung, Sie bekommen ein handschriftliches Exemplar der Übers. sobald dieselbe fertig ist. Die wissenschaftliche Arbeit geht eben leider etwas langsam vorwärts, weil es z. Z. für uns physisch sehr schwer auszuhalten ist und auch sonst die meiste Zeit des Tages für lebenswichtigere Dinge in Anspruch genommen ist. Übrigens

36 Vermutlich die Besprechung von Alexandra David-Neel, Lama Yongden: La vie surhumaine de Guésar de Ling. Paris 1931. *OLZ* 1933, 578-581.

37 Einheimische Quellen zur völkerkundlichen Erforschung Tibets und des chinesisch-tibetischen Grenzgebietes. *Tagungsberichte* .... Leipzig 1937, 75-84.

arbeite ich zusammen mit Herrn Dr. Köhler an einem allgemeinen Tibetbuch[38]; sozusagen einem Nachschlagebuch für angehende Tibetologen usw. Das Buch soll eine Einführung in die Probleme geben, soll das, was man bisher weiß, zusammenfassen, Literaturverzeichnisse, Karten, Tabellen und nicht zuletzt gute Bilder enthalten. Sollten Sie etwa für diesen Zweck einige bisher unveröffentlichte Photos (charakteristische Landschaftsbilder aus Osttibet, Tempel- und Gebäudeabbildungen oder Ähnliches) haben und mir dazu zur Verfügung stellen können, wäre ich Ihnen zu weiterem Dank verpflichtet, als es jetzt schon der Fall ist und dem Buch bezw. dem Leserkreis würde es wesentlich nützen. – Was Ihre neue Arbeit (*The ancient Na-khi kingdom of SW China*) anbelangt, so werde ich wenn die beiden Bände gut hier ankommen, darüber etwas schreiben, wie ich Ihnen ja schon einmal mitteilte. – Die Zeitschrift *Artibus Asiae* erscheint erst neuerdings wieder – in der Schweiz, Basel. Herr Prof. Hadl ist, wie ich vor einiger Zeit zufällig erfuhr, vor einigen Jahren gestorben.[39] Vielleicht interessiert es Sie in diesem Zusammenhang zu erfahren, daß ich Ihren diesbezüglichen Brief von seinerzeit nicht weitergeleitet habe! – Und es war wohl gut so? Ihre anderen Briefe, die alle erhalten sind, bilden für mich eine wissenschaftliche Kostbarkeit, weil wir z. Z. hier in Deutschland ja von all diesen Dingen so abgeschlossen sind und nur durch Korrespondenz hier und da einmal einen Blick in die internationale gelehrte Arbeit tun können. Ich danke Ihnen auch für die Mitteilung über P. Hermann[40] und dessen Anschrift, mit

---

38 M. Taube: "Nicht publiziert; Verbleib eines vielleicht existierenden Manuskriptes unbekannt."

39 Hadl starb 1944.

40 Vermutlich P. Matthias Hermanns SVD (1899-1972), Verfasser von u.a. *Die A mdo pa-Großtibeter; die sozial-wirtschaftlichen Grundlagen der Hirtenkulturen Innerasiens.* (Freiburg 1948?) XVI, 325 S. (Dissertation); *The Indo-Tibetans. The Indo-Tibetan and Mongoloid problem in the Southern Himalaya and North-Northeast India.* Bombay: K. L. Fernandes 1954. XVI, 159 S.; *Himmelsstier und Gletscherlöwe. Tibetische Mythen, Sagen und Fabeln aus Tibet.* Kassel, Eisenach: Röth 1955. 260 S. (Das Gesicht der Völker.); *Mythen und Mysterien, Magie und Religion der Tibeter.* Köln: B. Pick (1956). 400 S.; *Die Familie der A mdo-Tibeter.* Freiburg, München: Alber 1959. 403 S.; *Das National-Epos der Tibeter Gling König Ge Sar.* Aus dem Tibetischen übersetzt. Regensburg: Habbel 1965. 962 S., 1 Taf. Vgl. auch Dominik Schröder: P. Matthias Hermanns SVD (1899-1972).

deren Hilfe ich versuchen will, mit ihm eine Verbindung aufzunehmen. –

Und nun, lieber Herr Dr. Rock, komme ich zum zweiten, nichtwissenschaftlichen Teil Ihres Briefes, dem Certificate, das ich erhalten und vorschriftsgemäß auch schon an die betr. Stelle weitergesandt habe, mit meinen Wünschen. Welche Hilfe Sie meiner Frau und mir damit geleistet haben und welch große Freude Sie damit angerichtet haben, ahnen Sie nicht im geringsten! Sie glauben gar nicht, was schon allein diese Nachricht in Bezug auf das körperliche, persönliche Wiederaufleben bedeutet! Wie wir Ihnen das wieder gutmachen können, ist uns zunächst ein Rätsel. Auf alle Fälle danke ich Ihnen aber herzlichst für die Zusendung des Gutscheins und wir hoffen, daß auch die Waren selbst nun gut hier eintreffen möchten!

Mit den herzlichsten Grüßen
Ihr sehr ergebener
Johannes Schubert

Die beiden Sonderdrucke folgen mit gleicher Post, aber extra.
Bei Ihrer Operation begleiten Sie meine besten Wünsche. Ihre Gesundheit und Arbeitskraft ist besonders für die Wissenschaft unentbehrlich! Was durch Sie erschlossen worden ist, ist so ergebnisreich und eine solche Fundgrube für den Spezialisten, daß es nur Freude ist, darin zu schöpfen. Hoffentlich kann das ganze Material doch noch publiziert werden!
[Nachschrift:]
Sehr geehrter Herr Doktor!
Ich kam gerade von einer kleinen Tagestour müde, traurig, und ohne Erfolg heim, als mir mein Mann Ihren lieben Brief vorlegte. Vorerst konnte ich es gar nicht fassen, welch' große Freude für uns zu erwarten war, und kam mir beim Studium der Kiste vor wie ein Kind, das den Blick in ein Märchenland tun darf. Wie kann ich Ihnen, verehrter Herr Doktor, nun danken für Ihre Freundlichkeit? Es ist mein Wunsch, daß Ihr Leben bestrahlt werde von viel Freude und Erfolg, und Sie auch gesundheitlich bald wieder völlig hergestellt sein mögen.

Von ganzem Herzen dankt Ihnen mit freundlichen Grüßen
Maria Schubert

Anthropos 67.1972, 1-8.

15. Leipzig, den 27.6.48

Lieber Herr Dr. Rock!
Gestatten Sie mir, daß ich heute einmal den materiellen Teil an die erste Stelle rücke: denn das Paket mit den Lebensmitteln ist dieser Tage gut hier eingetroffen und ich bestätige Ihnen mit diesen Zeilen den Empfang der Sendung. Wie wir Ihnen danken sollen, wissen wir z. Z. nicht im geringsten und ich hoffe nur, daß ich alles doch wieder gut machen kann! Nehmen Sie also bitte vorläufig unseren herzlichsten Dank entgegen. Sie ahnen nicht – wie ich wohl schon einmal schrieb – welche Hilfe Sie uns angedeihen ließen. Meine Frau wird Ihnen am Schluß des Briefes selbst noch einige Zeilen schreiben.

Nun zur Wissenschaft! Am 15. Mai dss. J. habe ich Ihnen nach Arnold Arboretum die gewünschten Sonderdrucke als Einschreiben und einen Brief in einfacher Art abgeschickt. Wenn alles normal und gut abgelaufen ist, dann haben Sie, was ich sehr hoffe, Ihre Operation[41] inzwischen gut überstanden und sind wieder in Ihrem Heim in Likiang. Vorsichtshalber sende ich diese Notizen aber noch einmal nach Arnold Arboretum. Meine Tibetstudien nehme ich nun doch wieder in größerem Maße auf, und ich bin auf die Sruṅ.ma-Artikel gespannt. Der Dgra.lha-Text schreitet vorwärts in Bezug auf Übersetzung, und ich darf hier vielleicht einige Namen daraus anführen, mit denen ich noch nichts anfangen bezw. die ich nicht lokalisieren kann: ཀླུ་བཙུན་བུ་ལེ་འདགོཾ་བཀྲུང་བྲིལ, soll im Norden liegen; ferner Rdo.rje.brag.dmar; dbu.ru.ka.tshal; 'ol-dgâ; A.la.ṙnod (ཡ་ལ་རྔོད་); Te.chu; 'O.de ral. gsum; Ljaṅ.ri.smug.po, Ba.reg.'o.de. Alles sollen Orts- bezw. Länder- oder Berg-Namen sein. Sind Ihnen diese Ausdrücke bekannt? Für evtl. Mitteilung bin ich Ihnen sehr dankbar. Überhaupt, je intensiver man sich mit solchen Texten befaßt, um so deutlicher zeigt sich unsere Unkenntnis in diesen Sachen, und nur die Forschung an Ort und Stelle kann hier eingreifen und helfen. Wollen wir hoffen, daß Ihr Material noch voll und ganz ausgewertet werden kann! Außerdem hat nicht jeder soviel mitgebracht wie Sie! In welcher Arbeit soll denn der Amnyi-Machen-Text, den ich übersetzt habe, publiziert werden? Inzwi-

41 Es handelte sich um die Behandlung einer Gesichtsneuralgie, wozu Rock eigens nach Boston gereist war.

schen ist ja übrigens eine amerikanisch-chinesische Flugzeugexpedition in diesem Gebiet gewesen, hat die höchsten Gipfel überflogen und allerhand Luftbildaufnahmen gemacht. Wenn Sie Näheres darüber erfahren oder evtl. auch einige Aufnahmen (in Zeitschriften etc.) bekommen können, dürfte ich dann um einige Mitteilungen bezw. Zusendung von Literatur oder Abbildungen bitten? Ich bin – wie ich schon des öfteren Ihnen wissen ließ – so sehr hier auf solche Mitteilungen angewiesen. – Doch ich glaube, ich habe Sie schon wieder zu sehr mit Wünschen belästigt und will deshalb heute schließen. Andermal weiter!

Mit den herzlichsten Grüßen bleibe ich mit den besten Wünschen für Sie
Ihr dankbarer
Johannes Schubert

Sehr geehrter Herr Dr. Rock!
In welcher Weise wir die Dankesschuld abtragen sollen, weiß ich ja nicht zu sagen; es ist mein Wunsch, daß ein gütiges Schicksal Ihnen all' die Freundlichkeit tausendfach vergeben möge. Beim Eintreffen des Paketes trug der Tag (es war Sonnabend) den Stempel des höchsten Feiertages; jedenfalls wurde er dazu proklamiert. Schon das Lösen der äußeren Hüllen war beinahe eine kultische Handlung, und als dies geschehen war, haben wir beide erstmal ganz tief Luft geholt, und – ja, ich schäme mich nicht einzugestehen – jeder von uns beiden hatte nasse Augen dabei. Ihnen, verehrter Herr Dr., als dem Geber dieser guten Gaben, galt und gilt noch unser allerherzlichster innigster Dank. Bei einer, d.h. mehreren Tassen Tee haben wir dann einen recht fröhlichen Tag verlebt in einer überaus glücklichen Festtagsstimmung.

Mit freundlichen Grüßen und guten Wünschen für Ihr Wohlergehen und Gesundheit dankt Ihnen Ihre
Maria Schubert

16. Leipzig, 14.10.1948

Lieber Herr Dr. Rock!
Heute habe ich Ihnen für 2 Briefe, vom 24. Juli und vom 30. August 1948 zu danken. Die Post geht 8-9 Wochen. Ich danke Ihnen für Ihre Mitteilungen herzlichst. Ihre Ankündigung eines care-Paketes gibt mir – wie auch meiner Frau – neuen Lebensmut,

der sich natürlich auch auf die wissenschaftlichen Arbeiten auswirkt. Ihre Nakhi-Arbeiten begeistern mich außergewöhnlich, und ich freue mich jedesmal, wenn Sie mir von der Fertigstellung einer neuen diesbezüglichen Abhandlung berichten. Sie sitzen tatsächlich an der Quelle des alten Tibets. Meine Nakhi (Mo-so)-Studien haben mich zu der Erkenntnis geführt, daß eigentlich jeder Tibetanist, der sich über den Lamaismus hinaus auch mit dem nichtbuddhistischen und vorbuddhistischen Tibet beschäftigen will, keineswegs mehr an der Nakhi-Literatur vorübergehen kann, sondern diese eingehend zu studieren verpflichtet ist!

Um übrigens auf die Stelle zurückzukommen, die meine Bereiterklärung in punkto Erschließung des Mo-so-Schrifttums ausdrückt, so ersehe ich aus Ihrer Notiz, daß sie doch mehrdeutig aufgefaßt werden kann. Selbstverständlich ist dabei an eine Übersetzung noch unbekannter Texte hier „vom grünen Tisch aus“ gar nicht zu denken und die Frage ist auch nicht so erörtert worden. Es handelt sich vielmehr darum, das publizierte Material auszuwerten und in der Weise zu erschließen, daß – nach dem Vorbild der philologischen Wissenschaft überhaupt – lexikalische Zusammenstellungen angefertigt, Stellen verglichen, sowie Register aufgezeichnet werden, und alles Sachliche mit anderen (bes. tibetischen) Erscheinungen ähnlicher Art in Verbindung gebracht wird. Soweit die Texte mir durch Publikationen zugänglich sind, habe ich diese alle durchgearbeitet. Im Weiteren sehe ich den Artikeln entgegen, die Sie mir zu übersenden gedenken. Alle diese Dinge sind für wissenschaftliche Arbeiten hier dringendste Desiderata. Ich erlaube mir, Ihnen eine Liste Ihrer Schriften beizulegen, die ich besitze. Sehr viel gelegen wäre mir natürlich an dem großen 2-bändigen Werk über die Nakhi, das Sie mir schon mehrmals nannten! –

Für Ihre Erklärung des Bergnamens Ljaṅ.ri.smug.po ལྗང་རི་སྨུག་པོ་ danke ich verbindlichst. Es gibt auch einen Lcaṅ.ri.smug.po ལྕང་རི་སྨུག་པོ་ in Sog.yul སོག་ཡུལ་, der im Dgra.lha-Text erwähnt ist.

Als Lokalgottheit von Te chu (ཏེ་ཆུ་, das, wie Sie mir einmal mitteilten, Atuntse ist) wird im selben Text brag dmar bya rgod བྲག་དམར་བྱ་རྒོད་ genannt. Für das Gebiet des Amnye Machen (Rma.rgyal.spom.ra) ist Kha.rag.khyuṅ.btsun.ma ཁ་རག་ཁྱུང་བཙུན་མ་ mit dem gyi.liṅ. kha.dkar གྱི་ལིང་ཁ་དཀར་ (Einhorn mit weißem Maul) erwähnt. Eine 9-köpfige Schildkröte (rus-sbal.mgo.dgu རུས་སྦལ་མགོ་དགུ་) wird mit einem Land Sña.phyi.de.groṅ སྣ་ཕྱི་དེ་གྲོང་ in Verbindung gebracht. Auch wird der Berg 'Od.de འོད་དེ་ guṅ. rgyal གུང་རྒྱལ་ (in 'Ol.kha? འོལ་ཁ་) aufgezählt. Vielleicht ist es mit Ihrer Hilfe möglich, hier manches aufzuklären und festzustellen. – Bei dem Nakhi-Ausdruck Muan-mi-bo-lo kommt mir der Gedanke, ob etwa *bo-lo* dem bu.le im tibetischen Bergnamen lha.btsun.bu.le ལྷ་བཙུན་བུ་ལེ་ u.a. entsprechen könnte. Das tib. bu.le wird chines. *pu-la* 布拌, mong. bu-la umschrieben. Im Text des Amnye Machen (wird dieser übrigens auch bald im Druck erscheinen?), zu dem ich jetzt manche Verbesserungen bieten kann, (die ich Ihnen nach und nach mitteile), sehe ich im Ausdruck duṅ. khra. klaṅ. na. l daṅ. zhiṅ. stag. dmar དུང་ཁྲ་གླང་ན་ལྡིང་ཞིང་སྟག་དམར་ eine Anspielung auf oder besser Gleichsetzung mit Nakhi boa-du-la-hö mit dem Prädikat dtu[-p'er] (wie in dtu-p'er ssi-ngge); nur daß hier nicht weiß (dkar), sondern tib. khra bo *gefleckt* gemeint ist. In diesem Sinne muß meine Anmerkung 90 des Amnye Machen-Textes ergänzt werden.

Bei den Erläuterungen der Nakhi-Texte und vor allem der Bilderzeichen ist von großem Wert, bei jedem dieser Zeichen zu wissen, welcher Gegenstand dem Bilde zugrunde liegt und wieweit derselbe auch noch jetzt im Volke Verwendung findet: denn erst wenn die wirklichen, den Bildzeichnungen zugrundeliegenden Gegenstände bekannt sind, kann die Anordnung der Bildschriftzeichen nach gewissen Gruppen – ähnlich denen der Wörterbücher für die ägyptischen Hieroglyphen – erfolgen und mit Nutzen gebraucht werden.

Übrigens noch eine andere Frage: Die Nakhi-Bücher gewöhnlichen Inhalts sind dreizeilig, während die Tso-la-Texte vierzeilig sind. Gehört nun das Desgodins-MS (das nach Ihrer Bemerkung <Nichols Mo-so manuscr.[42], S. 229> den Ursprung der *ch'ou*-Dämonen erzählt) demnach auch zu den Tso-la-Texten, da es –

---

42 M. Taube: "Siehe Nichols Mo-so manuscript of the American Geographical Society, translated by J. F. Rock. *Geographical review* 27.1937,229-239; Berthold Laufer: The [Francis H.] Nichols manuscript. *Geographical review* 1.1916, 274-285."

wie die Abbildungen zeigen – immer vier Zeilen hat?

Sehr wertvoll erscheint mir Ihr Gedanke der Ikonographie eines Nakhi-Pantheons, ja zunächst der Zusammenstellung (Aufzählung der Gottheiten) des Nakhi-Pantheon's überhaupt. Es ist auf diesem Gebiete eben noch außergewöhnliches zu leisten: Dann jedoch ist die wahre Grundlage für das Studium Alt-Tibets und der Bon-Religion bis zum Auftreten Padmasambhava's (einschließlich!!!) geschaffen. Solange wir diese Unterlagen nicht haben, bleibt alles Forschen auf dem genannten Gebiete nur Stückwerk!

Bevor ich schließe, erlaube ich mir, Sie nochmals an Notizen über die Sruṅ.ma's und ähnliche Erscheinungen – die mich stark interessieren – zu erinnern, sowie um gelegentliche Angabe Ihrer Artikel (mit Titel) im *Journal of the West China Border Res. Soc.*[43] oder Ihrer Publikationen überhaupt – soweit sie nicht auf meiner Liste stehen – zu bitten. Von Herrn Rolf Stein habe ich bis jetzt noch keine Nachricht erhalten. Ich sehe seiner Mitteilung gern entgegen, da ich Lust habe, mit ihm eine große Gesar-Sache starten zu lassen.

Mit den herzlichsten Grüßen – zugleich im Namen meiner Frau zugleich auch an Ihren Herrn Neffen
bleibe ich
Ihr
Johannes Schubert

1 Beilage

Liste der in meinem Besitz befindlichen Schriften des Herrn Dr. Joseph Fr. Rock.

1. Hunting the chaulmoogra tree (NGM 41.1922: März)
2. Banishing the devil of disease among the Nashi of Yünnan Province, China (NGM 46.1924: Nov.)

---

43 The story of the flood in the literature of the Mo-so (Na-khi) tribe. J. F. Rock. *Journal of the West China Border Research Society* 7.1935, 64-80, Taf. 1-7, 1 Ill. – The origin of the Tso-la books, or books of divination of the Na-khi or Mo-so tribe. [By] J. F. Rock. *Journal of the West China Border Research Society* 8.1936, 39-52, 6 Ill., 1 Taf. – Hä-la or the killing of the soul as practiced by Na-khi sorcerers. [By] J. F. Rock. *Journal of the West China Border Research Society* 8.1936, 53-58, 2 Taf., 1 Ill.

3. The land of the yellow lama (NGM 47.1925: April)
4. Experiences of a lone geographer (NGM 48.1925: Sept.)
5. Seeking the mountains of mystery (NGM 57.1930: Febr.)
6. The glories of the Minya Konka (NGM 58.1930: Okt.)
7. Konka Risumgongba, holy mountain of the outlaws (NGM 60.1931: Juli)
8. Sungmas, living oracles of the Tibetan church (NGM 68.1935: Okt.)
Demon-possessed Tibetans and their incredible feats (NGM 68.1935: Okt.)
9. The origin of the Tso-la books ... (*Journal of the West China Border Research Society* 8.1936,39-52)
10. Hä-la or the killing of the soul ... (*Journal of the West China Border Research Society* 8.1936,53-58)
11. The Nichols Mo-so manuscript of the American Geographical Society (*Geogr. Rev.* 27.1937: April)
12. The birth and origin of Dto-mba Shi-lo ... (*Artibus Asia*e 7.1937)
13. Studies in Na-khi literature (*BEFEO* 37.1937)
14. The Zher-khin tribe and their religious literature (*Monumenta serica* 3.1938)

17. [Postkarte]
Herrn Dr. Jos. Franc. Rock
Likiang
Yün-nan Province
China

Absender: Johannes Schubert
10b Leipzig C 1
Lampestraße 6
Germany
sowjet. Besatzungszone
[gestempelt: Kunming 16.12. 1948; Likiang 29.12. 1948]

Leipzig, den 27. Okt. 1948

Lieber Herr Dr. Rock!
[...] als Ergänzung zu meinem Brief [...] noch diese Karte mit einer

[...] Bitte. Die Frage: Sind die in der [...]s befindlichen und im Report of the Librarian of Congress (Division of Orientalia) in den Jahren 1924, 1930, 1934 und 1936 beschriebenen bezw. erwähnten Na-khi-Texte noch erhalten? Oder sind auch diese MSS mit Ihren Übers. usw. untergegangen? Die Bitte: Darf ich Sie bitten, mir eine Photographie von Ihnen (Porträt etc.) mit Ihrer Unterschrift zu übersenden?

Für die Beantwortung der Frage und die Erfüllung des anderen Wunsches bin ich Ihnen schon im Voraus sehr zu Dank verbunden.
Ihr
Johannes Schubert

18. Dec. 17th, 1948

Lieber Herr Dr. Schubert:
Ihren lieben Brief vom 14ten October mit Beilage erst heute erhalten, besten Dank dafür. Leider hat CARE mich benachrichtigt, daß keine Care Pakete in die russische Zone nach Leipzig geschickt werden können, so bitte schreiben Sie mir, wie es möglich sein wird, Ihnen Proviant Pakete schicken zu können, eventuell durch welche Organisation. Es tut mir sehr leid, daß ich Ihnen deshalb zu Weihnachten nichts schicken konnte.

Herr Rolf Stein hat Paris am 5ten November verlassen und ist nach China abgereist, er wartet auf Ihren Brief, seine Adresse ist c/o The French Consulate, Peiping, China. Ich glaube aber nicht, daß er in Peiping bleiben kann, da alle Ausländer schon evacuiert sind, das heißt nicht Deutsche oder Schweizer, aber wohl Engländer und Amerikaner und andere.

Ich glaube, ich habe mich geirrt, wenn ich Ihnen sagte, daß Te Chu Atuntze sei, diese Stadt, die jetzt auf Chinesisch Te ch'ing heißt, heißt auf Tibetisch hJol, ausgesprochen Dschöl. Die anderen Namen kann ich vorläufig nicht identifizieren, da ich nicht meine Notizen bei mir habe. Wie ich Ihnen schrieb, habe ich an den [!] Naga Cult der Nakhi gearbeitet, auch habe ich die Ch'ou na gv Ceremonie fertig übersetzt, die Erste ist sehr mit Indien verwandt, sie ist viel umfangreicher wie die der Bön, siehe Schiefner das Bonpo Sutra[44], es hat aber viel damit gemein. Laufer seine Naga

---

44 Anton Schiefner: *Über das Bonpo-Sûtra: ‹Das weisse Nâga-Hunderttausend›*. St. Pétersbourg: Académie 1880. 86 S. (Mémoires de

Hundert Tausend[45] sind nicht Bön, sondern gehören der Gelben Sekte an. Die Ch'ou na gv Ceremonie wird vor der Ssu gv oder Naga Ceremonie aufgeführt, sie dient zur Reinigung des Platzes, der Leute, etc., ehe die Naga eingeladen werden können. Die Ms., die dieser Reinigungs Ceremonie angehören, sind von solcher Natur, daß ich kaum glaube, daß sich jemand finden wird, sie zu veröffentlichen, sie lesen sich wie *Psychopathia sexualis*[46] von Kraft Ebbing [!], sie geben einen Einblick in das sexuelle Leben der Nakhi, sind aber hochinteressant. Jetzt befasse ich mich mit den Begräbnis Ceremonien, man kann ja eigentlich von Begräbnis Ceremonie nicht reden, denn damals als diese Ceremonien aufgeführt wurden, die Leute ja nicht begraben, sondern verbrannt wurden. Es gibt deren 16, einige davon haben über 100 mss. eigen. Meine Absicht ist, bis Juni 1950 hier zu bleiben, aber ob das möglich sein wird unter jetzigen Verhältnissen, weiß ich nicht, denn die politische Lage ist eine sehr gefährliche.

Sie wissen ja wahrscheinlich mehr, wie es in China zugeht denn wir, hier erfahren wir gar nichts, nur Gerüchte, daß es schlimm aussieht, das wissen wir, das zeigt uns ja auch der Kurs des Geldes an. Jetzt kostet es 18.oo Gold Yüan für einen Brief nach Amerika, um so einen Brief aufzugeben würde man 648 Marken von dem Werte von je 50.000 Dollaren aufzukleben, und so viele Marken hat ja das Postamt nicht, und der Brief müßte von ganz ungemeiner Größe sein; neue Marken gibt es noch nicht, nur noch die alten, darum schicke ich diesen Brief nach Hong Kong, um ihn dort aufzugeben.

Ja, ich glaube, es wäre notwendig, eine Ikonographie über das Nakhi Pantheon zu schreiben, das kann ich thun, ohne hier bleiben zu müssen, denn so viel kann ich doch lesen ohne Hilfe der Dto mba, ich möchte gerne alle Gottheiten, Geister, Naga, Dämonen,

---

l'Académie impériale des sciences de St. Pétersbourg. VII, 28, 1.)

45 *Klu 'bum bsdus pai sñiṅ po. Eine verkürzte Version des Werkes von den Hunderttausend Nâga's.* Ein Beitrag zur Kenntnis der tibetischen Volksreligion. Helsingfors 1898. V, 20, 119 S. (Suomalais-ugrilaisen seuran toimituksia 11.) – M. Taube: "Laufer selbst bezeichnet das tibetische Werk, das er von dem 1891 verstorbenen Herrnhuter Missionar Karl Marx erhalten hatte, als ‹eine tendenziös lamaistische ... Fälschung.›"

46 Richard von Krafft-Ebing (1840-1902): *Psychopathia sexualis.* Eine klinisch-forensische Studie. Stuttgart: Enke 1886. VII, 110 S. (u. ö.)

etc. darin einbegriffen sehen, und alles, das man von ihnen weiß, angeben, es gibt ja mss., die von deren Abstammung erzählen. Wahrscheinlich haben Sie schon meinen neuen Aufsatz über das Sacrifice to Heaven in *Monumenta Serica*[47] erhalten, ich bat Dr. Rahmann[48], den Rector der catholischen Universität von Peking, Ihnen denselben gleich von Peiping aus zu senden, Sie werden darin wohl manches Interessante finden, speciell die Verbindung mit dem Ch'iang Stamm. Was das zweibändige Werk über die Nakhi anbetrifft, so bin ich leider nicht in der Lage, Ihnen ein Exemplar zu senden, ich erhielt nur sechs freie Exemplare, und diese waren schnell vergeben. Man hat 114 Frei Exemplare vergeben und da die Druckkosten auf 22.000.00 U.S. Dollar kamen, verweigert sich Harvard Yenching, weiters Frei Exemplare zu verschicken. Die zwei Bände kosten U.S. $ 25.oo, vielleicht wird es mir möglich sein, späterhin Ihnen diese zwei Bände schicken zu können.

Vorläufig habe ich keine hier, nur diese, die ich hier selbst gebrauche. Sie wissen wohl, daß Tucci in Lhasa ist, er gibt ein Werk heraus in Englisch betitelt: *Tibetan painted scrolls*[49], the subscription ist jetzt offen, der Betrag für das Werk kommt auf U.S.$ 300.00, wohl nur wenige werden sich das leisten können, ich habe mir vorgenommen, nichts mehr zu kaufen, ich bin jetzt 66 Jahre alt, und für wen sollte ich das kaufen, ich werde mich wahrscheinlich mit tibetischen Bildern nicht mehr befassen. Erinnern Sie sich auf das Bild des Ma chhen pom ra, ich glaube ich habe Ihnen eins geschickt, da ist der Berggott mit einem Garuda auf der Hand abgebildet, vor kurzer Zeit begegnete ich einer Beschreibung des Garudas in einem Nakhi ms., da stand geschrieben, er halte einen Garuda auf seiner flachen Hand, der Dto mba, der den Ma chhen pom ra (Ma mi bpo lo) abgebildet hat, hatte wohl nicht dieses Ms. gesehen. Daß das Nakhi Bpo lo in Muan oder Ma mi bpo lo oder bpa lo zusammenhängt mit dem Lha btsun bu le glaube ich sicher nicht, die Nakhi können keinen Endkonsonanten aussprechen, ein tibetischer Endkonsonant wird oft bei den Nakhi

---

47 The [2]Muan [1]Bpö ceremony or the sacrifice to heaven as practiced by the [1]Na-[2]khi. By Joseph F. Rock, Research Fellow, Harvard-Yenching Institute of Cambridge, Mass., U.S.A. *Monumenta Serica.* 13.1948, 1-160, Taf. 1-4.

48 Rudolf Rahmann S.V.D., 1902-1985.

49 Roma: Libreria dello Stato 1949. 2 Text- und 1 Tafelbd. 2°

der vorangehenden Silbe angefügt. Die Nakhi, die den Tibetern näher wohnen sprechen den Namen Ma ni pom ra aus, das *m* ist nasal.

Ich hatte die Absicht, als zweites Werk die Nordwest Gebiete von China zu veröffentlichen, eingenommen das Amnyi Machhen Gebiet, das Choni Reich und das Muli Lama Königtum, inclusive das gangs dkar la und Chia la Gebietes. Ich hatte alles übersetzt und fertig gemacht, nur das Ganze nicht coordiniert, es würde mich nicht drei Monate brauchen, um das Manuscript zum Druck fertig [zu] machen. Nach meiner Rückkehr nach Amerika von China während des Krieges arbeitete ich an der U.S. Army Map Service als Consultant, für ein einhalb Jahre, und während dieser Zeit hatte ich alle Karten von meinen Aufnahmen über das Amnyi Machhen Gebiet und alle anderen obigen Gebiete zusammengestellt, das Army Map Service in Washington wollte alle diese Karten für mich drucken, tatsächlich druckten sie 4 Karten (in drei Farben), die das Nakhi Gebiet umfassen und diese Karten sind meinem zweibändigen Werke beigegeben. Dann kam der Waffenstillstand, und der Congress gab kein Geld mehr, um die anderen Karten zu drucken. Es war kein Interesse mehr da. Sobald ich nach Amerika zurückkehre, werde ich mich an die Arbeit machen und das Werk druckfertig machen. Da Harvard Yenching so viel Geld ausgab für den Druck des Nakhi Werkes, konnte ich nicht verlangen, daß sie sofort das Andere auch drucken lassen sollten, da sie ja auch all ihr Geld für Veröffentlichungen verbrauchen dürfen. Mein Kontrakt mit Harvard Yenching verpflichtet sie, alle meine Arbeiten, die ich hier zum Druck fertig mache, binnen Jahresfrist zu drucken, und zwar über die Nakhi Literatur, etc. (Das heißt, nach meiner Rückkehr.)

Leider habe ich meine Tagebücher, in denen alle meine Notizen über die Srung ma sich befinden, nicht bei mir, ich habe 34 Bände dieser Tagebücher, die ich während meiner Reisen führte, da steht viel darin, aber sie kommen zu lassen, ist soviel als sie zu vernichten. Man kann hier nie wissen, was geschehen kann, und man muß vielleicht alles liegen lassen und fliehen. Nein, ich habe schon viel verloren, aber ich will diese Bände nicht auch verlieren. Vor einigen Tagen erhielt ich einen Brief aus Wien von einem Studenten, der sich René v. Nebesky Wojkowitz[50] nennt, er

---

50 Nebesky-Wojkowitz, René Mario von (1923-1959); vgl. die

schreibt, daß er an seiner Dissertation arbeitet über die Entstehung der tibetischen Schrift[51]. Er befaßt sich auch über die Bön Religion, und auch speciell über die Srung ma oder Oracle (Orakel) hatte er eine Arbeit fertig gestellt[52], er schickte mir eine Brochüre über das Böntum, die aber nichts Neues bringt, es ist eine Zusammenfassung von allem, das über die Bön Religion geschrieben wurde.[53]

Ich sehe in der Liste meiner Arbeiten, daß Sie Manches nicht haben, das, was Sie am meisten interessieren würde, ist der Aufsatz in *The [Nat.] Geogr. Magazine*: Life among the Lamas of Choni[54] mit vielen farbigen Platten von Lama Tänzern und Butterbildern vom Butterfest. Ferner: The great River Trenches of Asia[55], und eine Arbeit in the *Roy. Geogr. Soc. of London Journal*, The Land of the Tebbus[56]. Ferner erschien in *B.E.F.E.O.* 1939, vol. 39, fasc.1, The Romance of K'a mä gyu mi gkyi. A Na khi tribal love story, 155 pp., 32 Tafeln, 2 in Farben. Dann *The ancient Na khi Kingdom of Southwest China*, 500 pp. 257 Tafeln und 4 Karten[57]. 2 Bd. Das Letzte ist die Arbeit, die ich Ihnen von Peiping schicken ließ. Leider habe ich nicht die Referenzen hier über die anderen Arbeiten.

Nun muß ich schließen, ich werde Ihnen bald wieder schreiben und bis dahin verbleibe ich mit den besten Grüßen an Sie selbst wie

---

Publikation: *Oracles and demons of Tibet. The cult and iconography of the Tibetan protective deities.* By René de Nebesky-Wojkowitz, Ph. D., lecturer at the University of Vienna. Taipei: SMC Publishing (1990). 666 S. (Nachdruck der Ausgabe Den Haag: Mouton 1956.) Vgl. auch J. W. de Jong: René von Nebesky-Wojkowitz. *Indo-Iranian journal* 3.1959, 306-309.

51 M. Taube: "Der Titel der Dissertation lautet: *Schriftwesen, Papierherstellung und Buchdruck bei den Tibetern.* Wien 1949. 109 Bl. Phil. Diss., Univ. Wien."

52 Möglicherweise Das tibetische Staatsorakel. *Archiv für Völkerkunde* 3.1948,136-155.

53 M. Taube: "Gemeint ist vielleicht Nebesky-Wojkowitz: Die tibetische Bön-Religion. *Archiv für Völkerkunde* 2.1947,26-28; nicht gemeint sein kann der erst 4 Jahre später erschienene Aufsatz: Bon – The pre-Buddhist religion of Tibet. *Aryan path* 23.1952,509-513, 553-557."

54 *NGM* 54.1928, 569-619, 34 Taf., 16 Farbtaf., 1 Kte.

55 *NGM* 50.1926, 133-186, 47 Taf., 1 Kte.

56 81.1933, 108-127.

57 Cambridge, Mass.: Harvard Univ. Press 1947. (Harvard-Yenching Institute monograph series 8.)

an Ihre werte Frau Gemahlin,
Ihr stets ergebener
[Rock]

19. Dr. Johannes Schubert
Leipzig C 1
Lampestraße 6 Am 21. Dezember 1948

Lieber Herr Dr. Rock!
Vor einigen Tagen erhielt ich von der Catholic University-Peiping eine eingeschriebene Drucksache, die den Sonderdruck Ihres Artikels über die ^{2}Mùan-^{1}bpö-Zeremonie (aus: *Monumenta Serica* XIII, 1948) enthielt. Für die Zusendung danke ich Ihnen herzlichst und bitte Sie, die Ihnen mit meinem letzten Schreiben übersandten Liste der in meinem Besitz befindlichen Veröffentlichungen aus Ihrer Hand entsprechend zu ergänzen.

Ich bin gerade beim Durcharbeiten dieser recht gut gelungenen Arbeit. Sie ist ein sehr guter Beitrag, der das für uns hier am „grünen Tisch" so schwierige Einarbeiten in die Nakhi-Schriften außerordentlich erleichtert. Es gehörten nur noch, wie ich Ihnen schon einmal schrieb, zu jeder derartigen Publikation, Spezialvokabulare mit den entsprechenden Stellenangaben: dann ist das geschaffen, was der Wissenschaftler braucht! Bisher habe ich mir diese Verzeichnisse immer selbst erarbeitet, was zwar recht viel Mühe macht und noch mehr Zeit erfordert, für das Studium jedoch von größtem Nutzen ist. Bei dieser Gelegenheit fällt mir auf Seite 32 auf, daß im Ausdruck ^{2}mùan-^{2}ssu* die Silbe muan durch ** und nicht durch *** wiedergegeben ist, wenn der Begriff als „*nicht* wissen", „*nicht* verstehen" zu deuten ist? Sehr interessiert hat mich (auf S. 141-142 Ihres Artikels) der ^{2}Gyi-Dämon ^{2}K'wua-^{1}dtv-^{1}mber-^{2}ddv (-^{2}gyi) mit dem Gyag-Kopf und den flatternden Haaren: ich glaube beinahe nun doch, daß sich im Na-khi-Pantheon bezw. in den Na-khi-Texten fast alle Gestalten des Bon-Pantheon's (vor allem jedoch die Tierköpfigen) finden, von denen wir bisher nur Abbildungen, aber weder Namen noch sonst etwas kennen.
Vor kurzem hatte ich Gelegenheit, ein tibetisches Rollbild [...]

[Schluß liegt nicht vor]

* 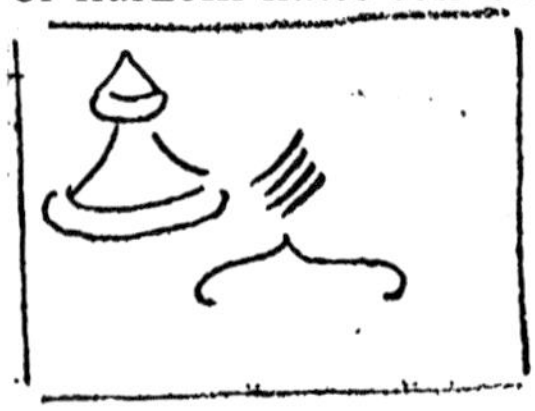

** 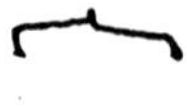

*** 

20. January 7th, 1949

Dear Dr. Schubert:
Meine Arbeit geht nur langsam vorwärts, unter gegenwärtigen Verhältnissen fehlt einem wirklich die Stimmung. Wie ich Ihnen schon schrieb, befasse ich mich vorläufig mit den Übersetzungen der Funerarien Litteratur; die erste Zeremonie, die am Tage des Begräbnisses stattfindet, ist sehr interessant, denn die Mss. berichten von sehr alten Gebräuchen, von denen man jetzt nichts mehr weiß. Ich glaube, daß die ganzen Funerarien, von denen es 16 gibt, und die dazu gehörige Litteratur mich wohl 1 und 1/2 Jahre befassen werden. Ob ich alles fertig bringen werde, das wissen die Götter, denn ich bin ja nicht mehr jung, mein Alter ist 66 Jahre, und ich bin immer kränklich, doch habe ich eine gute Constitution.

Es ist sehr interessant, daß die Nakhi Litteratur sehr viel mit der West Tibets zusammenhängt. Von vielem, was in der Gesar Litteratur vorkommt, ist in der Nakhi Litteratur auf den Dto-mba Shi-lo (gShen-rabs-mi-bo) übertragen, dieselben Situationen kommen vor, in der sich Gesar befindet, wie sein Kampf mit dem Teufel, sein Gekochtwerden im Kessel, etc., nur ist hier der Dto mba Shi lo darin verwickelt. Es wäre sehr wünschenswert, mehr von den Ladak Legenden zu erfahren. Speciell interessieren mich die Übersetzungen von A. H. Francke, leider habe ich nicht alles, das von ihm publiciert wurde. Könnten Sie mir sagen, wo ich seine *Hochzeits Lieder*, die im Verlag Folkwang erschienen sind[58], und auch etwas über Hochzeitslieder in der *Orientalistischen Litteratur Zeitung*, 1923, S.428[59], erlangen könnte? Ich möchte auch gerne S. H. Ribbach's Vier Bilder des Padmasambhava, in *Jahrbuch der Hamburger Wissensch. Anstalten*, Band XXXIV, 1916[60], kaufen.

---

58 Übersetzt nach Handschriften von Tag.Ma.Cig. Mit einer Einleitung über die Mythologie der tibetischen Sagenwelt und Bildern, meist nach Aufnahmen des Verfassers. Lieder in die ursprünglichen Versmaße übertragen von Anna Paalzow. Hagen, Darmstadt: Folkwang-Verlag 1923. 71 S., 3 Taf., 15 S. Abb. 4° (Kulturen der Erde. Abt. Textwerke.)

59 Dieser Hinweis zielt auf Franckes «Laufers Milaraspa». *OLZ* 1923, 426-429 (Zu: Milaraspa. Hagen, Darmstadt: Folkwang-Verlag 1922. 78 S., 14 S. Abb.). Die Hochzeitslieder Franckes sind besprochen in der *OLZ* von Johannes Nobel, Sp. 514-515.

60 Hamburg: Lüttcke & Wulff 1917. 53 S. Mit 5 (3 farb.) Taf. und 69

Wenn Sie mir da helfen könnten, dieselben zu kaufen, wäre ich Ihnen zu großem Dank verpflichtet.

Mit den besten Grüßen an Sie wie an Ihre werte Frau Gemahlin, verbleibe ich,
Ihr stets ergebener
[Rock]

21. Prof. Dr. Johs. Schubert
Leipzig C 1
Lampestraße 6 Leipzig, den 29. April 1952

Lieber Herr Dr. Rock!
Vor mir liegt Ihre Karte aus Meran v. 15. April und Ihr Brief aus Zürich v. 21. April. Für beide Mitteilungen danke ich Ihnen herzlichst. Die letztere erhielt ich am vergangenen Sonnabend. Es ist nicht nur selbstverständlich, sondern vielmehr für mich eine Ehre und eine Pflicht zugleich, daß ich Ihre Naga-Kult Arbeit bespreche. Diese Aufgabe ist für mich jetzt um so angenehmer, als inzwischen hier in der D.D.R. die weltbekannte und erstklassige *Orientalistische Literaturzeitung* wieder erscheint. Auf diese Weise erhält die Rezension des Werkes zur richtigen Zeit den richtigen Platz an der richtigen Stelle! Ich glaube und hoffe, daß Sie darin mit mir übereinstimmen. Wenn es möglich wäre, mir zum gleichen Zweck zur Besprechung in der OLZ auch noch ein Exemplar Ihres *Ancient Nakhi Kingdom* zu beschaffen, so wäre dies mehr als günstig! – Wenn auch das Werk schon etwas früher erschienen ist. Es wäre dann gleichzeitig das erste und einzige Exemplar in der D.D.R., ein Buch, dessen die Forschung hier dringendst bedarf! Wir könnten auch daran festhalten, daß ich Ihre künftigen Publikationen sofort nach Erscheinen in der OLZ bespreche. Damit ist dann ohne weiteres uns beiden geholfen. Wenigstens ist das meine Meinung!

Das Studienjahr ist nun nächste Woche hier in der D.D.R. zu Ende. Damit ist mir eine schwere Last von der Seele genommen, und ich kann mich ganz dem Abschluß-Strich unter den Texten

Abb. im Text. (Mitteilungen aus dem Museum für Völkerkunde zu Hamburg 5; Jahrbuch der Hamburgischen Wissenschaftlichen Anstalten. Beiheft 5.)

widmen, die ich für Sie übersetze und kann damit zugleich endlich mein äußerst unruhiges Gewissen entlasten. Wenn ich des öfteren gerade diese Arbeiten, die mir persönlichst sehr am Herzen liegen, wieder beiseite legte und beiseite legen mußte, weil sich neben schwerster Belastung noch ganz andere Arbeiten nötig oder nötiger machten, so geschah das aus Gründen, die Sie heute nach Empfang dieses Briefes vielleicht eher verstehen: denn, wie Sie schon aus dem Stempel am Briefkopf ersehen, habe ich nun – fast möchte ich sagen endlich – meine Professur bekommen! Es ist mir deshalb mehr als eine angenehme Pflicht, als erste Arbeiten, die ich in meiner Eigenschaft als einziger Inhaber einer Professur für Tibetisch in der D.D.R. leiste, Ihre Texte zu bearbeiten und in bester wissenschaftlicher Form der Fachwissenschaft vorzulegen!

Was lange währt, muß letzten Endes doch gut werden! Und die Länge der Zeit werden Sie mir, hoffe ich –, jetzt verzeihen. So kann ich Ihnen denn auch mit um so größerer Freude für die angekündigte Sendung danken, einer Sendung, die meine Frau sowohl als mich natürlich höchst angenehm überrascht.

Also, lieber Herr Dr. Rock, seien Sie mir nicht böse; Sie bekommen nun wirklich die Übersetzungen und wir werden doch noch – denke ich – zum Nutzen und Vorteil der Fachwissenschaft gemeinsam arbeiten können.

In diesem Sinne verbleibe ich mit den herzlichsten Grüßen und den besten Wünschen für die Zukunft – zugleich im Namen meiner Frau –

Ihr dankbarer
Johannes Schubert

22. z. Z. Rathen / Elbe, 24.8.'53

Lieber Herr Dr. Rock!

Anbei den Rta-thugs-dkar-Text[61] und die Anmerkungen zum kleinen Ritual für Amnye Machen. Ihr Telegramm wurde mir ganz zufällig (es wußte sonst gar kein Mensch, daß ich mich einige Tage hier aus Gründen der Erholung aufhalte) nachgeschickt. Ebenso zufällig hatte ich diese Texte bei mir und sende sie Ihnen. Nach meiner Rückkehr sende ich die Verbesserung des großen Rituals

---

61 M. Taube: "Das Ritual für Rta Thugs.dkar.po. Eine Beschwörung der Bon.po. (Mit 8 Taf.). *MIO* 8.1961,51-63."

sofort ab. Bis 15. September – also kurz vor Semesterbeginn - haben Sie alles (auch den Dgra-lha).

In der *OLZ* erscheinen demnächst auch die Besprechungen des Nakhi-Nâga-Kultes und des Ancient Nakhi-Kingdom's.[62] Von Leipzig aus sende ich Ihnen Durchschläge der Rezensionen mit. Bitte lassen Sie mich immer Ihre Anschrift wissen.

Mit den herzlichsten Grüßen
Ihr
Johannes Schubert

23. St. Anton a/Arlbg. 30.Aug. 53

Lieber Herr Dr. Schubert:
Soeben habe ich Ihren eingeschriebenen Eilbrief vom 24. Aug. erhalten, nebst bei auch den rTa-thugs-dkar Text und die Anmerkungen zum kleinen Ritual für Amnye Machhen mit Dank erhalten. Da Sie mir schrieben, daß Sie am 14. Aug. von Köln nach Leipzig zurückkehren werden, telegraphierte ich Ihnen nach Leipzig und hoffte, alles hier vor meiner Abreise, welche morgen am 31. Aug. stattfindet, vorzufinden. Ich begebe mich auf einige Tage in die Schweiz und werde von dort direkt nach London fliegen. Meine englische Adresse ist:

25 West Park Road,
Kew Gardens, Surrey

wo ich wahrscheinlich bis 20. September verbleiben werde. So bitte ich Sie dringend, das mir schon lange versprochene Material endgültig nachzuschicken. Ich habe auch Verpflichtungen und muß bei einer gewissen Zeit wieder in Seattle sein. Und da schon manche Briefe etc., die an mich dorthin gerichtete waren, verloren gegangen sind, würde ich Sie dringend bitten, alles an die englische Adresse abzusenden, wo alles sicher ankommt.

Mit freundlichen Grüßen an Sie sowie an Ihre werte Frau Gemahlin verbleibe ich
Ihr stets ergebener
[Rock]

P. S. Ich glaube, ich werde Ihre Arbeiten auf Deutsch drucken lassen wie sie sind, anstatt sie ins Englische zu übersetzen; ich habe

62 Vgl. Appendix.

jedoch Ihre erste Arbeit schon im Jahre 1946 ins Englische übersetzt, und da meine eigene Arbeit über die Geographie etc. auf Englisch gedruckt wird, so würde es vielleicht doch besser sein, Ihre jetzige Arbeit auch ins Englische zu übersetzen. Wenn Sie wollen, werde ich Ihnen eine Copie der Übersetzung der ganzen Arbeiten zur Übersicht senden. Bitte lassen Sie mich wissen, ob Ihnen das gefällig ist.

24. Prof. Dr. Johannes Schubert
Leipzig C 1
Lampestraße 6 Am 3. September 1953

Herrn
Prof. Dr. J. F. Rock
25 West Park Road
Kew Gardens
Surrey, England

Lieber Herr Dr. Rock!
Ihr Brief vom 29. August, für den ich Ihnen herzlichst danke, erreichte mich gestern. Zu Ihrem Postskriptum muß ich bemerken, daß ich von Rathen, Sächs. Schweiz, aus am Dienstag, d. 26. August, einen eingeschriebenen Eilbrief mit einem Teil der Manuskripte an Ihre Anschrift nach St. Anton geschickt habe. Ich hoffe aber, daß Sie inzwischen in den Besitz dieser Sendung gelangt sind.

Ich bin jetzt also beim 4 (!) durcharbeiten des großen Amnye-Machen-Rituals und sende es Ihnen schnellstens. Jetzt ist das Lesen des Textes für mich eine Freude, früher war es eine harte Nuß. Ich habe allerdings auch sehr vieles gegen früher geändert.

Zu dem mir in Abschrift übersandten Brief an Herrn Prof. [Serge] Elisséeff[63] habe ich folgendes zu sagen bezw. vorzuschlagen:

Daß Sie Ihr Nakhi-Wörterbuch nicht alphabetisch, sondern nach Gegenständen anordnen wollen, ist durchaus richtig und in Ordnung. Ich schrieb es Ihnen schon einmal vor Jahren, daß als

63 Serge Elisséeff, 1889-1975, Direktor des Harvard-Yenching Institute; vgl. E. O. Reischauer: Serge Elisséeff. *HJAS* 20.1957, 1-35.

Vorbild für diese Dinge die sachlich geordneten Wörterbücher für die Hieroglyphen und die Keilschriftzeichen-Lexika (die auch Silben- und Ideogramm-Werte besitzen) gelten können. Es sollte jedoch ein alphabetisches Register aller Wörter mit Hinweis auf die vorkommenden Stellen des Wörterbuches das Werk begleiten.

Ganz außerordentlich wertvoll ist es, wie Sie schreiben, in einem Appendix die Na-khi-Buchtitel mit Übersetzung, soweit es möglich ist, als auch die Lehnwörter mit ihren Original-Äquivalenten zu bringen. Ebenso wesentlich ist die von Ihnen angedeutete Beigabe der Namen in Na-khi-Silbenschrift (ggo-ba).

Meines Erachtens ganz außerordentlich wichtig wäre es – es würde allerdings die Arbeit wesentlich verlängern – wenn den Wörtern eine Stellenangabe aus den Na-khi-Büchern beigegeben werden könnte. Die Buchtitel, von denen Sie sowieso eine Übersicht liefern wollen, könnten dann durch Abkürzungen oder Nummern wiedergegeben werden. Das Ganze wird eine entschieden erstklassige Arbeit, die sonst keinem Menschen mehr möglich sein wird. Für den Druck dieser Arbeit muß auf alle Fälle gesorgt werden. Ich hoffe, von Ihnen über den Verlauf dieser Angelegenheit weiteres zu hören.

Mit den herzlichsten Grüßen
Ihr ergebener Johannes Schubert

25. Leipzig, den 20. 9. 53

Lieber Herr Dr. Rock!
Anbei sende ich Ihnen das große Amnye-Machen-Ritual in verbesserter Form und Übersetzung. Besonders die Anmerkungen sind ganz neu gestaltet.

In 14 Tagen sende ich den Dgra-lha-Text ab, so daß Sie dann alle Übersetzungen haben und nun endlich weiter arbeiten können mit diesen Materialien.

Ich nehme aber alle Schuld auf mich und werde Ihnen noch entsprechend schreiben, wann alles in Ihren Händen ist.

Mit den herzlichsten Grüßen
Ihr
Johannes Schubert

26. Prof. Dr. Johannes Schubert
Leipzig C 1
Lampestraße 6 am 5. Oktober 1953

Herrn
Prof. Dr. J. F. Rock
Kew Gardens, 25 West Park Road
Surrey, England

Lieber Herr Dr. Rock!
Für Ihren Brief vom 28. September danke ich Ihnen herzlichst. Ich freue mich, daß diese eine Übersetzung gut in Ihre Hände gelangt ist. Was, wenn ich gleich auf Ihre Notizen zu sprechen komme, die Änderungen anbetrifft, so bitte ich Sie, dieselben vorzunehmen, d.h.

in Anm 1 wäre zu ändern: die Photographie in „ein Photostat"; Chicago in „Washington, D. C., Library of Congress",

in Anm. 312 ist am Schluß hinzuzufügen: „und nochmals mit ausführlicherer Beschreibung in J. F. Rock: The Na-khi Nâga Cult and related ceremonies (Roma, Istituto Italiano per il medio ed estremo Oriente 1952) = Serie Orientale Roma IV, Part I, S.132/133, Anm. 116 und dazu das Bild in Part II auf Taf. XXXIX. – Die Darstellung des Berggottes nach diesen Bildern ist eine ganz andere als sie in der tibetischen Beschreibung geschildert wird!"

Ich bitte Sie auch höflichst, bei Erwähnung von Ortsnamen etc. (z.B. Co.ni, Ra.rgya [auch: Rva.rgya] die jeweilige chinesische oder tibetische Schreibung (also 卓 泥 རྭ་རྒྱ) hinzuzufügen.

An die Veröffentlichung des *ganzen* tibetischen Originaltextes hatte ich von Anfang an gedacht. Es wäre aber vielleicht gar nicht schlecht, wenn der Text (außer der Original-Wiedergabe!) nochmals in Umschrift und da abgeteilt wie in der Übersetzung gebracht würde. Wenn Sie damit einverstanden sind, müßte ich Ihnen noch den Text in der vorgeschlagenen Weise umschreiben.

Von der Dgra-lha-Text-Übersetzung sende ich Ihnen hier die ersten 2 bezw. 3 Seiten nebst den Anmerkungen 1-40 mit der Bitte, mir zu schreiben, ob Ihnen der auf Blatt 2 (letzter Abschnitt) ff. beschriebene Aufbau etc. bekannt ist bezw. richtig erscheint. (Es kommt in der weiteren Beschreibung noch einiges dazu, nämlich 12 Stäbe (3 x 4), 3-eckige Gtor.ma's und Kopf des

Gyag, des Schafes und der Ziege usw.) Die Blätter dieses Textes sende ich Ihnen so einzeln, weil es z. Z. nicht anders geht. Von *diesem* Text sollte unbedingt das Original mit publiziert werden, da vieles unleserlich ist!

Ebenso erhalten Sie beiliegend die Durchschläge der Besprechungen Ihrer beiden Bücher, wie ich sie an die *OLZ* gegeben hatte.[64] Die Rezensionen werden *im nächsten Heft* dieser Zeitschrift abgedruckt, soviel mir bekannt geworden ist.
Mit den herzlichsten Grüßen
Ihr sehr ergebener Johannes Schubert

Anlage: 10 Blatt

27. Prof. Dr. Johannes Schubert
Leipzig C 1
Beethovenstraße 6 am 22. Oktober 1953

Herrn
Prof. Dr. J. F. Rock
25 West Park Road
Kew Gardens, Surrey, England

Lieber Herr Dr. Rock!
Für Ihren sehr lieben Brief vom 13. Oktober 1953 danke ich Ihnen herzlichst. Ich freue mich, daß die Sendung gut angekommen ist und hoffe auch, daß Sie bald in den Besitz eines Abzuges der gedruckten Besprechungen Ihrer beiden Arbeiten kommen werden.

Der Dgra-lha-Text gibt zu Anfang sehr harte Nüsse zu knacken, soweit er sich auf den Aufbau der Ritualgegenstände etc. bezieht. Ich hoffe aber noch Klarheit hineinzubringen und sende Ihnen die Teile nach und nach. Es ist übrigens ein großartiger Vorschlag Ihrerseits, den Dgra.lha-Text und die Arbeit über die Sruṅ-ma zusammen heraus–zubringen. Ich teile Ihre Meinung ganz und gar.

Zuerst soll aber das Amnye-Machen-Buch erscheinen und die Texte sollen in Umschrift (in richtiger Abteilung) und Originaltext beigegeben werden. So sende ich *heute* beiliegend die Umschrift des Textes zum *kleinen* Ritual und lasse die Umschrift des Textes zum

64 Das Manuskript der beiden Rezensionen wird im Appendix wiedergegeben.

großen Ritual baldigst folgen. Wenn alles soweit ist, wäre es mir sehr lieb, wenn ich die *Korrekturen* erhalten könnte und dann vielleicht auch noch einige Verbesserungen anbringen kann.

Das Wörterbuch (Na-khi dictionary) ist eine mehr als dankenswerte, wenn auch schwierige Aufgabe und ich möchte Herrn Dr. Bornemann[65] gratulieren, daß er sich entschlossen hat, diese Arbeit zu drucken.

Natürlich bin ich Ihnen außerordentlich dankbar, wenn Sie mir Ihre Publikationen senden wollen; aber kann ich das überhaupt noch von Ihnen verlangen? Den *Anthropos* habe ich selbst nicht, so daß mir die von Ihnen darin veröffentlichten Aufsätze[66] schon willkommen wären.

Mit den herzlichsten Grüßen
Ihr sehr ergebener
Johannes Schubert

Beilage: 7 Blätter tibetischer Text [liegt nicht bei]

28. Prof. Dr. Johannes Schubert
Leipzig C 1
Lampestraße 6 Am 16. September 1954

Herrn
Prof. Dr. J. F. Rock
c/o Far Eastern Inst.
Univ. of Washington
Seattle 5
Washington

Lieber Herr Dr. Rock!
Ihren Brief vom 31. August habe ich erhalten und danke Ihnen dafür herzlichst. Kurz vorher hatte ich Ihnen bereits einen Luftpostbrief geschickt, den Sie vermutlich noch nicht bekommen

---

65 Fritz Bornemann (1905-), Herausgeber des *Anthropos*.

66 Excerpts from a history of Sikkim. *Anthropos* 48.1953, 925-948; später erschienen: The D'a Nv funeral ceremony with special reference to the origin of Na-khi weapons. *Anthropos* 50.1955, 1-31, 5 Taf.; Contributions to the shamanism of the Tibetan-Chinese borderland. *Anthropos* 54.1959, 796-818, 6 Taf.

haben und in welchem ich Ihnen auch den Empfang der Bemerkungen über die tibetischen Grammatiken bestätigte. Ebenfalls teilte ich Ihnen mit, daß sich meine Abreise nach China – Tibet bis 10. Oktober verschoben hat. So will ich Ihnen heute nur kurz Ihre Fragen aus dem letzten Brief beantworten und bitte Sie, meine Notizen den betreffenden Texten hinzuzufügen:
Amnye Machen Text – *Kurze Fassung*

Zu Anm. 7: siehe Helmut von Glasenapp: *Buddhistische Mysterien.* 1940.

Zu Anm.19: Das Rta thugs dkar po-Ritual habe ich übersetzt.[67] Ich sandte Ihnen einen Durchschlag davon mit. Falls dieser Text nicht mit dem Amnye-Machen-Text zusammen gedruckt wird oder werden kann, bitte ich mir dies mitzuteilen. Ich lasse diesen Text dann hier drucken. Die betr. Notiz in meinem Text (Anm. 19) kann auf alle Fälle stehen bleiben. Der eigentliche Dgra-lha-Text, dessen Übersetzung Sie noch von mir bekommen, wird davon nicht berührt.

Zu Anm. 25, 2. Zeile: Erster Buchstabe (nach der Klammer) ß ist ein Tippfehler. Bitte diesen Buchstaben zu streichen und durch = zu ersetzen.

Zu Anm. 32: Bitte hinzufügen: Vor allem aber J. F. Rock: *The Na-khi Naga cult..*. (Roma 1952, Serie Orientale Roma IV), Bd 1, S.94, Anm. 49!

Zu Anm. 67: Bitte hinzuzufügen: Notiz des Herrn Dr. Rock: In den Na-khi Texten wird von A mi geschrieben, er sei eine mächtige göttliche Verkörperung. Er soll früher ein ^{2}Bpö-^{2}mbö gewesen sein. Wenn die Na-khi ihn anrufen, d.h. wenn sie ^{2}A-^{1}mi (oder auch ^{2}Ä-^{1}mi) ^{1}ndsher ^{3}tsa singen, dann hängen sie das Bild der Gottheit des Amnye-Machen (^{1}Muan-^{3}mi-^{2}bpa-^{2}lo) auf. Es gibt mehrere ^{1}ndsher-^{3}tsa (= mit Macht begaben)-Rituale zur Austreibung von Dämonen. Im erwähnten Ritual ^{2}A-^{1}mi ^{1}ndsher ^{3}tsa ruft man diese Gottheit an, damit sie ihre Verehrer mit Macht begabt und sie dadurch in den Stand gesetzt werden, die Dämonen zu vernichten. Eines dieser sehr seltenen Ritualbücher befindet sich unter Nr 1071 (der Na-khi-Literatur) im Harvard-Yenching-Institute.

---

67 M. Taube: "Text und Übersetzung wurden 1961 in den *MIO* publiziert (siehe Anm. zu Rocks Brief vom 30.8.1953)."

Amnye Machen Text – *ausführliche Fassung*

Zu Anm. 7: Bitte hinzufügen: F. W. Thomas: *The Nam language...* behandelt den Namen gründlichst. Das Buch konnte ich leider nicht einsehen.

Zu Anm. 98: Bitte hinzufügen: Notiz des Herrn Dr. Rock: In dem Na-khi-Ms 1khü 3mä „Regen gewünscht" (vgl. J. F. Rock: *The Na-khi Naga cult...* (Roma 1952, Serie Orientale Roma IV), Bd 2, S. 609, page 7) ist auch vom Amnye Machen die Rede. Die Gottheit reitet ein weißes Pferd; der 2Bpa-2wùa-1ts'o-1bpö führt einen Tiger an einer Kette und die Gottheit hält einen perlmutterfarbigen Falken (Garuda) auf der Hand.

Vielleicht könnte diesem Zusatz auch die betreffende Zeile in Na-khi-Bilderschrift, die Sie mir mitteilten und die ich noch einmal in einem Abzug hier beifüge, beigegeben werden.
Hoffentlich erhalten Sie meine Zeilen noch rechtzeitig, bevor Sie nach Rom fliegen, um den Druck in die Wege zu leiten.

Mit den herzlichsten Grüßen
Ihr
Johannes Schubert

29. Leipzig C 1, Lampestr. 6, den 26. XI. 55

Lieber Herr Dr. Rock!
Mit herzlichstem Dank bestätige ich Ihnen den Empfang Ihres Schreibens vom 11. Nov. und freue mich Ihnen persönlich berichten zu können.

Zunächst einmal danke ich Ihnen für Ihre beiden Schriften, die Sie mir zugesandt haben und deren Empfang Ihnen meine Frau auch bestätigt hat. Beide waren mir ganz besonders wichtig, zumal ich mich nicht nur für die Herkunft der Na-khi, sondern auch vor allem für die der Tibeter interessiere und die Gegend im Süden des Kukunor (d.h. einschl. des Südteiles des Sees) Südwest-Kansu und Nordwest-Szu-ch'uan, das sind die Chiang, Rgyarong- und Amdo-Gebiete (einschl. des gesamten Amnye Machen-Raumes) für das Ausstrahlungszentrum halte.

Seit einigen Wochen bin ich aus China zurück. Ich habe sehr

viel gesehen und kennen gelernt.[68] Von den bedeutenden Städten sah ich außer Peking: Sian, Tschungking, Tschöngtu und Lantschow. Von letzterem Ort aus besuchte ich Sining, Kumbum und Lussar, den See Kukuṅor (Südspitze) und die dort nomadisierenden Tibeter (in den schwarzen Zelten), besonders die Arig in der „Goldebene“ (gser.thaṅ), direkt am Südostzipfel des Sees. Dann reiste ich auch südwestlich von Lantschow über Linhsia nach Labrang mit Hsiaho und ein Stück weiter von dort südwestlich in die Steppe Richtung A.mchog (ཨ་མཆོག་). In Kumbum wie auch in Labrang habe ich die Klosterbibliotheken als auch die Druckereien besonders besichtigt, hatte allerdings nicht die Zeit, um ausgedehntere Forschungen vorzunehmen. In Labrang ist übrigens seit einigen Jahren eine neue Inkarnation des Jambyangshadpa gefunden; derzeit ist der Junge 7 Jahre alt und macht einen überraschend intelligenten Eindruck. Mein nunmehr auszuwertendes Material ist hauptsächlich sprachlich (Lhasa-, Batang-, Derge- und Amdo-Dialekt) und phonetisch (Tonbandaufnahmen). Meine wissenschaftliche Arbeit ist nun keine Arbeit mehr, die nur am grünen Tisch geschieht, sondern eine fachwissenschaftliche Tätigkeit, die durch eigene Anschauung von Land und Leuten eine wesentlich festere Basis erhalten hat. Was an mir liegt, so werde ich auf alle Fälle sehr darauf bedacht sein, daß unser Briefwechsel nun doch wieder gut und regelmäßig funktioniert und Sie noch manches über meine Reise und von meinen Arbeiten hören werden.

Was Sie mir von Ihrer Tätigkeit berichten, so erfreut mich vor allem die Arbeit am Na-khi Lexicon. Das ist bestimmt das Wichtigste, was gebraucht wird und demnach das dringendste, was zu erledigen ist. Sehr wesentlich wird der alphabetisch angeordnete Teil sein, der alles erschließen muß und Hinweise auf die Stellen, an denen das Wort in der Bilderschrift zu finden ist, enthalten sollte! Die Beigabe der Manuskripttitel, wie Sie schreiben, wäre eine vorzügliche Ergänzung des Wörterbuches. Es sollte auch noch eine kleine Bibliographie aller auf die Na-khi-Sprache bezüglicher Bücher und Texte enthalten.

Sehr neugierig bin ich auf das Amnye-Machen Buch! Hoffentlich ist mir die Übersetzung des Textes aus dem Tibetischen

---

68 M. Taube: “Vgl. dazu Schubert: Vom Gelben Meer zum Blauen See (Miszellen meiner Reise 1955). *Wissenschaftliche Zeitschrift der Karl-Marx-Universität* 7.1957/58,1-12 (mit 13 Abb.”

einigermaßen gelungen! Wenn man nämlich selbst unter Tibetern war und auch mit hochintelligenten tibetischen Übersetzern, fabelhaften Literaturkennern usw. gesprochen und mit diesen Texte gelesen hat, dann fragt man sich, ob die ganze Arbeit so fern vom Schuß noch einen Sinn hat?
Mit den herzlichsten Grüßen
Ihr
Johannes Schubert

30. Leipzig, den 27. Februar 1956

Herrn
Prof. Dr. J. F. Rock
Honolulu 7 T. H.
c/o 497 [!] USA

Lieber Herr Dr. Rock!
Für Ihren Brief von Anfang Januar ds. Js., der mir nach Schierke, wo ich mich einige Zeit zur Erholung aufhielt, nachgesandt wurde, danke ich Ihnen herzlichst. Eigentlich müßte ich Ihnen ganz ausführlich und alle möglichen Einzelheiten schreiben, aber ich hoffe, gemäß Ihrer Mitteilung, auf den zeitigen Frühling, der Sie wieder kurz nach Europa führen soll, und wir dann Gelegenheit haben werden, uns persönlich zu unterhalten. Auf diese Weise würden wir uns nach sehr langer Zeit nun doch einmal wiedersehen.

Daß unser Buch (Amnye-Machen usw.) bereits im Druck ist und vermutlich noch dieses Frühjahr erscheinen wird, ist mir eine besondere Freude. Selbstverständlich würde ich mir die Korrektur-Bogen, die den tibetischen Teil und meine Übersetzung enthalten, gern noch einmal ansehen, ich glaube aber, daß die Zeit dazu viel zu kurz ist. Im positiven Fall allerdings wäre die direkte Zusendung vom IsMEO wohl das Beste.

Ich bin mit Ihnen derselben Meinung, daß im Abschnitt über Labrang auch den letzten, jetzigen ’Ajam-dbyaṅs-bzhad-pa (VI.) འཇམ་དབྱངས་བཞད་པ་ mit beizufügen, und es ist, wie ich in Klammer hier schon mit hinzugefügt habe, der 6. der Reihe und heißt Blo-bzaṅ-’ajigs-med-thub-bstan-chos-kyi-nyi-ma བློ་བཟང་འཇིགས་མེད་ཐུབ་བསྟན་ཆོས་ཀྱི་ཉི་མ་ Sie können das gleich noch mit einfügen. Tucci ist bereits aus Japan, Nepal, Afghanistan wieder nach Rom zurückgekehrt.

Hoffentlich macht Ihr Na-khi-Lexikon gute Fortschritte, denn so ein Hilfsmittel ist das Dringendste, was wir brauchen. Ich merke das jetzt ganz besonders zu einer Zeit, wo ich mich mit den letzten, von Ihnen herausgegebenen Na-khi-Texten eingehender befasse.

Mit den herzlichsten Grüßen, zugleich im Namen meiner Frau,
Ihr stets ergebener
Johannes Schubert

31. Prof. Dr. Johs. Schubert
Leipzig C 1
Lampestraße 6 Leipzig, den 4. VI. '56

Lieber Herr Dr. Rock!

Vorgestern erhielt ich Ihren Brief vom 26. Mai aus Rom. Ich danke Ihnen für die Zeilen herzlichst und bin froh nun wieder die Verbindung zu haben. Nach Honolulu hatte ich Ihnen geschrieben, aber bis jetzt noch keine Antwort erhalten. Deshalb schrieb ich auch an Herrn Professor Tucci.

Bezüglich des Amnye-Machen-Buches, ich meine jetzt den tibetischen Text und dessen Übersetzung, bestätigt mir Ihr Brief gewissermaßen nochmals das, was eigentlich schon Gegenstand unseres Briefwechsels war, nämlich die Revision der Übersetzung! Wie ich nun ebenfalls bereits nach Honolulu schrieb, halte ich eine nochmalige genaue Durchsicht für sehr nötig, wußte aber nicht, ob dazu noch genügend Zeit vorhanden sei. Ich wollte ja den Druck des ganzen Werkes damit nicht aufhalten. Wenn Sie mir allerdings heute schreiben, daß das Buch in zwei Teilen erscheinen soll oder kann[69], dann ist damit der Fall geklärt, und ich bin sehr gern zur Revision bereit und nehme mehr als gern auch die Hilfe des Herrn Professor Tucci, die er mir durch Ihren Brief anbietet, in Anspruch. Eine Schwierigkeit besteht dabei allerdings noch: meine vermutlich bevorstehende Mongolei-Reise[70], die evtl. schon An-

69 Der geplante 2. Teil ist nicht erschienen. Teil 1: *The Amnye Ma-chhen range and adjacent regions.* A monographic study. Roma: IsMEO 1956. IX, 194 S., 82 Taf., 5 Ktn. (Serie Orientale Roma.12.)

70 1957, lt. *Asienwissenschaftliche Beiträge*, 6. – M. Taube: "Siehe dazu

fang Juli beginnen kann. Schlimmstenfalls sehe ich den Text während meines Aufenthaltes in der Mongolei durch. Ich würde nur um baldige Zusendung bitten.

Ihr Na-khi-Lexicon wird, nach allem, was Sie mir darüber mitteilten, nunmehr bestimmt sehr gut geraten, und es wäre nur zu wünschen, daß es sehr bald publiziert werden könnte. Ich halte diese Arbeit für überaus wertvoll und wichtig!

Wer kann denn tibetische Zeremonialtexte, ohne daß man den Ablauf der Zeremonie an Ort und Stelle gesehen hat, richtig übersetzen?

Mit den herzlichsten Grüßen – zugleich auch von meiner Frau –
Ihr Johannes Schubert

Bitte bestellen Sie Herrn Prof. Tucci meine besten Empfehlungen.

32. Prof. Dr. Johannes Schubert
Leipzig C 1
Lampestraße 6 den 8. Mai 1958

Lieber Herr Dr. Rock!
Ihr Brief aus Honolulu, der vor wenigen Tagen hier einging, hat eine außergewöhnliche Freude bei mir ausgelöst! Allein ich fürchte, Sie glauben mir das nicht, weil Sie sicher sehr schlecht von mir denken – und dabei bin ich tatsächlich völlig schuldlos und habe mich immer auf unsere Zusammenarbeit bezw. die diesbezügliche Korrespondenz gefreut, genau so, wie ich mich jetzt freue, wo ich nun wieder ein Lebenszeichen von Ihnen habe und nun vor allem auch wieder schreiben kann. Nach der seinerzeit „verunglückten“ Italienreise – ich brauchte sehr viel Mühe, Zeit und auch Geld, um alles Nötige an Visa, Pässen etc. zu erhalten, und konnte Ihnen vor dieser Entscheidung nichts Positives oder Negatives mitteilen – für die ich schließlich alles zusammenhatte und dann die Absage bekam, ging ich ja in die Mongolische Volksrepublik. Von dort bin ich Ende 1957 bezw. Anfang 1958 zurückgekommen. Am 23.7.1957 schrieben Sie meiner Frau aus München[71]. Da meine

auch Johannes Schubert: *Ṛitt zum Burchan Chaldun. Forschungsreisen in der Mongolischen Volksrepublik.* Leipzig: F. A. Brockhaus 1963. 230 S. (mit 48 SW- und 8 Farbtafeln).”

71 Rock scheint aber nicht am Internationalen Orientalisten-Kongreß,

Frau während meiner Abwesenheit nicht immer zu Hause war, sondern bei Verwandten, erhielt sie Ihren Brief etwas spät und schrieb Ihnen am 10. September aus Torgau. Natürlich an die Münchener Anschrift. Den Erfolg bezw. Mißerfolg ersehen Sie bitte aus dem beiliegenden Brief. —

Bevor ich weiteres schreibe, möchte ich Ihnen auf alle Fälle noch den Erhalt Ihres Amnye-Machen-Buches bestätigen und natürlich herzlichst dafür danken. Daß das alles unentbehrliche Hilfsmittel sind und sehr wertvolle Schriften für die Zentralasien-Forschung, brauche ich wohl nicht besonders zu betonen. Was den von mir übersetzten Text über das Amnye-Machen-Ritual anbetrifft, so bin ich eigentlich sehr traurig, daß dieser noch nicht publiziert werden konnte, da Herr Tucci seinerzeit wohl nicht recht wollte. Ich habe dazu jetzt eine ganz andere Einstellung und traue auch Tucci nicht recht. Außerdem bezweifle ich, ob das alles wirklich so ist, wie er es übersetzen will. Es ist gar kein Zweifel, daß ich sehr gern den Text so veröffentlicht hätte, wie Sie ursprünglich wollten, nämlich als zweiten Band Ihres A.M.-Buches. Deshalb ist auch von mir bisher noch nichts weiter geschehen und hier keine Publikation des Textes erfolgt. Ich lege das in Ihre Hände! So, wie Sie das wünschen, soll es geschehen, selbst wenn auf die Zugeständnisse an Tucci eingegangen werden müßte! – Mit großer Spannung erwarte ich die Ausgabe Ihres Nakhi-Wörterbuches: ich halte das auf alle Fälle und immer noch für die wertvollste Veröffentlichung! – Wir sollten uns doch einmal sprechen! Rom ist aber nicht möglich, da ich sicherlich (?) im Juli (Mitte bezw. Ende) nach Nordindien reise[72]. Dauer etwa knappe drei Monate. Vielleicht wäre eine Zusammenkunft in Delhi (New Delhi) möglich? Ich würde mich wirklich sehr freuen!

Mit den herzlichsten Grüßen

Ihr

Johannes Schubert

Viele herzliche Grüße auch von meiner Frau.

München, 28. 8. bis 4. 9. 1957, teilgenommen zu haben. Zumindest ist er im Teilnehmerverzeichnis nicht genannt.

72 Assam, lt. *Asienwissenschaftliche Beiträge*, 6. – M. Taube: "Vgl. dazu: Johannes Schubert, Manfred Schindler: *Roter Fluß und blaue Berge. Durch Dschungel und Urwald von Assam*. Leipzig: F. A. Brockhaus 1960. 104 S. (mitt 22 Farb- und 83 SW-Aufnahmen)."

33. Prof. Dr. Johannes Schubert
Lampestraße 6

Leipzig C 1, den 6 Februar 1961

Herrn
Dr. J. F. Rock
P. O. Box 497
Honolulu 9
(Hawai[i])

Lieber Herr Dr. Rock!
Für Ihren Brief vom 3. Februar, den ich soeben erhielt, danke ich Ihnen verbindlichst. – Und ich habe mich gefreut, wieder von Ihnen ein Lebenszeichen zu erhalten! Natürlich und leider sind viele Jahre vergangen, ohne daß wir voneinander etwas hörten. Wenn Sie glauben würden, daß es Stolz oder sonst etwas von mir sei, daß ich Ihnen nicht schriebe, so würden Sie mir sehr wehe tun. Die Dinge liegen aber anders! Ich freue mich deshalb auch sehr, daß Herr Dr. Voigt[73], mit dem ich im August 1960 in Moskau zusammen war, mit Ihnen gesprochen hat. Es ist schade, daß Sie schon wieder zurückfliegen, sonst hätte sich bestimmt eine Begegnung in Berlin oder auch in Frankfurt/Main ermöglichen lassen (ich muß das eben ca. 6-8 Wochen vorher wissen). Auf die Herausgabe des Na-khi Wörterbuches werden viele warten; es ist tatsächlich eine dringende wissenschaftliche Notwendigkeit!! Da ich nun genau Ihre ständige Anschrift kenne (das war mir bisher wirklich nicht bekannt (nur Honolulu)), sende ich Ihnen mit gesonderter Post alles das, was ich von meinen letzten Veröffentlichungen noch habe; es ist nicht zuviel.

Ich denke, wir sind gute Freunde gewesen – und werden es auch immer bleiben! Mit den herzlichsten Grüßen
Ihr alter Freund
Johannes Schubert

---

73 Wolfgang Voigt (1911-1982), Direktor der Orientabteilung der Staatsbibliothek Preußischer Kulturbesitz Berlin (vorher: Westdeutsche Bibliothek, Marburg). Voigt initiierte das großangelegte Projekt Katalogisierung der Orientalischen Handschriften in Deutschland, in dessen Rahmen Rock später seinen Katalog der Na-hsi-Handschriften bearbeitete.

Im Sommer reise ich wieder in die Mongolische Volksrepublik und im Oktober (etwa) wieder in den Assam-Himâlaya.

34. Prof. Dr. Johannes Schubert
Leipzig C 1
Lampestraße 6 den 7. Mai 1961

Lieber Herr Dr. Rock!
Heute kann ich Ihnen einmal wieder etwas Tibetisches – einen Sonderdruck meiner Arbeit *Das Ritual für Rta Thugs-dkar po*[74] senden, der Sie vielleicht interessiert. Gleichzeitig habe ich mich noch für Ihren Brief vom 21. Febr., den ich erhalten habe, zu bedanken. Von Tucci's Unfall wußte ich nichts, ich habe das erst aus Ihren Zeilen ersehen, bisher aber auch nichts gehört, ob seine Gesundheit wieder hergestellt ist. Daß meine Haare ziemlich verschwunden sind, stimmt; es ist allerdings schon seit längerer Zeit so; vielleicht fällt es aber jetzt doch mehr auf. – Sobald wieder von mir etwas im Druck erschienen ist, sende ich Ihnen den betr. Sonderdruck. Ich hoffe, daß auch wieder bald etwas Tibetisches fertiggestellt werden kann. Von Juli bis Ende September bin ich wieder in der Mongolischen Volksrepublik auf einer Expedition (archäologisch-historisch). Ich gedenke jedoch noch vor meiner Abreise gewisse Sachen abzuschließen, so daß sie zum Druck gehen können.– Sollten Sie wider Erwarten doch einmal in Berlin sein, dann teilen Sie mir es doch bitte mit, auch wenn die Zeit nur kurz ist. Gesundheitlich geht es mir glücklicherweise gut.
Mit den herzlichsten Grüßen
Ihr Johannes Schubert

Herr Dr. Géza Uray[75] in Budapest bat mich kürzlich um Ihre Anschrift, die ich ihm gegeben habe.

---

74 Eine Beschwörung der Bon-po. Mit 8 Tafeln. *Mitteilungen des Instituts für Orientforschung* 8.1961, 51-63.

75 Führender ungarischer Tibetologe, 1921-1991; vgl. die Festschrift: *Tibetan history and language: Studies dedicated to Uray Géza on his seventieth birthday.* Hrsg. von Ernst Steinkellner. Wien: Universität 1991. XXXV, 536 S. (Wiener Studien zur Tibetologie und Buddhismuskunde 26.)

Appendix
Zwei Rezensionen Schuberts aus der *Orientalistischen Literaturzeitung.*

Rock, Joseph F.: *The ancient Na-khi kingdom of Southwest-China.* Vol. 1.2. Cambridge, Mass.: Harvard University Press 1947. XX, 554 S., 256 Abb, auf Taf. 8° (Harvard-Yenching Institute monograph series.8.)[76]

Als J. Siguret im Jahre 1937 seine (französische) Übersetzung des *Yün-nan pien-ti wen-t'i yen-chiu* veröffentlichte, war die Fachwissenschaft um ein gutes Quellenwerk zum Studium der Fragen, die die südwestlichen Grenzgebiete der Provinz Yün-nan betreffen, bereichert worden. Um wieviel mehr Ergebnisse aber liefert das in jeder Hinsicht einzig dastehende erstklassige Werk J. F. Rock's!

Der Verfasser hat sich das von den Na-khi bewohnte Gebiet und diese selbst, von den Chinesen mit dem Ausdruck Mo-so benannt, für seine Forschung ausgewählt und daher in dem Werk ein Gebiet beschrieben, das bestimmte Teile von Nordwest Yün-nan, von Hsi-k'ang, Tibet und Südwest Szu-ch'uan umfaßt. Der gesamte Raum ist im ungefähren Maßstab von 1: 500.000 (engl. Meilen bzw. chines. Li) auf 4, dem Werke beigelegten Kartenblättern (Te-ch'in, Wei-hsi, Li-chiang und Yen-yüan) festgehalten. Während des Aufenthaltes von etwa einem Menschenalter in Li-chiang und K'un-ming (Yün-nan fu) hat der Autor neben vielen Reisen, die er in diesem Gebiet unternommen hat, so gut wie alle chinesischen Schriften wie auch sämtliche in europäischen Sprachen publizierten Bücher über das westliche China und das östliche Tibet gesammelt und studiert, wobei an chinesischer Originalliteratur vor allem den Lokalchroniken der Grenzdistrikte besondere Aufmerksamkeit gewidmet wurde. Ergänzt wurde die Forschungsarbeit durch das Studium der Na-khi-Schriften, einer Tätigkeit, welcher J. F. Rock den größten Teil seines Lebens widmete.

Das Werk gliedert sich gewissermaßen nach den Mittelpunkten der auf den erwähnten Karten festgelegten Gebiete in der Reihenfolge Li-chiang (Teil II und III = Vol. I, S. 45-272), Wei-hsi

76 Publiziert in *OLZ* 48.1953, 465-466. Bei beiden hier gegebenen Texten liegen Schuberts Typoskripte zugrunde, so daß sich kleine Abweichungen zur Druckfassung ergeben.

(Teil IV = Vol. II, 294-319), Te-ch'in [A-tun-tzu] (Teil IV = Vol. II,320ff.) und Yen-yüan (Teil VI = Vol. II, 435ff.). Teil V (= Vol. II, 355-434) beschreibt das Yung-ning-Territorium mit dem berühmten See, der auf dem Kartenblatt Li-chiang noch besonders in einem Nebenkärtchen (Maßstab 1: 110.000 engl. Meilen bzw. chines. Li) dargestellt ist. Teil I enthält die Einführung (S. 3-44), die allgemeine Notizen über die Provinz Yün-nan, d.h. über die Gebiete zwischen K'un-ming und Li-chiang und besondere über den letztgenannten Distrikt enthält. 256 ganz vortreffliche Bildtafeln, die auf den S. XIII-XVII (Vol.1) verzeichnet sind, illustrieren die beiden Bände.

Das Werk liefert in erster Linie geographische und historische Daten, indem es jeweils alles nur Erdenkliche, was die einheimischen Lokalchroniken usw. bieten, exakt auswertet und größtenteils in einer vorzüglichen Übersetzung darstellt, wie sie eben nur jemand geben kann, der Land und Leute Jahrzehnte lang kennengelernt hat. Der Historiker findet durch die Übersetzung aller Inschriften (z.B. die Steininschriften von Shih-ku vom Jahre 1548 und 1561, Vol. II, S. 280-285), die sich in dem genannten Raum finden, und durch die Auswertung der Na-khi-Schriften (z.B. Vol. I, S. 76-86, die Na-khi-Überlieferung von der Abstammung der Menschen usw.) und der „Genealogien" (z.B. Vol. 1, S. 87-162) ein erstklassiges Quellenmaterial beieinander. Der Religionswissenschaftler trifft auf eine Fülle an Material bei Beschreibung der Lamaserien (vor allem der der Karma pa's) und findet auch die tibetisch geschriebene Liste der Hierarchen dieser Sekte (Vol. 1, S. 202). Vor allem steht aber der Philologe, der sich mit der Na-khi-Literatur befaßt – leider sind das sehr, sehr wenige! – in Verbindung mit dem Geographen vor einer unerschöpflichen Quelle hinsichtlich der geographischen Namen. Diese sind – das gilt aber teilweise auch für Personennamen – sowohl in chinesischer als auch in Na-khi-Form gegeben und ihrer Bedeutung nach übersetzt. Der Index (Vol. II,S. 475-539) und das von B. Armstrong Claytor von der orientalischen Abteilung der Library of Congress zusammengestellte Verzeichnis der geographischen Namen der Karten (Vol. II,S. 543-554), die beide bei chinesischen Namen auch die Schriftzeichen – wie im ganzen Werk überhaupt – geben, machen nach dieser Richtung hin den wichtigsten Bestandteil aus und erleichtern die Erschließung dieses so überaus reichhaltigen Werkes ungemein.

Man weiß nicht, was an der Publikation besser ist, der vorzügliche, tiefschürfende und exakt bearbeitete Text oder die außergewöhnlich guten Illustrationen. So zeigt sich die Arbeit als eine wirklich reife Frucht lange währender und langwieriger Studien und somit als das einzig dastehende Standard-Werk auf diesem Fachgebiet, das wohl kaum jemals überboten werden kann.

Rock, J. F.: *The Na-khi Naga cult and related ceremonies.* Roma: Istituto italiano per il medio ed estremo Oriente 1952. (Serie Orientale Roma.4.) Pt. 1: XII, S.1-383, viele Taf.; 2: XII, S. 385-806, viele Taf. 4°[77]

Es ist nunmehr eine anerkannte Tatsache, daß sich Formen des tibetischen Bontums unter den im chinesisch-tibetischen Grenzgebiet der Provinz Yünnan lebenden ^{1}Na-^{2}khi's erhalten haben, welche durch das hier zu besprechende Werk Rock's weiterhin stärkstens unter Beweis gestellt wird (vgl. S. 15). J. F. Rock, der sich ein Leben lang im Zentrum dieser Gegend aufhielt, hat vielfach davon berichtet und eine große Anzahl von ^{1}Na-^{2}khi-Schriften übersetzt.[78] Die Texte, die weniger in der eigenartigen Silbenschrift[79] als vielmehr in der noch komplizierteren Bilderschrift[80] geschrieben sind, sind zum überwiegenden Teil Ritualliteratur. Das Buch führt mitten hinein in das Na-khi-Schrifttum.

Der Herr Verfasser nennt 122 religiöse Zeremonien der ^{1}Na-^{2}khi (S. 25-30), die er in 15 Gruppen (z. B. Zeremonien frühester primitiver Naturverehrung, Gruppe 1; Zeremonien zugunsten des Familienglücks, Gruppe 2; Zeremonien zur Besänftigung der Bergdämonen und Schutzgeister, Gruppe 3 usw.) gliedert. Zur Abwicklung einer jeden Zeremonie ist das Absingen einer ganzen Anzahl verschiedener Texte nötig. Von den 15 Zeremonial-Gruppen zählt die vierte Gruppe 7 Zeremonien auf, die zur Besänftigung

---

77 Publiziert in *OLZ* 50.1955, 79-81.

78 Eine Übersicht über die betr. Arbeiten Rock's und über das Na-khi-Schrifttum findet sich in meinem Aufsatz Na-khi-Piktographie ... (*Buch und Papier.* Leipzig: Harrassowitz 1949, 114-142).

79 Eigennamen, z.B. hier Namen der Naga's sind neben den Bilderschriftzeichen noch in Silbenschrift geschrieben (MS No. 1020, vgl. S. 512).

80 Vgl. dazu S. 17-20 und auch meinen in Anm. 77 genannten Artikel.

der Wasserdämonen[81] dienen. Eine davon (S. 26. VI,30) ist die sogenannte „große Naga-Zeremonie“ (^{2}ssu ^{1}ddü ^{1}gv), deren Bon-Formen man bisher nur schleierhaft aus tibetischen Schriften[82] kannte und die bei Regenmangel, Versiegen der Quellen u.ä. Vorfällen ausgeführt wird. Diese Zeremonie bildet den Hauptteil des Rock'schen Werkes (S. 1-606), der neben zwei kleineren Ritualen zur Besänftigung der Naga's (S. 607-624; vgl. S. 26, VI 32 und 33) noch ein Reinigungs-Ritual (^{3}ch‘ou ^{1}na ^{1}gv; vgl. S. 26, VII, 36) angeschlossen (S. 625-806) ist. Die Grundlage der Arbeit bilden eine große Anzahl handschriftlicher Texte in Na-khi-Bilderschrift – ich habe 86 gezählt (erwähnt sind natürlich noch viel mehr) –, deren Originale sich zum größten Teil in der Bibliothek des Harvard-Yenching-Institute's in Cambridge, Mass., befinden.[83] Mit Hilfe der ^{1}Na-^{2}khi-^{2}dto-^{1}mba's (ob = tib. ston.pa?) konnte Rock zum Verständnis der Texte gelangen und sich auch die Zeremonien mit allen ihren Phasen vorführen lassen (das ist auch die Quelle der vorzüglichen Photobeigaben, auf die ich noch einmal zu sprechen komme).

Nach einer Vorrede aus der Feder des berühmten Tibetanisten Giuseppe Tucci's (S. XI) folgt die Einführung Rock's (S. 1-20). In dieser ist auf die ^{2}ssu ^{1}ddü ^{1}gv-Zeremonie, d.h. Rituale (^{1}gv) zur Besänftigung der Naga's (^{2}ssu = tib. se, bse usw., vgl. S. 11), das dem tibetischen klu. 'abum. nag. po (schwarzes Naga-Hunderttausend) entsprechen soll (S. 13), ganz allgemein hingewiesen. Sehr willkommen ist die bisher ausführlichste Darstellung der Aussprache der ^{1}Na-^{2}khi-Wörter auf S. 21-23. Den eigentlichen Texten voran geht noch eine übersichtliche Beschreibung über den von 4 ^{2}dto-^{1}mba's in der Zeit von 3 Tagen auszuführenden „großen Naga Kultes“ (S. 33-37).

Es folgen dann die Texte. Jedes dieser Rituale gliedert sich in viele (je nach dem betr. Ritual verschiedene) Phasen, Einzelhandlungen, für die besondere Texte vorgeschrieben sind. Nur manche für Einzelhandlungen vorgeschriebene Gesänge werden

---

81 Die sich aber keinesfalls nur im Wasser aufhalten. (S. 8)

82 Das Sa bdag klu gnyan gyi byad grol (bzhugs so), von Berthold Laufer hrsg. und übers. u. d. T. *Ein Sühngedicht der Bonpo.* (Denkschriften d. Akad. d. Wiss., Wien 1900, Phil.-hist. Kl. Bd 46, Nr 7) und das Klu 'abum bsdus pai snyin po, hrsg. und übers. von Berthold Laufer (Mémoires de la Société Finno-ougrienne 11. Helsingfors 1898).

83 Der Herr Verfasser hat Tausende solcher MSS gesammelt.

nicht nach einem geschriebenen Text, sondern aus dem Gedächtnis vorgetragen, wie z.B. hier (S.198) die Handlung, die sich auf den sogenannten „Schlaf der Naga's (^{1}ssu ^{3}haw) ...“ bezieht. Zur Zelebrierung des „großen Naga Kultes“ ist das Absingen von 58 Texten erforderlich. Die 58 Texte (z. B. das Austreiben der Naga-Feinde, S. 73-96, das Auswerfen der Körner, S. 124-159, das Herbeirufen der Naga's, S. 172-186, das Verbrennen von Wacholder, S. 230-253 usw.) sind nach den zugehörigen MSS, die nach Nr und Standort stets sorgfältig notiert sind, übersetzt. Den Originaltiteln der einzelnen übersetzten Texte sind die entsprechenden ^{1}Na-^{2}khi-Bilderschriften hinzugefügt. Von 10 Texten der 58 Zelebrationshandlungen (und zwar – wenn ich nach den Übersichten S. 41 und S. 632 zähle – die Texte 3, 17, 27, 29, 31, 36, 43, 56 und 57 des „großen Naga Kultes“ und die Texte 2, 4 und 20 der ^{3}ch‘ou ^{1}na ^{1}gv) ist jeweils die erste Seite in Bilderschrift, Umschrift und Übersetzung wiedergegeben. 14 Seiten des Textes Nr 36, der den Kampf zwischen den Naga's und ihrem Feind, den Garuda, behandelt, sind außerdem, weil es sich hier um das sogenannte „Schlüsselbuch“ der ganzen Zeremonie handelt (S. 385), auf den S. 385-486 vollständig in Bilderschrift, Umschrift und Übersetzung dargeboten. Zu jeder ausführlicheren Zeremonie gehört ein Buch bzw. Text (^{2}ddu-^{1}mun), der die Anzahl und Titel der abzusingenden Schriften aufzählt und die zur Ausführung der einzelnen Handlungen benötigten Gegenstände nennt und abbildet. Dieses, von Rock sogenannte „Index-Buch“ ist (S. 39/40) sowohl für die ^{2}ssu ^{1}ddü ^{1}gv als auch (S. 633) für die ^{3}ch‘ou (^{1}na) ^{1}gv beschrieben. Die S. 607-619 behandelte Zeremonie zur Herbeiführung von Regen (^{1}khü ^{3}mä) und das Ritual zur Austreibung des Schlangendämons (^{1}zhi ^{1}ts‘u ^{3}bpu), das auf S. 620-624 ausgeführt ist, haben kein „Index-Buch“.

Neben der Übertragung der Texte, die nur derjenige geben kann, der den Schlüssel dazu hat – und das ist einzig und allein der Herr Verfasser –, sind das Wertvolle an dem Buche die 1048 Anmerkungen, in denen eine ungeheure Masse an Stoff verarbeitet ist. ^{1}Na-^{2}khi-Namen mit ihren tibetischen bzw. chinesischen Gleichungen oder Äquivalenten (alles auch mit tibetischen bzw. chinesischen Typen und in Umschrift), geographische Namen sehr wenig bekannter Gegenden mit ihrer Deutung und nicht zuletzt die in den Texten vorkommenden Naga- und sonstigen Eigennamen usw., die hier erläutert sind, bieten der weiteren Forschung reichlichen Roh- bzw. Grundstoff.

Außer 11 Tafeln, die die bereits erwähnten Seiten der Bildertext-Originale wiedergeben (eins davon, das Titelbild zum 1.Band, farbig) ist das Werk mit 58 hervorragenden Photos (als Anhang zu jedem Band) ausgestattet, an deren Hand sich der Ablauf der einzelnen Ritualhandlungen bestens verfolgen läßt, weil die dazu benutzten Gegenstände, sei es einzelne oder im Gesamtaufbau der Objekte, vorzüglich abgebildet sind.

Diesem für die Fachwissenschaft unentbehrlichen und einzig darstehendem Werke (es wäre noch ein ausführlicher Index dazu erwünscht) sollten weitere Bände folgen! Das ist ein Desiderat, um dessen Erfüllung man den Herrn Verfasser nicht dringend genug bitten kann.

DIARY OF JOSEPH F. ROCK
1921–1922

From December 30, 1921 to May 10, 1922
Journey from Chiengmai to Li-chiang, N.W. Yunnan

December 30,1921

Left Chiengmai by motor from Doi Saket, 12 miles. At Doi Saket Haw caravan stopped for the night of December 29-30, 1921. Left together (29 ponies, 5 bearers, cook, boy, pony boy). Went through Dipterocarp forest consisting of *Dipterocarpus tuberculatus, Dipt. obtusifolius*, *Randia tomentosa, R. spinosa*, several species Lagerstroemias, *Strychnos nux vomica, Str. potatorum, Str. spinosum*, several species of *Zizyphus* (a small one – *Zizyphus oenoplia*), a red-flowering *Sterculia* (very handsome flowers in large terminal panicles), *Xylia dolabriformis, Caesalpinia* (climbing), two species of *Tectona*, a species of *Albizzia*, *Hopea odorata,* small Ardisias, also *Pinus Merkusii* (two-needle), a bignoniaceous tree, *Mayodendron igneum*, a climbing asclepiadaceous plant with long fruits, Orchidaceae on trees, no. 1551 – *Bridelia monoica* (Laur.) Merr., several species of *Desmodium*, *Cassia tora*, *Spatholobus Roxburghii, Phyllanthus* trees, a flowering bamboo called Mailai (Lao) (*Oxytenanthera albociliata* Munro), a *Canavalia*, *Mangifera* sp.?, several species *Ficus* (strangling figs), a large species of *Amoora* (*A. gigantea* – collected previously along Meh Ping last year), *Irvinga malayana* (very common), *Terminalia arjuna, Palaquium* (a milky sapotaceous tree), *Spondias mangifera, Bauhinia* sp.? (undershrubs), *Barleria cristata, Homonoia riparia* (in streambed) and *Grewia*. Followed along the Me Lai stream and arrived at the Chow-luang's place, Ban Me Wan, at 6 p.m. Elevation of Ban Me Wan 1,450 feet; laterite formation.

December 31, 1921

Left Ban Me Wan at 8 a.m. after a miserable night, people talking, burning fires etc. Glad to leave the barking dogs behind. Followed the valley of the Meh Kuang or Lampoon River to its headwaters and crossed the divide at Doi Nang Keo Hi Luang. Elevation 3,350 ft.;

distance to Chiengmai – 55 km; to Chiengrai – 117 km. The legend says that the above-mentioned princess, Nang Keo Hi Luang (Vagina magna) was unable to find a man to satisfy her strong sexual desires and consequently went mad and is said to now haunt the hill named after her, for a subduer. The natives make offerings of penises of all sizes made crudely of wood, leaning them against telegraph poles, trees etc. along the trail. This is done to appease the goddess as otherwise she would bring disaster upon any man passing her way. Most of these penis-offerings are made by married women in order that their husbands may escape the vengeance of the unsatisfied princess. We arrived at Phang kai camp at 6 p.m.

Saw the following trees and plants. Several species of bamboo – one small-leaved one in fruit called Mai Sang, and Mai-lai; also a larger species called Mai bong (*Dendrocalamus strictus*) with lateral horizontal branches. Bamboo was very common for quite a distance – several miles. After that we came into laterite country with *Hopea odorata, Dipterocarpus*, Compositae, a grove of very narrow-leaved trees of beautiful shape and large trunks, *Ficus* sp.?, *Duabanga* sp., along streams *Dillenia* sp.? (probably *indica*) and tall elephant grass. All the trees are very tall, some of them bare of leaves as *Bombax malabaricum* with red flowers (a very tall tree), a smaller *Bombax* with white flowers, *Erythrina suberosa* just coming into flower, also *Butea* sp.?, *Congea* sp.?, *Bauhinia malabarica*, *Bridelia retusa, Glochidion velutinum* trees, *Phyllanthus emblica* (very common), *Caesalpinia, Toona, Magnolia* (or *Michelia*), *Quercus, Castanopsis indica*, a species of *Prunus* with beautiful pink flowers, and *Anneslea fragrans*. Along the stream were ferns, also *Brainia insignis*; on the trees, *Pleopeltis, Platycerium* sp.?

The camp is situated at the headwaters of the Meh Lao which flows into the Mekok and the latter into the Mekong. Elevation at Pang Kia (Pine-tree Camp) 2,900 ft., – the highest point between Doi Nang Keo Hi Luang and Chiengrai. Also saw *Leea, Rhus, Solanum verbascifolium, Thunbergia laurifolia, Loranthus, Callicarpa arborea*, etc.
Specimen nos. 1562-1592.

January 1, 1922
Pine ridge or Pang Kié. Pang Kié is in a small valley in the headwaters of the Meh Lao. To the left, looking southeast, is a ridge covered with two species of pines, *Pinus Merkusii* (with a dark deeply

furrowed bark) and *P. Khasia* (with a smooth, brown, flaky bark). With them grow species of *Quercus, Anneslea* sp.?, a small bushy bamboo which the natives called Mai Phio. In the scrub grew a species of *Phoenix* much resembling *Ph. Robelini* and tall grasses as are found on the pine ridges of Doi Saket Chang near Chien-mai. At the foot of the ridge grew *Randia* sp.?, leguminous vines (probably *Mucuna*), yellow-flowered Compositae, a tree called Mai Pao with red leaves (very soft wood), *Loranthus* sp. and (on it) a species of *Viscum*. Photographed (1606) trunk of *Pinus Merkusii* with Dr. Mason – a view of the mountain. Doi Naw ngün (Silver sprout) facing Pang Kié with *Castanopsis acuminatissima* (Bl.) Rehd. and *Pinus Merkusii* in foreground, also a small *Pinus Merkusii* with *Quercus* and a Lao boy. It is said that when a boy is tattooed, the spaces between the tattoo marks grow wider as the boy grows older and then the spaces are filled in again with tattoo marks. The boy photographed is a native of Chieng-mai district.

January 2, 1922
Left Pang Kié at 8:15 a.m.; following the valley of the Meh Lao, passing the Pong nam ron – hot sulphur springs. The trails led through Dipterocarp forest. Passed through several villages and finally emerged on to the Wieng Papao plain with low hills ranging north and south. Passed through a village called Ban meh chedi, Crossed the plain to the village of Meh Kha chan elevation 2.000 ft., arriving at 4:15 p.m. Stopping in a new Sala.[1]

The valley extending around the pine ridge at Pang Kié is densely wooded with tall trees of many species. Huge *Bombax* trees occurred, many *Erythrina suberosa*, *Engelhardtia spicata* Bl. (1618), several species of *Ficus, Pipturus*, etc., long stretches of bamboo forest, followed by Dipterocarp or Eng forest – Mai hiang being the most common tree –, with it grew several species of *Strychnos* (1646 – *Strychnos Nux-vomica* L.)(3 species), *Terminalia, Desmodium, Spatholobus Roxburghii* Benth. (1645), *Entada scandens*, etc. see collection.
Specimen nos. 1611-1658.

---

1 *Sala* ist das Thai-Wort für (Holz)-Haus, Gebäude. Rock gebraucht das Wort wohl für das Amts/Empfangshaus des Ortes.

January 3, 1922
At Ban Tung Ha, 6 km from Wieng Papao, and 23 km from Ban Meh Kha Chan.

We crossed dry forest composed of *Strychnos spinosa*, *St. potatorum*, *Lagerstroemia macrocarpa*, also a rubiaceous tree, very common, resembling a *Nauclea* – *Monauclea* (Nauclea) probably *purpurea*), several species of *Ficus*, a few clumps of Bamboo, *Hoya* sp.?, *Dipterocarpus* sp.? (Mai hiang), and *Terminalia*. Finally emerged onto the Wien(g) Papao plain and ricefields. We passed through the latter to Ban Tung Ha where we arrived at 5:15, making camp under a *Phyllanthus* covered with *Loranthus*, along the bank of the Meh Lao.
In dry forest, collected specimens of various species beginning with no.1659-1669 inclusive.

January 4, 1922
Left Ban Tung Ha („Village of 5 rice-plains"), at 8:30 a.m. for Meh Soi (pretty water) – distance 32 km. The path led through dry forest over a spur and gradually passed through damp valleys with huge trees and underbrush. The trees were mostly so tall that it was impossible to collect them or make out what they were. Many species of strangling figs, *Engelhardtia spicata, Albizzia, Markhamia stipulata, Dolichandrone,* Milaceae [!], *Quercus*, etc. The forest gradually changed into Eng forest with *Dipterocarpus* sp.?, *Terminalia* etc.

Elevation at Meh Soi 1,600 ft. Arrived at 6 p.m. Passed several villages, last one called (Meh) Ban Nam Prik.
Specimen nos. 1670-1703.
1681 – *Oxytenanthera albociliata* Munroe in flower, a bamboo (Mailai), negative no. 22935 and 22936.

January 5, 1922
Left Meh Soi at 8:30 a.m. passed over a hill where we met 10 elephants of which 8 were wild. One of them was partially white. We also met a Lao who had been pinned down by a tusker and one tusk going into his side near the arm pit, tearing a hole in his shirt and penetrating his nipple. One of the elephants had a baby. After passing over the hill, we descended into a valley with bamboo and tall trees, many *Erythrina suberosa*, huge Alpinias, *Maranta* sp.?, *Quercus,* several species of ferns, *Clerodendron*, tall Pampas grasses and others as are found on Doi Chom Chang.

Later we passed through secondary jungle, mostly bamboo forest, the Mai Pung, Mai Sang (with glaucous stems) and Mai Lai. Thence we passed over a nice plain where we stopped for lunch at Ban Pong Pu Fuang. Later we passed into Eng forest and stands of *Pinus Merkusii*. Opposite the village where we had lunch is a large ridge covered with pines.

We arrived at our camp after passing through more or less uninteresting forests with bamboo and huge species of strangling Ficus. Later we descended to the village of Ban Hue San along the Hue San creek. Elevation 1,650 ft. The pine ridge where I collected *Pinus Merkusii* was 1750 ft. above sea level.
Specimen nos. 1832-1854

January 6, 1922
Left Hue San at 9 a.m. Crossing over small hills and Eng forest, composed mostly of *Dipterocarpus obtusifolius*, *Strychnos* (a large leaved species – a small tree), several species of bamboo (as Mai lai, Mai Sang, etc.), many climbers as *Dioscorea* (nearly all in fruit at this time), *Randia* (the very large fruited variety); *Schima* sp.?, and various other plants and grasses. Especially common are the *Ficus* species (strangling figs and non-epiphytic species), also *Garcinia*, vitaceous vines, Convolvulaceae, etc.

We reached Sun Dawk at 11 a.m. and had lunch under a huge fig tree *Ficus* sp.?, with cauliflorous figs. From there, the road passed through rice-fields and scrub forest of *Lagerstroemia* etc. We reached Chieng Rai at 4:30 p.m. and went to Mr. Bachtell's house.
Specimen nos. 1808–1831.

January 7, 1922
Dried the plants and unpacked.

January 8, 1922
Developed films taken at Pang Kia. Changed blotters and repacked. Called on the Governor of Chiengrai in the afternoon and visited several temples.

January 9, 1922
Left Chieng Rai for Soon Dawk and Hue San for the trip to Doi Chang. Had 6 horses and 5 carriers, cook, boy, and Pao Puer. At Soon Dawk I got a guide for Doi Chang, through the kind office of Kun

Phra, Governor. Collected a few plants on the way, stopped again at Soon Dawk for lunch under the big fig tree along the river. Between here and Hue San, in the Eng forest, I found a large bamboo in fruit but immature – the flowering stalk about 45 ft. or more in height, the culms short-jointed about 8 inches, glaucous with horizontal branches at the lower nodes. The Lao boys called it Mai Sang. Arrived at Hue San at 4:40 p.m., elevation 1,650 ft.
Specimen nos. 1855-1859.

January 10, 1922
Left Hue San at 8 a.m. for Doi Chang. The old guide was absolutely useless. He wanted to take us along a new road but missed the trail and we were wandering over paddy fields and water buffaloes' camping places which means mud. Finally we struck the trail crossing the stream Hue San. We passed through evergreen forest, tall trees 150 ft. in height with enormous boles. The valley of the Hue San was full of *Musa*, gingers, undershrubs and vines. We passed then into bamboo jungle which lasted to over 2,500 ft. elevation ó the natives called it Mai lai. It grows rather crooked and slender, reclining and forming dense thickets. the culms lying almost horizontally. A species of *Phlogacanthus* (*curviflorus*) with large leaves was found in the marshy places at the foot of Doi Chang. The trail went through open country along the outskirts of the forest till finally we entered jungle again with huge Mai Hok bamboo 100 ft. in height with large culms of considerable diameter, (elevation 3,500 ft.).

Ficus spp. were very common also many leguminous vines, especially *Mucuna* with purple flowers. The trail was rather steep. *Callicarpa* trees were common also *Erythrina suberosa* (but lower down). Large areas of the thatch grass was encountered which Laos were collecting and tying into cylindrically shaped bundles, four to one bamboo pole. We came to one small village of Miao and finally passed through the main village a little higher up. The dirt and filth is undescribable. The people wear dark blue home-spun cloth with a long narrow blue apron in front. The women wear a much pleated, short skirt reaching to the knees; nearly all wear heavy silver rings around the neck. The pattern of the skirt is of white figures and the basal part is ornamented with red stripes. They never wash their faces or bodies and the accumulated filth of years adheres to them. Several of them were insane or idiots. I took several pictures of the village; of a group of men in the jungle, several men standing near a house, and

Dr. Beech took one with myself standing with a group of dirty Miao near a house. Photographed a very primitive corn grinder and took another picture of a spinning wheel. The girl working the spinning wheel ran away on our approach as did all the women. They hid themselves in their houses and refused to come out. I managed to snap an old woman hunting lice on the head of a little child. We are camping above the Miao village where we pitched our tent.

We practiced shooting with my revolver and Dr. Beech's gun and two Miao came running up to see the shooting, and eagerly picking up the empty cartridges.

Above our camp is a deserted cornfield. Their corn bins are full of corn infected with weevels, red corn, white and mottled.

The moon is straight overhead. It is cold and I have to stop writing. Tomorrow we are going to climb to the summit of Doi Chang. Elevation at the camp is 4,200 ft.

Specimen nos. 1704-1746.

1721 – *Dioscorea rockii* Praisn & Burk. type.

1722 – *Bauhinia tenuiflora* Watt. cited in Brittonia 4:92.

1744 – *Celastrus hindsii* Benth., see Brittonia 4:103.

January 11, 1922

Left for Doi Chang (Elephant Mountain) summit at 9 a.m.. went through clearing and secondary forest and passed through a jungle of wild banana – Koi pâ. Thence through jungle of tall trees and Zingiberaceae, *Maranta*, *Carex baccans,* a few ferns etc. Finally we ascended a very steep ridge coming to the foot of the neck of the Chang (Elephant) where I took two photos. The trees were mainly *Quercus Junghuhnii*? We climbed to the summit of the ridge through undergrowth of *Amomum, Alpinia*, Rubiaceae (with red fruits resembling *Psychotria*), *Smilax* (a very spiny species with small leaves and fruits). Gradually the slope became bare of undergrowth and only species of *Castanopsis* and *Quercus* (1755 – *Quercus daichangensis* Camus) covered the ridge. There was some undergrowth, myrsinaceous plants, *Carex,* blue flowered Acanthaceae, *Evodia triphylla* (but only small plants 3 feet in height and fruiting), *Schima* sp.?, *Osbeckia* sp.?, and a shrubby *Pithecolobium, Gleichenia linearis* and *Brainea insignis* were about the only ferns save for no. 1757 – *Hymenophyllum exsertum* Wall., on trunks of *Quercus daichangensis* (see negatives in U.S. Dept. Agr.) which covered the trunks of *Quercus* etc., also the roots of a Polypod (on trees) and many orchids (also terrestrial ones).

I took two photos of the top without undergrowth. The trees are mostly *Quercus* and *Castanopsis*, also two tree trunks covered with the above *Hymenophyllum.*

We left Doi Chang summit, which according to aneroid was 5,350 feet, and descended to the neck of the elephant and climbed to the head through much spiny *Smilax* (1752). At the top of the head (5,125 feet), there was a beautiful view of the plain below and we could see Chiengrai in the distance. A Miao with two Miao boys appeared with bows and poisonous arrows. They were out for deer and birds. I tried to photograph them but every time the shutter of the camera clicked, they put their heads together and looked into the lens. Result – no picture. I bought a bow and arrows but he would not sell the poisoned ones. He said they were hard to get. They bought the poison from the Mo-so tribe.

We returned to the camp in the late afternoon. Dr. Beech and I talked till the midnight hour; fireflies came into the tent and, outside in the jungle their light made one believe that it emanated from the eyes of lurking tigers. We finally went to sleep, with curious noises around us, to wake up at 6 a.m.

Specimen nos. 1747–1807.

1755 – *Quercus daichangensis* Camus.

1757 – *Hymenophyllum exsertum* Wall., on trunks of no. 1755, see negatives in USDA.

1791 – *Castanopsis armata* Spach, negatives 22955 in USDA.

January 12, 1922

Owing to my indisposition (having the bother of a boil) and to the condition of the forest on Doi Chang, the Miao having cultivated the mountain slope and destroyed the forest, I decided to return to Chiengrai. I left the Miao village about 8 a.m. collecting on the way *Pterospermum acerifolium* with peltate large leaves was common, species of *Macaranga*, a white flowering climbing *Bauhinia,* also *B. purpurea* or *variegata* (it was not in flower but the tree had the aspect of either species), a beautiful *Argyreia* with whitish-lilac petals and deep purple center about 4 inches across, several species of *Ficus*, a *Bignonia* tree with reddish corolla tube, leaves with narrow oblong lanceolate leaflets. The tall timber bamboo called Mai Hok (Mai Hooak as Kerr spells it and which, according to him, is *Thyrsostachys siamensis* Gamble). It is of the latter that the Miao construct their

dwelling houses. Lower down we found again the Mai Lai, dense tangles of this species found flowering on the road to Chiengrai.
Met Mr. Lee going up to Miao village.

We returned by the Sun Dawk trail which follows a gorge with tropical forest emerging into Eng Dipterocarp forest and on to the Chiengrai road at a telephone pole bearing the number 2461.
Reached Chiengrai at 4:30 p.m.
The thatch-grass cut by the natives for their houses is *Imperata arundinacea.*
Specimen nos.
1793 – *Senecio acuminatus* Wall. forma *breviligulata* H.-M., type.

January 17, 1922
Left Chiengrai at 4 p.m., crossed on ferry over Me Kok River and rode as far as Ban Du where we found the Haws (Chinese caravan men) waiting. We have now 15 horses and 8 carriers.

We are sleeping in a Wat. A small Buddha sits enthroned on a bamboo altar. Our table is right under some Buddhist draperies. A lot of boy-priests are standing around looking and watching our doings. They are amused at our gas lamp. No collecting on this trip (day). Beyond the river is mostly teak forest and dry scrub. The village we are staying in is in teak forest. We are waiting for Mason (tomorrow). Mr. Baer is now with me.

January 18, 1922
Early this morning the priests came into the Wat to bring water to the Buddha who was dressed in yellow cotton. A little later a woman came in with a wash basin full of rice covered with banana leaves, banana skin and 2 little candles. A priest came and took the wash basin, stuck one lighted candle into the rice, the other before the Buddha. He began to recite Pali texts Namu amida Buddha, etc. The woman kneeling behind. It was cold and the priest was shivering and pulled his yellow robe over his shaven face. I watched the whole procedure from my camp bed and from behind the mosquito net. They did not mind at all. My table and chair was in front of the altar, set ready for breakfast.

We left at 9:30 a.m., passing through dry jungle and paddy fields. Had lunch at Ban Hue Pû and finally went on through more paddy, and Eng. forest as well as bamboo and then entered a beautiful bit of forest on the spur of a mountain. There we stopped and collected

ferns, *Mussaenda* etc. Rattan (*Calamus*) and Zingiberaceae ae well as Araceae are common. The trees were all *Dipterocarpus alatus*, Meliaceae etc. After passing through teak forest, we entered the village of Ban Meh Ki on the Meh Chan, the village of Ban Meh Chan is on the other bank and both are known as Chieng Sen Luang. The Ampur had sherry and dinner with us and I as usual, worked till late putting away my plants.
Specimens nos. 1860–1879.

January 19, 1922
Left Ban Meh Ki (Chieng Sen Luang) at 9:30 a.m., hunting for horses. The trail passed paddy fields and dry secondary savannah forest composed of *Nauclea sessilifolia*, species of *Randia, Tectona, Strychnos nux-vomica, Str. potatorum, Zizyphus* sp.?, *Imperata arundinacea*, or Elephant grass, *Randia* with large fruits, several huge *Ficus* (one I photographed), also large bamboo called Mai Pio at Ban Ba Sang, and *Ficus*. Among the large trees found there, were members of Sapindaceae and Anacardiaceae. We passed thence through teak forest of small growth, *Garcinia* sp.?, *Elaeocarpus*, *Amoora Rohituko*, *Pterosperum semisagittatum*, etc.

We stopped for lunch on the banks of the Nam Kham where I packed my plants. The village on the other side of the river was Ban Nam Kham, followed by grassy Savannah jungle with much *Nauclea* and *Randia.* All this plain was formerly under cultivation (about 75 years ago or more). The natives were carried off by a former King of Chiengmai as prisoners and made slaves of them in the latter place.

We followed along the foot of Doi Tum and the whole mountain range all afternoon. The mountains are very steep and intersected by deep valleys. Limestone out-croppings are visible. On Doi Tum there is a celebrated Wat and place of pilgrimage. We are stopping now under Mai Yang trees at a village called Hue Bong. We are now 12 km from the Shan border, altogether 60 km from Chieng Rai.
Specimen nos. 1880–1897

Januar 20, 1922
Left Ban Meh Pong at 7:30 a.m., where we camped in our tents for the first time on our trip. The dew was exceedingly heavy as it sounded like heavy rain from the teak leaves. The Haw horses had a free fight, turning their behinds to each other and kicking for all they were worth. A bullock caravan came along and caused a great

commotion – the Haws yelling and spluttering. We passed through a big swampy plain with numerous rushes, *Cyperus, Scirpus, Clerodendron, Desmodium*, Scrophulariaceae and *Carex*, etc. We then passed through a paddy field with the mountain range of Doi Tum to our left, before reaching the last village with a gendarmery, Ban Chong. I took two photos of the last mountain, Doi Chong (Umbrella Mountain), in Siamese territory. We passed the gendarmes and entered quite a tropical grove of large trees. The mist had risen and the whole range of mountains was clear. In the grove of large trees were several *Ficus*, yellow-fruited Meliaceae, a tree with catkin-like inflorescence was very common , also many *Strychnos* trees, Randias, *Imperata arundinacea*, etc. We came then to a small river called Hue Sai which flows into the Me Huak and both into the Mekong. The Hue Sai forms the boundary between Siam and Burma or the Keng-tung State proper.

The first village in Shan territory is Taki-lek where we stopped for lunch and where I dried my blotters. Here in Shan territory, the natives put up permanent paddy field fences, while in Siam they take them up after harvest and pile them way up in the forks of tall trees to prevent them from being stolen. The people here wear mostly top-notches or long hair but Lao is spoken.

We passed a large village called Hong-luck. At Ban Wieng Kean, I bought a coconut just before getting into Hong-luck. Priests are very scarce here but they carry swords which is not to be found in Siam. We reached Ban Meh Huak, crossing the stream of the same name and are now camping on an island in the Meh Huak. The vegetation is mostly rattan, *Caesalpinia* and ferns. Elevation about 1,500 ft.
Specimen nos. 1898–1922, still in Siam.
Specimen nos.1923–1928, collected in Burma. S.E. Shan States

January 21. 1922
Left Ban Meh Huak at 8 a.m. passing through paddy fields and villages. We passed by a mountain range at the foot of which was said to be hot springs called Muang Ko. At 11 a.m. we stopped for lunch, crossing the Meh Niu, and lunched at Ban Ta Kaw. This was the last village.

We entered much of bamboo jungle – many species from very tall ones to the Ma Lai and very slender ones. The glaucous Mai Sang was common as well as Mai Bong Pa. I photographed one fine clump of bamboo called Mai Pio (two exposures). After crossing several streams, we entered evergreen rain-forest with tall trees – species of *Calamus* or Rattan, *Caryota* palms, Zingiberaceae, tall Marantaceae and ferns. The trail led over a mountain range inhabited by Akas or Kaw people (their language is said to be akin to Chinese?). The mountains have been much cleared by these people – many of the various hillsides being absolutely barren, cleared for planting, others abandoned and taken possession of by tall grasses. We passed what seemed to be a surveyed road on which we found many Akas working. We pitched camp at a place crowded by bullocks. A large caravan had already taken posession of the place – the open space in the jungle. There is much yelling and shouting and calling. A large crowd is sitting not far from our tent gambling. We are peacefully situated along a babbling brook at Ban Ki Hat, in the midst of the jungle. The Akas are passing with torches through the forest and the shouting continues. One single woman is with that whole crowd of men. She looks for the world like an American Indian, wears a sort of tam-o-shanter decorated with silver, bone, etc., wears a short skirt and, in front, a sort of kilt. They are all around with bows and arrows and also with long knives, and are a cruel-looking people. I shall be glad when this night is over. There may be a free-for-all fight for they have drunk liquor and are now gambling – some sixty of them and we are only three. There are no police nor gendarmes nor officials of any kind in this region and all is so different from peaceful Siam. One would expect a lot of lawlessness in this region. Three Shans from Muang Yang paid us a visit and said they wanted to talk. They had been on the road one month and came from Muang Yan. They reported a band of 500-600 bandits working on the Shan-Yün-nan border at Muang Len.

To-morrow is Sunday and we are going to stop and visit a Kaw or Aka village. The collection on today's stage, was rather poor.
Specimen nos. 1929–1968, Burmese territory.

January 22, 1922
Sunday – Today we stopped all day at Bang Ma Ki Hat. Early this morning we were awakened by a lot of noise. Near our camp, there sat under a bamboo shed a Shan chief surrounded by his counsellors, holding in his hand a rattan stick. Before him sat a crowd of Kaws. In their front rank there were three who did the argueing. They yelled and shouted at the top of their voices, The chief struck one of them with the bamboo rod, and instantly he kept quiet. The chief visited

our camp and asked for a cathartic, which I gave him. He was surrounded by a group of Kaw onlookers. They wear blue turbans and short jackets, are armed with knives, wear thick silver bracelets and smoke out of silver pipes. I photographed them as they were at our camp. I also photographed a group of Kaws who passed by to work on the road as the British Commissioner is said to come next month. Kaw women passed and I took their pictures at the order of the Shan official. They carried paper-bark (*Broussonetia papyrifera*) on their back. Some of the Kaws ran away when I took the black cloth and tried to focus the camera. I had to discard the focusing cloth as they were afraid of it.

About noon they began to make music with drums and gongs while the chief weighed opium in a scale, the weight used being a silver rupee. They seemed to be all contented.

In the afternoon Baer and I went for a walk into the jungle following a small stream where I found ferns, and urticaceous undershrubs. Much bamboo was seen and very tall, buttressed trees between huge lianas. I photographed a tree trunk (2 views) also vegetation along the stream looking down and up with huge bamboo (Mai Bong Pa). We followed the creek bed till we came out on to a clearing or Hai made by the Kaws, of which clearing I took photos. It must take an enormous amount of work to clear such tremendous forests as we passed through with trees 150 ft. in height, enormous bamboos, *Calamus* and other spiny palms as well as *Pandanus*. There was not much in the collecting line, but we spent a very interesting day in this big evergreen jungle among the most curious people.

Many came for medicines, one wanted to buy quinine for a rupee. He was going to make money by so doing as he would sell each pill for twice the amount he would pay for them. We refused to sell any.

One Kaw came with bamboo-reed organs and played and danced to his tune. It was very amusing. Their hair is cut rather short, save a small patch on the very center of the head which they let grow long. Their jackets are short, blue, and have red and white borders. They are filthy but not nearly so as the Miao. Many of them have goiters. The women are very much decorated with beads and millet seeds which are sewn on the cornucopia-like (conical) headdress. Much silver ornament is worn.

To-morrow we leave for Muang Len.

January 23, 1922
Monday – Left camp at Pang Ki hat at 8:30 a.m. after saying good-bye to the chief of Muang Len, who still held a conflab with the Kaws.

Our road passed through much of Mai-lai bamboo and interesting scrub and weeds. After an hour's march, we came into dense bamboo jungle of Mai pong? where we met some Kaws working on the road. I stopped them and took three photos of them. One old man with long hair was exceedingly dirty. They all smoked large pipes. Shortly afterwards, we passed through a large grove of magnificent trees of which *Dillenia* formed a large proportion, many smaller shrubs and trees forming the undergrowth. We also passed many Hais, i.e. cleared land, which had been cultivated by the Kaws and, according to Dr. Mason, was in opium two years ago. Much thatch-grass occupies it now with *Broussonetia papyrifera* of which the people of this region make paper. We had a difficult crossing over the Meh Len River, the bank on the other side being very steep and exceedingly muddy, with a pool of mud at the foot of the bank. We crossed over paddy fields and entered Muan Len at 1:30 p.m., halting near the entrance of the village in a neat little sala or rather a little bot with an altar and a wooden Buddha in a corner where priests sit in meditation. The posts are red and gilded, the walls are red and the roof is of heavy tiles. After getting settled, we tool a walk through the village. Today was market day and the filth on the village road is indescribable, mangy dogs lying about with pigs and waterbuffalos. Two Kaw men, as filthy as can be with their long hair full of maggots, were driving two pigs through the village trying to sell them – 30 rupees for two large sows. We went to the headman's house and found a group of men playing the coconut game. They were all sitting on the mat and shaking coconuts listening to the water in the nut and weighing it in their hands, while a pile of rupees and French piasters were lying in the center of the circle which the players formed. The one who guesses the largest coconut inside the husk wins. The nuts are opened and the meat weighed, the one who guessed the heaviest nut gets the money. We left ere the game was finished. On the way back to the Sala we stopped at a native drug vendor and asked him for Kalaw (*Taraktogenos kurzii* Chaulmoogra). He had very old seeds and said that they were called Maw Kalaw in this region and that it occurred two days before reaching Keng Tung. We entered two wats, very poor ones with cheap-looking, smiling Buddhas wrapped in yellow cloth.

The building is of mud bricks, stuck together with mud. Old wooden figures were lying on the ground in the temple yard which reminded one of two lepers we met in the village street. In the evening I developed the pictures taken at Pang Ki hat and in the bamboo jungle on the road.
Mosquitos are bothering me so I am going to bed.
Specimen nos. 1969–1989.

January 24, 1922
Left Muang Len with a guide and three Kun-haps (carriers) for the Mekong River. We passed between lanes through the village of Muang Len out into the paddy fields and then entered a grove of tall trees and huge bamboos (the Mai Hok) with enormous glaucous culms. We then came to a small village on the Meh Len which river we crossed in a boat, the horses swimming. Thence we went through more paddy fields and gradually came to the foothills covered with low lying Mai lai bamboo. The ridge was rather open and covered with much *Scleria scrobiculata*. The trees were a *Schima*?, *Quercus truncata, Quercus fenestrata* (the small seeded variety), two species of *Castanopsis* and Mai lai bamboo in flower. The trees on the very steep, upper ridges were mostly *Quercus* (several species) and theaceous trees (*Eurya groffii, Eurya acuminata*) etc. Ferns were common as a species of *Cibotium* (2004 – *Cibotium barometz* (L.) Thn.) of which I took a photo (no. 30056). It is trunkless – the leaves coming directly from the ground. A simple-frond fern, with creeping rhizome, was common on the sandy quartzy ridges (*Oleandra cumingii* var. *longipes*) while a *Gleichenia* covered the exposed slopes. After climbing to about 4,000 ft. elevation, we descended into the valley of the Mekong. *Livistona* palms were very common on the slopes of the ridges of the Mekong valley but not on the Meh Len side. The forest changed from the deciduous *Quercus*-type to evergreen rainforest with tall tress such as *Dillenia*, Meliaceae, many aroids (*Pothos scandens*) one of which I took a photo on a *Dillenia* tree, also a general forest view of another aroid, probably a species of *Philodendron*. There were Zingiberaceae, Marangaceae, *Musa*, *Calamus, Caryota* and *Arenga* palm (with small, narrow leaflets), *Pinanga* probably *Pinanga gracilis* (the common terrestrial fern found in wet low forests), also *Asplenium nidus* on trees, and many other fleshy plants as *Pandanus* etc. Gradually the forest gives way to Mai lai bamboo until we reached the streambed of the Mekong. The

riverbed is about a mile wide but at this time contained little water. Enormous rocks fill the whole riverbed which produce terrific rapids. The vegetation in the streambed is *Homonoia riparia* and a low stout shrub (2002 – *Syzygium abortivum* (Gagnep.) neg. 22988), bent in the direction of the current, this plant which I photographed is submerged in many feet of water for the greater part of the year. I took a photo of the plant, also 5 of the Mekong up and down stream and one of the guide in the Mekong river against a rock. The heat was terrific. We found large tiger tracks in the muddy riverbed. They led to the river and were quite dry; they must have been there quite a time ago. We ascended again the steep ridge and returned the same way we came. Photographed a Kaw village on the top of a hill in the distance. We reached camp about 7 p.m.

The photos of the Mekong were taken near the deserted village of Tang Oh or Tah Aw, a trail leads from there to Muang Sing in French Territory.

Specimen nos. 1990–2030.
1995 – (*Quercus*) *Lithocarpus fenestrata* (Roxb.) Rehd.
2002 – *Syzygium abortivum* (Gagnep.), neg. no. 22988
2004 – *Cibotium barometz* (L.) Thn., negative 30001. USDA

January 25, 1922
Left Muang Len about 10 a.m. stopped at the Paper-village and took photo of their paper drying racks. Also bought 33.75 rupees worth 2250 sheets of paper made of the bark of *Broussonetia papyrifera* for the plant specimens. The trail led through uninteresting bamboo jungle onto a surveyed road which follows the Meh Len. The scenery along the Meh Len is very beautiful, evergreen rainforest. We stopped about half way between Muang Len and Pang Kah Luang. Many of the trees are very tall and have pinnate leaves either Sapindaceae or Meliaceae. *Duabanga sonneratioides* is very common. On one tree of that species I found a large fern (*Drynaria coronans*, neg. 30059 in USDA), which I collected. Bamboo of the large timber variety is also common besides *Pandanus*, Zingiberaceae, Marantaceae, and of palm, *Calamus, Caryota* and *Arenga*, besides many climbing ferns (*Lygodium*). In the streambed are many shrubs, and *Ficus* (2034 = *Ficus gibbosa* Bl.; 2035 = *Ficus comosa* Roxb., large trees, 2037 = *Ficus pyriformis* Hook & Arn., a shrub, is also very prevalent but mostly large trees). Leguminous vines and aroids cover the trees and form huge entanglements. *Musa* may be seen everywhere. *Homonoia*

riparia forms large shrubs with thick knotty trunks growing in the riverbed between the rocks in the water. They are submerged for a great part of the year. I took several (3) photos of the Meh Len showing the vegetation of its banks, also the *Homonoia* shrubs in the water. Much of the trail had been cleared owing to the Commanding Officer visiting along the road. We met him in the Sala at Pang Kah Luang. He is Col. Shelton Smith, a very dry and droll individual. I asked him how many white people there are in Chieng tung and he answered „One and he is away“. This Sala is quite a substantial building with a verandah of hard wood and a room built in the center. Taung gyi, whence he came, is 25 stages from Chieng Tung with a beautiful road and Dak Bungalos all along the way. Taung gyi is 20 miles from the railway.
Specimen nos. 2031–2039
2031 – *Pteris ensiformis* Burm., negative no. 30059, USDA
2034 – *Ficus gibbosa* Bl. (large treses)
2035 – *Ficus comosa* Roxb. (large trees)
2037 – *Ficus pyriformis* Hook. & Arn., shrub

January 26, 1922
Left Pang Kah Luang (the big thatch camp) at 8:30 a.m. We saw the Colonel off and then started ourselves. The trail led through dense rain forests along the bank of the Meh Len. Enormous trees towered for 150 feet above the other smaller trees, many ferns grow along the small rivulets emptying into the Meh Len, as well as on the banks of the trail – collected many plants among them no. 2040 – *Laportea crenulata* Gaudich. var., with stinging hairs which are still smarting me. We stopped for lunch on the banks of the Meh Len. The trail led away from the Meh Men upwards to 2,000 feet elevation into dry bamboo jungle (2053 – *Melocalamus compactiflorus* B. & H., 2066 – *Dendrocalamus birmanicus* A. Camus) with *Bauhinia malabarica*, bignoniaceous trees with spathaceous calyx (*Dolichandrone cauda–felina*), etc. We arrived at Muang Hpyak at 5 p.m. We looked for the house built especially for the Colonel and when we arrived, we found it occupied by Shans, who bolted with all their belongings as fast as they could go. We took possession of the place and are quite comfortable. We are sitting under a tamarind tree. The air is cool and damp. Muang Hpyak is on a fairly large plain between two mountain ranges. The plain is well cultivated. All the rice has been harvested and the fields are brown. The village is scattered over the plain rather

more detached than one village. Across the creek is a temple rather primitive, where we found a number of Buddhist priests – their chests tattooed with red, their legs and abdomen black. They are all a very civil lot. So far we have had nothing stolen.

Today I took one photo (neg. no. 22995) of a species of fern (2041 – *Alsophila glabra* (Bl.) Hook.), with a trunk 6 feet in height; the frond measured 10 feet. It grew along a creek which emptied into the Meh Len.

The scenery along the river was magnificent.

Specimen nos. 2040–2071.

2040 – *Laportea cranulata* Gaudich. var.

2041 – *Alsophila glabra* (Bl.) Hook., neg. no. 22995

2051 – *Pittosporopsis kerrii* Craib. cited in Brittonia 4:67.

2053 – *Melocalamus compactiflorus* B. & H.

2066 – *Dendrocalamus birmanicus* A. Camus nov. spec.

January 27. 1922

Early this morning I took a photo at Muang Hpyak of the girls sitting and spinning on the platform in the middle of the road of the village. These girls are eligible and expose themselves to the view of boys looking for wives. The boys select the girls after inspection and take them home.

We left Muang Hpyak about 9 a.m., the trail passed through dry scrub vegetation and, just outside the village on the edge of a paddy field, I found a large *Castanea* tree with old fruits. It has entire leaves but is a true *Castanea*. Bamboo gave way to scrub forest such as is found on the outskirts of the villages. We soon entered the gorge of the Meh Len River and dense forest with tall trees and lianas. The trail is very winding and it is a continuous up and down. The forest became denser and ferns very numerous; tree ferns were observed and Nai In swam across the Meh Len to cut down one large tree fern (2097 – *Alsophila glauca* (Bl.) J. Sm., neg. nos.: 22950, 22996, 22998, 22999). He succeeded and brought it toward the rapids where he secured it before reaching the current. The trunk was 35 feet tall about 6-7 inches in diameter, blackish, the fronds about 15 feet long with 36 pinnae or 18 alternating pairs of pinnae, and an additional tip of 3-4 inches in length. The scenery is grand. Ferns became very numerous and the trail reached an altitude of 2,300 feet. It began to get dark when we reached our camping place, called Pang Sop Lao (The mouth of the Meh Lao) which we crossed just before entering a

small village. It is pitch dark as I write. Lao Puer is playing his flute and the water of the Meh Len is rushing madly over rocks toward the Mekong.
It is time to go to bed.
Specimen nos. 2072–2112.
2097 – *Alsophila glauca* (Bl.) J. Sm., neg, nos, 22950. 22996. 22998. and 22999,

January 28, 1922
Stopped at Pang Sop Lao for the day. After breakfast we climbed the ridge facing our camp to investigate the Pine trees on the top of it. We had to climb through tall grass and over a burnt area till we reached the pine trees which proved to be *Pinus Khasia.* With it, we found growing a species of *Andromeda* (2118) – *Vaccinium sprengelii* (G. Don) Sleumer (cited by Merrill in Brittonia 4:156, (1941) and by Sleumer B. J. 71: 462, 1941, but by Sealy in Curtis Bot. Mag. vol.164: pl. 9658, 1944, it is discussed as similar to *V. brachybotrys* (Fr.) H.-M. Rehder) again determined it as *V. mandarinorum* Diels det. Rehder. *Gleichenia* was common and covered the ground; also Compositae, *Rubus, Carex, Rhynchospora*, and the common grasses found on Doi Sootep with the pines. I took two photos, one looking towards the tall mountains on the other side of the Meh Len, the other looking across through pines down into the valley. After lunch I went along the trail back to the tall tree ferns which are the same as the one found on the other side of the Meh Len (2097 – *Alsophila glauca* (Bl.) J. Sm., a tree fern attaining a height of 35 ft., frond 15 ft., 36 pinnae in dense rain-forest of the banks of the Meh Len, photo, USDA neg. no. 22950, 22996, 22998, and 22999.) I took four exposures, also a *Phlogacanthus pubinervius*, neg. no. 30003. We followed the creek by the tall tree ferns and found quite a number of species not collected yesterday. On the return trip to the camp, I took a photo up stream with the mountains in the distance. After a bath and swim in the cool waters of the Meh Len, we walked along the banks and collected quite a number of species of ferns. Our camp is under a *Dillenia indica*, the fruits have been dropping frequently but no one got hurt.
Specimen nos. 2113–2144

January 29, 1922
Left Sop Lao at 8:30 a.m. The trail led through dense forest along the

Meh Len and just before crossing the river over which the bridge was washed away, we struck a grove of pine trees – *Pinus Merkusii* which I photographed. The whole hillsides on both sides of the river were clothed in pines. After crossing the stream, the trail ascended to the top of the ridge which was mainly covered with huge pines (2155 – *Pinus Merkusii* Junghuhn & De Vriese, USDA neg. no. 30008) and Mai Sang bamboo (2154 – *Dendrocalamus strictus* Nees, see above neg. Oaks (*Lithocarpus spicata*) and *Castanopsis* were also common. All the pines on the north side of the ridge were *Pinus Merkusii.* The trail led to an elevation of 3,000 feet at the divide, but the mountains above it were all covered with pines up to about 3,600 feet elevation. The mountains opposite, which I photographed yesterday, must be about 6,000 feet elevation. The bamboo we photographed was collected (2154 – *Dendrocalamus strictus* Nees., see neg. no. 30008 – Mai Sang). Accompanying the pines was *Phyllanthus emblica*, *Quercus* (2172 – *Lithocarpus spicata* (Sm.) R. & W.), etc. The trail descended again into the valley of the Meh Len. The south side was covered with pines but *Pinus Khasia* with only very few of *Pinus Merkusii.* The trunks of *Pinus Merkusii* are deeply furrowed but those of *P. Khasia* are covered with a flaky or flat, scaly bark. Although very old, *P. Merkusii* show their furrowed-bark characteristic only at the base. The cones of *P. Merkusii* are longer, as are also the needles, than those of *P. Khasia.* From the summit of the ridge a wonderful view could be obtained over the valley. The Kaws have ruined much of the forest by clearing the land which, set off by commas after abandonment, is taken possession of by tall grasses and *Rubus* (2151 – *Rubus elongatus* Sm.). The trail descended into dry Mai Sang bamboo jungle and gradually rose again into dry forest with *Strychnos*, and spiny *Randia.*

We had lunch at a most awful place full of bull dung and dry and dusty – most of it burnt-over area. It is known as Bang Tong. We passed afterwards through more or less open jungle, the Meh Len getting smaller all the time. We entered a little valley with an affluent of the Meh Len where we collected many ferns. The altitude at our camp which is Ban Hka is 3,100 feet. On our way to the camp we found *Prunus* trees (2177 – *Prunus salicina* Lindl. in full flower. They were gorgeous to behold and reminded one of home. We were all very angry on account of a package of dry plants having disappeared.

We encountered much *Engelhardtia* (2174 – *Engelhardtia spicata* Bl.

tree 80–100 ft,. tr. 3 ft in diam.) – one very tall tree, and climbing *Mussaenda*, Compositae etc.
Specimen nos, 2145–2191
2154 – *Dendrocalamus strictus* Nees, see neg. no. 90008 – Mai Sang
2155 – *Pinus Merkusii* Junghuhn & de Vriese; USDA neg. no. 30008
2172 – *Lithocarpus spicata* (Sm.) R. & W.
2174 – *Engelhardtia spicata* Bl. tree 80–100 ft., tr. 3 ft. in diam.
2177 – *Prunus salicina* Lindl., see Brittonia 4: 89,

January 30, 1922
Before leaving Ban Yang Kha, I went down the road and photographed a lovely tree of a Yün-nan pear – it was all in flower with ascending branches. I also photographed a branch spread along the trunk (2191 – *Pyrus pashia* Ham.). The trail led along the Meh Len which here was rather small, resembling a brook; along it grew many pear trees, one with small fruits which we collected. Further along, we came across two other *Prunus*, one with small flowers (white), and one with deep pink flowers and stout ascending branches (2199 – *Prunus persica* (L.) Batsch). It looked as if cultivated, as it grew in a deserted field. We entered rather dry forest with *Ficus* trees, *Strychnos*, spiny *Randia* with yellow fruits and a real *Prunus* (2195 – *Prunus puddum* Roxb.) with pink flowers resembling a Japanese cherry. The trunk was clean and not attacked at the base as are the cherry trees in Washington, D.C.

The Meh Len became smaller and smaller, the hills were bare having been cleared by the Kaw people and everything looked like desolation. The Kaws are ruining the state of Keng Tung by their Hai-methods. We passed several huge chestnut (*Castanopsis*) trees with large spreading crowns and large trunks 3 feet in diameter; the branches were drooping (2201 – *Castanopsis hystrix* A. DC., neg. no. 22977 USDA). We could see Loi Mwe on the hills. Bare hills they are with brick buildings. A little beyond the tall chestnut trees, the Meh Len divided and we followed a branch known as Pi Bang – the Ghost divide.

One hour's ride along the banks of this small branch brought us to our noon stop at Pang Bi Bang. I decided to stop and dry the numerous plants collected. We built a large fire and put up a frame scaffolding to use for hanging the plants over the fire. 15 parcels are drying on the frames over a ripping fire.
It is time to change plates and go to bed.

Specimen nos. 2192–2204
2192 – *Pyrus pashia* Ham.
2195 – *Prunus puddum* Roxb.
2199 – *Prunus persica* (L.) Batsch.
2201 – *Castanopsis hystrix* A. DC., neg. no. 22977 USDA

January 31, 1922
Shortly before breakfast this morning, two Kaw women came to the camp – spinning while they were walking. One (an old one) we persuaded to pose for a picture in a spinning attitude; the younger one refused. The old one carried a baby on her back and every now and then she shook as if she were cold – it was meant for the baby.

We left and passed through oak and chestnut forest – a rather semi-dry region. I photographed a *Castanea* tree (2252 – *Castanopsis hystrix* D.C.) with Noi Lot standing at the foot of the tree. We followed a small streambed up to its source (elevation 4,450 feet) through dense and magnificent forest; the air was glorious cool, and the scenery wonderful. We left Loi Mwe behind us encircling the ridge on which it is situated. We found wild peaches with deep pink flowers (2240 – *Prunus persica* (L.) Batsch) growing near the summit of the pass. The trail descended rather steeply through drier jungles and at about 3,500 feet elevation, *Quercus* and *Castanopsis* species became numerous. It became very hot. The lower slopes were mostly covered with small oak trees with large spikes and large leaved oaks (2218 – *Lithocarpus fenestrata* (Roxb.) Rehd. & Wils.; 2219 – *Castanopsis indica* A.DC.). *Phyllanthus emblica* was very common, also *Buddleia asiatica* and numerous trees as *Schima* (2220 – *Schima wallichii* Choisy (det. Rehder)) etc. We stopped for lunch at Muang Chen where I dried blotters etc. We left nearly as late as four p.m. The trail passes through paddy fields and open timber known as Mai Ki Lek (Fire wood). It looked, for the world, like an orchard. The branches are cut for fire-wood and all the trees had a pruned appearance. Very common is a large tree with very stout and crooked (twisted) branches which had very light green foliage – just appearing, although it was flowering (specimen nol. 2248 – *Quercus acutissima* Carruth.).

We came to a village called Peng Sai and there we stopped at the temple – a very crude affair made of sun-baked mud-bricks. The abbot is a deformed chap but very kind. We made our beds in the temple before the altar. The priest came and recited Pali texts while

we were listening. A tiny home-made candle burnt on bullock horns in front of the altar.
Bed time.
Specimen nos. 2205–2252.
2214 & 2221 *Quercus brandisiana* Kurz
2218 *Lithocarpus fenestrata* (Roxb.) Rehd. & Wils.
2219 *Castanopsis indica* A. DC.
2220 *Schima wallichii* Choisy (det. Rehd.)
2236 *Clematis acuminata* D.C., cited by Nurrile
2237 *Celastrus monosperma* Roxb. In Britt. 4: 104 but det. as *Celastrus Benthamii* Rehd. & Wils. Other name for it is *C. championi* Benth.
2240 *Prunus persica* (L.) Batsch
2248 *Quercus acutissima* Carruth.
2252 *Castanopsis hystrix* D.C. photographed this spec.

February 7, 1922
We left the Dak Bungalow about 10:30 went to the post office and thence to the big market (neg. no. 30026, Keng Tung market). There were not quite so many people as five days ago, however, the place was crowded. Many Chinese were there, Kaw women, Shans, Wa, Lahus, etc. The Lahus wear jackets with red and green triangular strips of cloth sewn onto them. On the right, below the shoulder, they wear a silver rosette-buckle resembling the Japanese chrysanthemum crest. The Kaw dress in dark blue, without any ornaments, though some of their women are overloaded with silver rings around the neck. Their cap is covered with millet seeds and those of *Coix lachrima jobii* but the elongate variety. I photographed two Kaw women who were absolutely ignorant of what was going on. They looked straight into the camera. I also took photos of the various temples – one of the temples is situated within the Sawbwa's palace (walls) with a Wa standing in front (neg. no. 30044). In the temple opposite I photographed the Library; masonry covered with stucco; the priests standing in front (neg. no. 20037); an interior view of the temple with many Buddhas, all covered with yellow cloth. We said good bye to Mrs. Thelford, the Baptist Missionary, and went to see the old stone, covered with old Siamese characters.

The trail led through paddy fields over the circular plain of Keng Tung. We stopped for lunch on the banks of the Meh Lap – having lunch rather late (3:20 p.m.). Not far from the Meh Lap, there grew a large *Pyrus pashia* Ham., full of fruits and covered with white flowers. We collected the fruits and specimens (2253 – *Pyrus pashia* Ham.). The native name is Mai Kaw Ki Kai. The trail led through several villages and paddy fields. We passed a Haw caravan loaded with tea from Muang Hai, Yünnan. They said that they met with no robbers. The sun began to set and the hills to the left were a rich blue which was strangely contrasted with the dull yellow paddy fields; the red hills to the right bathed in the red fading light of the setting sun. Large bamboo groves announced the site of the village Ban Saa where we are stopping now in the Wat Ban Saa. The abbot of the temple, a rather intelligent individual, came to interview us and admire our benzine gas lamp – in fact all the women came from the village to gaze at this marvelous lamp. We had to turn it on and off to show them; they gazed at it for hours and could not understand the brightness. We told them the cost of it and they said for that price they can build a house with a lot of wood in it. The cost would be 30 rupees – today's exchange (358 to 100 dollars) as per Rangoon cable. A nice old woman who said she was seventy years old, had eye disease and Dr. Mason treated her with blue-stone (copper sulphate). The electric torch-light amused them greatly and, shortly after the women left, the men came and wanted to see the light which was not hot – also the photographs of Mason's family. They are a very nice lot of people. They all watched us eat and sounds of surprise went up as we used forks and knives. They could not get over it. It was like a theater. We poured brandy on our pudding and lighted it and the fun was great for them. Later, some more old women came – one a shrew with a hard mouth of the thickness of a string and straight, and with eyes that pierced. She gazed at the lamp. Another woman was ugly beyond description. We finally told them that it was bed-time but they wanted to see us go to bed – how we did it. Our beds are in the temple, banners for departed spirits are hanging down and paper decoration on the wall representing elephants, cows and horses. Buddha, a very ugly statue covered with gold-leaf and yellow cloth, sits enthroned and surrounded by numerous little ones. On the sides are white paper umbrellas and on the ground (in front) are small elephants made out of wood, carrying on their backs banners covered with tinsel. The posts, which are evidently chestnut wood, rest on stones to preserve them from the termites. It is real cold.

Mason told us that, when he went up last time, they were told to sleep in a Wat but the place was so dirty that they slept under a large

tree. This happened to be a spirit tree and in the morning the people came gazing and remarking „Why they are still alive". Mrs. Thelford said that they would encircle the whole town with a string to keep the spirits out and place offerings outside the string for the spirits. Her Wa tribal servants, she said, once remarked when they slaughtered sheep, that that was the way they cut the heads off humans when they still lived in the hills.
We are going to stop tomorrow as I want to send mail back to Keng Tung.
Specimen nos. 2253–2257
2253 *Pyrus pashia* Ham.
2270 *Castanopsis tcheponensis* Hick & Camus (see neg. no. 30071)

February 8, 1922
Stopped all day at Ban Saa. Developed plates exposed in Keng Tung of market, wats etc.

Afternoon climbed the dry hill to the left of the village. The whole hill has been cut over and is now occupied by secondary jungle – mostly Mai Sang bamboo, some *Quercus, Castanopsis, Schima*, a beautiful leguminous shrub (*Indigofera elliptica*) with pink flowers; one of the most common shrubs was rubiaceous with large panicles of cream-colored flowers and simple leaves, the old ones of which turned red (no. 2269 – *Wendlandia glabrata* D.C.), *Desmodium* and a few pine trees was about all. The *Pinus* was *P. insularis* but some had four needles as well as two and three on the same tree. Along the creek-bed below the village which is really situated on the hill-top, the vegetation is mostly bamboo, and a rubiaceous tree with fragrant flowers which we also found planted as a fence near and at Muang Len. I sent a large package of the fruits of *Pinus insularis* to Washington. Also photographed the village of Ban Saa, where there is a small bronze Buddha in the Wat of the same name.
Specimen nos. 2258–2270
2269 *Wendlandia glabrata* D.C.
2270 *Castanopsis tcheponensis* Hick. & Camus, USDA neg. no. 30017.

February 9, 1922
We left Wat Ban Saa early this morning, the night was very cold and it was hard to get out of the warm bed. The trail led along a dry hill with bamboo and oaks and paddy-fields on the other hand, and

gradually ascended to a ridge on which we found a fine bamboo in flower – they called it also Mai Sang but it was much smaller, the culms thinner and less tall (specimen no. 2280 – *Dendrocalamus strictus* Nees). The whole hillsides had been cut over for fire wood and now bamboo and scrub-oaks have taken possession of the country. The view from the top of the ridge was magnificent over the Keng Tung (valley plain) towards Ban Saa and again up the valley of the Meh Lui. The hills were rather bare except the steep ridges, fires were burning on the steep slopes and in the rarified air Kaw villages could be seen distinctly on the top of the hills.

I took two photos – one over the plain towards Keng Tung, the other into the distant Meh Lui valley with pine trees in the distance. *Pinus insularis* was quite common but as isolated individuals. Near a bog at 4,500 feet elevation we found a species of *Pyrus (pashia)* in flower, it was about 30 feet tall. I photographed the tree and a flowering branch pinned against the trunk. From there the trail descended gradually, winding through bamboo jungle and much *Quercus* and *Castanopsis*? and *Schima* with here and there a pine. We had lunch under a large species of *Ficus* with *Quercus truncata* (specimen no. 2295 – *Lithocarpus truncata* (King) Rehd.) and *Vaccinium* (2296 – *Vaccinium sprengelii* (G. Don) Sleumer) trees near by. We met much Mai Lai bamboo also Mai Pok, Mai Sang, Mai Hok and others. At a bend in the trail we came across a family of Sam Tao, a Buddhist Tai hill-tribe, who have adapted hill costumes and also part of the hill language especially much of Chinese, although they all understand Tai. The dialect spoken in this district is Khu(r)n but Chieng-mai dialect is understood everywhere in this region. We met several Haw caravans with camphor and tea from Muang Rai. They all carried Winchesters and Mauser pistols.

The trail led through everlasting bamboo jungle with *Quercus* and *Schima* and very few ferns. It descended into the valley of the Meh Lui at a steep grade through much bamboo, *Musa* and tall trees. The forest near the banks of the Meh Lui is lovely consisting of large trees, tall and erect with many lianas. On the very banks of the swift flowing Meh Lui, which is larger than the Meh Len, there were huge species of *Ficus* which made a beautiful effect. We crossed the Meh Lui on two dug-out canoes which were tied together. The pack loads were taken over in the boats while the horses swam the stream which is very swift and deep in the center. I took several photos of the banks and upstream. The ford where we crossed and where we are staying for tonight is called Ta Ping. The house is a brick building, white-washed and covered with figures and writings. The roof is nearly caved in and dirty straw is on the dusty floor.

Bauhinia variegata is very common on the ridges and valleys at elevation of 4,000 to 4,500 feet. I photographed (neg. no. 30048) a tree in flower (2272).

10 p.m. It is cold and while I was putting away the specimens, noise like gun-fire was and still is heard every few minutes. It is the exploding of green bamboo joints in camp fire. This is mostly done in order to keep wild animals away.

Specimen nos. 2271–2296.
2272 *Bauhinia variegata* L., neg. no. 30048
2274 *Lithocarpus spicata* (Sm.) R. & W.
2295 *Lithocarpus truncata* (King) Rehd.
2296 *Vaccinium sprengelii* (G. Don) Sleumer

February 10, 1922

We left our prison-like brick-building camp on the banks of the Meh Lui early this morning. The whole valley was filled with fog. The trail passed through a large grove of beautiful tall trees along a little brook where there grew many ferns, some very tall ones (but not trunks) – the leaves measuring about ten feet. We climbed the rather steep trail to the top of the ridge at an elevation of (about) 3,600 feet only to descend again to the banks of another stream, the name of which we could not learn. The jungle was composed of several species of oaks and much bamboo, also *Pinus insularis* was very common (with male flowers). The undergrowth was formed mostly by *Gleichenia, Braynia insignis* often with trunks 3 feet in height, also a stout fern which we collected on the ridge between Muang Len and the Mekong. A *Cibotium* was very common. I photographed one chump (specimen no. 2298 – *Cibotium barometz* (L.) Thn.; neg. no. 30056) – the only photo taken today. Bamboos increased in numbers – several species, especially Mai Sang, and a small one which they called Mai Lai (2322 – *Dendrocalamus hookeri* Munro; 2304) Where we had lunch under a bamboo chump, I found a peculiar species which seemed not to die after flowering – small branches of the inflorescence appeared from the axils of the leaf-bearing culms. They drooped in long whip-like slender branches from the lower axils. The stipules have a very long tip even longer than the actual sheath. We saw a great deal of *Cibotium,* also many large species of *Ficus*

obtusifolius (one 12 feet in diameter), also smaller species *Ficus hirta* Vahl var. *roxburghii* King, growing in company with araliaceous trees, etc. The trail descended again to 2,600 feet elevation through much bamboo, over swampy ground where we found again *Pyrus pashia* with *Smilax prolifera* Roxb. and much *Congea velutina*. We finally struck a paddy-plain and a small village with a miserable temple. We did not stop but moved on to the larger village of Muang Mah – one stage from the Chinese border. We are stopping in a bamboo Salah with wooden walls. The floor is flattened-out timber-bamboo (split). After arrival, we called on the Kun (or headman) of the village, a congenial old man. He was making thatch for roof while his daughters were drying some yarn which they had dyed blue with the leaves of an acanthaceous bush (*Strobilanthes flaccidifolius*). The leaves are steeped and the yarn dipped into the blue stew of leaves and young twigs. After several dryings and soaking the result is a nice deep blue. The women wore nothing above the waist but short skirts below. Around their arms, above the elbow, they had many silver rings for about five inches. The rings were evidently placed there when they were still small girls as they constricted the upper arm considerably. After a few pleasant remarks, we left again to make ourselves as comfortable as possible in the dingy Salah surrounded by dust and pack-animals. About a dozen of the Sawbaws soldiers were being drilled by a Siamese ex-soldier from Chieng Rai. They are dressed in natural-brown cotton trousers (knee-breeches), coats of the same material, yellow turbans and blue cotton puttees.
Tomorrow we shall sleep in Chinese territory.
Specimen no.
2288 & 2289 *Oxytenanthera albociliata* Munro (a bamboo)
2298 *Cibotium barometz* (L.) Thn., neg. no. 30056
2304 small clump Mai Lai and 2322 *Dendrocalamus hookeri* Munro.

February 11, 1922
We left our Bamboo salah at Muang Mah early and followed on the edge of the paddy-fields through jungle. The forested hills were to our left and the terraced plain to our right. The hills had been much cut over and covered with bamboo, on the upper and middle slopes. Along the margins of the paddy-plain the forest is left undisturbed, only along the actual path (or mule-trail), the vegetation is stunted. Meliaceae are quite common as *Toona* sp.?, also a larger number of

Ficus of various species (as *F. Benjamina*, and several small fruited species) and many bamboos (as the Mai Hook and Mai Sang). We passed through several villages and paddy-fields. Found a fine specimen of *Pleopeltis* on a tree trunk which I photographed with N. Keo standing beside the trunk. *Quercus* was met with, also Garcinias, etc. The trail followed practically all along the Nam Mah – a tributary of the Nam Lam which flows by Chieng Law.

The dry jungle gave way often to beautiful groves of large trees with innumerable vines and huge woody climbers forming dense tangles. In these large groves we met with large specimens of *Cibotium barometz*, also of the other dark-scaled fern *Alsophila glabra* with trunks of 6-7 feet in height. I photographed one specimen (two exposures); neg. no. 30057, 40058 USDA). Also photographed an orange (Mandarin) grove with tall slender trees, not at all pruned or otherwise taken care of.

At a village which we passed, all the women came out dressed (or rather undressed) similarly as the women at Muang Mah, with silver rings above their elbows. They sold us Mandarins but when I tried to photograph them, they all ran away like deer. At about 5 p.m. we reached the Chinese-Burma border at Chieng-law. A much faded Chinese flag was implanted almost in the center of the road. To the left of it was a bamboo-wooden shanty where the Chinese official, a small unwashed fellow but with a kindly smile, gave us rather a nice reception. He had a few soldiers with him dressed in dirty white, with narrow red borders along the seams of the trousers and brass buttons. Several Chinese were sitting about dressed in blue. They brought out stools for us, and entered into a lively conversation. The official whose name is Wai Kia-ki (like the beach in Honolulu) knew Mason and, without asking for any passports or anything, we were shown the way to Chieng Law. He gave Mason some money to send down quinine. He had been on this border for seven years. I took a photo of Mason and Baer standing by the flag, while Mason took one of me standing with the official near the Chinese flag. He also gave us a letter to take to the Chow Fah of Chieng rung (Keng hung). We followed a narrow trail through bushes of *Vitex* (specimen no. 2325 – *Vitex trifolia* L.) and spec. no. 2347 – *Justicia adhatoda* L. to the ford of the Naw Law. We are staying at a poor miserable temple with two little alabaster Buddhas and a small wooden one in the back. The name of the Wat is Ban Ta Ngua („the crossing of the buffalo“). The Wat was crowded with Lüs a tribe of the Tai people. They were again much interested in our lamp and the electric flashlight. They watched us eat and when I put my plants away, they were so crowding around me, coughing and belching all around and on top of me – I could stand it no longer. Many wanted medicines – one for an epileptic boy, the other for asthma, and many other ailments. We finally drove them all off telling them we were now going to bed.

The moon rose shortly before our arrival at the Wat. The sun was just setting and the hills in the distance were purple, the sky slightly hazy and the full moon was gently riding on the pale lilac haze, herself a pale silver disc, with the land-masses clearly showing. It was a wonderful sight. It is now very cold and I must stop writing. All the rest are asleep.

Specimen nos. 2325–2348
2325 *Vitex trifolia* L.
2332 *Ficus altissima* Bl.
2335 *Ficus obtusifolius* Roxb.
2341 *Alsophila glabra* (Bl.) Hook., neg. USDA 30057-30058
2347 & 2348 *Justicia adhatoda* L., Chinese plants from Chieng-law border

February 12, 1922
Left Chieng Law while the heavy fog was still lying over the valley or alluvial plain. We are traveling without escort. There is a noticeable difference in the road. There is no grading and a trail, simply made by the pack trains, goes in every-which direction. We passed over paddy-plains and then entered beautiful forest – tall trees, much under-brush, many *Cibotium* ferns, and also the *Alsophila* tree fern. The trail led over a steep ridge with dense rainforest – a regular tiger-jungle, dark and mysterious. We passed several natives (Lür) with cross-brows and arrows, also with guns. They were going out to hunt deer. One beautiful glen we examined, it was full of *Cibotium* and tree ferns (several terrestrial species), many climbing aroids and terrestrial Marantaceae, *Pinanga gracilis* was one of the most common palms. In the same glen, there was a tall *Castanopsis tribuloides* (2360) with old burs on the swampy ground. *Pandanus* was also not uncommon. Today we had one of the most delightful jungle-walks. We passed one small village (the tail end of Muang Pan), crossed many paddy-fields till we reached the main village of Muang Pan which had been burned down completely before the harvest some months before. Eighty-four houses were destroyed and the people are living in

provisional bamboo houses built on the ground. We bought a large pig for fourteen rupees – the pig weighed more than 100 lbs. It is for the Haw train and the Lao carriers. We are sleeping in the Wat of the place. The altar is small and only a few small wooden Buddhas are on it. Before reaching Muang Pan, I photographed a beautiful *Engelhardtia* tree (specimen no 2382 – *Engelhardtia spicata* Bl., USDA neg. 30063).
Specimen nos. 2349–2385
2382 *Engelhardtia spicata* B., neg. 30063

February 13, 1922
Left Muang Pan early. The whole village saw us off. They became curious in the morning, for the night of our arrival they left us alone. Only one or two priests came and watched us eat our supper..

We crossed the paddy-plain, leaving the pine-covered mountains to our left – the western mountains being devoid of pine. We soon entered dense jungle. The road was full of rocks and muddy in places. It was a continuous climbing up and down, the trail leading on the crests of the ridges. We followed a valley and then climbed and climbed the steep ridge ahead. There were many oaks, *Engelhardtia, Castanopsis, Bauhinia*, and a very very stout spreading bamboo like *Dendrocalamus Hamiltonii*, *Cibotium* ferns became very common – whole slopes were covered with the huge fronds. In the shady ravines, ferns were plentiful, many aroids of various types, Marantaceae, *Dracaena,* several species of araliaceous trees, Acanthaceae, Leguminosae etc. *Pinanga gracilis*, *Wallichia caryotoides, Calamus*, *Arenga* sp.?, *Philodendron* – several species climbing and covering the trees. There were many very tall trees, 100 feet or more in height, mostly with pinnate fronds.

The trail rose higher and higher. The vegetation changed gradually. On top of the ridge at an elevation of 4,500-5,000 feet *Quercus* became very common, also a very large *Castanopsis* with very large burs. One tree, about 130 feet to 150 feet (neg. 30064 USDA) had very large chestnuts (spec. 2405 – *Castanopsis diversifolia* King). One I photographed with Dr. Mason beneath. There were other species of *Castanopsis* but without fruit. The undergrowth was mostly Compositae – one a fine small tree with large panicles of lilac flowers (spec. 2406 – *Vernonia Volkameriaefolia* D.C.). There were other Compositae with white flowers (spec. 2414 – *Vernonia papillosa* Fr.), also a yellow *Senecio* (spec. 2424 – *Senecio scandens* Buch.-Ham., ex D. Don). The trees were all very tall. There were magnificent oaks with large leaves, and smaller ones – the acorns being arranged in long spikes. Even as high as 4,900 feet, there were Hais or clearings made by Kaw Hill people. We saw some wild peacocks strutting about but the gun was with the caravan, which was far ahead.

Still the trail ascended and descended only to rise again till finally we saw the plain of Muang Hun. Lower down (at about 4,000 feet) we found several *Quercus* – one with very crooked trunk (spec. 2407– *Quercus acutissima* Carr.) on which the flowers and young leaves had appeared simultaneously. On the Muang Plain, there were hundreds of trees of *Pyrus pashia* all in flower, the snow-white appearing trees being very conspicuous against the dark foliage of the other trees. The yellow Composite (spec. 2424 & 2436 – *Senecio scandens*) was very common climbing over shrubs, as well as a large flowered *Clematis* (spec. 2408 – *Clematis heracleifolia* D.C.?), *Rubus* (the yellow fruited species, *trifoliata*) being very common with other lobed species which were not in flower. We reached the plain at sunset and made straight for a small village called Ban Nâ. The village was stockaded and on inquiry about our caravan, we were told that they had not seen it at all. It was getting dark and a Haw caravan which had pitched a tent outside the village informed us that they had gone on to another village – so on we went in the dark, the Haw horse-boy once in a while striking a match to see if we were on the right trail and if there were hoof-marks. We yelled and yelled and finally reached the Wat of the village. we are now comfortably settled in the village temple. The mosquito curtains are being tied to the altar on one side, and to a spirit-flag on the other. The elevation of this village is 4,500 feet.

Took only one photo of a large *Castanopsis* (spec. 2405 – *Castanopsis diversifolia* King – neg. no. 30064), at an elevation of 4,900 feet.

February 14, 1922

We spent the whole forenoon at the Wan Ban Ta Woa Tuan at Muang Hun, the Chinese Meng-hun 猛混. I labeled the numerous plants collected the day before. The people became exceedingly curious and crowded around us. They belched and coughed right on my head as I was sitting at the table; their garlic breath scented the air. They were filthy beyond description, especially the boy-priests. We had to tell them to go out of the temple as we sat down to lunch. After lunch we left the Wat (where, by the way, I bought a native embroidered towel – home-spun and embroidered – all for a rupee), encircled the beautiful plain of Muang Hun keeping to the left. Not far out we passed through a large grove of *Pyrus pashia* trees. They were lovely to behold in their white attire of beautiful flowers. I took three photos there. The trail led by the main village of Chieng Hun (or The Walled City, although there is no wall now). There are two distinct sections – the Tai part, with high roofs, and the Chinese part, on the other side of the brook, with low houses. In the Tai village there was a peculiar Wat with many roofs, and a pracheedy[2] on the side of the hill near by, resembling much Wat Kutau of Chieng Mai.

We passed through a large grove of chestnut trees with small fruits arranged in spikes (no. 2439 – *Castanopsis hystrix* D.C., neg. no. 30071 USDA). The trees were small but a few were rather large. I think it is the same as we collected at Hue Phi Bang in the Keng Tung State. There was much of the flowering *Clematis* and numerous trees of *Pyrus pashia* all in flower. In the distance there is a little hill in the midst of the paddy-plain with a Wat – called Wat Ban Nowng Kam – is to be our shelter for the night. On the way, just on the slopes of the hills encircling the paddy-plains, we found a large tree of *Pyrus pashia* in fruit. We reached the Wat just before the sun disappeared behind the hills over which we came the day before. Several maidens with rosy cheeks came up to the Wat, after having spied us, to sell vegetables for a few coppers. The priest of the Wat, a nice fellow, blind on one eye, told us that this part of the country is much infested by white tigers – perhaps the blue tiger hunted for by Chapman Andrews[3] in Yünnan some years ago. He said that six people were carried off by tigers during the last four months. Mason tells us of having passed through here four and a half years ago, just the day after two men were killed by tigers at high noon on the road while they were smoking opium.

There are tea trees on this hill where I am staying, also peaches with large deep-pink flowers, *Sapindus saponaria*, orange trees and

---

2 Auch *prachedi*, Rock erläutert den Begriff in *NGM* 41.1922, S. 243 als «votive spires».

3 Roy Chapman Andrews, 1884-1960, zoologischer Forschungsreisender. 196-1917 leitete er die Asiatische Zoologische Expedition des American Museum of Natural History (dessen Direktor er übrigens 1934 wurde) nach Yünnan und anderen chinesischen Provinzen.

Pyrus. A beautiful landscape spreads out to one's feet: the brown paddy-plain encircled by forested hills on which smoke rises as from volcanoes – fires kindled by passing caravans.

Tomorrow we must pass through tiger-infested jungles and hills. It is said that they are especially active near Muang Hai, our next stop. Specimen nos. 2428–2440.
2439 *Castanopsis hystrix* D.C., neg. no. 30071

February 15, 1922
Left Ban Nowng Kam early this morning. The road led directly into the hills. We looked out for tigers. I carried my colt-automatic, ready for use, but fortunately we did not meet any. The trees were mostly oaks and the large fruited chestnut, also several *Elaeocarpus* and the large composite with the purple flowers. The ridges were comparatively dry and Fagaceae formed the major part of the vegetation. I photographed one very large oak (no. 2453 – *Quercus acutissima* Carr., neg. no. 30069), also a beautiful *Vaccinium* with white and pink bell-shaped flowers (no. 2455 – *Vaccinium sprengelii* (G. Don) Sleumer – neg. no. 30068 (see also 30090). It grows with no. 2453 – *Quercus acutissima*). The highest elevation reached was 5,050 feet. The oak I photographed was at 4,800 feet. The native name was given as Mai dim. An *Alnus* (*Alnus nepalensis*) we also collected called Mai Wai. *Pyrus pashia* was exceedingly common along the plain (4,200 feet). The most common tree was a *Wendlandia glabrata* D.C. (no. 2269). We descended again onto a plain – the plain of Muang Hai. *Pyrus* was again common, also a tree, an alder tree whose leaves were all dry and the branches seemed to have been cut many times and used for firewood. We came on to a grove of pear trees in flower. It was an orchard and was said to be the large-fruited variety. The flowers are much larger than of the wild species but they seem not to be such prolific bloomers. I photographed the grove (elevation 4,300 feet). We passed through a large grove of enormous oak trees (no. 2407 – *Quercus acutissima* Carruth.), and presently came upon the village of Muang Hai, the Chinese Meng-hai 猛海. There are two distinct villages one Chinese on the plain, and the Tai village on the hillside. The temple (Tai) is a rather large one and is situated overlooking the plain – somewhat isolated from either village. It is by far the largest temple we saw since leaving Keng Tung. There are fifty priests in this temple but mostly boys only four real priests. We settled at the temple and then

called on the Chinese official whose residence is on the top of the hill. He is a real Tai but goes by the name of Li Fu-tin. He was with two rather nice Chinese men (officials) and a lot of children and women. We were seated to the left drinking tea and smoking cigarettes. We conversed with him for some time and then took our departure.

Arrived at the temple, we were surrounded by the whole congregation of priests who admired the gasoline lamp. Shortly afterwards they all went on their knees and the changing commenced – rather beautiful. The abbot explained to us that the prayer offered was to the coming lord.

The Wat in which we are settled is a rather large one, surrounded by a mud-wall with Chinese gate-ways. Many banners hang from the ceiling and a large childish-looking Buddha sits enthroned on the altar.

After dinner we went to see the Chinese theater which has been erected in the middle of the street of the Chinese village. We were directed by the young Chinese official to a store where the chief was sitting with a fire burning under the table. We afterwards went out in front of the theater where we were again made comfortable by the Chinese officials. As it was rather cold, we soon made our departure thanking our hosts. The theater was very primitive and the play equally so. Tea is sent from here in large caravans in baskets. The name of the Wat here is Wat Luang.

The Chow-fa reports *Pinus khasia* or *Pinus insularis* a little west of Muang Hai – a few were planted at this place. Where we had lunch at the foot of the hill on the edge of the Muang Hai plain, there was a large grove of *Castanopsis* (no. 2439 – *Castanopsis hystrix* D.C.) with tall climbing *Mucuna* (seeds collected).

Specimen nos. 2441–2461

2407 *Quercus acutissima* Carruth.

2439 *Castanopsis hystrix* D.C.

2453 *Quercus acutissima* Carruth., neg. no 30069

2455 *Vaccinium sprengelii* (G. Don) Sleumer, neg. no. 30068 (see also 30090). It grows with 2453 *Quercus acutissima.*

2269 *Wendlandia glabrata* D.C.

February 16, 1922

Left Muang Hai 猛海 about 8:30 a.m.; before leaving I took three photos (neg. 30072) of the temple and one of the caravan in front of the temple near the big Bho tree. The priests ran after me, and the abbot wanted to see the pictures at once, begging me to open the plate holder.

It was a very cold night and the fog was lying over the Muang Hai plain like a heavy blanket, only the gables of the Tai houses on the hill appearing above the fog. It was a wintery scene, the crows flying about in the temple ground and picking up the remnants of food left by our carriers. Dr. Mason went to say good-bye to the Chinese official while we turned again toward the hills and northeast to Chieng Rung (Keng Hung), the Chinese Ch'e-li 車里. At first we passed through thick groves of a *Castanopsis* with the small bur and gold-brown leaves – magnificent trees. There were some very tall and at least three to four in diameter. I took a photo of the large grove through which the trail led with Noi Loh standing under the big trees. The most common undershrub was the tea tree (2469 – *Thea sinensis* L., neg. no. 30074)(*Ton Miang*) the leaves of which are chewed in Siam and Laos. Very numerous was the epiphytic *Drynaria coronans* (Wall.) J. Smith which preferred tall trees. Here in Yünnan the leaves are made into tea which is drunk. At the temple where we were staying, there were a few Chinese carriers with their wooden carrying-frames for tea. The leaves are very coarse as I could observe while they were selling some to the Abbot of the temple. I photographed a wild -growing tree of *Thea* (no. 2469 – *Thea sinensis* L., neg. 30074). It grew in the jungle with *Castanopsis* and *Pyrus*. Further along the trail it seemed to be cultivated or trimmed and the underbrush cleared out. It was very common all the way to the last village which we passed about noon. At a small place called Yam Kham, I took a photo of the large-fruited *Castanopsis* – a tall tree on the edge of the village. The *Quercus* with crooked limbs and young leaves and flowers was very common. At the last (next) village, I photographed a flowering bamboo (no. 2467 – *Dendrocalamus giganteus* Munro) – Noi Keo standing at the foot of the culms. It is a large timber-bamboo with glaucous stems and large leaves. Before crossing the Nam Ha I found another species of bamboo in flower (no. 2462 – *Oxytenanthera felix* Keng, det. Keng). It is a smaller species only about 25 feet tall – or 30 feet at the most. After crossing the river at the ford, we entered a narrow valley through which the Nam Ha flows. The heat was intense. In addition to that, forest and grass fires were burning on both sides of the trail and in passing through them at a gallop, my horse's tail got singed. Oaks and *Castanopsis* as well as *Schima* form the tree vegetation. Much

Cibotium is found in shady places together with the shaggy fern we found on the Meh Len ridge. Willows grow on the banks of the Nam Ha (no. 2747 – *Salix araeostachys* Schneider), especially at a small islet where we had lunch. *Psilotum triquetrum* grew on the trunks of these willows which were overhanging the banks of the Nam Ha. The trail became very narrow and led alternatingly through open burning grass country (thatch grass) and again through beautiful forest of oaks with large acorns (no. 2477 – *Lithocarpus spicata* var. *brevipetiolata* (A. DC.) Rehd. & Wils.). We are stopping in such a grove with the trunks of a large *Ficus* as a wall for our tents. This is tiger jungle and we must be careful. The elevation at this particular place, about half way between Chieng Rung and Muang Hai, is 4,200 feet. A very common tree on the valley slopes is a species of *Alnus* (no. 2443 – *Alnus nepalensis* Don.). No pines are visible.
Specimen nos. 2462–2477
2443 *Alnus nepalensis* Don.
2462 *Oxytenanthera felix* Keng. Det. Keng
2467 *Dendrocalamus giganteus* Munro
2469 *Thea sinensis* L., neg. no. 30074
2474 *Salix araeostachys* Schneider
2477 *Lithocarpus spicata* var. *brevipetiolata* (A. DC.) Rehd. & Wils.

February 17, 1922
We left our camp under the large oak trees at 3,900 ft. elevation real early. At 7:30 we were on the way. There was no fog. We followed the Meh Nam Ha down its course, the drop of this treacherous river is very considerable. Its bed is full of huge boulders and the water rushes in cascades and rapids over them. The vegetation becomes quite tropical. *Garcinia* trees are very common but below the *Quercus* zone, *Bauhinia variegata*, the same as photographed in the Keng Tung State), is very plentiful. A *Eugenia* but not in flower (2479 – *Eugenia*), and numerous trees form this dense tropical jungle. The Mai lang bamboo was all dead throughout the valley with exceptional flowering clumps. The trail was absolutely awful, full of huge rocks and continuously steeply ascending and descending. The numerous landslides made traveling very difficult and detours necessary. We walked nearly the whole way. It may be stated that *Pinus insularis* was observed in the Meh Ha valley just above our camping place. *Podocarpus* (no. 2495 – *Podocarpus neriifolius* D. Don) occurred lower down at an elevation of 2,800 feet. It was more a large climber

than tree, overhanging the waters of the Meh Nam Ha. The small *Castanopsis* was also quite common. Araliaceous trees, *Pipturus*, *Boehmeria* (no. 2502 – *Boehmeria platyphylla* Don.), huge *Milletia* climbers with enormous beans trailing from tree to tree, many of the trees observed in Siam and in the Keng Tung State, with an occasional *Pyrus,* and *Erythrina suberosa* occur along the slopes of this valley. Several species of *Smilax*, leguminous vines (the yellow-flowering species – large racemes), Eurphorbiaceous trees (no. 2434 – *Mallotus philippinensis* Muell.), Acanthaceae, and ferns form the plant ensemble. In the dense shade of this tropical jungle, I found a species of *Cycas*. It was trunkless and without flowers or fruit. It has a creeping thick rhizome about five inches below the surface of the soil. In a distance of about five miles the river drops about a thousand feet, until it reaches the plain of Chieng Rung (Keng Hung), the Chiu-lung-chiang 九龍江 of the Chinese where it flows placidly but winding considerably, almost making complete circles. We crossed the paddy-plain leaving to our right, a tall mountain mass inhabited by Haw people. The mountain, called Doi Nun Nang, must be about 8,000 feet in height. The paddy-fields were lined with dry jungle of *Ficus*, Euphorbiaceous trees and one tree especially common (which I found plentiful near Raheng and on the Meh Ping (no. 2511 – *Streblus asper* Laur.)), *Clematis* (no. 2512 – *Clematis kerriana* Carib.), *Jasminum* (no. 2507 – *Jasminum* sp.?), etc. A *Salix* with broad leaves (no. 2514 – *Salix wilsonii* Seem. or a related species) was not uncommon lower down. In the upper part of the Nam Ha, a Salix (no. 2474 - *Salix araeostachys* Schneid.) formed small islands in the river, with ferns and grasses. The Meh Nam Ha flows into the Mekong not far from the Mission Compound.

We reached the Mission House at Chieng-rung about 7 p.m. having had a march of about eleven hours over a most difficult trail. Dr. & Mrs. Park welcomed us. They haven't seen a white face for over two years. They are all alone with a little boy, John, born in Chieng Rung and a little girl.

This finishes the journey with Dr. Mason. On our trip we met many caravans (both mule and ponies) carrying salt, lead, tea and camphor. Usually two caravans going together, every man carrying rifles and knives. The Yamen is situated on the plain just across the Mission compound. It is enclosed within three walls with watchtowers. The Chow Haw (Chinese magistrate) is away at present. He has been here for many years, having been called to assist the

Chou Fa (the native Tai ruler) in a rebellion. He has remained here ever since and is now absolute ruler over life and death – the Chow Fa being without power.
Specimen nos. 2478–2516
2434 *Mallotus philippinensis* Muell.
2474 *Salix araeostachys* Schneid.
2479 *Eugenia*
2494 *Podocarpus macrophyllus* Don.
2495 *Podocarpus neriifolius* D. Don
2502 *Boehmeria platyphylla* Don.
2507 *Jasminum* sp.?
2511 *Streblus asper* Laur.
2512 *Clematis kerriana* Craib.
2514 *Salix wilsonii* Seem. or near?

February 18, 1922
Keng Hung
We are stopping in the mission compound, the house is built of round stones from the Mekong River-bed, held together by mortar. It looks like a prison. Baer and I went down to the edge of the mighty Mekong, the present level of the river being about 35 feet lower from that of the rainy season. It does not look so treacherous as at Tang Oh. The sandy bed is covered with bushes of *Homonoia riparia* which for months are under 20 feet of water.
We paid off all our carriers today.

February 22, 1922
We left Chieng Rung early for the Mekong. Two canoes were to be ready but instead, of course, only one was ready. We made ourselves comfortable as best we could. The boat leaked terribly and I bailed most of the time. We had four men poling upstream with oars. I walked along the shore or rather river-bank collecting – a blue *Campanula* (no. 2541 – *Campanula mekongensis* Diels), *Rosa* sp.?, *Homonoia* (no. 2520 – *Homonoia riparia* Lour.), several Compositae, *Spiraea* (no. 2540 – *Deutzia corymbosa* R. Br.), a medium sized tree with drooping large foliage and large terminal panicles or cream-colored scented flowers. We poled among huge rocks, in swift current, turning to the left where we stopped under huge bolders (see photo). We climbed the embankment and followed

a small creek-bed to near the top of the range. Most of hillside had been cut over but I found some ferns I had not previously collected, also a few trees. We proceeded by boat across the river and lunched on a sand bar. After lunch I walked along the stream-bed collecting plants, *Phoenix* (no. 2531 – *Phoenix roebelenii* O'Brien) was quite common and I photographed a few specimens (neg. nos. 30079, 30085, 30086). They form dense clumps but some have stems several feet in height as could be seen from specimens brought down by the boatmen who cut timber for the A. P. Mission, and of whom I took pictures on the raft.

We then followed a small creek-bed to the left where the natives (Tai) had erected a small library and where they hid all their temple treasures during a Haw[4] (Chinese) robber invasion. I found several other species of ferns and a Bignoniaceous tree in full flower (no. 2556 – *Mayodendron igneum* Turcz.), the Dak dip of the natives, I photographed a flowering branch (neg. no. 30084). The tree is deciduous in the cold season when it flowers, usually in February and March – the beginning of the hot season. We returned to the Mekong and took a swim in the icy waters.

Specimen nos. 2517–2559.

2520 *Homonoia riparia* Lour.

2531 *Phoenix roebelenii* O'Brien – neg. n os. 30079, 30085, 30086

2540 *Deutzia corymbosa* R. Br.

2541 *Campanula mekongensis* Diels

2556 *Mayodendron igneum* Turcz.

February 23, 1922

Chieng Rung

After lunch we proceeded by boat to Chieng rung village to pay a visit to the Chow Fa („the Lord of Heaven") who was ruler over the Hsip Song-pan-na but plays now the second fiddle – the Chow Haw (Chinese magistrate) being the Chief.

---

4 Haw wird in verschiedenen Bedeutungen gebraucht. Vgl. James McCarthy: *Surveying and exploring in Siam*. London: John Murray 1900 (Nachdruck: Bangkok, Cheney: White Lotus 1994), 44: «They [the Haw] were, in a word, Chinese brigands. At one time, Chinese traders, known in Luang Prabang as Haw, came down from the north in great numbers to traffic with the inhabitants, and when the peaceful traders gave place to brigands of the same nationality, the name Haw was naturally transferred to these.»

He (the Chow Fa) is a very ignorant coolie-like individual who lives in a long barn-like structure with pigs and horses living under the house. Manure is thick everywhere. He offered us tea in dirty cups which we nearly had to empty for politeness sake. We went to the largest Wat of Chieng Rung (Wat seri Chawm muang) where we found a fine old priest who had been in the temple for over forty years. He gave me a nice bronze Buddha in a graceful and pleasing way.

The second temple is a little lower down. We also visited the priest there, who was quite cold and more or less uncivil. We rode horseback to the Mission Compound. On arrival we heard that the Chow Haw had returned to the Yamen and so we called on him immediately. He received us in a very pleasant manner and promised to send us passports and a letter to the Tao Tai of Szu-mao.

February 24, 1922
We had the Chow Haw for dinner tonight. He stayed till nearly ten o'clock p.m. talking through a very pleasant interpreter, a young northern Tai. He pleasantly said that we were friends now, having eaten rice together, and that he would send food along with us on the road. He also promised an escort, and a letter to the Tao Tai of Szu-mao 思茅.

Today I tried to send two parcels by letter mail. The postmaster, a young Chinese, said he could not take them as he had not stamps enough to cover postage.

February 25, 1922
Left Chieng Rung at 11 a.m. in the mission boat across the Mekong. Dr. Mason going with us as far as the trail separates. Before leaving Chieng Rung we received a ham and Chinese sausages from the Chow Haw which were brought over by a soldier bringing two cards with him from the Chow Haw. Our caravan (fifteen packhorses and two riding ponies) went across by the regular ferry for which we paid $ 5,00. Our carriers are Lü people who wear large dark-blue baggy trousers which they roll up way above their knees when not walking, the crutch is near the ankles, otherwise they would remind one of French zouaves. They wear long hair which is tied like a top notch and wear a pink turban. Their blankets are spotted like a leopard's skin.

We ascended a rather steep hill covered with dry shrub-jungle composed mostly of oaks (no. 2588 – *Lithocarpus spicata* (Sm.) R. & W.), anacardiaceous small tree or shrub, *Spondias mangifera*, *Hoya* sp.? (no. 2568 – *Toxocarpus* sp.) and (no. 2566 – *Bauhinia variegata* Kinn.), with snow-white large flowers, save one petal which is pinkish striped. This tree is the most common, the whole hillside being white with these flowering and leafless *Bauhinia* trees. Other trees are the large fruited *Gardenia*, large trees of *Albizzia* (probably *procera*), climbing *Mussaenda* (no. 2591 – *Mussaenda hossei* Craib.) and a climbing spiny Mimusoid, known as Sum Poi by the Lao people who use the pods for washing their hair. The trail led down across the ridge at about 3,900 feet elevation, and followed along the mountain slope at a slightly lower elevation, overlooking a deep valley. *Bauhinia variegata* was scattered all over the hillsides to the top of the ridge. Mai Sang bamboo was very common – many clumps just past flowering. *Nauclea polycephala* (large trees), *Erythrina* sp.?, *Ficus*, *Thunbergia,* etc., formed most of the jungle. We crossed another mountain spur which seems to cut off the deep valley. Owing to our late start from Chieng Rung, we stopped on the top of a ridge with a streamlet considerably down the hillside. This camping place is called Pang Khun. I went collecting into the jungle and down to the stream-bed, where I found quite a number of trees in flower and several species of ferns not previously collected.

Tonight it is quite warm – a considerable difference from cold, damp, foggy, Chieng Rung.

Specimen nos. 2560–2605
2566 *Bauhinia variegata* Linn.
2568 *Toxocarpus* ?
2588 *Lithocarpus spicata* (Sm.) R. & W.
2591 *Mussaenda hossei* Craib.

February 26, 1922
Pang Khun, although higher than Chieng Rung, was quite warm during the night and morning. No fog. We left the camp rather early climbing higher and higher. The small species of *Engelhardtia* grew with *Castanopsis* (no. 2615 – *Castanopsis indica* A. DC.) and a *Sterculia* (no. 2606 – *Sterculia fulgens* Wall.) with bright sienna-red flowers and leafless, *Bauhinia variegata, Erythrina* (no. 2633 – *Erythrina stricta* Roxb.), *Cycas siamensis*, etc. Near the top of the ridge we looked over the Chieng Rung plain although a deep valley and another mountain range lay between us and could see a blanket of

fog over the Mekong. At the top of the ridge about 4,600-5,000 feet there were the large fruited oak same as (no. 2477 – *Lithocarpus spicata* var. *brevipetiolata* (A. DC.) R. & W.), and pinnately-leaved trees but without flower or fruit. The hillside was one huge bed of *Cibotium* – same species as previously collected. The trail led over the hills rather rapidly down to the Muang yang 猛養 plain. *Andromeda* and *Castanopsis* (no. 2615 – *Castanopsis indica* A. DC.) was quite common with *Schima* (no. 2622 – *Schima Wallichii* Choisy) and here and there a *Phyllanthus emblica*. We found a large liana climbing from tree to tree of enormous size with racemes of large purplish-green flowers on the huge stems same as (no. 2389 – *Mucuna*). On the Muang Yang plain we found again *Pyrus pashia* quite common, also a huge *Ficus* tree (no. 2631 – *Ficus altissima* Bl.) which I photographed with Bun Mah standing under the tree (neg. no. 30087). We met with a procession of Shan or Tai people beating gongs and singing. They were going into the jungle to cut logs for the temple. We lunched on the plain near Muang Yang, and afterwards crossed the paddy-field to enter jungle again near a village, Ban Sai Kau – an exceedingly dreary affair. All were eager to stop but it being very early in the afternoon I went on following the new road which the Chow Haw (Chinese magistrate) is making. We entered a narrow valley after climbing over a steep hill. This valley we followed for a considerable time, trying to find a place for a camp. Finally we reached a spot where there was some grass for the horses. It is known as Pang Nam Wah, The camp on the Nam Wah stream. This place has a bad reputation for tigers and the men are out building fires and letting big bamboo joints explode and burst in the fire. It sounds like a cannonade. This is to scare off the tigers. Baer has been shooting off his revolver but the bamboo is more effective and cheaper. We hope nothing will happen.

Specimen nos. 2606–2641
2389 *Mucuna*
2477 *Lithocarpus spicata* var. *brevipetiolata* (A. DC.) R. & W.
2606 *Sterculia fulgens* Wall.
2615 *Castanopsis indica* A. DC.
2622 *Schima Wallichii* Choisy
2631 *Ficus altissima* Bl., neg. 30088
2633 *Erythrina stricta* Roxb.

February 27, 1922
Left our camp on the Nam Wah rather late. No tigers visited our camp as far as we know. The trail follows the valley above the stream-bed of the Nam Wah. Vegetation similar to yesterday; *Mayodendron,* Mai Sang, Mai Hok, but palms made their appearance on the steep valley slopes. Tall specimens were observed high up. It is a *Livistonia* (probably *speciosa*). Much of the valley slopes is burnt over but near the stream-bed can be found *Caryota* sp.?, *Melia azedarach, Musa, Wallichia caryotoides* ?, *Ficus benjamina*, Marantaceae in numbers, here and there a *Pyrus*, red flowering *Bombax, Randia*, and tall *Cycas* (no. 2680 – *Cycas* sp.?) with trunks 8 to 10 feet tall, I photographed one specimen (no. 30092, 30093). They are very common on the grassy slopes of the Nam Hpang Koh. Also two exposures (neg. no. 30092, 30098) of a *Cycas* (no. 2680 – *Cycas* sp.) with tall trunks quite common on the grassy slopes. After following the valley for a considerable time we came to a small plain with six houses. This is Kuang-p'ing 廣玶, a Chinese village. We went on a little farther till we came to the Chinese soldiers camp where we are staying in our tent. It is very cold again, a starlit night and bedtime has come.

Tomorrow we shall follow the regular stage to Ching-tung-chai, this means getting up early. The soldiers are playing cards and are smoking opium, which is very disagreeable.

Specimen nos. 2642–2660
2453 *Quercus acutissima* Carr.
2654 *Cycas pectinata* Griff.? (perhaps)
2680 *Cycas* sp.?, neg. no. 30092, 30093

February 28, 1922
We left our camp early at 7 a.m. The soldiers who came with us from Chieng Rung bade us good-bye and two new ones from the soldiers camp came with us. The trail leads from the valley directly to the top of the left hand ridge (Szu-mao way) and follows it on the other side into another valley, winding back and forth near the top from one valley into the other, finally to stay on the very crest of the ridge (5,000 feet or more elevation). The trees on the valley slope were *Randia, Phyllanthus emblica, Bauhinia variegata, Gardenia, Mayodendron* and *Pyrus*. We passed a little hamlet – the houses of which were made of sticks of wood and plastered together with mud. The roof is of thatch grass. Higher and higher leads the trail among oaks, and chestnuts (no. 2663 – *Castanopsis tribuloides* A. DC.). I

photographed (neg. no. 30094) a specimen pinned against a trunk, flower and fruit. *Vaccinium sprengelii* (G. Don) Sleumer (no. 2671) was very common, with large *Schima* and small-fruited species, also the ubiquitous anacardiaceous tree *Buchanania latifolia* Roeb. Small bamboo lines the trail. The trail then passed through this forest on the bare hills. On the right hand side for miles the hills were bare and brown, not a vestige of green. On this plateau – like hilly range Ta-tu-kang 大渡岡 is situated while Muang Meng, the Chinese Meng-mang 猛忙, is in the deep valley to the left. The mountains in that direction are densely wooded. From Ta-tu-kang where we had lunch, we went through very dry forest, shallow valleys and dry rounded hills, bare but with an occasional oak and *Schima*, thence we passed through burnt areas; to the right there extended a ridge with *Pinus insularis* (two trees) which species I also observed at the camp near Ta-tu-kang. The forest gave way to bracken-fern in spots, but later we entered again dry jungle in pockets till we reached Ching-tung-chai which is situated on the slopes of a bowl-shaped hollow. We made our camp near the stream where there are trees and shrubs. I collected quite a number of ferns and vines, also of the trees etc. We met a man with tiger's skins on the trail. He said they poisoned one with opium. He killed two men and a horse in a village not far from Szu-mao. They bought the skins and bones to make medicine for 20 piastres.

Specimen nos. 2661–2690

2662 *Quercus helferana* DC., or 2659

2663 *Castanopsis tribuloides* A. DC., neg. no. 30094

2671 *Vaccinium sprengelii* (G. Don) Sleumer

2689 *Benettiodendron brevipes* Merrill, n.sp. (here a new species is recorded)

March 1, 1922

We left our camp near the bank of the little stream at Ching-tung-chai quite early. This path led out of the little valley to the top of the ridge. The vegetation was at first stunted and scattered – here a little clump and there a little clump of *Symplocos* (no. 2707 – *Symplocos paniculata* (Thunb.) Mig.) with white flowers, oaks (2696 – *Quercus acutissima* Carr.), chestnuts and the anacardiaceous shrub. To the right is a deep valley with a mountain of over 5,000 feet – densely wooded. High mountains are ahead. We descend to a small stream called Nam Ching-tung-chai after the Chinese village where we stopped. Near the stream there was a beautiful grove of oaks and pines (*Pinus khasia*). I photographed (neg. no. 30095) the oak grove (2696 – *Quercus acutissima* Carr.) with its pale green foliage in the morning mist, and another view with pine and oak. From the streambed on the trail led steeply to the top of the ridge. There were many *Pinus insularis* trees with a *Phoenix* with trunks of one to two feet high as underbrush. *Cycas* sp.? was also quite common, especially tall thick-stemmed and branching species were seen.

Higher and higher led the trail through beautiful forest of oak with *Schima* (2704 – *Schima wallichii* Choisy). A deep valley was to our left and a high mountain ahead – densely wooded with *Pinus insularis*. The haze and heat was intense. We stopped for lunch on the top of a ridge where there was a little shed, under the roof of which we ate our lunch of rice, chicken and cabbage.

A very strong wind sprung up and the dust and our drying-blotters were blown down the hill side. The wind increased and also the heat, the sun was burning down through haze which made the sky gray. From now on, the trail led downhill at a steep grade. At one turn we found a cane with a cross-handle such as is used by Chinese carriers to rest their heavy load on and, lo and behold at the turn of the trail – just a foot by the side of the trail – there was the last resting place of the man to whom the cane belonged. He was not buried but dirt had been heaped over him and a few branches, myriads of flies were buzzing over his built-up grave. Such are the things one encounters on the trail through Yünnan. Perhaps I also will rest somewhere in Yünnan in similar fashion. The wild beasts will soon discover him and then he will be dug out. We got a glimpse of the plain of Muang Hing and soon spied the Tai village of the same name. We are now encamped near the village temple on the banks of the Nam Tü. The temple is a miserable affair – dirty and filthy, the sides are open and the dust of the plain gathers, chickens gather and scratch in it in front of the ragged altar with its paper umbrellas. We decided to pitch our tent rather than sleep in the filthy holy of holiest. The priests have gathered around us. The head priest is dressed in yellow, pinkish-purple and orange cotton cloth. He looks perfectly silly and acts likewise. He wears a white cornucopia cap with a red border. His teeth are black as that of any ordinary Tai. The sky is now pinkish-grey from the afterglow of the sun which has set over the hills whence we came. The scenery is rather dreary and I, for once, am longing for a comfortable room and the ordinary comforts of civilized men. The mountains are a bluish-ash-gray in the thick haze. Crows are noisily flying about. We took a walk through the filthy Tai village with its pigs and carabaos. The dust is a foot deep and in the center are blocks of stone – the path in the rainy season when the whole place shall be a pig-mire. I took two photos of this filthy place, one with our Chinese escort.

Specimen nos. 2691–2723

2696 *Quercus acutissima* Carr., neg. no. 30095

2704 *Schima wallichii* Choisy

2707 *Symplocos paniculata* (Thunb.) Mig. det. Rehder

March 2, 1922

Left our dreary camp at Muang Hing or P'u-ting 普定, or P'u-wen 普文 as it is called in Chinese. At 2 a.m. we were wakened by thunder and lightning. We expected a rain storm but luckily nothing came of it. We were real glad to leave that awful dirty and dusty spot, the filthy village the inhabitants of which keep ready coffins under their houses as it is rather far to fetch wood from the jungle. It is quite gruesome to sit beneath their houses with coffins around. The trail led along the paddy-plain due west, hills are on both sides, the vegetation is common oak, *Quercus acutissima, Engelhardtia spicata* Bl. (no. 2728), *Ficus, Schima, Pyrus*, etc. One of the paddy-plains seems to have been abandoned. We encountered *Castanopsis*, climbing *Mussaenda, Randia* (2751 – *Randia wallichii* Hook. f. ?), *Rubus* (2743 – *Rubus elongatus*), *Bauhinia, Alnus nepalensis* and *Syzygium rockii.* We crossed the Nam Tü over a covered bridge, rather nice for this part of the world. A little farther on we came to a hamlet of Chinese houses, it is known as Ta Ka Haw. Thence we crossed the Nam Tü again in dense forest and afterwards passed again into open jungle of bamboo, *Phoenix*, purple composite tree, etc.

Along the stream-bed in the sand I found a species of *Equisetum*. The trail led through grassland with pale green-foliaged oaks the common species. Today it was not so hot nor so hazy as yesterday. We had lunch on the edge of the jungle. In the afternoon we continued our journey along the paddy-fields, crossed the stream again where *Alnus* and *Salix* and *Pyrus* were quite common. To the left we passed the village of Ma-li-p'ing 麻栗坪 a small hamlet of a few houses on a terrace near a small valley, some 50 feet above the paddy-plain. We turned to the right towards the village of Ban Man where we are going to stop for the night. We found a newly erected Wat of bamboo – dirt floor. It was at least clean and so we decided to occupy it. The priest sat on the ground, sewing his dirty once-yellow robe together. He is rather friendly and bade us welcome. I took a walk into the hill back of the little Wat, with its bamboo shrine and tiny little Buddha, to collect and, as I roamed about, I stumbled onto a graveyard in the dense jungle. One grave was new, covered with bamboo sticks and branches, also the ubiquitous broken pot was present. It was quite a lovely spot and I felt inclined to leave the silent company and returned to the temple. A company of natives gathered in front of the temple with drum and cymbals and made a joyful noise unto the Buddhist Lord. They were celebrating the opening of the new Wat. It was a monotonous noise of two tones. Everything is quiet save the dirty chief priest is watching me.

The *Pyrus* along the paddy-plain on the edge of the forest were beautifully contrasted with red-flowering Erythrinas and the reddish foliage of various trees. *Cycas* plants were also quite common.

Specimen nos. 2724–2752
2728 *Engelhardtia spicata* Bl.
2735 *Primula malacoides* Fr., det. W. W. S. (first primula on journey in Yünnan)
2742 *Syzygium rockii* sp. n. type Merr & Perry
2743 *Rubus elongatus*
2749 *Olea densiflora* H. L. Li type
2751 *Randia wallichii* Hook. f. ?

March 3, 1922

It was a lovely cool morning when we started on our last stage to Szu-mao. The trail from Ban Man To led around the little hill and followed the paddy-plain. We passed through groves of *Quercus* (the common species) and many flowering *Pyrus* also *Albizzia* trees. The ridges on both sides of the narrow plain were covered with *Pinus khasia* along the creek-bed there grew *Salix, Erythrina, Mayodendron*, etc., *Ficus hispida* (the *Ficus* with semi-stemclasping leaves), and large Mai Yom trees (*Melia Toona*?). The trail led up the ridge and down a valley again with tropical forest, only to lead out again into a grassy spot with *Pyrus*.

We crossed the stream and saw ahead of us a high ridge covered with *Pinus insularis* with scattered Vacciniums, small-leaved *Castanopsis*, etc. Down we went along a small paddy-field with pine-covered ridges on both sides – the lower slopes being taken possession of by *Quercus* (light green leaves) *Pyrus* and *Alnus*. We came to another paddy-plain which merged into a long narrow swamp in the center of the valley with pine-ridges on both sides. Narrow plains shut in by pine covered hills seemed to be the order.

We lunched under a large araliaceous tree (no. 2757 – *Tupidanthus calyptratus* Hook. f. & Thoms.), the same species as we observed in the valley of the Nam Ha before reaching Chieng Rung. The trail now led to the top of the ridge – a deep valley is to the left and high mountains ahead. The trees are mostly the small chestnut, pines, oaks, *Schima, Pyrus*, species of *Smilax* and others. From the top of this ridge (which must be 5,500 feet), the trail descends. A beautiful rhododendron (no. 2762 – *Rhododendron stamineum* Franch.). Trees 15 feet tall (or shrubs) decorated the banks along the trail – this is the first one collected on this journey north. The trail descends to a broad plain surrounded by low, deforested hills but covered with young growth of some sort. They are too far distant to allow guessing as to the nature of the growth. The road is now paved with large stones which makes traveling rather difficult. *Castanopsis*, the small-leaved species, and *Pyrus* form the commonest trees. We pass a rather large Chinese temple, more an ancestral shrine, with a few figures in the courtyard. A little village come to view, mud houses, some better, others very poor. An open restaurant along the road where Ai In was sitting refreshing himself. Extensive paddy-fields spread out before us and the stone-paved trail led through the midst of the paddy-field. The white-walled houses and blackish tiled roofs of Szu-mao glistened in the distance in the evening sun. A broad stone-paved road leads into the suburb of Szu-mao. We stop as we see the sign of Fu-yin-t'ang 福音堂 – Happy Tidings Hall. Here lives the Danish Mission. We knock and a very pleasant Chinese opens and calls Miss Peterson, the only lone missionary in the whole of Szu-mao. She bade us welcome and there we are now, very comfortable and well taken care of. We arrived at Szu-mao at 4:45 p.m.

We sent our cards to the Custom House people – Mr. Cassati and Mr. Pattison, expecting to call tomorrow.

Specimen nos. 2753–2776
2757 *Tupidanthus calyptratus* Hook. f. & Thoms.
2762 *Rhododendron stamineum* Franch.

March 4, 1922

Today Miss Peterson's teacher took me to the Custom House where I met Mr. Cassati and Mr. Pattison, the former went with me to the Tao

Tai, but as he was just then at lunch, we called later. At three o'clock we went to call on the Tao Tai in the Yamen in the inner walled city. He received us very cordially at the Yamen gate, leading us into an upper room where he served tea, champagne, cigarettes and cakes. He informed me that my Yünnan passport had come from the Viceroy and as to the way to travel. After a very pleasant interview we bade him good bye, he accompanying us to the Yamen gate, bowing low.

March 5, 1922
Sunday – Had lunch with Mr. Cassati.

March 6, 1922
Today there was a large parade in which the idols of Szu-mao were taken about, the whole procession ending in a large Buddha being carried under an embroidered umbrella. The great Buddha was preceded by the musicians and singing priests who were walking under a tent (or baldachin). Took many photos. Boys danced gracefully with a long blue dragon marching in spirals and gyrating the dragon swiftly through the air. The procession was in honor of the great Buddha of Szu-mao, soldiers and officers took part in the procession, the latter keeping the soldiers in step by blowing a whistle.

March 8, 1922
Szu-mao.
Went out this afternoon with Miss Clara Peterson and the Chinese Evangelist, Mr. Leng and his little Chinese girl, into the low hills surrounding Szu-mao. We went over large graveyards – in fact the whole country side is one huge graveyard. There are several large groves of trees, mostly *Castanopsis* (the small-leaved species) and the common oak with the light green foliage. The rest of the vegetation is mostly scrub (see nos. 2777-2798 inclusive). We visited a small village (Chinese) and a little temple, the lady caretaker (an old maid – a rare thing in China) taking care of it. She was ill but received us very kindly. We went back to Szu-mao by a circuitous route entering the city through a narrow lane and finally emerged into the main street which leads to the mission. It was a glorious evening, the hills were clear, and Szu-mao sparkled in the sunlight.
Specimen nos. 2777–2798.

March 11, 1922
We left Szu-mao for a trip into the Eastern Hills – Miss Peterson, Mr. Pattison, Baer and I. We went through the main street of Szu-mao and thence through a small side-street into the open; huge graveyards, much neglected with their circular graves, here and there a stone (gray) column with a mythical animal (Liogryph) on top, denoting a rich man or official. These graveyards extend all over the lower hills surrounding the Szu-mao plain. Shrubbery of various kind, like *Smilax, Jasminum,* etc., is very common. Not far from Szu-mao, the pines commence – first small, stunted, then larger trees, up to 6,400 feet elevation, being one of the most common trees. I think they are not *Pinus khasia* (three needle) but *Pinus insularis*, however, the cone looks very much like that of *Pinus khasia. Pyrus pashia* just covers the hill-sides with a *Symplocos* (2804 – *Symplocos chinensis* (Laur.) Drude) with beautiful white flowers clustered around the branchlets. High on the ridges we found a *Docynia* (2803 – *Docynia delavayi* (Fr.) Schneid.) with large flowers, and large leaves dark green above and white beneath (*Pyrus indica*?), the branches are reddish, the trunk about a foot in diameter and sending out large spiny shoots from the base. A glorious wind blew, which made the air cool and fresh. On the top of the ridge Compositae were quite common, the one with the large panicles of purple flowers (2812 – *Vernonia volkameriaefolia* D.C.), and a smaller tree (bare) (2808 – *Leucomeris decora* Kurz) with sessile clusters of white tubular flower heads, very fragrant, a leguminous shrub 2 feet high (2806 – *Milletia pulchra* Benth.) could be found in the open gravelly places; there were several (2816 – *Symplocos paniculata* (Thunb.) Mig.) but only one was in flower. At about 6,000 feet we found a *Leontopodium* (2813 – *Leontopodium andersoni* C. B. Cl.), Edelweiss, about two feet high quite common on the dry hill sides, many stalks springing from the same root stock. Lower down on the dry slopes, an *Engelhardtia* with pubescent leaves was quite common, also the ubiquitous Anacardiaceous shrub *Buchanania* with panicles of cream-colored floweres. *Lobelia* (2825 – *Lobelia pyramidalis* Wall.) was very common on the top ridges (6,200-6,400 feet), also *Smilax* and a small tree with globose fruits arranged in drooping spikes, collected also in the Meh Len valley near Loi Mwe. *Rubus* (2226 – *Rubus ellipticum* Sm.) was very common as well as the undivided species (2791 – *Rubus alceaefolius* Poir.?), and many *Schima* and small oak trees. We returned to Szu-mao by the Mang-li trail, finding on the plain a beautiful Pink

Shower, *Cassia* sp.? (2827 – *Cassia javanica* L. var. *indo-chinensis* Gagnep.), also *Melia* (2818 – *Melia azedarach* L.) and the Mai Yom trees, the Chinese Ch'un-shu 春樹 (2964 – *Cedrela toona* Roxb.).
Specimen nos. 2799–2827
2226 *Rubus ellipticus* Sm. (ellipticum)
2791 *Rubus alceaefolius* Poir.?
2803 *Docynia delavayi* (Fr.) Schneid.
2804 *Symplocos chinensis* (Laur.) Drude
2806 *Milletia pulchra* Benth.
2808 *Leucomeris decora* Kurz
2812 *Vernonia volkameriaefolia* D. C.
2813 *Leontopodium andersoni* C. D. Cl.
2816 *Symplocos paniculata* (Thunb.)
2818 *Melia azedarach* L.
2825 *Lobelia pyramidalis* Wall.
2827 *Cassia javanica* L. var. *indo-chinensis* Gagnep.

March 13, 1922
We left Szu-mao at about 10 o'clock after saying good-bye to our dear friend Miss Clara Peterson, the Danish Missionary at Szu-mao. We went to the post-office, mailed a package of plants and films and seeds, said adieu to Mr. Cassati and Mr. Pattison of the Custom House service. Four soldiers accompany us to P'u-erh Fu 普洱府, the present Ning-erh 寧洱. Passing through the east gate (or rather by it) and the temple called Wen Miao 文廟 (which I photographed), we followed the Szu-mao plain in a northern direction over a stone-paved trail, through scrub vegetation of *Jasminum primulifolium, Pyrus pashia, Prunus puddum*, etc. On ascending into the hills, pines become very common. It is however not the pine of the south (*Pinus insularis*) but a different species. The vegetation has been much destroyed on these hills and *Pyrus pashia* and a variety of *Pyrus indica* with the ubiquitous anacardiaceous shrub are the main plants with *Smilax, Leontopodium*, etc. To our left is a valley and to our right a very deep one. The top of the ridge is reached and *Quercus, Castanopsis* (the small-fruited and small-leaved species) are found with *Vaccinium, Prunus* (2836 – *Prunus puddum* Roxb.) a cherry with red fruits and *Alnus* sp.?, the latter the most common tree. The trail descends rapidly through similar scrub vegetation. We met many men and women carrying salt in larges chunks, some little girls carrying as much as 75-80 lbs. on their backs with a strap over their

forehead. The trail follows a stream (but rising and falling) passing through pine groves, *Rhododendron, Smilax, Clematis, Schima*, and *Castanopsis*. We reached the village of Na-k'o-li 那柯里, sixty li distant from Szu-mao after 6 p.m. We could find no place to stay, the only inn (a horse stable) was already full so we pitched our tent near the stream. Bandits are said to be about but the soldiers are watching.
Specimen nos. 2828–2845
2836 *Prunus puddum* Roxb.

March 14, 1922
We left our camp at Na-k'o-li near the river; the trail winding through the valley following closely an irrigation ditch. The vegetation was extremely poor, everything having been cleared and the hillsides burned as we could see last night by a huge fire which was raging on the mountain above us.

The trees are mostly *Pyrus pashia, Prunus* sp.? and the ubiquitous anacardiaceous small tree. Pines were quite common. It certainly is a different species. *Alnus* was sparingly represented but *Smilax, Jasminum* and leguminous climbers (2849 – *Mezoneurum sinense* Hemsl.) common.

We stopped for lunch at an inn where we met a number of other soldiers returning from an escort. We followed a valley surrounded by numerous separate hills covered with pines, and with temples nestled among the trees. The paddy-fields were full of water and ready for planting.

We climbed over a high hill only to descend again. The hillsides are barren and only such pine-trees have been left standing to which the telegraph wires have been attached. Over some places there is no support for the wires for 400 meters. We go through a pass with a high hill of black rocks to our right. From a little shrine between two large *Ficus* trees, where we stopped for a rest, we could see the P'u-erh plain. To the left a huge mountain loomed up, a mighty black mass in the blue mist and haze. At the foot of the pass some women selling sugar cane. Being thirsty, we sat down and bought a few, dividing them among the soldiers and the boys.

The trail leads along the edge of the plain and across rice fields, still paved with these awful rocks. We skirt around a hill, and follow the edge of the paddy-plain to our right. The Chinese were already planting their fields. The telegraph wire hanging so low almost touching the water. Palm trunks of *Trachycarpus fortunei* are used as

telegraph poles on this plain. After passing through small villages, we reached the south gate of P‘u-erh with an open veranda on the top. The streets are better paved and not steep. The soldiers wanted us to go to an inn, but we went to the Fu-yin T‘ang where we met the Chinese evangelist who escorted us to a Buddhist temple but in every room were coffins, some packed in straw, the bodies ready for shipment. We decided not to sleep in a morgue and went to the Mission where we are occupying a small room on the ground floor. We paid off the soldiers and tried to get comfortable but in vain. Three Chinese called on us, one a teacher in the Normal School, the other the chief of police and the third, the chief of the telegraph service. We chatted till late at night, the teacher agreeing to take us to the Magistrate next morning.
Specimen nos. 2846–2850
2849 *Mezoneurum sinense* Hemsl.

March 15, 1922
The soldier came today for medicine for his sores. I painted him with iodine and some itch salve. The teacher then came and we went to the Normal School.

Mr. Sha called in the morning and we went to the Normal School where we were greeted by the various teachers and the headmaster. It is certainly a credit to P‘u-erh Fu to have such a school. It is supported by the Government of Yün-nan Fu, and furnishes books, clothing and food to the students, free of charge. They are taught even English grammar and conversation. They are erecting a science building for chemistry and natural history. We had tea after listening to the singing-music classes, a Chinese teacher playing the organ. The song was a poem about Spring and Cherry trees. They wanted to have their picture taken and we agreed to come after having called on the magistrate.

After the lunch the magistrate sent for us and we were escorted to the Yamen. The old gentleman was exceedingly cordial, shaking his own hand and bowing. He wanted to send us to Ta-lang which would be out of our way. We produced a map and he seemed to be ignorant of the lay of the land, so he called in a young official who was very keen and pleasant and knew the condition of the road and all the stage. So he agreed to give us an escort to Hsin-fu which is on our way. He wished to have his picture taken and the vainglorious old magistrate began to make toilet for the occasion like a prima donna.

He stood there like a statue but, having had much liquor, he swayed a bit and I am afraid the picture will be blurred. He suggested to have his picture taken with his staff and ourselves. So we went into the courtyard and there arranged the sitting, he, yelling for chairs at the top of his voice. After shaking hands, each one with himself – Chinese fashion, we were bowed out of the house. From the magistrate we went to the new General Liu 劉. At the gate there was a soldier on guard. A courtyard where they drilled, then a door, then another door and we were in the inner courtyard. We were cordially received by the general who also had his picture taken (neg. no. 30128, USDA) with the old General and then again with ourselves. He enquired about our trip and said he had much to do with America where he was stationed previously. He promised to give us a large escort, saying four soldiers not to be enough.

After having received an invitation to dinner for the next evening we went to the Normal School where we took photos of the graduating class and also of the student body (over 300 boys) on the drill ground.

The headmaster was dressed in European clothes which he obtained in Japan and which he had outgrown – his hat was about three sizes too small and his collar had evidently never been laundered. His hair was cut like that of a Japanese doll. He invited us in the name of all the teachers to a Chinese dinner for the 17th of March.

March 16, 1922

We left in company with Mr. Shao [!] of the Normal School and the evangelist for the top of Mt. Hsi-men ngai-tzu 西門崖子 or T'ien-pi-shan 天壁山. We passed through the west gate across paddy-fields to the foot of the mountain. There are large black limestone boulders scattered over the lower slopes for Hsi-men ngai-tzu is an extinct volcano with superimposed limestone in the middle of the P'u-erh plain. It is about 7,400 feet in height. A zigzag trail leads to the small plateau planted with pine trees which is about two thirds the height of the mountain. All the vegetation is burnt and only black rocks and burnt grass stumps remained. *Dracaena* sp.? trees are quite common growing among the rocks, on steep cliffs, for the mountain rises nearly abruptly from the plain. On the little plateau, there is a small rest house and a cave which leads into the mountain but only a short ways. There is a very narrow entrance with two figures carved in the rock, while to the right (in the main cave) there sits a Buddha also carved from the same rock and a part of the cave-wall, although gilded and covered with a shrine of wood. A stunted specimen of *Ficus* grows among the rocks. On ascending the steep slope a few hundred feet higher, we reached a shallow crater with steep walls on the northwest side. This wall was covered with sharp limestone rocks among which grew a *Lithocarpus carnea*? (2558) and orchids (2859 – undetermined) as well as ferns. *Dracaena* (2869) was very common, especially on the very steep walls of the north side. Southwest there is a very steep wall (limestone rock) and covered with vegetation. It forms the wall of another crater-like depression, the western wall of which does now not exist but the northwestern is still standing. Near the main crater, there is a big riff in the rocks and a hole (limestone sink-holes) extending to the depths below. The caves in the mountain are evidently limestone caves. On enquiry I found out that hot springs are found in the neighborhood, which should prove that volcanic activity is not altogether extinct. I took several photos of the crater and of P'u-erh as well as Dracaenas growing on cliffs.

We returned to P'u-erh and got ready for General Liu's dinner. He is a native of P'u-erh and has recently returned from Szu-ch'uan where he was wounded, they wishing to take his life. He however escaped, coming to his native place where he reigns supreme and taking the place of the old Mohammedan General Ma 馬. Mr. Sha came for us and we went to dinner to the General's house. The soldiers saluted and called out a welcome, opening the dragon doors wide and we marched into the large courtyard. We were escorted to a sitting room, the floor of which was covered with fresh pine needles. Tea was served, water chestnuts, small cakes, melon seeds and pine seeds. We exchanged compliments and talked about the political conditions, Washington conference, etc. Soup was served and, after some five minutes, we were escorted into the dining room, the floor of which was also covered with pine branches. He served liquor first to myself (the guest of honor) and then to everyone according to his rank. He motioned to each while pouring the wine thereby indicating his seat. The dinner was excellent consisting of sixteen courses. Four dishes were placed first on the table. They were cold. Then dish by dish was brought each helping himself from the large dish. There were whole stewed chickens, frogs legs, and many other things which tasted very good but I could not make out what they were. A white sweet liquor was served in small glasses and an orange colored one in small silver cups. Sweet litchi stew finished the main dinner, but rice

and some delicious soup was served in the end. After dinner we returned again to the sitting room where tea was served, cigarettes and fruits. The host, general Liu is a very quiet delightul gentleman who speaks very quietly, in contrast to the very nervous and hysterical district magistrate, Mr. Li 李. We were escorted again to the outer door where everybody bowed, the soldiers saluting, two of them escorting us to our place of residence at the poor mission. He was very anxious to have a doctor come to stay and he would furnish land free if he could get a hospital built, and assist in the construction of the same by furnishing workers.
Specimen nos. 2856–2871

March 17, 1922
Early this morning I was called by servants of the Normal School and so I went to teach an English class. This was the first time in the history of the school that an American conducted a class in it. The boys were very interested, and we also had visitors of other schools. I told them about America, Hawaii, Java, India, Burma and our trip to Yünnan from Siam.

From the Normal School, I was escorted to the Higher Primary School by the principal of that institution. Mr T'ang was a very pleasant gentleman in fact the nicest I have met. I made an address in English to the student body which was translated into Chinese. Everybody expressed his sincere friendship to America, the Boxer indemnity being mentioned very freely.

Tea and cigarettes were served. I took a photo of the student body, they were quite a patriotic group of men, flags were arranged and the school banner, there was a fine spirit exhibited in that school which was started in the 31st year of Kuang-hsü 光緒, 1905. There were 149 boys although 160 are enrolled. The school was built by the magistrate Hu Pi 胡壁. It has six teachers who give their services for very little money, $ 17.00 Mex. Yünnan a month.

The motto of the school is truth and labor. Food and shelter is furnished to boys of outlying villages but P'u-erh children go home. I was indeed surprised to find such an alive institution and such progressive and delightful people. The teachers were a fine set of men. After lunch I went out to take a few pictures, with the usual mob behind me. I photographed the South gate also the West gate.

In the evening we went to dine with the teachers in a temple without coffins. Much meat was served and a pink or purple, very strong liquor. Flies were present galore and sat on the cut sugar cane and water chestnuts. The dinner was of many courses but far behind that of the General's. However they meant very well and we appreciated their kind of hospitality. About 20 teachers being present, we talked much about China and I asked questions about P'u-erh. It was founded 200 years ago by Chinese who found only tribespeople living in the hills. Szu-mao was also founded about the same time Yung-cheng 雍正. The tribes to be found now are the O-ni, Lolo, Ma-hei or Pu-kong, Tai, Yao and P'u-man. P'u-erh contains about 1,100 families, average 5 in a family. The old name of P'u-erh is Chali.

There are two pears found in this part, the T'ang li 棠梨, a large pear and the common small pear which is known as K'o-i chieh-pen 可以接本 the pear used for grafting on.
The wild persimmon is called Mao-shih 毛柿 or hairy persimmon.

March 18, 1922
Left P'u-erh at 8:30 a.m., Mr. Liu, the evangelist, accompanying us to the city gate. The trail followed the valley northwards by terraced rice-fields, the hills being covered with pines. We ascended a steep hill covered with pines, *Pyrus* (the two species being very common), also several species of *Prunus*, some bamboo, the ubiquitous anacardiaceous shrub (same as at Ban Saa, Keng Tung State), also *Vaccinium.* We saw *Keteleeria* (2888 – *Keteleeria davidiana* Beiss.) trees for the first time, *Rhododendron* several species, a *Myrica* (2889 – *Myrica esculenta* Buch.-Ham.), a *Symplocos* (2876 – *Symplocos crataegoides* Ham.) etc. We lunched at a small Chinese rest-house half way up the hill – the six soldiers spitting everywhere and smoking their bamboo water pipes, producing an ugly gargling sound. The soldiers were otherwise clean and pleasant boys, full of fun. By the way, before we left P'u-erh the Magistrate, Mr. Li, called on us. He is a very congenial soul and very kind and pleasant. He sent us canned fruits, Muang Hai tea and two tins of Magnum cigarettes.

After lunch we climbed up to the top of the ridge over 6,000 feet. Pines almost ceased but *Keteleeria* (2888 – *Keteleeria davidiana* Beiss.) trees were found and *Gordonia* (2885 – *Gordonia chrysandra*?), the Chinese Shan-ch'a-hua 山茶花. On the top there was only shrub-vegetation. On both sides, was a deep valley with high hills on the other side (some 7,500-8,000 feet in height) but all sparingly covered with bushy growth. Many landslides were noticed; only on the right hand valley there were trees along a streambed.

Melastoma plants were common, the ordinary *Osbeckia*?, *Rubus* (three species – one with entire ovoid-leaves, the other one 3-foliate and one lobed-leaf species).

The trail, still paved with large slippery rocks of all sizes, led steeply down into the deep valley in which Mo-hei is situated. The slopes were barren, and in places terraced for rice-planting. Mo-hei is in the bottom of the valley along the Mo-hei stream. The soldiers led us through the dirty filthy town swarming with flies and loitering pigs which revel in the gutter with small kids in the quagmire. The smell and filth and buzzing of myriads of flies in the horse-manure (all over the street from passing salt caravans) does not give a pleasant picture of Mo-hei 磨黑河 to remember. We were first led to a Ma-tien 馬店 or horse inn, but I refused to go there asking for the Salt office where we found Mr. K'e who is the acccountant and who with two other young men spoke English quite well. He asked us to stay at the Salt office quarters and showed us a room with a small courtyard. The former was occupied by chickens also a chicken coop. He had it cleaned out and we moved in. Mr. K'e had dinner with us.

Specimen nos. 2875–2891
2876 *Symplocos crataegoides* Buch.-Ham.
2877 *Rhododendron mackenzianum* Forr., this is a synonym of *Rhod. stenaulum*
2885 *Gordonia chrysandra*?
2888 *Keteleeria Davidiana* Beiss.
2889 *Myrica esculenta* Buch.-Ham.

March 19, 1922

Today Mr. Keh guided us to the Saltmine which is a new shaft about 400 feet deep (that is Chinese feet – 8 inches Chinese to 1 English foot). This shaft is right in the city a little west of the main street. 26 boilers are scattered throughout the district. We were taken into a small room above the entrance to the mine where we divested ourselves of some clothing as it was said to be very hot down there. The entrance is rather small, a steep trail leads into the depth of the earth, the roof and sides are supported by timber. Steps lead down to about 100 feet depth then the trail turns, corkscrew fashion water standing in pools and the air is exceedingly foul as no sanitary arrangements are made for the miners who work absolutely naked in the mines. We met some of these naked creatures loaded with salt-ore carrying two bags, one in front and one behind. One strong man is

said to be able to carry 200 catties in one day of 6 hour work, one cattie being four English pounds. Their wages is on the average of 8 dollars Yünnan per month and those of the miners who dig from 7 a.m. to 1 p.m., 15 dollars per month. They wear no clothing not even a gee string. The working hours in the wet season are only 4 hours. There are employed 100-200 miners, but the majority of them work only half a year and tend to their fields and rice crops the other half. Their pay is also increased according to the output of ore. The ore we saw was purplish-red or brownish with glittering salt, it was rather poor ore. At the end of the trail there was what looked like a huge cave in which men were working; we could hear the muffling sounds of picks. The ore is weighed and taken by women to the boilers. Brine is pumped from the shafts into vats after 1 p.m. when other work ceases. The ore is placed in these flat vats – just depressions in the earth, and there it is dissolved. It is left for two days and then drained and pumped into cast-iron, shallow basins which are arranged in pairs in a terraced fashion, 24 in number. It takes 2 days and one night to make one cake of salt which weighs about 180 lbs. Each cake represents a value of 10 dollars. The boiling of the salt is farmed out to private capitalists under Government supervision. For each cake of salt the Government taxes $ 3.50 and 25 cents extra which goes to the boilers. There are several shafts but some have been abandoned as hey are too dangerous to work. We took several photos of the miners in front of the entrance of the mine, also of the boilers, ore store room, brine pumpers and a general view of Mo-hei.
Tonight we had Chinese dinner with Mr. K‘e at his home.

March 20, 1922
Left Mo-hei at about 8 a.m., Mr. K‘e accompanying us. We looked at the ridge northeast of the Mo-hei Ho 磨黑河, known as the Dragon, it winds through the valley of Mo-hei and is densely wooded, the reason why the beautiful forest has not been destroyed is superstition; the dragon would squirm. We had considerable trouble with our caravan and carriers this morning who were all late. We followed the stone-paved road till we came to the streambed from there we walked along the irrigation ditch which feeds the terraced rice fields in the Mo-hei valley; the trail led through a small hamlet and there we were attacked by a bunch of ferocious dogs. Baer shot one, it was the only way to keep them off. Seven soldiers accompanied us but they were behind still in the streambed having started late. We met many Lolo women

carrying fire wood on their backs supported by a strap which is passed over the forehead. This wood is carried for several li (Chinese miles, 3 English to about 10 Chinese) to the salt boilers. The hills around Mo-hei are quite barren having been denuded for fire wood to keep the boilers going. Not a single tree is left all are cut down, so natural reforestation is impossible. We followed the Mo-hei River through a beautiful valley the ridges covered with pine trees and other foliage trees; *Pyrus pashia* is still very common, one *Pyrus* with velvety leaves and large flowers we saw only three trees. *Schima* is still common, and so are *Prunus* but more of the red fruited variety (2884 – *Prunus puddum* Roxb.), *Ligustrum, Osbeckia, Castanopsis* and *Quercus* make up about the rest of the vegetation. Terraced rice-fields on the steep slopes are the mode in this part of Yünnan. We passed two small villages on the way to Mo-po 磨波. The scenery became grand, the mountains on each side of the valley rise 3,000-4,000 feet or even higher, while the narrow valley bottom is in itself about 4,500-4,800 feet elevation. The mountains are covered with dense forest and it is quite pleasant to behold these forested hills in comparison to the naked hills around Szu-mao, P'u-erh, Mo-hei. The forest takes on a tropical nature, *Cibotium* occurs several other ferns, *Engelhardtia, Castanopsis*, even *Wallichia*, aroids and lianas, the whole ensemble taking us back to the forests of Chieng Tung and Chieng Law. For lunch we stopped at a small house. The owner looked more like a Tai than Chinese but he only spoke Chinese but could not read it. We found a little girl with severe fever and I administered quinine and to an old man iodine for some sores.

We arrived at Mo-po at about 6 p.m. after skirting or rather passing some rice fields. Mo-po is on a small slope in the Mo-po valley. I looked into one inn but had enough with one look, so the soldiers went into a home where an old lady with bound feet – she looked for the world like old pictures of the devil with goats-feet – objected to the soldiers requisitioning an empty room. She agreed finally and I sat down. Alone with me in the room sat a funny looking young man quite as a mummy and all at once he said, „From where did you come from Sir“, in perfectly good English. I was much astonished to find, in such a place, an English speaking Chinese. He told me he was the teacher of the primary school of Mo-po and had studied English in Yünnan Fu with an American teacher. He made us welcome in the school house, where we are staying. He wrote a verse of one of Byron's poems on the blackboard, much to my

astonishment. Mo-po is still very small consisting only of a few houses. There are 30 pupils in his school, but some of them do not belong to Mo-po.
I am tired and it is late.
Specimen nos. 2892–2921
2884 *Prunus puddum* Roxb.

March 21, 1922
We left Mo-po 磨波 early at 7:10 a.m., with five soldiers vested in various garbs, only one was able to read and write Chinese. The trail led along the Mo-po Ho; there was little forest at first. The tall mountains, over 10,000 feet in height, still loomed up majestically below Mo-po, on the west side of the valley. As far as I could see it was densely wooded. There were two sharp ridges, the southern one was very steep and abrupt, while the northern one was sharp but of a more gradual slope. Pines were quite common. Red cherries, *Schima*, two species of *Pyrus*, the yellow *Rubus*, and *Alnus* trees grow on the slopes. Mai yom trees, *Ficus, Engelhardtia* trees, roses, *Ligustrum* and the stiff reddish *Jasminum* were quite common. We passed a small village called Fen-shui-ling 分水嶺 and followed the Man-pa-ho 蠻巴河 to Ma-li-shu 麻梨樹, thence to the village of Lan-ka ts'un 蘭嘎村 where we had lunch. near here the Lan-ka Ho (river) flows into the Man-pa Ho. The hills are rather steep and forested, with the exception of a chain of hills running east-west, the tops of which were bare. We saw several valleys leading off the main valley. In these deep valley branches fire was raging. The wind started again, as usual at about 11 a.m., and lasted till about 2:30-3 o'clock p.m. We lunched under a bamboo clump at Lan-ka ts'un. The heat was intense although the wind made it bearable. To the left, in the valley, looking north, there is a small mountain, the top of which is wooded with pine trees. It was before reaching Lan-ka ts'un that we found many *Abies* trees in fruit – quite a grove of them. The valley near Lan-ka ts'un leads northward but is a blind valley, a trail leading at the foot of the hills along the terraced paddy fields. The Lan-ka ts'un stream divides and flows through a gorge at right angles to the Lan-ka ts'un valley towards Man-pieh and from that point it is called Man-pieh Ho 蠻別河.
We left our horse caravan behind at Lan-ka ts'un and went down the trail to the streambed, three soldiers preceding me. As we came to the Man-pieh Ho over which there is a single log for a bridge, I saw them

kneel down and put cartridges into their guns and aiming at a small mountain which controls the two trails, one leading up to the Lan-ka ts'un stream, and the other the Man-pieh Ho stream, they yelled at the same time, and I saw several men disappearing behind a wall on the mountain top. The soldiers recognized them as robbers. I went back to Baer and the carriers informing them of the presence of robbers who controlled the two trails from a very strategic position. They looked down quite unconcerned but made for cover when the soldiers aimed at them. We got our caravan to join us, two merchants, a little boy and his father, being very frightened, joined us, too. Between us we had six rifles and two revolvers. The bandits watched us for a long time but did not attack us, we were too many for them. It was quite an exciting time, the soldiers said there were 15 men up there. They would have charged their nest had we gone with them, but I thought best to let them alone as they did not attack us. They played a hide and seek game. We had to skirt the hill on which the bandits had the vantage point. A steep rock wall with a narrow trail leading in its face with the Man-pieh Ho roaring at the bottom of the valley. We looked back watching their every move, we saw several heads over a wall watching us. We went on and nothing happened. This part between Lan-ka ts'un and Man-pieh is very wild and dangerous, the hills rise steeply for several thousand feet and are covered with pines and foliage trees. There is one very narrow gap with steep rocky bluffs. One soldier told me that a short time ago there was a fight between soldiers and robbers at this particular place and one man was killed. We scanned the rocky hillsides for robbers but saw no more. *Engelhardtia* trees both the large and small species are quite common also pines, *Schima* and the other theaceous trees with drooping fruits and parge persistent calycine lobes. The trail which is very rocky leads down to the streambed and to Man-pieh which is situated on the Man-pieh Ho. We were glad to reach this small hamlet at 4,500 feet elevation. The soldiers took us to the schoolhouse where we are now located for the night. Many boys came to see me and brought their English primer asking me to teach them to read. They are all eager to learn although they are far removed from any railway and English speaking people. They were all very friendly and kind.
Tomorrow we are going to Man-lien 蠻連.
Specimen nos. 2922–2929

March 22, 1922

We left Man-pieh at 7:45 a.m. saying good bye to the school boys who were coming into the schoolroom to prepare for their lesson. We had a miserable night – at first we were awakened by some people quarreling and yelling calling each other awful names, they kept it up till midnight to the top of their voices. After midnight the dogs gave a concert barking continuously. It was a restless night after a long day's march. The only pleasant thing at Man-pieh was the cold bath in the river. We followed the rice fields and climbed a ridge and then descended into a deep lonely valley flanked by high steep mountains, at the bottom a wild stream flowed in a circuitous way around a pine covered bluff; deep gorges and steep rocky mountain slopes clad with *Pinus insularis*, a 3-needled species, could be seen dimly in the mist.

We kept to the left side of the mountain slope also clad with *Pinus, Schima*, the Rubiaceous *Wendlandia* shrub, *Quercus* (2930 – *Quercus Mongolica* Fish., neg. no. 30150), with the light colored foliage and oblong acorns in a very long scaly cup; we descended to the streambed, the trail was still paved but there were other trails less awkward to the bottom of the valley. We had to cross the stream and ascend again to the top of the ridge opposite about 5,500 feet elevation, where there was a gap. Two high, cone-like mountains appeared in full view, they must be over 7,000 feet in height; we saw them from Man-pieh. The hills are covered with pines and *Quercus, Schima, Ixora* white-flowered, *Clematis* (2937 – *Clematis armandii* Fr.), with Smilax-like leaves, the yellow *Jasminum* (2934 – *Jasminum lanceolaria* Roxb.), and a yellow spiny *Caesalpinia* (2933 – *Caesalpinia sepiaria* Roxb.). We descended into a narrow valley with steep slopes the top of which were covered with pines. The large oak with light green leaves was *Quercus acutissima*, again very common. The heat was intense although a little breeze blew. We lunched near a streambed where some stark naked boys repaired a ditch. We descended again dropping down to about 3,000 feet in a valley where we passed the village of Chien-law where we should have stopped for the night but we depended on Major Davies' map which marked Man-lien as a big place, but we found out that the latter is now deserted and only one family lives in it. We climbed the steep hillsides north of Chien-law and reached the top of the ridge where we found a gate near a large fig tree with large (cherry-size) yellow fruits (2939 – *Ficus altissima* Bl.). It is a queer and lonely looking scenery, the hill-slopes are broad and covered with scattered pines but no underbrush only the dry brown needles which gave the whole landscape a very sombre aspect; not a breath of air was stirring, the tops of the peaks hidden in haze and mist which gave them the color of the sky. Not a bird not an animal of any kind, quiet reigned among those mighty pine clad mountains. We descended on a gradual slope, where we found *Schima,* oaks, *Albizzia stipulata* and bamboo clumps, *Ficus* sp.?, and even Guava which showed human habitation. We crossed the Ah-chien Ho (stream) climbed a ridge again and looked down at what was once Man-lien. The hills were bare, only pines on the very top and a lonely grave on the brown hillside; if anything looked like a graveyard it was Man-lien. Only one house inhabited, the others fallen down and empty. There was nothing to do but to stay and occupy a ramshackle house which served once for soldiers. Fleas are plenty. We put up our baggage under the leaky roof and made ourselves as comfortable as possible.

This is a very lonely place, the one family living here said that 10 days ago there was a band of robbers on the hills here but since then they have not seen any. Four new soldiers have come but only one has a gun; so we decided to keep the other four who, however, are simply villagers armed with old style rifles, old flint lock, they carry their powder in a gourd on their belt. In case of trouble they would not be of much use, but from the distance any gun looks alike. A bluff goes a long way here. We barricaded our door which we found on the floor, placed stones and bamboo behind it as a prop. So we will go to bed and sleep lightly.

I could not believe that this place was Man-lien, so I went on ahead alone to the top of the hill to see if there was not a larger place ahead – only lonely hills – and the Black River 黑河 or Pa-pien Ho 把邊河 winding in the bottom of the valley, flowing into a gorge at right angles to this valley.

Elevation of Man-lien 3,350 feet.

Specimen nos. 2930–2937.

2930 *Quercus mongolica* Fish., neg. no. 30150
2933 *Caesalpinia sepiaria* Roxb.
2934 *Jasminum lanceolaria* Roxb.
2937 *Clematis armandii* Fr.
2939 *Ficus altissima* Bl.
2943 *Wendlandia paniculata* Roxb.

March 23, 1922

We left Man-lien at 8 a.m. glad to leave the ruined and forsaken place behind. The scenery was weird and wild. We followed the direction of the wild Black River or Pa-pien Ho for some distance, then climbed a ridge following the hill side skirting the mountain. A deep valley lay at our feet, pine trees only are to be seen with here and there a lonely grave, the only remnants of the former inhabitants of Man-lien. On the higher slopes we saw some lonely houses, people that had fled from the poisonous air that rises from the ground and hovers over and in the valley. We came to the crest of a ridge again after having passed through a shallow depression. Ahead of us is a long narrow valley, to the right extends a narrow deep gorge through which a stream has cut its tortuous way; rock filled the streambed and the sombre atmosphere, cloudy sky and mist, the graveyard stillness of the air, the dark green scattered pines over a brown hillside made the scenery rather awsome and weird. The picture was a little brightened by terraced paddy fields over which water trickled from terrace to terrace. The people who tend these rice fields have taken to the hills and only come down to plough, plant and harvest. We found the same species of *Ficus* with its semi-sagittate semi amplexicaul-leaves, which we met first in the Meh Len valley of the Keng Tung State. The yellow, double-flowered *Jasminum* is also common but the most common plant is the red flowered *Woodwardia* [!] (2945 – *Woodfordia floribunda* Salisb.) which I saw last year also in the Chin hills and upper Burma. The Mai Yom trees called ch'un-shu 春樹 in Chinese are also met with as are *Ligustrum, Clematis*, tall grasses etc. The hillsides on the right are burnt-over and only pines have survived the fires. At the head of the valley, which is closed in by a ridge, there is a depression, a sort of pass guarded by two sentinel trees (*Schima* sp.?) through which the road passes. By the way we have now left the awful paved road and once more it is comfortable going over a smooth dirt road. Clouds gathered blacker and blacker and thunder began to roll. We descended to the level of the river called Ma-pa Ho, it comes from the west through a side branch of the long narrow blind valley, makes a big curve, and disappears through the narrow gorge already mentioned to join the Black River, which in this region has another name: Ching-lai Ho 景來河. There is a little village at this streambed Chiu-tu-ching 舊毒菁. As the weather looked threatening we decided to stop and lunch here. The air was terribly depressive and everyone suffered from headache. I myself being badly affected by nervousness and palpitation. We entered a poor widow's house built of mud, the two sides half open to let in wind and rain. She was brewing a liquor made of corn and paddy in a huge shallow, circular iron pan. She was alone with her little girl whose father had died when she was three years old. While we had lunch it rained and the thunder rolled. The air was still bad, but we left shortly after the rain had stopped, the first since we left Chiengmai December 30, 1921. With ears aching and a depressive feeling in the temples, we started on our way climbing the hillside on the east bank of a small stream which joined the Ma-pa Ho, we reached the top of the hill and there the air was fine, the scenery was still wild it must have been over 5,000 feet. The oppressive feeling we all experienced in the valley was due to poisonous gases rising from the ground. We descended again into a narrow valley which leads down to Hsin-fu 新拊 or 新撫 The trail becomes paved again near houses and there it is generally steep and for horses almost impassable. By the vegetation we could see that we have descended to the level of the tropics, as we found Mango trees, guava, *Oroxylum indicum* and a number of plants found in Siam. After crossing another ridge we passed down the Ta-pu-tzu Ho 大舖子河 which we followed for some distance walking over the rocks in the streambed. As our soldier escort looked the most miserable we have had, being simple, dirty villagers without uniform and carrying muzzle-loader rifles with the powder in a gourd on their belts, they were mistaken for robbers and many of the people we met made haste to get out of our way. Some, after seeing them running back whence they had come, and that was within a few hundred yards of the gate of Hsin-fu. We entered Hsin-fu by the south gate a whitewashed miserable affair. Inside there was a lamp post of bamboo which was set in a pile of rocks and bricks held together by earth and surrounded by bamboo. An oil lamp is hoisted to the top of the bamboo by means of a rope like on a ship's mast. It was market day and the one street of Hsin-fu was crowded; a large caravan of Tibetans was on its way to P'u-erh to buy tea. The tea comes from Muang Hai where I photographed the wild tea trees. The dried leaves are brought to Szu-mao in large bamboo bales. There the leaves are sorted by hand (picked out) and packed in circular packages about 8 inches across and five inches high. Once a year these Tibetans, some 300-400 come to P'u-erh and Szu-mao to buy tea. We looked for a place to stop and after taking a glance at a dirty temple decided to try the schoolhouse. We found a spacious courtyard surrounded by a wall within which were four buildings. The rear wall of three of the houses formed the surrounding wall of the whole compound. The building on the west side was the largest and was a temple erected to Confucius, his soul tablet rested in the center, pine branches were on the floor and everything looked quite cosy. We decided to occupy the Confucius temple, while the rest took possession of the other buildings. Facing the shrine on the East side was the classroom, but no school was held and it was by far the quietest and most peaceful place we yet occupied. We decided to stay one day but as it rained all afternoon of the following day and night and all forenoon of the next day we stopped another day. We called on the official a Mr. Wang, a very pleasant man whom we had met on the road to Man-lien. He called on us the next day. We took some photos the whole of Hsin-fu from our place on the hill, one of the north gate, and one of the street (neg. 141050 Nat. Geogr. Soc.), also one down the Black River from the other side of the stream (neg. 30150) which we crossed on a bamboo raft.

The afternoon of the 24th was clear and pleasant.

Specimen nos. 2938–2947

2945 *Woodfordia floribunda* Salisb.

March 26, 1922

We left Hsin-fu about 8 a.m. There are two roads one over the hills on the left bank of the river and one narrow trail on the other side following the river. We decided to take the one following the streambed as we had enough of mountain climbing. The hills enclosing the valley of the Black River are not high, although further on we met with black rocky cliffs covered with pine trees. The vegetation in the river bottom is very scarce; *Engelhardtia* trees the small species, *Bauhinia variegata*, a Cactus-like *Euphorbia*, Mango trees in flower, fruiting bananas, *Smilax*, *Jasminum* the stiff white variety, no *Pyrus* nor *Prunus*, nor *Schima*, the elevation being too low. Bamboo was common, also *Caesalpinia*, etc. We followed the right side for some time but in the afternoon it became unbearably hot, so we crossed over to the left, fording the stream, and travelled through lovely pine forest till we reached Kuan-ni a tiny village on the slopes of the hillside. We are camping in the tent on the bank of the Black River.

Specimen nos. 2948–2959. Along black River.

2949 *Trevesia palmata* Vis.

2950 *Ehretia dicksoni* Hance

2952 *Diospyros yünnanensis* Rehd. & Wils.
2953 *Ficus ischnopoda* Miq.?

March 27, 1922. Monday
Left Kuan-ni after seven a.m. The mosquitos were very bad in the tent, hundreds of them, so I moved out and put the mosquito curtain over the bed trusting that it would not rain. The east bank of the Black River here called the Ching-lai Ho 景來河 where we camped is flat and sandy. The village, just a couple of houses, is a little higher up above the terraced paddy fields on the hill side. The other half of the village is on the other side of the river but has been abandoned. The trail leads along the riverbed then climbs the steep slopes. I passed a newly placed coffin among the trees, with a large stone on top, evidently the method of burying people in this part of the world. There was a slight odor but I was soon on top of the ridge. The trail was very narrow and very winding along the steep gorge it entered. No human beings could be seen for miles. The small velvety *Engelhardtia* was quite common, also *Quercus* (2959 – *Quercus mespilifolia* Wall.); the light green foliaged *Quercus acutissima* and another species without fruit or flower. Pines covered the upper slopes to the very tops of the ridges. *Phyllanthus emblica* is very common all along from Siam up. A few *Pyrus pashia* trees were observed but only near the stream. Bamboo clumps were quite common. We came on to a few terraced paddy fields but no houses. The mountains are not very high but covered with pines. Along the trail we found *Mayodendron igneum*, the common purple tree composite, *Woodwardia* [!] and along the streambed *Homonoia riparia* as well as *Ficus* sp. (2957 – *Ficus nitida* Thunb.). The leguminous vines like *Mezoneurum* (collected at Chieng Rung) was very common. An occasional bamboo, an occasional *Bombax malabarica* tree with red flowers, also the papilionaceous shrub-like vine found on the hills of Szu-mao was also common. Yellow *Jasminum* grew along the trail with spiny yellow flowered *Caesalpinia* (2933 – *Caesalpinia sepiaria* Roxb.), tall grasses etc. (2962 – *Neyraudia madagascariensis* Hook. f.). The country here looked very wild and weird. *Bauhinia variegata* made its appearance on the opposite bank of the stream and *Albizzia* sp.? (2960 – *Albizzia julibrissin* Durazz.) is a common tree along the Black River. We lunched at the small village of Hui-po, and then followed the broad valley of the Black River which is here cultivated. Sweet peas being

the principal crop, also opium poppies, mustard and a coarse bean. Further on we saw large plantings of sugar cane and a most primitive sugar mill, the grinder consisting of two upright posts between which stick by stick of cane is shoved in by hand, the posts being turned by a carabao. A wooden vat is beneath the upright grinders which catches the juice. In the rear are five shallow circular pans like those used in boiling salt, five in a row here, with a brown dirty mass boiling in it. This is stirred or rather the sides of the pan are brushed with a bunch of dirty straw. The latter is handled by semi-naked men whose skin is so encrusted with filth that it is black. With a large braided bamboo dipper they skim the foam off the boiling mass, and with a wooden dipper it is poured into earthen pots behind the boilers to dry. The little brown cakes resulting from this process are about five inches across and one inch thick and of a deep chocolate color. They are then packed in large bamboo stipules tied with bamboo, and sent to Szu-mao and elsewhere. I took several photos of the whole outfit. The cane itself is rather juicy, long jointed, and of a yellowish brown color.

Shortly after leaving the sugar mill which is an open shed consisting only of a straw roof on posts, we reached En-lo 恩樂 where we are located in a newly built Chinese house, mud floor and walls. The valley is very broad and a high mountain is visible in the distance, part of the village, about thirty houses, is on the other side of the river.

Specimen nos. 2955–2983
2957 *Ficus nitida* Thunb.
2959 *Quercus mespilifolia* Wall.
2960 *Albizzia julibrissin* Durazz.
2962 *Neyraudia madagascariensis* Hook. f.
2933 *Caesalpinia sepiaria* Roxb.

March 28, 1922
We left En-lo rather late, 8 a.m. The soldiers had not appeared and soon we found out that we are to have the same men who accompanied us from Hsin-fu. Those of the village of En-lo having gone out chasing away robbers who infested the neighborhood of En-lo. The trail follows the Black River for a short distance through a rather narrow valley; a large broad valley back of the real village of En-lo on the other (west) side of the Black River is to our left. We enter dreary country with pines, the trail winding around the hill

sides. We had not gone very far till we met a large caravan of 500 horses and mules coming from Tibet. The men were all dirty Tibetans wearing queues and turquoise earrings, most of them had rifles, the muzzles covered with red cloth. The men wore woollen clothing, others yak jackets with the long fur inside. Their animals carried their own rice, also provision for the Tibetans, half smoked pigs the hairy tail sticking up into the air. We had to wait a long time till they had all passed, as the trail was too narrow to permit the passing of two caravans. The trail leaves the river and leads over a high hill; high mountains are to the right in the distance. Pines form the only vegetation. We crossed the top of a ridge onto a spur around which the Black River flows placidly. The water of the Black River is dark and muddy hence the name. The sky is overcast and haze enshrouds the whole country. We cross a dry gulch, the hills are bare and brown. We follow again the Black River at the banks of which cultivation is carried on, sweet peas are grown, sugar cane and beans. Mai Yom trees (2964 – *Cedrela Toona* Roxb., the Chinese ch'un-shu) are very common in the valley especially near the banks of the river and about the villages. We lunched on the river bank under Mai Yom trees at a semi-forsaken village called Hsin-kai 新街. The sky looked threatening, black clouds gathered but it did not rain. We followed again the river, crossed over the Lo-ma Ho near cultivated rice fields (terraced), here tall bamboo is very common also the light green foliage *Quercus acutissima* of which there are nice groves on the banks of the Lo-ma Ho. We climbed a pine-covered ridge with *Schima* trees, the small village we had left behind was Tang-fa, while below us winds the now dry streambed of the Chin-chi Ho which finds its source in the broad valley to our right with high mountains in the distance. From here we climbed over a ridge leaving the Black River to our left and considerably below us. We are passing lonely groves of pine trees, the ground being covered with the dry brown needles. High blue hazy mountains are beyond the distant west bank of the Black River are visible between the sombre pines. Parrots called ying-ko 鸚哥 screeched and flittered about, while down below along the streambed, brownish gray birds kept close to the ground jumping from grass culm to grass culm like toads.

We followed again along the Black river, the banks being low, the valley broader and the mountains becoming lower. We stopped and took a swim in the stream where big fish (cat fish) swam lazily looking on in horror of a decent bath. On the lower hills to the left is the village of Che-hou. From this side it appears as quite a good-sized place with rather decent looking buildings.

Only a short distance from here and we are in Che-hou 者后 on the east of the Black River. It is composed of one very short street, with a number of tumbled-down empty houses. An old temple is on a hill with large *Ficus benjamina* trees, and a few houses inhabited by a few families. We are stopping in a newly built house rather larger and more comfortable than the one we occupied at En-lo. A shower of rain fell from a black sky and the high hills to our right almost touch the black clouds. Today we heard the cuckoo call among the pines. Vegetation, *Sapium sebiferum*, bamboo, *Erythrina* and large Mango trees.

March 29, 1922

We left our comfortable dwelling at about 8 a.m. The sky was cloudy and stormy looking. The trail passed between some deserted houses and soon we crossed a broad dry streambed which found its source in a broad valley to our right. We followed along paddy plains, we crossed another stream, and followed again some paddy fields. It would be considerate of the Yünnanese farmers if they would not convert part of the rottenly paved road into an irrigation ditch which makes travelling over such stretches almost impossible. *Bauhinia* trees are still flowering here and there. To the left is a deep valley whence an affluent of the Black River, the Chuan Ho issues. The trail follows in the straight valley ahead leaving the Black River for some time until we reach Nan-Ma-kai 南馬街. The trail leads into the streambed of the Man-kang Ho 蠻綱河 which it follows amidst numerous rocks. There is no water in it at this time of the year. The valley through which this river flows is rather shallow being bordered by low pine-covered hills, some oaks can be seen among the pines and an occasional *Ficus benjamina*. To our right, a little further on, there is a valley with high mountains all pine-covered. The Man-kang Ho comes from this valley, a long spur juts out into the main valley, at the point of which there is a small tea house for travellers. Huts can be seen on the other side of the valley. From the valley ahead there enters a small stream called Ssu-ko-shu into the larger Man-kang Ho. This little stream we followed in its bed to a bend caused by a low spur to the top of which we ascend. It is clad rather thickly with sombre pine trees. To our right are higher mountains, the crests of which are clothed with pine, while most of the steep red slopes are bare or harboring an occasional pine or oak. The scenery is weird and at the same time it has a pacifying influence on one. The sky was black, not a breath of air was stirring, only the dark pines, their brown needle-covered ground with here and there a tomb or even a row of them, all gave the scene a rather picturesque appearance. The scenery was a little brightened by a few narrow terraced paddy fields with usually one or two seed beds of rice, sparkling like an emerald in even this sombre atmosphere. We still climb somewhat and soon find ourselves in front of a nine-storied pagoda, white with nine little roofs. Just a little lower down to the right are a few deserted houses, one of which looked as if it had been struck by lightning, it was burnt down. The pagoda like all others is called *wen-pi* 文筆. To the right of this picturesque pagoda is a little shrine built of stone and bricks, it contains some slabs of rocks giving the names of those who constructed the pagoda. Both, on the top of the pagoda, and on the flat roof of the shrine, there are *Ficus* trees. The latter looks as if it were a pedestal especially erected for the tree. On the ground in front of the pagoda there is a stone slab with an inscription but no year is given the heading reads K'ai hung ti chia 開宏弟甲. The trail leads past the pagoda which is surrounded by numerous graves, into a little valley and a streambed called Ch'uan ho 川河, which we follow for a short distance. Soon we crossed another stream coming from a broad valley to the right, it is known as Ch'u-shan-men 出山門 or Mountain exit. We now followed the broad plain with its paddy fields and tall thatch grass; we cross another stream called Ch'ien-shao Ho 錢少河. On its bank, to the right, is a small village called Hsiao-wan-p'u 小碗舖. A little ahead we pass through a stone gate in bad repair, it is the entrance to Nan-ma-kai. An inscription reads as follows, to the left K'uei-wen 魁文, in the center Yüan-chieh 元解, and to the right K'uei-erh 魁二. Another inscription says that it was erected during the Ta Ch'ing dynasty under the reign of the Emperor Chia-ch'ing 嘉慶 in the twelfth year and 3rd month of the spring, April 1807 A.D. A miserable stone-paved trail leads to the top of the little hill on which the dirty village of Nan-ma-kai or Man-ma-kai 蠻馬街 is situated. We looked for a place to lunch and found an old temple in a very bad state of repair near some large *Ficus* trees on the outside of this rotten town. It consists of a tumbled down wall and two buildings, one contains three gods or rather genii and eight coffins, the other a body probably the Kuan-yin sitting on a lotus flower and adored by devas, all carved out of wood. It is all very dreary, the sky is black and a cold wind is blowing. The country is bleak and dry and covered with brown grass. There is no food for the horses. Rain drops are falling and the whole outlook is far from cheerful. We are sitting outside the temple with the goddess, in which by the way there are six coffins, and eat our lunch, while a few people peep through the tumbled down gate in the wall to watch us eat. The trail leads down again through a large graveyard. We are bound for Man-wo 蠻窩 with our new escort, the other four soldiers returning to Hsin-fu. *Sapium sebiferum* is one of the most common shrubs here, also Mai Yom or Ch'un-shu (2964 – *Cedrela toona* Roxb.) trees are met with, with an occasional *Schima*, oak and pine. We follow paddy fields, the people are now transplanting their rice. The Black River is quite a distance west of us but we are following it up in the same valley. Suddenly at a sharp turn of the trail we found ourselves in Man-wo, Den of the Savages. There is a large house belonging to a Tai woman from Lampang who had married a Chinese and who had lived here for 13 years. We are occupying the tower of the school building, it is a ramshackle affair but at least it is airy and away from the curious eyes of the people. The school is in a state of utter decay, the walls are crumbling, the steps have fallen, the floors are sagging and the courtyard, although small, is filled with rubbish. A few hibiscus bushes (*H. rosa-sinensis*) Ta-hung-hua 大紅花 are in the courtyard also one Oleander tree and two crape myrtle trees. A stone slab is lying on the ground partly buried in rubbish, and stating that it was erected in the 57th year of the reign of Ch'ien-lung 乾隆 of the Ch'ing dynasty, in the third month and 13th day (April 15, 1792).

The teachers are very nice and civil and everybody is most kind. From the tower we have a beautiful view of the Black River valley and of the village of Ma-ching-kai.

During the night a fierce thunderstorm arose, the lightning playing on the high hills of this valley and lighting up our ramshackle wooden tower, the wind is blowing and the air penetrates through the large cracks and windows. If ever this tower is struck by lightning it will collapse as flat as a pancake. The walls on which it rests are literally split and lean in every direction. Above our tower-room is a smaller one in which there is an altar, but the god that once sat enthroned is now lying on the floor. The head is gone and the rats have pulled out the straw stuffing.

It is morning. It rained heavily but apparently it is clearing up and we are getting ready to leave for Ching-tung 景東.

March 30, 1922

We left our tower at Man-Wo rather late as it still rained in the morning; in fact, at first we had not intended to go at all but stay for the day. We followed the Black River in the rather broad valley, the hills being fairly low on both sides. The village of Man-ching-kai is diagonally across from the place we stayed. We crossed a streambed and then walked between sparingly pine-clad hills, to the right there were large landslides and deep washouts, one place was particularly interesting as a beautiful canyon was forming with towers and cathedrals and chasms, but all was very precipitous. We came to a very dirty village with a peculiar tower-like structure in the center, built of wood. A lake of mud on the other side, numerous flies, pigs, ducks and dirty children. The soldiers wanted me to stop there for lunch but I refused and so we walked on across paddy fields to the Black River which we crossed just a little ways from the village mentioned. We followed on the other side of the river over a very narrow miserable road between paddy fields; there was a ruined temple, the roof fallen in and as in some, the idols were buried in debris from the falling crumbling mud walls. Along the trail I noticed a meteorite about 1 foot in diameter placed between ordinary rocks, before it incense burners and incense was placed; it had evidently been deified. The Black River makes a big bend, the bed being rather wide and sandy. To the left there was a hillside absolutely bare only a single *Ficus* tree grew on the upper edge. On the sides there were numerous graves but not another tree. It was late after one o'clock and I decided to stop for lunch. We only gathered enough fire wood to boil our cocoa and our lunch we ate cold. From there we left for a village called Chi-liang Kai where we changed soldiers. From this dirty village the trail led up a hill much eroded but covered with scrub *Schima*, young pine trees and the anacardiaceous shrub. On the top of the hill was a small plateau with pine trees, graveyards and *Schima* trees. To the right there stood a seven-storied pagoda, another wen-pi, small trees growing on the top of it. To the left, a large stone slab overgrown by a large *Ficus* tree stated it was erected in the 45th year of Ch'ien-lung – 1780 A.D. From there we went down hill again passing the village of Wang-kung-tung. Many *Schima* and theaceous trees were in evidence also oaks. Near the pagoda I collected a large leaved cherry with light red fruits. We met several people on horseback and many people on foot carrying strings of brass cash on their shoulders and some vegetables or chickens. They had come from the market of Ching-tung. Many carried earthen pots which are manufactured in Ching-tung. We passed along paddy fields, after making a turn to the left we came on to the city of Ching-tung on the foot of the hills which are bare. Before entering the city we pass over a bridge and after a little climb reach the suburb of Ching-tung. To the left, on a small hill, is a formidable gateway which leads to the Yamen. We went on and after a short distance came to where the lone American missionary lady Mrs. Marston lived. She gave us a rather cold reception. A crowd immediately gathered and thronged the poor little courtyard where a parrot screeched. Mrs. Marston was dressed in Chinese clothes. After a few very pointed questions one of the first if we were Christians, asked us to come in. The police official came, he was rather uncivil but finally after hearing that I was not a missionary escorted us to the Szechuan temple at the head of the street, where we are now located. Unfortunately I had not my calling cards with me nor my credentials and the horse caravan had not yet come. Our kung-ssu or letters from the official of Hsin-fu had also not arrived so they thought we were ordinary travellers and wanted us to go to a miserable inn. The whole reception both on the side of Mrs. Marston and the officials was all but a friendly one. If I could have gotten an aeroplane I would have flown immediately into a more hospitable country.

March 31, 1922

This morning I called on Mrs. Marston who is all alone. She is rather queer being a Pentecostal Christian refusing medicine and trusting in divine healing. She said that the whole of Ching-tung is very unfriendly to foreigners and they had tried to get her out. She has been here for three years but so far has no converts. We called on the official who was very civil to me but cold and indifferent to Mrs. Marston who came along to interpret for me. He said that via Ta-li 大理 was the best way to reach T'eng-yüeh and he suggested that we go that way; he would furnish 20 soldiers. We are said to be the third European party in three years that passed through Ching-tung. The official seems very anxious to know when the foreign post offices are to go, the wireless, etc. and as a whole would be very pleased to see every foreigner out of China. This place seems especially hostile to Europeans or Americans or white people in general. It has been my experience that officials are usually discorteous in towns where there are missionaries.

April 1, 1922
The chief official called on me today at the temple; he said it was impracticable to go to the top of Wu-liang-shan 無量山 as it is the hiding place of many robbers and he said we could not go there. Wu-liang-shan is a high mountain (11500 ft.) west of Ching-tung which is 3850 ft. above sea level. The word liang, the second syllable in the name of the mountain is here pronounced liao and it appears so on major Davies' map. The official was very civil and said the best way for us to proceed would be to go to Ta-li. I questioned him regarding the production of sugar cane. He said that more than 70,000 acres (were) are under cane cultivation and that more than 30,000 people were living in the Ching-tung valley, all cultivators, and that the whole district comprised 200,000 men and women.

In the afternoon we climbed to half way of the mountain back of our temple, where we had a lovely view of the town and from which vantage point I took a photo of the surrounding rice fields and the town.

April 3, 1922
All morning a terrible wind blew and the dust in the temple was unbearable. We packed our plants ready to mail but the post office had no customs declarations so they would not take them.

In the afternoon at 3 p.m. we had Chinese dinner at Mrs. Marston's to which the official was invited. It was a poor dinner but rather boring as the Chinese will not eat unless they are continuously urged. It was also liquorless and no Chinese will eat without drinking. As usual soup to begin with, then all kinds of dishes, followed by dessert and after that soup and meats again etc., rice at last. As soon as the official had gone we had some coffee and cake and made up for the deficiency of the dinner. In the evening we packed our trunks etc.

April 4, 1922
The soldiers, 10 in number were ready At 7:30 a.m. We left our Szechuan temple dedicated to Wan t'ien ch'uan chu tsung 萬天川主宗 etc. and went to say good bye to Mrs. Marston. We had our picture taken with her in front of her sitting room and then we said good bye (neg. 30160). The caravan had already come down and started ahead of us through the street. The horses' bells were jingling, all the trunks and boxes followed in procession, our carriers, horseboy, assistants, the Lao cook and myself and Baer on horseback.

All the people came out of their shops and everybody looked. Four soldiers were walking behind me and two in front. I wished Washington could have seen this triumphant exit from Ching-tung through the old city gate. We followed the hillsides above the paddy fields over the old rocky road, an officer walking ahead of the whole procession. We passed the stream near Ching-tung, which issues from a deep valley at the head of which Wu-liao shan is visible. It was very dark and hazy, the sky dark gray and gloomy. We passed a few villages and single houses on the hillsides. We saw a beautiful pink *Cassia* in flower near a village, there was also a *Ficus* tree covered with large white herons which flew about in great numbers. At ten a.m., we arrived at a village Tung-pao and there the soldiers informed us that they will return and others will take their place. The new soldiers were already waiting and we marched on without delay. We crossed a ridge and left the Black River to our right, thence crossed other well forested ridges. The country looks very wild, pine trees are the predominating trees. Ahead is a peculiar mountain, the south slopes giving it the appearance of a perfect cone; it is quite separate from the surrounding mountain ranges, and as we approached closer I recognized it as being limestone. This is an ancient volcano with super-imposed limestone; it is known as Hua Shan 花山 or Flower Mountain. At the base of this limestone mountain winds the Black River practically encircling the mountain and flowing S-shaped to the right. The slopes of the limestone mountain are very steep and on its southern face a trail leads zigzag to the rocky top. The black rocks, huge boulders jut out prominently from the red soil which otherwise covers the slopes. We stopped for lunch on the southeast side of the Black River opposite the mountain, we crossed the foot of the mountain which must be about 5,000 feet elevation or perhaps 5,500 feet. In the crevices of the huge limestone rocks near the base of the mountain, I could see where the lava had squeezed through. From the overhanging cliffs the jutted out huge boulders having the appearance of stalagtites which formed in the process of dropping and remained there, quite free from the wall, only attached at the upper part. The trees around here are mostly *Cedrela toona.* I took two photos, one of the cone and one showing the canvas. Other trees found plentifully are *Alnus, Vaccinium, Schima, Pyrus pashia, Rubus flavus,* a candelabra-like *Euphorbia*, pines, etc. On the eastern side of the limestone mountain its whole appearance changed. There is no cone visible but the whole mountain extends into a range with three distinct divisions, through the last the Black River issues. The rocks are huge black boulders deeply fissured. The village of Hui-yao is on the opposite side. We cross a stream coming from a valley to the left and ascend a ridge. High mountains are to the left. We found *Pyrus pashia*, several species of oaks, the wooly one and the one with the light green leaves and oblong acorns; pines are of course very common. To the right is a deep valley with the Black River flowing at the bottom some 1,500 feet below. The trees are mostly *Phyllanthus emblica*, *Schima*, *Bauhinia variegata*, etc. From the top of the ridge about 5,000 feet, we descend into a valley and village called Pan-ch'iao 板橋. The soldiers intended to stop there for the night but as it was still very early I decided to go on. We crossed the stream and climbed to the top of another ridge amongst pines, *Abies, Schima, Prunus, Pyrus* and oaks. The famous Wu-liao shan, 11,500 feet high looms up high above us to the west; it is a continuous mountain range of which one point projects like a cone into the sky, the other peaks are less high.

Dark haze and mist hangs over the high mountains and the deep mysterious valley to the left. Range upon range tower into the air overtopped by Wu-liao shan, all the hills are pine clad. A deep gorge is to our right, with the Black River some 1,500 feet below.

We are camping among these high mountains on the top of a beautiful ridge with pines, *Keteleeria*, oaks, *Prunus*, etc. Near as small village called Cheng-yen-shao 正言少. The air is lovely, no mosquitos. We went to the top of a hill just ahead of our camping place, it overlooks the deep valley ahead. On the top then hill is surrounded by an earth-wall with a trench within and without. It looks like a stronghold against the invasion of robbers. The road passes below the hill.

It is getting late and I must quit, tomorrow we must reach Chu-kai 豬街.

April 6, 1922

We rose early this morning; although we were at about 5,500 feet elevation, mosquitos visited our tent during the night. Two horses came late and one had to have its load carried, it could walk no more. A very stony trail led to a gap and then down a narrow valley. The hill to left (west side) was one mass of large limestone rocks or boulders full of large smooth holes. Huge black crags, overgrown with *Ficus*, jutted out into the sky line. Enormous boulders covered the lower hillsides near the trail. We descended into this narrow valley along wheat fields with here and there a few houses. At the end of the limestone ridge there is a circular hill of volcanic origin. The ridge is geologically different from the other surrounding hills and country. All the way however, here and there to Chu-kai or Pig Market, are outcroppings of volcanic and limestone ranges on the east and west side of the Black River. We cross the stream at the bottom of the narrow valley, turn to the right and follow the creek bed of another valley for a short distance and ascend the ridge among *Keteleeria*, *Pinus*, etc. It leads through a gap, back into the same valley which makes a large bend. From the top we can see the Black River below. The ridges are high and steep and wooded as are the surrounding mountains. We find *Keteleeria, Pinus* in preponderance, two species of *Quercus, Castanopsis, Engelhardtia* the small pubescent species, *Schima, Ficus* large trees, Mai Yom trees, *Vaccinium* and the fragrant *Rhododendron.* Ahead, a few li distant, is the village of Lung-kai 龍街, pronounced Lung-gai or Dragon Market.

There is a two span bridge across the Black River. The soldiers stopped for breakfast there and while they were eating, I attended to my plants. We sat down at a small house and were soon surrounded by the usual crowd. One man, very civil, asked if we were missionaries and produced a calling card of a Rev. Mr. Hanna. He said he was a Christian; he brought us tea and was very polite and gave me the Chinese name of many of the plants collected. We left Lung-kai following the Black River in the rather narrow valley with its very high mountains. The trail is not so bad but occasionally becomes paved near villages or houses and then is almost impassable, especially for pack animals.

We stopped for lunch under some trees near the banks of the river whose waters are very clear now, in contrast to the water in its lower course near Hsin-fu where it is blackish muddy. We cross several streambeds and then follow again the lower slopes of the eastern range; the river makes a sharp bend flowing through a defile. Black crags, steep and fissured, extend into the sky line, huge blocks of fissured rocks are allover the hillside extending to the water's edge. High mountains are to the left covered with pines, one range beyond the other; here the valley becomes so narrow that the trail leads to the top of the eastern ridge, a steep ascent at first through pinewoods, we climb higher and higher, then follow the western slope of this ridge separating us from the Black River to the very top where we cross it through the gap. To the left are high mountains but their peaks are enshrouded in haze. We pass by poor wheat fields, the plants not higher than a foot and a half at the most and the culms about a foot apart. The trail leads us to a pine covered peak where we intended pitching our camp, but as there was no food for the horses we moved on to a small plateau on another hill, farther on, overlooking the deep gorge through which the Black River flows. The gorge is so narrow and deep that the river remains hidden. On Davies's map the valley appears broad. The surrounding mountains are about 7,000 feet in height. The trail continues on the right hand eastern mountain range. We can see Chu-kai 豬街 or Pig market deep down in the valley ahead. Later, after the sun had set, part of our horse caravan appeared after we had feared that we would have to sleep out in the open. Everybody seems to be played out, the horses cannot keep up, they are incapacitated; the men have either boils on their backs or feet, one has a tooth ache, the other sore eyes and so on. We are a travelling hospital. The soldiers came tonight each with a different ailment and wanted to be treated, some asking medicines for their family at home.

The moon (first quarter) shines placidly over our camp with the pine trees, with the deep gorge below us. A fire rages on the steep slopes of a mountain in the southwest.

We are camping at about 5,800 feet; the mountain opposite us is much higher and is cultivated in the upper third to the top, the lower slopes being densely wooded with evergreen forest; narrow trails lead to the very steep hillsides.

April 6, 1922

We left our camp on the top of the hill over 6,000 feet high, early. The night was cold and a strong cold wind blew up from the deep gorge below us through which the Black River winds its way. We descended for a short distance, skirting the hillside through forest of *Engelhardtia* good sized trees with gnarled trunks, *Pyrus pashia*, plum trees, the red cherry but small fruited and the white flowered *Memecylon* which turns yellow on drying. *Jasminum, Smilax,* the stipuled variety with large leaves, and a number of other climbers, *Pinus* and the *Keteleeria* make up the vegetation; there are narrow deep ravines with a tropical vegetation of *Musa* and Zingiberaceae as well as ferns. The trail leads higher skirting the mountain, it is very rocky and narrow, cut into the nearly vertical cliffs at the foot of which the Black River surges some 1,500 feet below. Ahead of us the valley broadens a little and to the left a broad and cultivated, more or less gently sloping hillside extends to the Black River. We turn in the direction of the Black River (now) and at our feet lies the villages of Chu-kai 鼠街 here called the Rat Town or Rat Market, while on Chinese maps it is called Pig Market. The word chu can mean either. The village is divided one part is below the Black River, the other on the slope some 200 feet higher up on the east range. We stopped at the lower village where we fed our poor horses with peas. It was here that I discharged the opium sodden Lao Wang the horse boy.

The caravan was long time in coming, three horses never came at all and the baggage was being carried up by the men. We did not want to get separated too much from our caravan and so we decided to stop at Chu-kai and buy or rent horses to enable us to proceed. We had already started on our way with a new lot of soldiers with a very nice leader who presented his calling card, when we returned and took possession of a temple with a school, in the former is a tablet to the great teacher Confucius. The people here are exceedingly kind and helped us in every way.

This afternoon came a letter from Nan-chien, brought by four soldiers, welcoming us to Nan-chien 南澗 territory. They presented it and saluted. We are being shown more attention now than was the case at our arrival in Ching-tung.

Tomorrow we leave early for Mao-kai with two new mules rented for for one dollar per day for each animal.

April 7, 1922

Left Chu-kai early this morning. The elevation of lower Chu-kai which a month or so ago was visited by a fire destroying 50 houses, is 4,800 feet. At the school house where we stopped the elevation was 5,000 feet. The Black River winds snake-like through the narrow valley; high mountains tower to the left, a continuation of the Wu-liang shan (range). The trail winds along the mountain side. The vegetation is about the same as met with the day before. We found several cherry trees some loaded with fruits of which we collected much seed. I also photographed a branch pinned to the trunk of a tree.

About 10 o'clock we reached a village called Chi-ma-kai. The trail was very good and on a level, except for a few ups and downs and few rocks, except near the village. From here the trail leads gently down and then ascends rather steeply at least 1,200 feet. The vegetation is scrub pine, *Schima, Pyrus pashia* and *Prunus*; wheat and the poppy are mainly cultivated. The hill we ascend is all slate and the slabs are lying about and project from anywhere. At the top I found for the first time the 5-needled *Pinus armandi*, it is of a bluish green color, the needles are thinner and the bark is smooth, the cones are much larger than those of the common pine found everywhere in southern Yünnan (2994 – *Pinus armandi*, the first collected in this region). The trail ascends to about 7,000 feet. From here on, the blue pine as I shall call it, is very common and forms groves covering the upper hillsides to 8,000 feet together with *Keteleeria*, a beautiful large white-flowered *Rhododendron* (3007 – *Rhododendron dendricola* Hutch.) with *Spyraea* grew on the edge of a gulch. *Alnus* forms extensive groves here; a *Lobelia* (white flowered same as collected in Keng-tung and Szu-mao) reaches here a height of over 8 feet. The young plants reminding very much of *Trematolobelia macrostachys* of Hawaii. *Vaccinium sprengelii* is common with another white-flowered Ericaceous shrub (3006 – *Pieris ovalifolia* Don.) also *Rubus flavus* with very stout branches, *Osteomeles* (2995 – *Osteomeles schwerinae* Schneid.) a shrub with white flowers is common, also *Oxyspora paniculata* DC, no. 2997, *Pyrus pashia*, the white flowered *Memecylon* which turns yellow on drying and *Ligustrum* shrubs with fragrant flowers and *Docynia* (3004 – *Docynia delavayi* (Fr.) Schneid.) trees. Near houses walnut trees are planted and also persimmons (3000 – *Diospyros lotus* Linn.). No oaks nor chestnuts are to be seen. Higher up we find *Lycopodium*, a *Rhododendron* with small pink flowers (3008 – *Rhododendron microphyton* Fr.), and a tree *Rhododendron* with deep red flowers (3012a – *Rhododendron delavayi* Fr.) and *Rhododendron pachypodum* Balf. f. S., no. 3012. This is said to have yellow flowers while the white one is also common among the rocks and pines. Large areas are here cultivated with the opium poppy, the women being busy in the fields making longitudinal incisions in the green fruit. Wheat is also cultivated but the poppy occupies larger areas much to the detriment of the population. It was a glorious morning, a cool breeze blowing through the pines and my feet felt light, it was easy to walk even over steep places. The trail descended slightly and a little lower we came across large beautiful trees with a dense round crown, by far the finest trees we have seen. I took a photo of the 5-needled pine with the head soldier at the foot of the tree. *Trachycarpus fortunei*, the Chinese fan palm was planted everywhere along the fields and near houses, the fibrous bast around the leaf-bases being used in the manufacture of raincoats. Lower down a large white-flowered single rose (3015 – *Rosa odorata* var. *gigantea* (Call.) Rehd. & W.) formed rambling shrubs flowering profusely. The hills are high on both sides of the valley, the main ridge, a continuation of the Wu-liang shan (range) appears to be about 10,000 feet in height judging from the other nearer mountains. The trail descends again to the Black River, follows in its streambed to a point where the Shih-tung-ssu Ho or Rock-cave Temple River flows into it. The Black River makes a bend to the left, the Shih-tung-ssu Ho 石洞寺河 coming from the right. The hill in the center is a sheer cliff of peculiar limestone rocks with caves. Two of these caves have a façade the entrance to the temple. The trail leads to the village of Shih-tung-ssu 石洞寺, the Rock cave temple over a bridge across the river of the same name. On the left side there are two temples built over the entrance of two caves, one being a more elaborate one with a second story. There are three divisions in this temple, the central part, a dirty black hole, where the caretaker sleeps, in the left one sits a fierce looking god with a black beard, an inscription says: Ts'ai shen tien 財神殿 – The god of wealth hall; the one on the right harbors several female gods but male ones are along the side walls. There are several small caves in the rock wall opposite, but do not amount to much. The houses of the village are built immediately under the cliff. The trail winds along the bluff under the temples and leads to the top of the ridge to the village of Mao-kai 貓街 or Cat market where we are located in the schoolhouse. We were beseeched for medicines of all kinds, heart trouble, itch, rheumatism, swollen faces, boils, eye trouble, etc. It is bed time and I got rid of all the patients.

For lunch we stopped at Niu-kai 牛街 or Cattle Market, a filthy hovel full of flies so that I had to go outside and drink my cocoa in the gale that was blowing.

April 8, 1922

We left Mao-kai (Cat Market) at 7 a.m. The mules and their drivers had run away the night before, that is those we hired at Chu-kai, evidently they had enough of high mountain travel. Animals were not obtainable in Mao-kai so we had to hire carriers, who were Lolos. There are quite a number of Lolos in Chu-kai by the way.

From the school house where we stopped the trail leads through the village – everybody being out to gaze at us and see us off – and then steeply over a very rocky road to near the top of the hill behind the dirty town. It is quite cool, a lovely breeze is blowing. The rocky hill is covered sparingly with scrub pine which grows prostrate here, also *Keteleeria, Rhododendron,* the blue-green 5-needled pine of which there are large groves on the top; *Pyrus pashia, Docynia*, etc. are common. At about 6,800 feet elevation we found Magnolias, oaks, one a tall tree with a straight trunk, round crown, and brownish leaves. It began to rain, and a cold wind blew which made an overcoat quite necessary. We crossed a bridge to the right leaving the Black River way in the left valley below. Here the Black River turns to the left and makes a big bend to the right. We descend again to the Black River but do not cross until we reach Hsin-kai 新街, New Market. On the hillsides Rhododendrons were all in bloom also pink and white ones, quite fragrant, among the pines. In the cold breeze they looked so fresh and pure as the air itself. *Rubus flavus* was very common, many pears, *Populus* a small species and scrub oaks; and among all this lovely vegetation were lonely graves. We descend and to the left we see a village called Hsin-kai. Here we cross the river for the last time; it is here a small little creek the very head waters of the river; a short log, a few feet long, is laid across it. The banks are so densely lined with a narrow-leaved willow that it gives a fine outline of the windings of this river which we are glad to see no more.

We are soon to pass into dry scrub jungle, no more 5-needled pines, only the common 3-needle species. The trail leads among wheat fields to the village A-k'o-t'ang 阿克塘, only a few li distant. The atmosphere is here quite different, the air is clear and the dry jungle mentioned above commences. Small pines, cover the many small hills for the country is very undulated, no high hills in the immediate vicinity. The soil is very gravelly and the whole country is weird and wild, no more houses or villages till we come to our noon stop.

We enter hills with a little denser vegetation but only scrub pines, *Rhododendron*, a Theaceous shrub (3045 – *Thea pitardii* (C. Stuart) Rehd.) with whitish pink flowers the size of a dollar, called Yu-kuo hua. Also a white *Rhododendron* with pubescent pedicels (3043 – *Rhododendron decorum* Fr.) it is called Niu-chin-hua. We arrived at our noon stop called Mieh-pa-shao 篾笆哨 named for a bamboo which grows here. This is a very poor village, nothing could be had only large beans for horse feed, which my yellow horse refused to eat. From here, where wheat is so poor that one can easily count the individual culms and leaving the myriads of flies behind, we cross over rocky slopes; nothing but rock, rock, rock. Among them grew, beautiful *Andromeda*, two species of oaks one a very wooly one, *Schima*, 3-needle pines, *Hypericum* (3030 – *Hypericum patulum* Thunb.), etc. We are winding through gulches and over rocky hills, oaks predominating. A yellow *Jasminum* (3026 – *Jasminum diversifolium* var. *glabricosum* (W. W. Sm.) Kobuski) with tubular flowers became very common. We cut through a gap across a ridge. Ahead is a very dry valley with bare hills and scattered pines. Way ahead, at the foot of a dry range lies Nan-chien. We follow this dry valley the lower slopes of which are deeply eroded due to deforestation. Scrub pines, an occasional *Ficus* is all that there is. The hills are bare and cut up into weird canyons, exposing red soil everywhere. We descend to the river bed, stony and without a drop of water in it, it is the Wu-chieh Ho. We follow in the bed of this stream for a distance, then turn to the right and up over an eroded hill. The trail is dangerous. We follow another streambed higher than the surrounding fields, the river being dammed in artificially on both sides. The Nan-chien 南澗 valley is here very broad and in the rainy season the whole valley must be one lake. We cross the bluish waters of the Nan-chien Ho, which is here about a quarter of a mile wide, but with very little water at this time of the year. It flows northeastwards at the foot of Nan-chien nestled on the hillside, and then enters a dry gorge ahead. The town of Nan-chien is on the north bank. We enter the town and at once are accompanied by all the youth of the place, a soldier leads us to the top of the town to a large and commodious Confucius temple, which, before our arrival, was as quiet as a graveyard, but is now teaming with curious people. We found the common coarse bean cultivated in the plain drying in the temple ground. The soldiers went out to buy chickens, eggs, have water carried up the hill and firewood. I sent my card to the official who called at the temple about an hour later, he is a very friendly and civil person quite young and pleasant. We decided to stop for a day to rest our horses. We made 80 li today and 70 the day before, quite long stages considering the high hills we crossed.

April 9, 1922
The morning we called on the official (magistrate) in the Yamen. He received us courteously, served the usual tea, etc. Today was market and crowds were on the open spaces outside of the town, peas, beans,

corn, and sugar cane were the only articles of which there were quantities, pigs wallowed in the large stinking pools surrounding the town, myriads of flies hovered and hummed about quagmires, Chinese dishes with food and sugar cane. Pigs and the dirty-faced children encrusted with filth and skin diseases cavourted about.

There is no fruit cultivated and for the last three years there has been very little crop. Drought and floods cause a great deal of damage in this bare poor country.

I climbed the hill back of the Confucius temple and took two views of the town of Nan-chien, and two into or across the Meng-hua 蒙化 valley where the Meng-hua Ho flows into the Nan-chien Ho. Both forming the Red River which flows by Hanoi in Indo-China.

Near our temple there is another smaller one with funny gods in them, at the gate way on both sides are two identical figures holding a horse, which I photographed. In another one there were large stone tablets some 8-10 feet high, others smaller, the largest dated back to the first moon of the first year of K'ang-hsi 康熙 1662 A.D. admonishing people as to learning and religious culture. A smaller one dated back to Tao-kuang 1821 A.D. and gave a description of the districts. This stone would be of great interest. Another dates back to the 30th year of Ch'ien-lung 乾隆 1766 A.D.

April 10, 1922
We left our temple early, but before leaving the town of Nan-chien, I called on the school as promised the night before. They served tea and after a short written conversation I left, going by our temple down into the valley of the Red River or Meng-hua Ho as the Chinese call it. We passed a small village in the bottom of the valley and then climbed to the top of the hill near by. The road was exceedingly bad and rocky, passing many eroded places. We are skirting the hillsides and climbing to an elevation of about 6,000 feet. Vegetation mostly pines, three species of oaks (3076 – *Castanopsis delavayi* Fr., and 3093 – *Quercus delavayi* Fr.), *Engelhardtia,* a little lower we found a pistache tree (3072 – *Pistacia chinensis* Bunge) with fine leaves. This tree is very common all around Nan-chien 南澗. Among the hills and forests there are some terraced fields where beans, the poppy and wheat are cultivated. There are numerous trees, of *Malus, Pyrus pashia*, white Rhododendron (3081a – *Rhododendron decorum* Fr.), *Vaccinium* (3075 – *Vaccinium fragile* Fr.), small and large species, yellow *Jasminum,* with long tubular flowers (3026 – *Jasminum*

diversifolium var. *glabricymosum* (W. W. Sm.) Kobuski) and *Hypericum* (3077 – *Hypericum patulum* Thunb.). the trail continues at a high elevation skirting the hillside till we reached a small hamlet called P'ing-an-shao 平安哨 or The Outpost of Peace. The trail follows a small side valley leaving the Red River to the right. Red Rhododendrons (3081 – *Rhododendron delavayi* Fr.), *Abelia*-like bushes (3085 – *Leycesteria formosa* Wall. var. *stenosepala* Rehd.) with green fleshy stems and a *Pteris* fern of the *Pteris cretica* type was quite common. Here and there an *Alnus* tree was met with. We passed through a small gap over a spur and found ourselves again in the main Red River valley. We passed the little village of Ta-p'ing-ti 大平地 where we should have stopped for lunch, but went on to a temple at the suggestion of the soldiers. A side trail led up hill to a beautiful grove of *Cupressus* trees under which the little temple of Pai-yin-ssu 白銀寺 or White Silver Temple was hidden. We sat down under these lovely trees and there, on the hillside, with a cool breeze blowing, enjoyed a rest after a strenuous hill climbing. We can oversee the valley of the Meng-hua Ho, with a village spread out over the hillsides on the opposite side of the valley visible from between the tree trunks, it is the village of Ko-chiu-mo 各舊麼. After lunch we descended from the lovely temple grounds to the dirty village of Hsiao-li-tse 小李澤 or Small Plum Marsh. We had hardly arrived when it began to rain. They opened the school house and wanted us to stop there for the night but as we had only gone 45 li, I insisted to continue as otherwise we would have to go 80 li tomorrow to Meng-hua. The trail ascends again to over 7,000 feet, pines and oaks form the only vegetation here with rhododendrons. We follow the Red River valley, skirting the hillside again, about 1,000 feet above the river bed. The trail descends again, crosses a stream, then ascends steeply for a short distance and we find ourselves at the village of Mei-hai-ti 梅海底. We have pitched our tent and I am about to go to bed after having treated most of the soldiers. It is quite cold and the full moon is looking down on this peaceful hamlet and our camp.

April 11, 1922

Left our camp on the paddy fields at about 7:30 a.m. The night and morning was cold and several blankets were necessary. We followed nearly all day along the Red River or Meng-hua Ho as the Chinese call it here. It has two other names one is Li-she Chiang 禮社江 and the lower end is Hung Ho or Red River. There are two trails from

Nan-chien one the Ta-lu 大路 which follows over the mountains, the other the river in the bottom of the valley.

Bare hills encircle the river, here and there scrub pines are visible and scrub oaks, *Osteomeles*, roses (3094 – *Rosa multiflora* Thunb. var. *legitima* Regel) and the beautiful dainty pink-lilac leguminous shrub (3084 – *Campylotropis polyantha* (Fr.) Schindl.), the latter being very common. The river is clear but full of green algae. The trail is rather bad only rocks and rocks, the lower trail is much better, but too narrow for horses and is used only by pedestrians. On the top of one of the hills is a pagoda, the sign of a larger town in the neighborhood. The trail crosses a bridge at one end of which is the temple, a nice place for a lunch stop. On the other side not far distant is the village of Hsi-tsao-t'ang 洗藻塘.The village consists of only one street and after passing through it we enter the plain of Meng-hua with its beautiful green fields; beans, wheat and opium poppy being the principal crop. To the left, nestled against the bare hills, is the village of San-ku-p'ang 三古旁. Ahead in the middle of the plain there is another village called Tien-wei-ch'ang 殿尾廠, and beyond another nine-storeyed pagoda is visible.

We are stopping on a little knoll in the shade of some pistache tree for lunch. After a quick repast, we follow the trail which leads on the eastern mountain slope along the river bed. We pass the big pagoda with large *Opuntia* cactus growing on the top of it; this pagoda is about 3 li from the town of Meng-hua which is situated in the center of the valley or rather plain, surrounded by a big wall, we can see the roof of some high building projecting.

We cross an absolutely dry but very rocky streambed. An iron chain bridge spans it but we did not use it. Outside the wall there are a few buildings and past them the main gate about 25 feet thick of solid stone; the streets are broader, the houses appear lower. In the center there is a tower-like structure. We turn to the left and pass the military quarters where we are given a military salute and are then escorted to the Wu Miao 武廟. Crowds are again gathering and making themselves a nuisance. It is supper time and I am going to get rid of them.

Most of the carriers we met on the way were Lolos, all have huge goiters and many are idiots. They carry a board over their shoulders, a sort of yoke part which the neck fits, a strap goes over the forehead and thus they carry their load over the head as well as on their backs. They are as filthy and dirty as our Li-su carriers.

All is quiet in the temple courtyard, the full moon is just behind the cypress tree to the left. Our soldiers, a nice lot especially one of strong build and joyful manner, are gone, so has the troublesome crowd and I hope to leave tomorrow. Robbers are said to be on the road but the soldiers are to protect us. Elevation of Meng-hua 5950 ft.

April 12, 1922
Left Meng-hua t'ing 蒙化廳 early. We passed through several gates in the inner city and then through a small gate out into the open. The north gate is the small one while the south gate is about 25 feet thick. The plain broadens out, to the left are beautiful terraced wheat fields, also beans and poppy fields. One li from the city is another village through which we pass, while to our left is a huge graveyard the population of Meng-hua of bygone days. The morning air is lovely and cool, it is a real pleasure to be out in the open again away from the dirty towns with their myriads of vermin and flies. The name of the village we passed near Meng-hua t'ing is Chiao-ch'ang 教場, here we met with a beautiful tree with clusters of pink flowers, it is *Catalpa duclouxii* Dode, 3097, called in Chinese *ch'iu-mu-shu* 秋木樹. It reaches quite a large size some 75 to 80 feet in height and 3 feet in diameter. We pass through many beautiful fields mostly poppy, the large white and purple flowers gave the landscape a beautiful mantle of color. About 3 li distant we pass another village called Chi-ma chuang 紀馬庄. The hills on both sides are bare and the broad plain is cultivated to the very foot of the hills. The road is lined with white flowering rose bushes and here and there large pistache trees (3072 – *Pistacia chinensis* Bunge) called Huang-lien-t'ou 黃練頭; another tree quite common (2998 – *Celtis* cf. *sinensis* Pers.) is called *Sha-t'ang-kuo* 沙糖果. To our right is a large village set in a field of white poppies called Ku-ch'i-ch'ang. The whole plain is one large field of poppy; to the left, opposite from Ku-ch'i-ch'ang 顧旗廠 is another village called Tsung-ch'i-ch'ang 宗旗廠. The villages are too numerous to record; in all there are 557 villages on the whole plain. The scenery is rather monotonous nothing to collect. A rose grows along the trail from village to village. The next village we passed is called Miao-kai 廟街, the next Meng-shih 盟石 is on Davies's map. We lunched on the road near a little ditch flowing by poppy fields. After a quick repast we left again following the trail through a sea of poppies to the village of Tien-chung-kai 甸中街 , thence to Ta-ts'ang 大倉 which seems to be quite a village. Market is

on and many hill people, mostly Lolos have brought fire wood, and other hill products to the market; flies are numerous. We stopped at a school in the center of the town. The reception was a cool one so I left and the head soldier, who in the meantime had come up with Baer, sent my card to the village magistrate who opened the girls school for us which was unoccupied and so we had a very quiet place to rest.

April 13, 1922
We left our girls school at Ta-ts'ang early following the rocky streets into the open; nearby is another small village and ahead loom up the Tali mountains called Tien-ts'ang Shan 點蒼山[5] Before us is a ridge partly bare and partly terraced which we are about to ascend. We cross the Hsün-chien Ho 巡檢河 and follow between poppy fields into Hsün-chien 巡檢. This village is rather small and is situated at the foot of the hills which separate the Meng-hua plain from Hsia-kuan, its elevation is 6250 ft. Near the foot of the hills on the other side of Hsün-chien we cross the Wen-yao Ho, a now dry streambed, but evidently a wild torrent in the rainy season. From here we ascend the foothills past the village of Ta-p'ing-ti 大坪地 into open fields of poppy and wheat. To the right are terraced hills with more poppy. We cross now another small rocky streambed issuing from a gorge. The inner slopes of the latter are covered with Rhododendron. To my left is a bleached skeleton of a man, a robber shot and left to rot. On the slopes to the right is the village of Wen-yao. The hills are bare and sparingly covered with small Rhododendrons and *Thea* (3105 – *Thea pitardii* (C. Stuart) Rehd.), scrub pines, *Malus, Pyrus*, etc. We ascend gradually over the otherwise bare hills. To the right a small gorge much eroded and exposing limestone formation. The range of which encloses the Meng-hua plain to the west is called Lung-yü-t'u Shan 龍于圖山. The mountains are very rocky and dreary, we met many salt caravans carrying grey dirty cilindrical blocks flat on one side, rounded at the other. We continue to climb the bare hills, the soldiers being constantly on the watch, keeping a sharp eye out for robbers for this is a favorite hunting ground of theirs. On the other side of the pass lies the commercial city of Hsia-kuan 下關 where most of the business in the west of Yünnan is transacted. A band of 100 robbers are roving over these bare rocky mountains doing much mischief. We

---

5 For description see *The ancient Na-khi kingdom of S. W. China*, p. 27, note 25.

reach a flat space near a creek, where the wind howls in gusts. There are some tea sheds, dirty and filthy; a short distance away is a military camp or fort surrounded by a wall which can almost be jumped; the place is called Wa-fang-shao 瓦房哨. Here we stopped for lunch under a shed built of brush wood. After a quick repast I visited the military camp. After passing the wall one enters another gate leading into the central building. To the left is a dingy room full of smoke. Soldiers are lying about smoking opium, others (ours) are cooking their noon meal in an iron cauldron. To the right is a small room where the sargent lives. Here also soldiers are lying about smoking opium, it is a dingy dirty place. The entire fort is badly situated in a hollow, from a small hill about 20 feet away it can be covered with rifle fire.

By the way, half way up to the fort, we found a few soldiers near a fireplace among the bushes on the lookout for bandits.

A better class of Chinese riding nice fat sleek ponies, dressed in gay colors arrived at the tea sheds where they tied up their horses. While they were at lunch their horses were fed beans placed on the ground of which a courageous stray pig took advantage, kicking the horse with its snout and squealing finished the lunch intended for the horse.

We left for the high pass which leads over the divide called Chung-ho Shan 中和山. Higher and higher we climbed. Vegetation was found only in spots. From Wa-fang-shao the trail skirts the hillside with scattered *Pinus armandi* the five-needled pine, of which a few tall ones were left standing near the military camp. We ascend higher till we come to a very narrow gap circular in outline. In the center of this pass soldiers were guarding this strategic spot. Here the wind blows like at the Honolulu Pali. The hillsides are covered with gorgeous lilac *Rhododendron* (3131 – *Rhododendron rubiginosum* Fr., see neg. no. 30192) 5-6 feet in height. From the top of the gap 8,700 feet elevation a beautiful view is obtainable over the mountains and the Ta-li lake. Unfortunately it is hazy and photographing is out of the question. The trail leads zigzag down the east side of the Hsin-kuan range. There are many Ericaceae, one a shrub, a species of *Pieris* with dark foliage and greyish white flowers (3100 – *Pieris formosa* (Wall.) D. Don.) in erect terminal racemes, abounds, also *Rhododendron*, roses (3112 – *Rosa omeiensis* Rolfe. and 3113 – *Rosa sericea* subsp. *eusericea* var. *Mairei* (Leid.) Herr.) and many other species of plants. The descent is very steep and rocky; we follow a dry streambed full of rocks past a small village called Chi-wu-ts'un and shortly afterwards enter the city of Hsia-kuan 下關 or the Lower Pass. After crossing the stream of the same name by a stone bridge. This stream is the outlet of the Ta-li lake called Erh-hai 洱海 and flows between the Ta-li Ts'ang Shan (mountain range) and that separating Hsia-kuan from the Meng-hua plain. Hsia-kuan is situated on the lower gentle slopes of the big range, overtowered by a temple, a Wen-miao 文廟 with two large *Ficus* trees in front. While entering Hsia-kuan it began to rain. We ascended the steep slippery stone road passing a filthy market with millions of flies, to the other end of the town where there is a school. The soldiers thought it would be best to stay here. I went in but received such a hostile reception, that I left at once, sending in my card on which I had written that they were absolutely ignorant of the rules of propriety as taught by Confucius. We left in pouring rain up to the temple mentioned first where we found a peaceful and happy spot to rest after a long and hard journey over high mountains.

The wind blew furiously all night thunder and lightning played and rain descended in torrents. We stopped one day at Hsia-kuan in our temple fortress, visiting only once the filthy town.

The spur between Hsia-kuan and Wa-fang-shao is called Chung-ho-shan 中和山, the highest point of which is 10,000 ft. elevation.

April 15, 1922

We left our Wen-miao on the top of the hills situated among innumerable graves rather late as we expected to get to Ta-li 大理 about noon; the distance is about 8 miles or about 25 li. The road is broad, paved and passes through fields of wheat and poppy, with graves beyond up the hillside. The poppy fields adjoining the graves a fit setting – one step from the opium to the grave. Many are resting there who, had they not been slaves to that awful habit, might today enjoy God's sunshine, the beautiful air, the singing of the birds, the mountains and all that is grand in nature. We meet many people, the traffic is indeed heavy on this road; it might have been on account of the Ta-li fair which is held once a year in April. We passed a 13-stories pagoda on the left, the apex was in ruin. The hills were bare and brown, their lower slopes being covered with graves. Towards the lake the scenery is more friendly, sails are visible and boats are crossing over the placid waters. We met many troops, the 18th regiment on the march, with soldiers looking as if they had been hatched from a Japanese cannon ball. A perfect image of a Japanese even to the loops projecting from the Japanese boots (foreign pattern) with rubber sides. Hundreds of people were on the road wearing (the men only) large hats covered with blue, yes sky blue oiled cloth and long blue ribbons which are tied in front over the chest, some letting them float down to their knees. Straw hat sellers were found all along the road side, offering plain straw hats for sale, but some were ornamented with gaudy colors of pink, blue and yellow. We pass a village called Ya-ma-t'ang and not far distant is the famous temple of the goddess of mercy Kuan-yin ssu 觀音寺. The entrance to the temple ground is crowded with worshippers. Mules are tied in front, restaurants or food kitchens – a better term, are doing a good business. We enter into the courtyard where there is a marble shrine in the midst of a pond, full of gold fish. Here I took a picture of our three soldiers. In the rear is a large temple with large Buddhas, and to the left of it is, the same courtyard, a small room in which sits enthroned the famous general Yang 楊, the greatest benefactor of Ta-li, of bygone days. I took a photo of this well-carved image, said to be a perfect likeness of him. Not far from the temple we spied the gates of Ta-li, for there are several massive gateways, some 25 feet thick. We pass through them and arrive before an enclosure, where soldiers stand guard – this is the general's yamen. It was the old palace of the Tai rulers, of ancient days. We pass the famous bell tower, Wu-hua-lou 五華樓 and follow the long narrow main street. The soldier who met us at the gate guiding us to the west gate to the temple of Mi-t'o ssu 彌陀寺 where we made halt.

There was little collecting on the way; dry streambeds full of rocks we passed, with *Oxyria* growing among the rocks, the same (*Oxyria sinensis*) that covers the walls of the city.

I had my bed made under the very eyes of Dharma, the mosquito net being tied to his staff while Baer's bed was made under the vigilance of some hideous devil of indigo blue color which recalled Mahâkala (it actually is Mahâkala) of the Tibetans.

We called on Mr. Fergusson of the Salt Gabelle with whom we had lunch and later returned to our temple; there were many Buddhist monks praying; women came with offerings and the place was always full of smoke from the constant incense offerings. In the morning they placed tea in front of the gods for a minute, only to be taken away and to be drunk by the temple guards.

The fair is on in earnest and in the afternoon I went out with two soldiers through the west gate. There were 1,000sands of people, many Tibetans with medicines and old rubbish perfectly useless which they hoped to sell. Many temporary teashops and kitchens had been erected where hundreds of people were fed with fly-bespeckled food, while pigs and dogs catched the drippings from the bowls out of which people sucked their food. There was nothing sold of any interest only cheap goods and junk.

I took several photos, one from the Wu-miao where soldiers were quartered in case of trouble. After the fair which lasted one more day, Sunday the 16th, rain set in and it rained pretty constantly.

April 20, 1922

After lunch we started from Ta-li for the Tsang Shan Range, Mr. Fergusson, Gregory, Cook, Baer and myself and 20 coolies carrying tents, provision and botanical outfit. We left by the west gate over the grassy plain between graves and memorial tablets to the foot of the mountain range. Our objective for today being the temple at 9,000 feet elevation on Chung-ho Ssu 中和寺. The trail at first is not steep, although we followed a shortcut by which we shall reach the temple in an hour and a half.

The first plants observed were a pink flowered bushy *Rhododendron* (3118 – *Rhododendron serotinum* Hutch.), 3121 – *Elaeagnus umbellata* Thunb. with yellow flowers, near the temple, little blue-flowered 3164 – *Polygala sibirica* L., the common yellow-flowered deep-rooted perennial growing in clumps, 3162 – *Stellera Chamaejasme* Linn. *Pinus armandi* forms here small trees. A few Rhododendrons (3123 – *Rhododendron oleifolium* Fr.), willows (3122 – *Salix tenella* Schneid. var. *luctuosa* Lév.), 3166 – *Rhododendron triflorum* series with pale lilac flowers, *Leptodermis* (3161 – *Leptodermis carissa* shrubs and various herbs, as a fleshy *Euphorbia* (3125 – *Euphorbia chrysocoma glaucophylla* Lév.?) are about the most striking plants. The trail is a very rocky one until we reach the temple from where a wonderful view is obtained over the lake, the plain and city of Ta-li, the latter a perfect square situated in the center of the plain halfway between the mountain and the Erh-hai 洱海. We made ourselves comfortable in the upstairs room of the temple at the western wall of which was an altar with a wonderful painting of a Dragon, a veritable masterpiece of the old school. Incense and offerings were reposing before it.

Next morning we started early for the top of the mountain or rather for a place suitable to pitch our tents from which the summit could be reached. The trail leads along a ditch lined with flowers, while the hillsides were covered with the common *Rhododendron decorum*; *Clematis montana* climbed over the common yellow-flowered *Rhododendron trichocladum* Fr. (3137), a deciduous shrub which occupied the open hillsides among rocks; with it grew the rose-colored shrubby *Rhododendron cyanocarpum* Fr. (3129), while the gorgeous lilac colored *Rhododendron rubiginosum* Fr. (3131, neg. USDA 30192) was less common being found only as scattered individuals at the foot of boulders. *Vaccinium delavayi* Fr. (3156) a lovely compact shrub occurred only among rocks while on rocks grew with the delicate purple orchid *Pleione yünnanense* Rolfe (3119). The orange-flowered *Pleione forrestii* was rather rare. *Abies delavayi* occurred here and there on ledges and cliffs along the main gorge which I photographed. Cane-brake *Bambusa arundinacea* made climbing very difficult. It grew so densely to the exclusion of nearly everything else that one was absolutely buried in it. The trail became rather steep and very rocky. An occasional *Primula* grew under rock ledges with the rubiaceous purple-flowered *Morina betonicoides* Benth (3132). One of the most gorgeous flowered *Rhododendron* was here very common and was at its best at about 10,000 feet elevation, *Rhododendron neriiflorum* Fr. (3146) with dark red brilliant fleshy flowers. Spiny *Berberis* were common, and on rocky cliffs the bright yellow-flowered *Rhododendron sulphureum* Fr. (3142) testified to being a distinct rock plant. At about 11,000 feet cane-brake gave way to the somber forest of *Abies delavayi* Fr. (3148; neg. 30181) with its dark blackish green needles. The underbrush was here formed by *Rhododendron haematodes* Fr. (3149). In little depressions in this *Abies* forest we found our first snow on the Ta-li range. Occasionally we met with *Juniperus squamata* Buch, & Hamil. (3147, neg. 30178) prostrate et the foot of giant *Abies delavayi*. The trail skirted a deep ravine, its slopes were covered with the gorgeous pink-flowered *Rhododendron cyanocarpum* Fr. (3129) which forms here trees 20 feet in height. Above it *Abies delavayi* Fr. reigns supreme with a dense carpet of moss at its feet and its trunks clothed in lichens and moss. In this mossy damp carpet just recently released from a blanket of snow and ice, there flowered the beautiful *Primula sonchifolia* Franch. (3133) of which I took a photo (neg. no. 30180, USDA). Here in a lovely grove of *Abies* near a running brook, somewhat sheltered by a steep rockwall, we pitched our tents. The summit of the range seemed not far away and we hoped to reach it the following day. Our camp was at 12,000 feet elevation. To the south on the other side of the rocky streambed there were low hills taken possession of by species of *Rhododendron* which was however not in flower. A very small leaved bronze-colored *Rhododendron* formed regular cushion plants in the ice and snow.

We had hardly settled down when it began to snow, and snow it did all night. I looked out of my tent and saw the snow flakes dancing from a grey sky, the *Abies* (fir) trees were soon covered and so was the ground, with a blanket of white. As we feared that the snow would not abate and obliterate our trail, we decided to break camp, but four of us and two coolies decided to make for the top. We followed the creek bed in the snow among the small leaved prostrate *Rhododendron fastigiatum* Fr. (3152), the last *Rhododendron* in the snow. *Juniperus squamata* Buch & Hamil. (3150) continued till we reached a bowl-shaped depression filled with water slightly frozen on the surface. This lake is called Hei-lung-t'ang 黑龍塘 or Black Dragon Pool. The slopes were still covered with *Rhododendron fastigiatum* Fr. but blanketed with snow. Low prostrate Rhododendrons led steeply to the summit of the ridge, but as it was foggy, no view could be obtained of the western plain and slope of Ts'ang Shan .

As it continued to snow heavily we decided to return to Ta-li the same day. After taking some photos of the snow peaks and vegetation, we returned the same way we came, collecting on the way a few more Rhododendrons, *Piptanthus concolor* Craib (3139), *Indigofera mairei* Pamp. (3127), *Corylopsis yünnanensis* Diels (3128) and others. We reached Ta-li about 6 p.m. on April 22, 1922.

May 4, 1922

Mr. Chu 朱 of the Salt Gabelle, I and Boomah went out to see the San-t'a ssu 三塔寺, or three pagodas. It was a beautiful summer day rather warm for Ta-li. We left by the north gate following the main road which leads to Li-chiang. The distance to the pagodas is only 3 li or about 1 mile. According to the Chinese calendar the 4th of May is a very lucky one this year and especially so for tailors. Numerous people were gathered together in the various temples with musical instruments, singing and bowing and burning incense. Men in one temple, women in another, each trying to outsing the other. We visited the Kuan-yin shrine with the large image of the Goddess of

Mercy, cast of metal in a most crude fashion.

There were many temples in nearly all of which worship was going on.

May 6, 1922

We left Ta-li at 8 a.m., Mr. Cooke of the C. I. Mission accompanying us. We followed the middle road which is not paved, but is very comfortable, sandy and smooth. We selected the lake road but always some distance from the shore. There is nothing to record in the plant line, except alder trees, *Alnus ferdinandi-coburgi* Schneid. (3184) different from those found in the south which is *Alnus nepalensis*, a willow, *Salix variegata* Franch. (3180), and cultivated apricots, pears, plums and peaches were everywhere as were a good many *Albizzia* trees. We passed Shang-kuan or the Upper Pass 75 li from Ta-li. We did not stop in Shang-kuan 上關 which is a dirty place with only one long road, but went on to Sha-p'ing 沙坪 to the north of the lake. There we went to one of the temples near the lake shore but finding it full of beggars we went to a hsioh-t'ang 學堂 or school, a former temple, and spent the night.

The hills are mostly bare but in the deep ravines there seems to be plenty of growth.

May 7, 1922

We left Sha-p'ing very early as today's stage is to be 90 li. The trail follows round the last bend of the lake which is crescent or ear-shaped, hence the name Erh-hai. It still extends a little further north in the form of a lagoon around which the trail skirts and then climbs over a rocky slope covered with graves. The hills are bare and all eroded exposing red soil containing oxidized iron. There are a few pine trees scattered over the hillside. The trail leads through an orchard of pears and peaches as well as plums. *Berberis pruinosa* Fr. (3178) is much in evidence, as are Ericaceae, the white flowered *Rosa banksiae* Art. var. *normalis* Regel (3197), the small flowered *Bauhinia densiflora* Fr. (3189). I found in Nan-chien, *Sophora flavescens* Ait. (3207), a purple flowered legume *Campylotropis polyantha* (Fr.) Schindl. (3191), and a small fruited wild pear *Pyrus calleryana* Decne (3206). We are now on a high plain surrounded by bare eroded hills some of them over 12,000 feet in height. Ahead is a very fertile plain encircled by absolutely bare hills. Opium is cultivated everywhere; the heads are beginning to ripen and the farmers are gathering the exudation from the longitudinally-incised capsules with a sickle-shaped knife.

Ahead is the village of Teng-ch'uan 鄧川. We enter the town through a very imposing gate, follow the main street and find at the other end an agricultural school and garden. The plain is wonderfully fertile, the wheat fields are the finest I have seen in Yünnan. Buckwheat is also grown but not so frequently. The crop being opium. To the left of us is the famous canal, the outlet of the Lang-k'ung 浪穹 River which passes through a narrow gorge from the Lang-k'ung plain. The canal is lined with large trees, mostly *Albizzia, Pistacia chinensis* Bunge (3203) and *Paliurus orientalis* (Fr.) Hemsley (3194), etc. The hills are bare and the pass ahead particularly so, with the exception *Opuntia indica* common everywhere. *Alnus* and *Pistacia* are the most common trees. The trail passes over a sort of causeway between two lagoons in which a species of rush is cultivated. It is said to be used in the making of the famous Ta-li hats. Willow trees, *Salix cavaleriei* Leid. (3193) are lining the road which is here paved. Ahead is the village of Yu-so 右所. We pass more villages and follow along the canal. There is a regular sea of poppies which extend to a village at the foot of the hills buried in trees. The trail now turns to the right (east) and ascends the canal embankment following under the shade of the lovely trees mentioned above. We cross a bridge near a village called Chung-so which is situated on both sides of the canal. Here I took two photos one up the canal, north, and one down south. The trail follows on the top of the eastern embankment. We met many salt caravans with the dirty grey cylindrical salt blocks; also many tribespeople carrying loads of firewood on their backs supported by straps over their forehead. Further on as we near the actual gorge into which the canal merges, the stream is hemmed in by one embankment only, the left one being formed by the mountain.

The trees are again *Albizzia, Pistacia, Paliurus*, and *Capparis indica* which form here trees 20 feet in height. The gorge is indeed narrow and the river rushes through in torrents. The drop in the short distance through this wild and rocky gorge is considerable. The rockwalls are black and overgrown with the yellow-flowered *Opuntia indica*. The name of the river is Lang-k'ung Ho. After a few ascents and descents in the gorge we pass out into an undulating plain. The mountains to the right are bare, but those to the left in the distance are covered with scrub forest, the black growth on the summits being masses of *Rhododendron*. It is the Lang-k'ung Range. Mr. Chu tells me that the Lang-k'ung River is now known as Erh-yüan Ho 洱源河, the source of the Ta-li or Erh-hai 洱海 lake. We are camping along a streambed under some willow trees. *Photinia prionophylla* (Fr.) Schneid. (3205) shrubs are not uncommon as is a *Castanopsis* or *Quercus*, and a shrubby species with large, white bell-shaped flowers, *Styrax langkongensis* W. W. Sm. (3198). There are wild pears, *Osteomeles schwerinae* Schneid., the same pinkish-purple delicate *Campylotropis polyantha* (Fr.) Schindl. (3191) so common north of Nan-chien and also south of it in the Black River valley. It is hot but a strong wind makes the heat bearable.

We cross a hill much eroded; on the top is a pond or reservoir, the ground is boggy and a yellow *Caltha* is common as is the blue flowered *Antiotrema dunnianum* (Diels) H.-M. (3199) and other herbaceous flowering plants. The mountains to the right are black and steep, but wooded only at the very top. They must be 12,000 feet high. The village opposite where we camped was Ying-shan-p'u 應山舖. High up on the very steep slopes of the high mountain known as Fo-kuang-chai shan 佛光寨山 there is a temple or several monasteries called Ling-kuang-ssu 靈光寺. We now follow the well cultivated plain, wheat being the most abundant crop, while opium is very scarce. Roses line the path on both sides. We pass the villages of Ch'ang-ying 長營, Wen-pi-ts'un 文筆村, and finally the village of San-ying 三營, where we will stop for the night. We are comfortably housed in a military temple. It is bed time.

May 8, 1922

We left our temple in which police were also camping, at 7 o'clock a.m. San-ying is situated at the foot of a hill which is entirely limestone. On the hill there is a small pagoda visible from quite a distance. The hill is full of limestone boulders, and honeycombed with caves. Hot springs are bubbling up everywhere and near to the boiling waters flowing across the road. The hills around here seem to be volcanic with superimposed limestone. The trail or road skirts the other side of the hill and about 5 li beyond reach the village of Niu-kai 牛街 or Cattle Market. In Niu-kai there is a large hot spring. Poppy is very plentiful and is interspersed with wheat. The volcanic hill is called Huo-yen Shan 火焰山 or Flaming Mountain. On the road we met many Min-chia women, marching in single file, each one carrying thick, heavy coffin lids, each two planks, they are made of a species of Juniper, *Juniperus saltuaria* which comes from about two days of Niu-kai. We follow the valley and skirt the hillside. At the head of the valley there is another volcanic hill white with huge blocks of limestone in which large tablets have been chiseled. It is the Kuan-yin Shan 觀音山. There are many climbing *Rosa sericea* subsp. *bisericea* var. *mairiei* (Lév.) Herr. (3222) and *Trachelospermum donieri* (Leid.) Woods. (3186). The small leaved *Bauhinia densiflora* Fr. (3189), the beautiful purple-flowered *Campylotropis polyantha* (3224), *Broussonetia papyrifera*, a narrow leaved *Salix* etc. The trail skirts the hillside, then passes through the small village of Hsia-chai 下寨 about 20 li from Niu-kai. Here there are two roads, one leading straight to Chien-ch'uan 劍川, the other at right angle to Ho-ch'ing 鶴慶 over a high mountain range 11,000 ft. high. I decided to follow the Chen-ch'uan road as it leads through the valley, although the high mountain road would have been much more interesting, but as Boomah had malaria and the horses were tired I thought it wise to follow the valley. The country from now on became very hilly, the trail leading over undulating land up to about 8,000 feet elevation. These hills were covered with scrub oaks and here and there a tall oak tree with golden yellow entire leaves *Quercus semicarpifolia* (3227), *Pinus armandi* and *Pinus yünnanensis* were quite common also *Rhododendron delavayi* Fr. (3241) with scarlet flowers; *Rhododendron decorum* (3257) the common white flowered species, then *Ligustrum sempervirens* (Fr.) Mansf. (3258) were found on the hillside but not in flower. Roses were quite common especially *Rosa Banksiae* with small white flowers, a prolific bloomer. *Osteomeles schwerinae* with small white flowers, and *Pteris ovalifolia* with long horizontal racemes formed the scrub vegetation. We lunched near a small village called Ch'ou-shui-ching 臭水井 where I found a large rose (*Rosa odorata*) with large pink flowers with a touch of yellow.

The country becomes very hilly and is covered with pines, both *P. armandi* and P. *yünnanensis*. I came across a large tree of *Docynia delavayi* (Fr.) Schneid. (3215) which I photographed. *Rhododendron chartophyllum praecox* (3269) of the *Rhododendron triflorum* series, a beautiful shrub with pink, lilac, or mauve colored flowers and with the lower petal beautifully marked, wild plums, roses and scrub oaks associated on the rocky hillside. The soil is red in color and full of iron. There are a number of oak trees with ilex-like foliage (*Quercus semicarpifolia*). Among the shrubs of which *Pyracantha crenulata* (D. Don) Roem. (3202) with stiff branches and white flowers is the most common, grew *Clematis chrysocoma* Fr. (3248) with magenta-colored flower, also *Berberis pruinosa*. *Jasminum beesianum* Forrest & Diels (3216) with wine-colored flowers was also found entwined with scandent roses. We descend into a plain called Sha-pa or Sand Plain, and follow a rocky watercourse lined with willows. We pass the village of Yeh-ya-t'ang 野鴨塘 or Wild Duck Plain, and another small village called Yeh-chi-p'ing 野雞坪 or Pheasant Plateau. We had intended to stop there but as there was no decent place we went on to Tien-wei 甸尾. The Sha-pa Ho 沙壩河 empties into this sand plain which appears to be a regular sink hole and must have been formerly a lake. Yeh-chi-p'ing is situated in this plain. Crossing over some hills and passing through a small gorge we pass into another plain with a lake to the right, it is the Chien-ch'uan Hai 劍川海 or Chien Hu 劍湖. High mountains are straight across this valley or plain. The surrounding hills are covered with pines. We pass irrigated fields, follow a ditch on the banks of which grow Forgetmenots, *Myosotis sylvatica* (Elrh.) Hoffm. (3220) and the dark blue *Cynoglossum amabile* Stapf & Dramm. (3221). The village of Tien-wei is situated on this plain but on the slopes of a hill. I stopped in a small temple overlooking the plain and the lake, all alone, having a very quiet rest.

May 9, 1922

Left my temple passing through the village of Tien-wei and descending to the plain and over a fine bridge across the Tien-wei Ho to Shang-teng 上登, a small village. At the entrance to the village opposite the bridge is a temple with two consumptive-looking guardian gods showing every bone and vein in their body. They are of huge proportion and each is brandishing a sword. The temple is 3 li from Tien-wei. The plain is intensely cultivated. On the hillsides grow *Catalpa duclouxii* Dode (3217). The air was fine, cool and brazing, a lovely spring morning. As we turn right I got my first glimpse of the Li-chiang Snow Range. As seen from here, there are two distinct mountain masses, one a lower composed of sharp peaks, and one higher about 21,000 feet covered with snow for about 5,000 feet. In the distance there is another lonely snow mountain, the Yangtze flowing between it and the Li-chiang Range. The trail follows the edge of the plain skirting lower hills between fields of wheat, beans and peas. We pass the villages of Han-teng 漢登 and Hsi-chuang 西庄 a distance of 10 1/2 li from Tien-wei, then Chuch'üan-ch'ang 豬圈場 or Pig-Pen Floor and from there to Shui-chai 水寨. From this last village it is not far to Chien-ch'uan a large village. We pass through an imposing gateway and before us lies the walled city of Chien-ch'uan, elevation 7750 feet. Entering through a massive gate Main Street displays a dirty road with pigs, dirty children and myriads of flies. This town is known for its carpenters. We met hundreds of women carrying large primitive furniture, resembling wardrobe cases with drawers. These can be bought for one or two dollars. We followed the plain between irrigated rice fields and wheat fields to Pan-teng ho 板凳河, then to Yung-p'an ts'un 永盤村, to T'ai-p'ing-ts'un 太平村, to Liu-chia-teng 劉家登, to Tien-hsin-ch'iao 甸心橋, to Mei-tzu-shao 梅子哨, to Chiang-tung 江東. We had lunch among graves in a pine grove with *Rhododendron decorum.* The trail was lined with roses especially *Rosa Banksiae* and *Rosa moschata* subsp. *brunonii* (Lindl.) Herr. (3238) with viscous calyces and tomentose leaves. From Chiang-tung we went to Wen-mu-ho 文木河, then to Chiu-ho-kai 九河街 (elev. 8150 ft.) still following a plain. Straight ahead was the wonderful mountain mass of Li-chiang covered with snow and the top enshrouded in clouds. To our right is a limestone hill with huge black boulders. After passing Chiu-ho-kai, we keep on following the plain. To our left is a large village Nan-su-mei 南穌梅. On the hill on the right grow *Salvia digitaliodes* Diels (3235) and *Androsace spinulifera* Knuth. (3232) with small pink flowers in terminal capitate inflorescences.

We did not stop at Chiu-ho-kai 九河街, as there was no decent place and having been told that 5 li further was a fine temple, we went to Pei-su-mei 北穌梅. On arrival there we found an unswept smoky school full of dirt, and so we decided to stop at the theatrical stage. A ladder was procured and we climbed up to the stage where we are now comfortably located, the caravan below. I attended to the Ma-fu (horseboy) who had very sore eyes.

May 10, 1922

We left our theatrical stage early enough so that there was no audience at rising time. We followed the rather narrow plain from our last village Pei-su-mei to Tu-wo 杜娥, which is 5 li distant, both villages are situated on the east side of the valley. The hills are covered with pines and numerous walnut trees both wild and cultivated can be found along the road; young fruits are now forming although male inflorescences droop from every branch. There were also numerous pear trees, plums, peaches, etc. We passed another hamlet of a few houses called Wu-li-p'ai 五里牌. Along the road *Cotoneaster linifolia* and *Lonicera* were predominant, the former with pubescent linear leaves and stiff spiny branches. The rosaceous shrub *Prinsepia utilis* Royle (3281) with purple drupes, long spines and dark green foliage was exceedingly common and formed regular hedges. It is said that it is very pretty when in flower (December) when the long arched drooping branches are one mass of small white flowers. The seeds were collected by the village women all along the road, they express an oil from the rather large drupe which they use for cooking purposes hence the specific name *utilis*.

There are a few more villages like Kuan-shang 關上, five li from Wu-li-p'ai, then P'o-chio where we had lunch up a creek bed among pines and oaks. There I found *Primula begoniiformis* Petetitm. (3299), a microform of *P. obconica* Hance and a shrub belonging to the genus *Decaisnea*, *Decaisnea fargesii* Fr. (3301). From here we climbed up to about 8,500 feet elevation or nearly 9,000 feet, over a steep slope and a very rocky trail or rather paved road. *Pinus yünnanensis* was exceedingly common also *Rhododendron chartophyllum var. praecox* of which I photographed (neg. no. 30244) a lovely shrub. Below at the end of the valley we had a view of a beautiful green little lake, like an emerald gem surrounded by dark mountains, the Na-khi call it Bbu-t'u-nder. Along our trail were many plants as the beautiful white *Clematis montana* Buch.-Ham. (3284), the lovely *Buddleia heliophila* W. W. Sm. (3298), a purple leguminous shrub, a fine cherry *Prunus conadenia* Koehne (3286) and lovely *Rhododendron delavayi* and *Rhododendron decorum*. On the road below we found *Jasminum humile* L. (3293) and *Chionanthus retusa* Lindl. & Paxt. (3294) with white flowers. Scrub oaks covered the hillside, mainly *Quercus semecarpifolia* Sm. var. *glabra* (Fr.) H.-H. (3291), one of its numerous forms. On the top of the hill there is a broad plateau somewhat undulating, the soil is a yellow clay but the whole plateau is of limestone origin, there are numerous sink holes called Lo-shui-tung 落水洞 (dropping water hole). The plain is cultivated down to the margin of the deep holes (sunken basins). Limestone outcroppings are numerous and the fields extend to their rocky margins, these limestone areas are taken possession of by scrub oaks and *Pinus yünnanensis*. *Rhododendron racemosum* Fr. (3274) a small shrub with deep pink flowers forms regular carpets with a prostrate form of *Quercus semecarpifolia*. On this plain we met a miserable specimen of humanity dressed in rags, a Chinese beggar – not Na-khi and looked so wild that I photographed him as a curiosity. Further on we found *Populus tibethicus* with pines, and more Rhododendrons. We descended from this high plateau to La-shih-pa plain. The whole plain of La-shih-pa 剌是垻 is one great limestone basin with sharp limestone outcroppings, hills and innumerable sink holes, the inner slopes of which are densely covered with trees. This plain is also under cultivation, but the sharp grey basaltic limestone rocks are covered with scrub oaks, *Pieris* and pines. At the head of this valley there looms up the enormous mountain mass called Yü-lung Shan 玉龍山, Jade Dragon Mountain, it is the Li-chiang Snow Mountain. The scenery was wonderful and the air cool and bracing.

Lower down on the plain is a small resting place called Shan-shen-miao 山神廟 or Shrine to the Mountain God. Here a robbery took place just before our arrival, an hour earlier. From here we descend to another more level plain densely cultivated and in which at the extreme north end, is situated a lake called Lung-hai 龍海 or Dragon Lake. The rather scattered village is situated on the very foot of the undulating southern end of the plain.

I am staying in a small temple on a hillside among roses and oaks, apricots and plums. It is very quiet and much more restful than the noisy inns which I detest.

Journey from Ta-li to T'eng-yüeh, September 23, 1922 to October 9, 1922
由大里至騰越

September 23, 1922
We left Ta-li on September 23, 1922 at 7 a.m. The sun was shining brightly and the lake sparkled in the early morning light. The road abominable as ever leads almost in a straight line to Hsia-kuan 下關, 28 li distant, through beautiful green rice fields, water is running everywhere rushing hither and yon. Our escort is way behind so we wait at the temple of the goddess of mercy where General Yang Yü-k'o 楊玉科 has also found quarters in a small room in the rear to the extreme left. I paid another visit to his shrine, this time deserted, no one in the whole temple as ourselves, for this is not fair time. Rice everywhere now where in the spring the fields were one mass of

poppies with white pink, purple and red flowers; these opium fields joined the thousands of graves beyond to the actual foot of the hills. Edelweiss is everywhere among the graves, pink *Anaphalis margaritacea* B. & H. f. as well as the silvery-white *Leontopodium sinense* Hemsl., the yellow flowered *Campylotropis trigonoclada* shrub and a pink flowered *Desmodium*. In the ditches and pools a purple *Pedicularis* holds forth, also tall grasses similar to those (or identical) we find around Chien-ch'uan, north of Ta-li. We reach Hsia-kuan at about noon, called at the B. A. T. and were informed that many robbers were on the road 50 li beyond Yang-pi, and that 250 soldiers had been dispatched from Ta-li and Yung-ch'ang 永昌 to drive them away. Traffic between Yang-pi 様鼻 and Yung-ch'ang had stopped for about a week owing to their hold ups and making the road unsafe. As we had only four soldiers as escort from Ta-li, we asked the military authorities at Hsia-kuan what they knew about the robbers. They confirmed what we were told at the B. A. T. and to our requesting a larger escort they replied that they would give us four more soldiers. We went to the Fu-lien Bank and exchanged 1800.00 dollars for a draft on T'eng-yüeh in order not to risk having all our money taken away by robbers. After a hasty farewell to dirty Hsia-kuan we followed the river which outside of Hsia-kuan flows rather rapidly towards the famous narrows with a fort on the right side. The outlet, I would say, is only about four feet wide and is arched by a stone, a natural stone bridge known as the T'ien-sheng-ch'iao 天生橋, which I should think no one would care to cross. The waters rushing madly forward find the narrow outlet too small and a back-eddy is formed. The water is mostly foam and as it finally forces itself through the narrow stony gate falls in a mad rush over boulders some 10 or 20 feet to the riverbed below, in its whole length it is one mad torrent. The banks are willow-lined, and how the trees can stand the terrific onslaught of the angry waters is more than I can understand. On the very edge, and jutting out into the riverbed, is a small stone house, its base is constantly washed by the roaring, dashing stream which some day will carry it with it on its mad rush.

The fort is old and the wall facing towards the lake fallen in. There are a few houses and a small tower within the enclosure, but no human life seems to inhabit it. On the left side is a stone gateway and beyond an arched thoroughfare, the steep rock wall having been broken through and arched over. The gateway is about 20 feet thick and a rocky path leds steeply down into the main road. Some stone tablets are in front of the entrance to the right.

The river here is not called the Hsia-kuan Ho but the Ch'ing-shui Ho 清水河 or Clear Water River which later becomes the Ho-chiang ho 合江河. The walls on either side rise steeply. The mountain being almost bare of trees but covered with shrubs and grasses and on the upper part some trees can be seen. The vegetation along the road is mainly shrubs of a pink *Desmodium* very common, a climbing *Aconitum volubile* Pall. (6598), *Delphinium* here and there a *Senecio*, with yellow flowered spikes; our old friend *Primula poissonii* is still to be found in flower; one or two leguminous shrubs, the common purple flowered one with velvety leaves *Uraria clarkii* found at Nv-lv-k'ö among the scrub oaks; the yellow flowered species, *Campylotropis trigonoclada*, and a white small-flowered climber found in the south is very common. A shrubby *Hydrangea* with purple lilac flowers and creamy bracts, a lilac-flowered Composite shrub, some 8 feet in height; the new *Quercus augustinii* var. *rockiana* Camus, *Castanopsis armata*? and *Castanopsis delavayi* Fr. (6584) can also be found. The yellow shrubby Composite, the *Inula cappa* (Hom.) DC. usually found on laterite with *Pinus khasia* in Siam, is here also present. A profusely climbing leguminous plant *Lespedeza bonatiana*, with long racemes of pinkish purple flowers we also collected, it occurs north of Ta-li beyond Chien-ch'uan. A purple flowered leguminous vine resembling *Dolichos*, a purple flowered *Strobilanthes* and the pink flowered *Begonia labordei* Lévl. (6605) are always found among damp boulders. One of the commonest shrubs is a scarlet-flowered *Colquhounia coccinia* Wall. *var. mollis* Prain (6566) shrub 5-6 feet along ditches with tall grasses, which we first observed around Chien-ch'uan. Walnuts grow wild down the steep valley slopes, also a wild persimmon (a small tree), nettles, *Debregeasia*, and one with long spikes, the same as I found along the Mekong in the south. One of the handsomest shrubs is 10 feet high, the one with pinkish white long tubular flowers, forming large terminal corymbs, *Luculia intermedia* Hutch., while a magenta mauve-colored *Paederia* is a common climber among small cane-brake, grasses and shrubs. Two yellow-flowered *Hibiscus*, *Hibiscus forrestii* Diels (6594) grow along the trail, also *Hibiscus venustus* Bl. (6587) not uncommon along Ho-chiang Ho. As we descend into the valley, leaving a well wooded narrow valley to our left, we cross a small affluent and soon reach the Ssu-shih-li ch'iao 四十里橋 or the 40 Li Bridge so called on account its being forty li from Hsiang-kuan.

From here it is 10 li to Ho-chiang.

The river is nothing but foam dashing against huge boulders. Here and there is a patch of rice, brown and ready for harvest. In some places it is being already harvested while on the Ta-li plain it is still green. Tall bamboos grow on the banks and we meet several caravans carrying large quantities of the large bamboo stipules used for wrapping tea and sugar. Also mules with coffin boards we encountered. We pass one or two villages till we come to the iron chainbridge, a rickety affair, the boards old and broken, and one can see the river roaring beneath. The village at the bridge is also called Ssu-shih-li ch'iao 四十里橋. The entrance gate from Hsia-kuan into the gorge is known as T'ien-sheng-ch'iao or Natural Bridge. I took two photos one showing the west the other the east side, including the fort. We passed the village beyond T'ien-sheng-ch'iao called Ta-po-chin, and the village before reaching Ho-chiang is called Hsiao-ho-chiang 小合江 or the small Ho-chiang. We are stopping in a delapidated old temple back of Ho-chiang, there are many lady and gentlemen gods one of them a small one with a bright red dress, called Tsu-shuai 祖帥, ancestral leader; this image is taken in procession around the village once a year to drive all devils away. The mosquito net is tied to the leg of a lady-goddess and I hope she will not be offended. It is getting late and tomorrow is a long journey.

September 24, 1922

Left our tumbled down temple at Ho-chiang at 7 a.m. The village street is straight, lined by dirty houses, pigs everywhere. There are rather large inns but they look so uninviting, filthy and dirty, that I always give them a wide berth. We follow the Ho-chiang River roaring madly on its onward course to the Yang-pi Ho 様鼻河. We climb a ridge and leave the Ho-chiang Ho to the left. It finds its way through a valley or rather gulch which opens out in a southern direction. Leguminous plants are common, tall grasses, wild pears, *Impatiens*, *Gordonia axillaris* (Don) Diels (6618) shrubs with large white flowers. Corn and rice fields are being prepared for harvest. The candelabra Euphorbias resembling *Euphorbia antiquorum* but with spathulate fleshy leaves, grow along the roadside. A flat hill which separated or rather diverts the Ho-chiang Ho juts out southward and westward, on its flat top is the filthy village of P'ing-p'o 平坡 named after the hill „flat hill“, it is about 5 li from Ho-chiang. Pigs are everywhere, the people are so accustomed to living with pigs that they have acquired piggy habits, there is not much difference in the habit of the two. After crossing the flat hill we turn to the right and see for the first time the Yang-pi Ho, it is brown and muddy; there is quite a contrast between the Ho-chiang and Yang-pi River, the former being one mass of white foam, its current is also more rapid than that of the brown Yang-pi River.

The road is absolutely abominable and is more a rock pile than a road; first mud and boulders, then it becomes a brook again used as a ditch for rice fields, then a pond or lake of mud with boulders protruding, others hidden. It is worse than some of the roads in the Black River. On the right hand side is the village of Li-tzu-yüan 栗子園 – Chestnut Garden. There are many chestnut groves – *Castanea mollissima* Bl. (6645) whence the name; as the fruits are gathered before they ripen there are now no fruits to be seen. The trail follows the muddy, but still rushing torrent of the Yang-pi to the small village of Chi-yi-p'u 雞一舖 – One Chicken Shop, and it certainly is worthy the name. There are many walnut trees, chestnuts, peach trees, wild pears with larger fruits than found north of Ta-li but the remainder of the wild vegetation reminds one more of the southern part of Yünnan and Burmah.

We had lunch under some old chestnut trees with few nuts still on the trees. I took a photo of a branch with open and closed fruits pinned against the trunk. The chestnut grove is near the village of Chin-niu-t'un 金牛屯 or Golden Ox Village. After lunch we were just about to proceed to Yang-pi, when we spied a deep and narrow gorge to the right, a steep rocky mountain side covered with trees in its upper part and a temple on the top, near the edge of the precipitous cliff. Lau-loo told me that it was here he found a wild chestnut; this decided me to stop at once and proceed to the temple. As it is however impossible to bring the loads up we enquired further and found that there is a large temple at the foot of the hill and east of the gorge, and hither we went. The temple is called Fu-kuo Ssu 福國寺 or Happy Country Temple; it is situated on a knoll covered with cypresses. As its name implies it is a Buddhist temple and in the room, the last one, third terrace, there are three huge images of lady-like Buddhas; a big drum if back of my bed, with a gong beneath, on the altar table with a fine bronze bowl giving forth a beautiful sound reminding me of the temple bells of Nikko in Japan.

We made ourselves comfortable and settled down, as the next day we are to climb to the top of the hill, where the wild chestnuts are –

Castanopsis concolor Rehd. & Wils. (6646).

In the late afternoon Ma Ting-an, Lau-loo and Li Ssu-ch'en and I went out behind the temple to find a trail into the gorge with its narrow entrance, but instead climbed the sides of the hill through scrub. At the foot of the hill we found *Corylus yünnanensis* (Fr.) A. Camus (6635), a shrub, 8 feet or so in height with tomentose leaves, and loaded with fruit of which we collected quite a large quantity. With it grew *Rhus cotinus, Leptodermis* (6640) melastomaceaous plant *Osbeckia chinensis* L. (6631), a herb with purple flowers; tall grasses, one a lemon scented one and the yellow composite *Inula cappa* DC. (6639). Of several wild pear trees we collected seeds (6636-6637 – *Pyrus pashia* Buch.-Ham.). Two species of oaks are common, *Quercus griffithii* and *Quercus glauca. Keteleeria davidiana* and *Pinus sinensis* are also common. We got a thorough soaking notwithstanding our trying to hide under a *Keteleeria* tree during a heavy downpour.

September 25, 1922

Left our temple for a trip into the mountains beyond. The trail lead down to the river bank, crossed the rocky streambed and ascended steeply on the other side. The lower slopes are mostly covered with weeds, but soon we enter pine woods the trees growing scattered, *Leontopodium andersoni* C. B. Clarke (6662) and *Anaphalis margaritacea* Benth & Hook. f. *cinnamomea* (Wall.) H.-M. (6648) are common, also scrub oaks, *Quercus griffithii* Hook. f. & Th. (6658) at 6,000 feet, trees 25-30 feet in height, *Quercus glauca* Thunb. (6659), at 5,000 ft elevation, tree 15-20 feet, and alder trees, *Alnus nepalensis* D. Don (6657) are not uncommon. The following plants were also collected: Of leguminous vines *Dumasia villosa* DC (6653), flowers yellow, elev. 7,000 feet, *Pueraria peduncularis* Graham (6663), lavender purple flowers in long racemes, 5,800 feet; *Apios carnea* (Wall.) Benth. (6669), fl. in lax racemes pinkish, 6,800 feet, and *Dollinera cinerascens* (6674), a bush 6-10 feet, fl. purplish-pink, 5,800 feet, also *Indigofera scabrida* Dunn. (6617), fl. reddish-purple. A white daisy-like composite *Inula nervosa* Wall. (6660) grows along the trail and the whole aspect is somewhat subtropical, large banana groves nestle in the ravines, Zingiberaceae like *Hedychium,* etc., *Keteleeria* trees are common, also *Diospyros lotus* L. (6670) of which I collected seeds. Water is rushing everywhere. Ferns grow on the banks, some Polypodia with creeping rhizomes cover mossy boulders. Oak trees become very common also here and there *Castanopsis concolor* Rehd. & Wils. (6647). *Rhododendron delavayi* being here the most common species. Li Ssu-ch'en climbed a tall wild chestnut tree and the seed collecting began, we filled several bags and proceeded to the temple above. The whole mountain side is one mass of enormous boulders; all the rocks are mica and glitter like silver. The fog which had filled the valley had now lifted and the sun shone brightly. We finally, after a very steep and slippery climb, reached the temple called Yu-hua-ko, where a wonderful view can be had over the whole of the Yang-pi valley, with the town of that name to the west, across a small ridge, on the banks of the muddy river. We climbed higher to the left and entered oak and rhododendron forest, all trees here 20 feet or so or more in height; one tree still bore a few large pale pink flowers which we collected – *Rhododendron decorum* Fr. (6681). The trail is very steep and from the edge, under large Rhododendron trees, we could gaze down the precipitous rock wall into the deep gorge below, several thousand feet down. Here were many trees of *Rhododendron delavayi*, and other species all with immature fruits, fine oaks and *Castanopsis concolor* Rehd. & Wils. (6647) with straight trunks, dark clean bark and horizontal branches.

The boys went ahead to gather more chestnut seeds, while we waited under large trees on mighty boulders covered with various ferns (6676, 6677, 6678). A heavy shower surprised us and so we hurried down to the temple where we awaited the boys. We had some roasted corn for lunch, which the temple keeper grows up there; being very hungry it tasted mighty good. The sun came forth again the atmosphere was clear and the whole valley below was bathed in sunlight. The descent was quite rapid, but still difficult owing to the slipperiness of the wet trail. The heat was intense especially in some of the corners of the trail, which seemed more of the temperature of a heated furnace. It must have been at least 100 degrees. A heavy thunder storm closed the day which began so brightly, accompanied with a tropical downpour. Tomorrow we shall go as far as Yang-pi, 20 li from here.

September 26, 1922

Left the Fu-kuo ssu temple at 8 a.m. in pouring rain, the trail down hill was a regular brook, the rocks slippery and the deep holes filled with mud and water. There was unfortunately a little quarrel before starting, two women came and complained that the caravan had eaten about 10 *sheng* of beans from her field so I gave her $ 1.50 as compensation but she was still unsatisfied. The main road was next to impassable, piles of rocks of all sizes, with holes filled with mud and water; every now and then these rock piles were inundated and made traveling for horses a terrible task and it was small wonder that no legs were broken. We followed the winding muddy Yang-pi Ho and finally reached the first part of Yang-pi, a dirty hovel, the gate of thick masonry was cracked in several places at the top and leaned inward, ready to fall any moment on passing caravans. There was one prop back of it but it certainly did not support the tons of rock leaning in every direction; such a gate would not be allowed to remain in any other country but China. Along the road to Yang-pi, which is rather a narrow one, we met many caravans, of 40-50 mules each, carrying cotton and cotton yarn to Ta-li, out of which the cotton garments are woven. All this comes from Burmah.

It was very difficult to pass and we were delayed considerably in the pouring rain. The valley was filled with mist and clouds, all the mountains being hidden in black vapor. After passing through the first part of Yang-pi we came to a river which joined the Yang-pi Ho on the right bank (coming from Ta-li) the river is crossed by a wooden bridge, and is known as Hsüeh-shan Ho 雪山河, or Snow Mountain River. This small stream separates the two Yang-pi's; the streets of the other half of the town are very narrow and dirty, it is one of the dirtiest places I saw. There were many mule loads of coffin boards lined up, a caravan was ahead of us, and headway was difficult. We passed through the town, crossed the iron suspension bridge which swayed considerably and climbed the hill on the other side to a temple which had just recently been vacated by soldiers who are now chasing robbers near T'ai-p'ing-p'u 太平舖. There was a man and a woman caretaker; the place was filthy beyond description, so we went to another one a little higher up and were met by a lousy beggar in rags who, with four other companions, had taken possession of the place. The walls were tumbled down and all in rot and decay, a fit habitat for the beggars, so we went back across the bridge and were led to a school and temple in the town. It turned out to be a Mohammedan Mosque, although its exterior did not suggest a mosque at all, only on closer inspection could one notice that the decorations were a bit different. The interior however was that of a mosque. An old man with distinct semitic features, although a

Chinese, waved his hand as greeting and made us welcome. This is the first time I ever am to sleep in a mosque. The old man a very congenial soul, a hadji, speaks Arabic with a peculiar Chinese accent, but we could converse quite well together and when a word failed he wrote it down in Arabic, although the Chinese characters would have done as well. My Mohammedan Chinese interpreter Ma Ting-an felt quite ashamed. We would talk about the Koran in Arabic, and he would be absolutely speechless, knew none of the terms as Imam or Mueddin which had to be explained to him. I think most of the Chinese Mohammedans are like him, the only thing by which they differ from other Chinese is in their refraining from eating pork or using pork fat. The old priest looked on while I had lunch, and I offered him a piece of raisin bread made with pork fat, he ate it greedily and said it was very good. I did not give it to him intentionally, but he looked longingly at it so without thinking I handed him the plate; all my men are cooking with pork fat in the mosque compound, and my food cooked with the same fat is served me in the mosque itself; there seems to be no objection and they seem to take their religion not so seriously. My bed is in one corner of the mosque and at the other end is a coffin, the bier etc., is exactly like those I have seen in Tunis and Algeria. We talked much of Mohammedanism and there was much which even the Imam did not know, I told him of Mohammed Sidi Okba, the barber of the prophet who lies buried in the sacred city of Kairouan as the French call it or Qaruan قيرٯوان in Arabic. We talked of the Kaba of Mekka, and many other things until he asked me in earnest if I was a Mohammedan; I answered him saying that I take the good of all religions, until he insisted that I was a Nusrani (Christian).

All the soldiers of this place are chasing robbers but 25 have just returned and so we will have an escort of ten tomorrow for T'ai-p'ing-p'u. We have to cross a range 8,300 feet in height and as there was heavy rain and a thunderstorm lasting all evening, the steep trail will be very slippery.

We found a pear here on the market which they call shui-pien-li 水遍梨; it is not much eaten raw, but as a cooking pear it is delicious and quite the best I have eaten. It has a delicious flavor, and is quite as good as canned bartlet pear. In shape it resembles a large apple is of a rich dark green with dark red cheeks, it appears to be very hard, but is very watery and sweet and slightly acid. We packed the seeds and cooked some of the pears. There were also a few wild pears along the road with rich yellowish brown fruits and we collected them. Tomorrow we start early for T'ai-p'ing-p'u, the Hsien-ch'ang 縣長 (magistrate) having promised an escort of 10 men.

The old man came and prayed in regular Mohammedan fashion looking towards Mekka, while I had my supper served in the mosque. I told him *leltak saide* (good night) and he smiled and tried to correct me by saying rer-tak saide, he talking Mandarin in Yünnan Chinese, found it difficult to say „l".

September 27, 1922

We left Yang-pi the filthiest hole I know of, at an early hour just to be rid of that rotten place. The dog at the mosque barked every now and then at the slightest noise so sleep was much interrupted. The friendly old man gave me a real handshake and with a salaam we parted.

The trail led up steeply on the other side of the Yang-pi River and town. The heavy rain of the day before had made the road exceedingly slippery. It certainly cannot be called a road, a continuation of mud-ridges with troughs of mud and water about a foot wide, but the depth varied from one to two feet. Riding was absolutely impossible.

The vegetation at first consisted of herbaceous plants, and low shrubby climbing rose-bushes as *Rosa moschata* subsp. *longicuspis* (Bertol.) Herr. (6668), full of fruit of a fiery red, the flowers are said to be large and white; other bushes were *Rubus cotinus*, *Rubus flavus* very common, a few *Pyrus* shrubs with *Pinus sinensis* and quite a distance up the hill, at about 6,500 feet elevation, *Zizyphus yünnanensis* Schneid. (6695), a tree about 20 feet tall with a round spreading crown, and loaded with seeds, of which we collected one large bag full. Further up the hills are clothed with *Pinus sinensis* only. On the left is a big gorge well-wooded and in parts cultivated, the hillsides being terraced and planted to rice. Among the *Pinus sinensis* trees there were small trees of *Pieris*, *Phyllanthus emblica* with small fruits, a species of *Docynia* which formed quite large trees, loaded with apple-like fruits larger than a crab-apple, with the fragrance of a quince, the color of the skin is greenish yellow and red on one side, others uniformly yellowish-green. The ground beneath them was literally covered with the fallen fruits. I took a photo of one tree with Li Ssu-ch'en standing beneath, also one of a branch pinned against the trunk, we collected fruits of that particular tree. *Alnus nepalensis* D. Don (6657) with very large leaves appeared to be very common in fact further on it formed pure stands, large trees of it, forming dense forests. The male catkins are yellow and drooping, in masses. It is quite different from the species found on the Ta-li plain (3184 – *Alnus ferdinandi-coburgi* Schneid.), also different from the species found south of Szu-mao.

The road became worse and worse and almost impassable. A splendid view could be had over the Yang-pi valley to the Tsang Shan range 14,000 feet, but the peaks were covered with clouds. We had a splendid view of the deep gorge with the temple high up on the cliffs where we collected *Castanopsis concolor* Rehd. & Wils. (6683) two days before. As we climbed higher, pines became scarce and *Alnus* trees formed pure stands with here and there *Malus* (6687, 6688, 6690) trees scattered between them, and loaded with fruits. Here I also met the Shan-li-kuo 山梨果, a species of *Meliosma pungens* Wall. (6836), a tree 30 feet high with the fruits forming clusters at the ends of the branches. The elevation here was about 8,200 feet. The trail climbs to the top of the ridge with magnificent alder trees, some bamboos (not in flower), and finally emerges into an open flat swampy place called Lan-pa. There we stopped for lunch. The swampy part was one mass of fruiting *Primula poissonii* with a species of *Pedicularis* and the labiate *Plectranthus striatus* Benth. with lilac flowers. On the margins grew the pink flowered *Impatiens forrestii* while in the drier places grew *Dipsacus chinensis*, a thistle, a sage, *Anaphalis busua* (Ham.) H.-M. and a species of *Leontopodium*, as well as a *Potentilla* with large leaves, white beneath, also found on the Li-chiang plain. On the hillsides grew, masses of *Rhododendron delavayi, Keteleeria davidiana;* a shrub belonging to the Theaceae (*Ternstroemia japonica* Thunb. var. *wightii* Dger) with leathery leaves, cream-colored flowers and red fruits; with it grew a small *Vaccinium* and a *Pieris*, here and there occured a small *Populus tibeticus.*

The trail led through a narrow pass between two hillsides along the left bank of a brook on the right hand bank of the trail grew *Lobelia erecta* Hook. f. & Thoms. (6839) with blue and purple flowers, a tall branching *Anaphalis busua* (Ham.) H. & M. with quadrangular stems bright green, and white flowers.

To the left there was a fine jungle composed of *Alnus*, roses, theaceous trees with large brown fruits *Thea pitardii* (C. Stuart) Rehd. (6846), and a new species of *Malus* (*Malus rockii* Rehd.) with long drooping branches bare of leaves at least the lower ones, and loaded with oblong-obovoid, apples about 1/2 inch long yellowish-green and dark red on one side; we collected large quantities of these. It was quite common along this brook. The large fruited apple trees were also common, an araliaceous tree with straight ascending branches, and digitate leaves but without flower or fruit was also observed. The trail descended somewhat, following the brook, in the bed of which we found a species of *Prunus* with yellow globose plums, of which seed was collected. Walnut trees *Juglans regia* L. (6694) became very numerous and covered whole hillsides together with large fruited wild pears all at an elevation of 8,300 feet. We secured about 500 of the wild walnuts; the nuts are very thick-shelled and consequently heavy and of all shapes, some long and pointed others globose, and again others ovoid and small. Small *Quercus* trees occur and with them also a large leaved species forming quite a tree, allied, if not identical with *Quercus delavayi*. *Rhododendron delavayi* and *Rhododendron decorum* are quite common but the former more so. Yellow flowered *Impatiens* cling to a grassy bank in the shade of *Malus* trees. *Pinus armandi* occurs with *Pinus sinensis* and walnuts become more and more common till we emerge into *Pinus sinensis* forest with here and there an *Alnus* tree. the trail now winds along the steep hillside; to the left is a deep gorge, with a small stream the headwaters of which is the aforementioned brook. Two villages come into sight the first is T'ai-p'ing-p'u 太平舖; here we stop for the night in a small school house above the village. While I write this the thunder rolls and lightning flashes on the other side of the valley.

September 28, 1922

Left the old school house after a rather sleepless night owing to continuous lightning and thunder which shook the hillside and the poor mud foundation of the tumbled down building. The storm was of the merry-go-round kind and reminded on a similar experience in Colorado Springs one summer night in 1914. The hillsides of this valley are rather steep and the vegetation is mostly *Pinus sinensis*, *Juglans regia* and *Castanopsis delavayi*; of the latter there seemed to have been a great many trees which had been cut down by the inhabitants of T'ai-p'ing-p'u, they used the logs for fire wood. Fortunately, the trees ratoon and a crop of young strong saplings grow from the old stout trunks. A few had seed and we had them collected. *Pyrus* trees with small yellow fruits are scattered here and there over the otherwise bare hillsides. Further on we met *Pinus armandi* trees, *Quercus delavayi* and *Rhododendron delavayi*, this latter tree seems to be much desired for firewood as most of the trees had been cut down within a foot or so of the ground, they also ratooned. At a bend of the road we found a hastily-made grave covered with pine branches it is that of a T'eng-yüeh merchant who was shot by robbers a few weeks ago. It certainly is a lonely country, I saw a lot of men in the distance and not knowing who they were I looked around for my escort but found none so I shot off all seven bullets in quick succession from my colt and soon saw them come running; they thought I had met with robbers. I gave them a good lecture and told them that they had no business to stay behind. I scared the people ahead of me as they also thought robbers had descended from the hillside and they waited till I approached. The trail follows close to the top of the ridge and we could overlook the valley of the Shun-pi Ho otherwise also called the Sheng-pei Chiang 勝備江 on the right, filled with clouds only one peak protruding. The trail finally descends rapidly, passing through pine forest with *Rubus flavus* and *Alnus* trees, with an occasional *Malus* of the large fruited type. We continue to descend passing through oak forest to the village of Ta-shan-hou 大山後 – „Behind the large mountain".There are three species of *Quercus*, the small-leaved shrubby species, and that with large leaves of the *Quercus delavayi* type and the most common of which we collected seeds and specimens. Wild persimmons are to be found with wooly leaves, and large cultivated species with very large fruits but not yet ripe. We now follow on the left hand side slopes of the Shun-pi Ho valley, densely wooded and very narrow. The forest changes and pines give way to pure stands of *Quercus variabilis* Bl. with oblong acorn half enclosed in a spiny-scaly cup, the acorns literally covering the ground. The trail goes down the streambed of the Shun-pi Ho a more or less sluggish red-brown muddy stream flowing along like thick soup. We cross a suspension bridge called Ch'ing-lien-ch'iao 清連橋 built in the 18th year of Kuang-hsü (1892) by merchants of T'eng-yüeh, Yung-ch'ang and other places, their names being engraved on the marble tablet. We now climbed up the other side of the narrow valley over a more abominable road as we came down; riding absolutely out of the question, mud holes, rock-piles with water rushing over them, narrow, hardly passable, landslides over the steep hillsides in places real precipices. The heat was intense, no wind, not the slightest breeze, only when we had climbed a considerable distance did it become cooler. The oaks gave way to pure stands of *Pinus sinensis* with *Phyllanthus emblica*, and an anacardiaceaous tree resembling a sumach; the undergrowth being grasses. At about noon we arrived at the hamlet of Huang-lien-p‘u 黃連舖 nestled against the steep hillside. Here the soldiers who were sent to chase robbers had their headquarters. We spied a temple and intended to have lunch in it but found it so full of coffins that there was absolutely no room in it to sit down; so we passed through the dirty place on to the banks of an affluent of the Shun-pi Ho and had lunch. Here we met a soldier who was a friend of Boon Mah, and who had lived 3 years in Chieng Rai. Being acquainted with most of the American missionaries of northern Siam. After lunch the trail led us up the steep hillside, through pine forest with which the surrounding hills are covered to the exclusion of everything else. We still can see the Ts‘ang-shan 蒼山 range of Ta-li in the distance, the peaks being covered with clouds. We follow a ridge with valleys on both sides, another ridge to the right had a square temple at the very tip top of a hill. I took a photo of the mountain ranges over which we had come, and climbed on through wonderful hilly country at an elevation of 8,000 feet, till we reached the hamlet of Pai-t‘ou-p‘u 白頭舖. On a small knoll overlooking the few houses forming this village stands a little temple in the innermost sanctuary of which (mud walls) I made my headquarters. There are 8 images all new and hideous, in one corner is Mahâkala blue as ever, and in the center Kuan-yin riding a fish with one foot on a frog, next to her a Lung Wang 龍王, and next to him a devilish brown figure with black round eyes he represents Shen-nung 神農, I think an insult to the famous emperor. One god, naked, only covered with a few leaves, hideous beyond description, deformed and evidently modeled after a choice specimen of their own human kind. Next to him a goddess with a baby with its legs spread out showing the genitalia, it is for barren women to be prayed to. Such are my companions for the night.

September 29, 1922

Left the small temple at a very early hour. The mountains surrounding us peeped above the clouds which filled the numerous valleys. The trail ascended to the top of a pine covered hill with *Alnus* and *Keteleeria* trees and *Cornus capitata* Wall. with strawberry-like red edible fruits. The common oak *Quercus variabilis* was plentiful. *Castanopsis delavayi* made its appearance, we found several trees loaded with burs. I photographed this species (trunk with small branches) and stopped to collect them. I took a photo of one branch pinned against the trunk of the tree of which we collected seeds. *Alnus* trees were common and so was another species *Lithocarpus viridis* Rehd. & Wils. Large *Pieris* trees became abundant with pines, and trees with stout trunks and branches of *Rhododendron delavayi.* The further we went the commoner became the *Castanopsis*, a small fruited species, and *Castanopsis concolor* (negative 30307) the same as collected back of the temple of Fu-kuo Ssu 福國寺.

We had hardly made 25 li when I decided to stop at 11 a.m. at a small village of only 4 houses, called Hsiao-shui-ching 小水井. We found an old tumbled down empty house and made our headquarters there, I pitching my tent outside. In the afternoon we went chestnut collecting and found one large tree about 80 feet tall loaded with fruits. Ma Ting-an, Li Ssu-chʻen and Lao Lee and I gathered, cut, and filled two huge baskets. In the evening we sat and cleaned them until our fingers were sore and black. There are many different plants found here, the beautiful purple *Impatiens delavayi* Fr. (6890), *I. clavicuspis* and *I. arguta*, and of trees *Ilex polyneura* (H.-M.) S. Y. Ha (6888), 25-30 feet high with very attractive bright red fruits. Also the red fruited *Anisadenia pubescens* (6689) very handsome and attractive. A large leaved *Rubus* covers the open spaces of the hillside. Oaks are common and tall *Alnus* trees; *Smilax*, etc. make up the rest of the vegetation with leguminous climbers of the Papilionatae group.

September 30, 1922

Today we went *Castanopsis concolor* collecting; we found one large tree loaded with fruit and filled two large baskets with the large spiny fruits. In the afternoon we went out again but did not meet with great success although we gathered two half baskets full, from several trees. A severe thunder storm broke and we hastened to our camp black and delapitated. Our soldiers are filthy with itch and I gave one medicine for his sores. Tomorrow we shall be off for Chʻü-tung 曲洞.

October 1, 1922

It rained all morning, in fact a thunderstorm broke early at about 7 a.m. We left camp at 11 a.m., the rain had just stopped. We had packed all our chestnuts both species in charcoal and ready for mailing. The trail led downward through beautiful groves of chestnuts and I photographed 3 trunks with one of my assistants standing in the middle. It was the small fruited species *Castanopsis delavayi* Fr., although *Castanopsis concolor* Rehd. & Wils. grew near it and its branches intermingled with the other. There was *Alnus nepalensis* D. Don. quite common, araliaceous trees with entire but deeply and acute, sinutae-lobed leaves, (*Lobelia pyramidalis* Wall., 6901) grew on the banks, wild pear and apple trees were also common. We finally entered pine forest and after a ride of another hour over the most miserable trail imaginable full of mudholes, rock piles, many wash outs etc., we entered a valley and followed a creek bed on the banks of which many pear trees, the wild species *Pyrus pashia* Buch.-Ham. here growing in abundance. The fruits were larger than any previously observed. There were also two trees of a very small fruited variety with stiff ascending branches while the large fruited species had long drooping branches. I photographed a tree of which I collected seeds also a branch pinned against the trunk. From there we continued following the stream which is known here as the Ch'a Ho 叉河 climbing and descending over a muddy slippery trail. We soon entered the bare hills mentioned by Davies in his book. Erosion everywhere only grass and landslides, not a tree visible. We soon had a view of the Yung-p'ing 永平 plain with little cone-shaped volcano-like hills on the margins of the larger bare hills, Ch'ü-tung 曲洞 (or 硐) is situated at the foot of one of the conical hills partly wooded.

The Ch'a Ho crosses the Yung-p'ing plain diagonally and is then known as the Yung-p'ing Ho 永平河. Yung-p'ing is to the right of the valley and Ch'ü-tung on the left hand coming from Huang-lien-p'u, and at the extreme north (west) end. There seem to be several small townships on this plain. We had to ford the Yung-p'ing River, the bridge having been washed away. Yung-p'ing, like every other Chinese town is very uninviting; many Mohammedans reside here, and as there was no decent inn and no temple, we went two li further to a place called T'ao-yüan-p'u 桃源舖 composed of few houses, but the whole village is Mohammedan. We were guided to the Mosque of the place, a clean looking place and were made welcome by a nice old man, the headman of the village, his name is Ma Chang-hsing 馬長興. To his credit it may be said that he is one of the kindliest men I ever met in Yünnan. He presented us with vegetables, firewood etc. and opened the mosque wide. I was astonished by the severe Semitic features of the people, they had nearly all large round eyes not the small slanting pig eyes of the ordinary Chinese. Many of the boys were of typical Arab appearance and I would never have taken them for Chinese. They all speak Chinese, of course, and it seems so queer to hear people of that type speak Chinese, Persian or Arabic would be less astonishing; there were three boys typical Arabs in features and bearing, bright, keen eyed and full of fire, quite different from the slow and lazy-looking opium smoking Yünnanese one generally meets. The plain of Yung-p'ing had a most sleepy and drousy looking appearance, as if completely lifeless. The clouds were hanging heavy over the ridges only the peaks of the Mekong range protruding black and uninviting. It was steaming hot, the atmosphere heavy, and certainly quite different from the cool hills we had left where a sweater was quite welcome. There is not much to be recorded of plant life. *Delphinium tatsienense* with Prussian blue flowers grows on the grassy hillsides with *Leontopodium sinense* and *Anaphalis adornata,* rose bushes were common, the purplish-red melastomaceous herb, and a ground orchid *Arundina chinensis* Bl. with lovely flowers, but not common.

October 2, 1922

We left the little village of T'ao-yüan-p'u rather late, the good old man of the village, kindness personified persuading us to remain at the mosque owing to the nasty rain weather. However we decided to go at least as far as Hua-ch'iao 花橋 should the weather not improve. We said farewell to all these kind people and bright children who gathered around us. The trail led between steep low hills perhaps only 400 feet high. At first the road is paved and only here and there bad; the hills are covered with shrub vegetation; they seemed to have been cut over and are just recovering; low shrubs form a dense cover. There were shrub oaks *Quercus griffithii* Hook. f. & Th. (6931), tea bushes *Thea yünnanensis Pitardi* (6936) with large white flowers; *Callicarpa rubella* forma *angustata* P'ei (6932) bushes with purple berries, *Pieris* shrubs, *Rhododendron delavayi* bushes and *Cotoneaster*. Especially common was the lavender-flowered composite shrub *Vernonia clivorum* found so common near Yang-pi and Ho-chiang, also the yellow flowered shrubby one common in pine forest. The trail leads up some few hundred feet, probably 600 or so, through pine trees with many wild *Pyrus* trees, Cotoneasters, *Keteleeria davidiana* and through a gap wooded on both sides, put into more or less open country with here and there a pine tree, pear and *Alnus.* The trail now descends to the valley or small plain on which Hua-ch'iao is situated. To the left a road leads into a valley called Pa-chiao-ching, Banana-bamboo, while we follow the main paved road into the well cultivated plain called T'ieh-chang 鐵廠 or Iron Mine. The hill sides are cultivated to corn (maize) while the plain which gradually rises towards the mountain in the northwest is one mass of rice fields to the foot of the hills, one beautiful sea if green. We pass a stream flowing through the plain, called Hsiao-ch'a Ho 小叉河 or Small Cross River, which flows into the Yung-p'ing Ho. At the Yung-p'ing Ho we descend to the foot of the hills and just before entering the plain we pass a small village called T'ieh-chang from which the plain derives its name. An iron mine existed here formerly but is now closed. There are peculiar *Salix* sp.? trees (6929) lining the streambed, and bushes (*Viburnum*) with long drooping branches loaded with red flat fruits borne in large terminal corymbs. We pass through the center of the plain which on the higher end divides into two branches, one ridge densely covered with pine trees separating it. The second village we pass, a rather dreary affair, is called Hsiao-hua-ch'iao 小花橋. We ascend gradually to the left head branch of the valley at the end of which is situated Hua-ch'iao or Flower Bridge village. We stopped for lunch at a temple converted partly into a school where a lot of boys made awful noises repeating pages of their study books. We were undecided as to going beyond, but the rain did not increase, at least while we were there, so we decided to go on after lunch. The temple was a so-called Buddhist one, but full of female gods in wire cages, dressed in gaudy colored costumes and white painted faces, To the right was the now commonly observed goddess for barren women with a little boy in her hand whose penis had gone, either by accident or done by some willfull boy as a joke, for this generation is not very religious. From this place the trail, well paved in spots, leads steeply up the hillside, between pines and oaks *Lithocarpus spicatus* (6939) and *Castanopsis concolor* Rehd. & Wils. trees being the commonest one. Steeper and steeper we ascend the air is cold, rain descends in torrents, clouds hang heavy over the mountain tops, and we march enshrowded in mist and rain. We ascend to over 8,000 feet again; the scenery must be grand on a cloudless day; the vegetation is dense, *Castanopsis*, several species of oaks, several *rhododendron*, stout trees of *Rhododendron delavayi*, *Thea* sp.?, several of them with large leaves some fruiting others with large fruits and shrubby but with large white flowers, *Elsholtzia rugulosa* Hemsl. (6935), leguminous climbers, *Dalbergia mimosoides* Fr. (6913) with spiny rambling branches, all these and a willow tree with stiff grey foliage hairy beneath, are very common near the top of the range. *Desmodium griffithianum* Benth. (6911), *Indigofera gerardiana* Wall. (6917), *Indigofera scabrida* Dunn. (6918) with other leguminous shrubs formed the main vegetation. *Pinus armandi* trees quite large ones occur with *Malus* (apples) with large fruits, oaks of various kinds some small shrubby ones, other large forming trees were also collected. The large flowered golden *Senecio clivorum* Maxim. (6915), a blue flowered Campanulaceae herb and the small flowered common blue-to purplish *Codonopsis* trailing and twining over *Gnaphalium, Leontopodium* and leguminous herbs were to be found among other plants already enumerated. There are many shrubs and trees which it was not feasable to collect owing to the muddy slippery road and heavy rain, but the collection of plants is quite a large one and many are in the press not here enumerated. Near the top of the range it is very cold, a strong wind blows in addition and drip drip goes the rain, and the splashing mud from slipping and sliding horses over an impassable road. On clear days it must be a joy to pass over this forested mountain range. Finally we descend through *Castanopsis concolor* Rehd. & Wils. forest, the trees being here rather small, the above mentioned willow is still common, and pines occur both *Pinus sinensis* and *Pinus armandi*. The rain still continued and as we unexpectedly came on to a rather large temple at about 7,500 feet elevation I decided to investigate it and if suitable to stop for the night, it being still 18 li to Sha-yang and the weather miserable. The temple called Yung-kuo Ssu 永國寺 is situated in a large grove of *Castanopsis* both species, the small fruited one and *Castanopsis concolor* with *Pinus armandi* of unusual size and a beautiful flowering tree near by – *Michelia lanceolata* Wils. (6919).

The temple, a very old one, over 400 years, is quite roomy, built in three terraces, in the last one I decided to make my headquarters. A huge golden Buddha said to be an Indian one sits enthroned, while 2 others one on each side, but hidden from view by screens. A black coffin is in the right hand corner, I hope empty, also a large drum, while outside is an old bronze bell cast during the Ming dynasty. On the left are two large stone tablets inscribed in Chinese. On the walls and wooden doors are maps of Mt. Omei in Ssu-ch'uan and Sanscrit prayers. Below on the second terrace are also images. To the left being Da-mu or Darma or Daruma as he is called in Japanese, the famous Hindu Buddhist who sat for six years in continuous meditation till his legs rotted off. On the right side a devilish-looking image the identity of which is unknown to me. In the courtyard between the second and third terrace are two large trees 50 feet tall of the genus *Michelia*, they are loaded with large brown fruits the size of a pear, and we collected all the seeds possible. The keeper of the temple says that the flowers are nearly 4 inches in diameter and white. The first terrace is occupied by a large gilded figure in a sitting position with two attending standing figures, one with whiskers which could well be employed as a broom. All my loads are stored there and the boys built a fire to dry the plants and blotters collected since we left Nv-lv-k'ö of blessed memory. Everywhere I look are shrines, tablets, gods small and large, and peace reigns, the golden image with half closed eyes sitting in peaceful meditation, since four centuries.

The mountain on which this temple is situated is called Po-nan Shan 博南山 and is famous in Chinese history. In the Han dynasty, during the rule of Ming-ti 明帝 A.D. 58-75 a road was opened over this mountain which led to a ford across the Lan-ts'ang Chiang 瀾倉江 or Mekong.

October 3, 1922

Rain, rain, rain, and a sleepless night. The tiny oil lamp before the great golden Buddha went out about 2 a.m., but the moon, although hidden from view by mist and rain, made everything quite visible in the temple, even the black coffin opposite me. The noise from the dripping water off the large *Michelia* trees in the small courtyard into the pools of water kept me from sleeping although dead tired. We left this peaceful spot descending the mountain range to Sha-yang 沙陽 but not so quickly as it is written. The road is paved and passable near the temple leading through *Castanopsis* groves and pine forest, but all at once it descends very steeply over a road which beggars description. The continuous rain made the clay-road, here not paved, so slippery that it was impossible to walk, we just slid down the hill, pine trees were scattered here and there. Finally we passed out into open grassy country, still hilly, the tops of the range hidden in clouds. Part of the trail was a ditch 4 feet deep, one foot wide, with water rushing down, on the edge boulders sticking out of the mud. And down such trails people are forced to travel. We soon spied the terraced fields of Sha-yang, passing first a stream and climbing a small hill on the other side and we soon are in Sha-yang. A delapidated gate stands alone in open country and leads into a small suburb with a few houses, we pass another gate and enter Sha-yang proper, one long narrow street dirty and filthy as usual. Many inns full of mud and standing water, pigs wallowed in the courtyards, thank goodness the flies have nearly all gone owing to the rains and cold weather. The rock of this region is a dark greyish black slate and all the smaller bridges over ditches and brooks are made of long flat slabs of this rock; many water troughs are about and big square tanks made of the same rock. The road through Sha-yang descends slightly, to the left, just before ascending a steep but rather well paved road to the upper gate, a trail leads to the top of a hill with tall trees surrounding a large temple. On arriving at the gate at the top of the hill the trail descends as steeply as it ascends and leads down into a small plain called Hsiao-wan-tzu 小灣子. The trail runs to the right following the narrow plain, and across a rapidly flowing greyish stream called Sha Ho, Sand River, over a bridge well built of stone called Feng-ming-ch'iao, Call of the Phoenix Bridge. We now enter the larger end of the plain called Ta-wan-tzu 大灣子, the river turning to the right through a gorge and flowing into the Mekong. The hills surrounding this well cultivated plain (rice being now the crop) are bare not a tree in sight.

There are two roads, the Ta-lu 大路 or main road, climbing the hill, the sensible or lower or inferior road leading on a level to the end of the plain (left branch) only to ascend over the most bare clay mud slopes at an angle of 75°.

To ride was impossible, not a rock or a pebble, just yellow hard clay as slippery as ice; the rain descending in a drizzle, every little furrow a rushing brook. To ascend such a hill with a well loaded caravan in such weather is really a feat and I thanked my lucky stars that no legs were broken or boxes smashed. Up and up we climb till finally we reach the crest of the ridge on which there is a temple, the road passing right through it. To the left is the real temple, to the right a sort of resting place with a small kitchen of mud bricks and a fire with an old kettle, an old woman selling a few eatables to passing coolies. This temple is called Chiang-yin Ssu 江音寺 or The Sound of the River Temple, it surely was as the road through the center of it was one mad rushing stream of yellow mud. But of course the name had no reference to that but to the Mekong flowing some 2,000 feet below in the gorge beyond. We stopped here for lunch building a fire with wood brought from the temple and sat down on some logs which were lying outside and partook of a meagre repast amidst mud, rain, beggars etc. Not at all inviting as a place to dine, although in pleasant weather it may be all right and the place for a lunch stop just before descending the steep and rocky road to the Mekong called here Lan-ts'ang Chiang. We waited for our caravan which was behind. On they came drenched to the skin, the mules paddling along in mud like ducks and splashing the trunks, boxes, bedding etc. until they appeared as if painted red. We told them we would go 15 li beyond to a place called P'ing-p'o 平坡 and would there spend the night. They seemed pleased as the prospect of going the regular stage to Shui-chai 水寨 in such weather was not a pleasant one especially after passing over such roads. The road, leading down the steep mountain sides obstructed by many landslides to the Mekong River 2,000 feet below, was one rushing stream. The road is paved but rocks have been washed out everywhere, and none were visible the water rushing over them like a waterfall, the grade being exceedingly steep. Over such slippery road of smooth stones, with rushing water transforming the whole into a veritable streambed, we descended to the Mekong. The tops of the mountain were heavy with mist, and the Mekong could be seen winding its way through this narrow much eroded gorge, like a brown snake. In all my travels this day October 3, 1922 was the worst of all experienced and I was sincerely grateful for arriving whole and sound at the iron bridge spanning the Mekong just as it emerges from a narrow gorge with steep precipices on both sides. I took a picture of this famous bridge with my mules and baggage crossing it. The road first enters through a gate and one arrives in a sort of courtyard, a small rest house for coolies is to the right, and to the left, opposite the entrance to the bridge, is a well built masonry or rather stone building. The bridge is called Hsien-hung-ch'iao 霰虹橋. The river must be exceedingly deep here as it flows between steep mountains forming a narrow gorge, many whirlpools can be seen everywhere and its current is rather swift. The height of the bridge is now about 20 feet above the water, although at the height of the rainy season it is said to come within very few feet (2 feet) of the bridge. On the other side is a very smooth steep rock wall, covered in its lower part with many Chinese characters. We ascend again over a paved road and land in the village half way up the hill on the edge of the gorge; P'ing-p'o, Flat Hill, where we are quartered in a temple with images of Confucius and a lady on his right, on the left 3 ladies, one the famous Sung-tzu-niang-niang 送子娘娘 or the Lucina Giver of Sons, with a baby in her arms this time decently clothed.

October 4th, 1922
Still cloudy but no rain; fleezy clouds drift through the valley of the Mekong, while heavy black ones enshroud the tops of the ridge. We leave our small temple at P'ing-p'o with the intention of reaching Yung-ch'ang tonight. From this small village, the stony road leads up and up the hillside which is only scrub covered. All we saw of interest was a *Wendlandia*, some *Thalictrum*, composites, *Desmodium* etc.; steeper and steeper the mountain side rose, and zigzag went the rocky road. We could hear the thunder of rushing, falling water in the distance. Riding was impossible so we climbed and climbed, meeting many loaded mules coming from T'eng-yüeh carrying cotton thread well packed in the fiber of the common fan palm *Trachycarpus fortunei.* After an hour and a half or nearly two we reached the top, crossing twice a torrent of a small stream tumbling at full speed into the Mekong. The gap is a narrow one and as we looked back into the prison of the Mekong all we could see was a sea of clouds and mist. A little beyond the actual gap there is a gate way and temple as in the gap before descending from the Sha-yang side. The defile is smaller and perhaps a trifle dirtier; at least the left side looked absolutely black and filthy, the abode of beggars. The trail is a sea of mud and covered with landslides from the bare slopes. All this made traveling over it an absolute torture for man and beast. We continued to climbing until we reached the small hamlet of Shui-chai 水寨. To the right somewhat ahead jutting out on a rounded hill is a little larger village called Ma-chia-ts'un 馬家村, a Mohammedan village. Shui-chai is, like every other Chinese village, no more than a filthy piggery; pig stables in America are beautiful parlors in comparison to a Chinese place of abode. Instead of cleaning their roads they throw all the rubbish on it, hold the behinds of babies out by the door over the roadside to relieve themselves and let the attentive dogs lick them clean, this instead of toilet paper. In the middle of the road a pig is cleaned out and all the dirt and hair and filth remains in front of the house or shop, people stepping over it or into it. We still climb from Shui-chai to another small village Cha-wan 茶碗.

A beautiful view could be had looking back; the black clouds had lifted and the black mountain range of the Mekong Valley could be followed till it was lost in a mass of ranges equally high. The hillsides were quite bare, here and there was a leguminous bush, and tea bushes were common. Corn was planted on the hillside while in the small valley below rice fields abounded. Further south the hills were terraced, the terraces being very narrow owing to the steepness of the hills. We follow along the side of a ridge leaving the valley to our left, climbing it soon afterwards; the trail here was sandy and rocky and with the exception of washouts was quite passable but only for a short distance. Here the hills are covered with the glaucous *Pinus armandi* which formed pure stands, taller trees than I had observed elsewhere with the exception perhaps of those found in the Lama temple near Pai-sha-kai on the Li-chiang plain. *Alnus* trees, the same as observed beyond Yang-pi were common again, *Rubus flavus*, *Hypericum patulum, Desmodium, Thea* sp.? and a few other shrubs and climbers.

The trail, terrifically muddy, winds along a hill side, the upper and lower part of it, although very steep, had been cleared for cultivation and in consequence landslides covered the trail in many places. Regular streams had cut beds into the soft earth, and made the road a sea of mud and consequently impassable.

We found many wild pears, trees with pears of all sizes and shapes, also a Shan-li-kuo 山梨果 tree with bright red fruits and a shrubby *Crataegus* with corymbs of bright red drupes. Walnuts were common and the small-fruited *Castanopsis delavayi* had again made its appearance, but only cut-over trees could be observed. We collected three varieties of pears and then followed the miserable trail up and up, partly paved; one would ride for 3 minutes and then dismount again owing to the road being left in mid air. A sea of mud 3-4 feet below the end of the pavement which in itself was composed of uneven boulders with mud holes between. One often had to turn back to find a place to descend into the ditch on the side of the pavement in order to pass over that stretch where it had been washed away. When a bit of straight road could be overlooked for some distance, one could see pieces of pavement left high and dry some feet above the mud road. Everybody avoided the paved part of the road and consequently the dirt road became a deep channel. We finally descended over mud, mud, mud to a small hamlet of three houses where we stopped for lunch. On the edge of the road grew a peculiar wild apple tree, with small oblong dark reddish black apples about 1/3 of an inch or less long; the tree was absolutely loaded with fruits. Here it began to rain again; the caravan appeared and put all our loads in the mud; after a miserable repast, my Tibetan cook buying nothing on the way, and giving me only boiled potatoes, we passed on again. The trail led between bare round hills, 400 feet or less high reminding me very much of the volcanic cones of the Waimea plain of Hawaii when the clouds swept their rounded tops. A strong wind blew and it was really cold. The road again was one vast morass. There I found *Delphinium, Aconitum venatorium, Anemone japonica* growing quite tall, an araliaceous shrub, *Nothopanax delavayi* with deeply divided leaves and another with Rhicinus-like leaves but sterile. Roses abounded, also *Dioscorea,* and further on *Senecio clivorum* Max. made its appearance with large golden yellow flowers, several Leguminosae, Elscholtzias, thistles, *Anaphalis*, the greenish large-headed *Leontopodium*, pink *Pedicularis* and others. From here on the road became worse and worse till it reached a small village the like I have never seen. From this village on the road was the most awful I have ever traveled over, there is no vocabulary in any known language adequate to describe the condition of that road. One may fancy a hillside perhaps 30 feet wide at one end narrowing down to 5 feet, a grade of 75° with furrows in it 6-10 feet deep leaving narrow ridges between; on the top of these ridges one had to balance oneself. Here and there a slippery boulder of 3-4 cubic feet would stick out the remnant of a pavement. Water rushing down, everywhere; this continued for about 150 yards or so, followed again by a sea of mud, a bit of pavement ending in mid air, then entering a chasm with a road or trail 3/4 of a foot wide, with water 2 feet deep running down from the enclosing banks. Rose branches which were hanging down on both sides from the banks with spiny raspberries did not ameliorate the condition of the road. All this at a grade of 59 to 60°, one long slide; with one word it was the worst road I have ever experienced in all Yünnan. This continued till we finally had our first glimps of the beautiful plain of Yung-ch'ang. As we left the narrow gap, the trail now more or less paved, slippery to a degree, led down a spur to the plain of Yung-ch'ang 永昌 but not without first passing a filthy village called Kuan-p'o 官坡. Graves made their appearance in the usual pine groves; many circular ones were noticed quite different from those so common in Yünnan. In one place, the banks had been washed out and 3 or 4 coffins were completely exposed. We followed the broad spur to a small village on the right, to the left was a huge temple the largest I have ever seen in Yünnan. As it looked so inviting we decided to stop here instead of going to Pan-ch'iao 板橋, 1 li beyond. An old caretaker bade us welcome; there are so many rooms, shrines, courtyards, and buildings that it would take many pages to describe this temple which is several hundred years old. It was repaired during the reign of Tao-kuang about 1821.

Tomorrow is a festive day for the Chinese it being the 15th of the 8th month; big fruit at Pan-ch'iao.

October 5, 1922

Before leaving the large temple where we spent a very quiet and restful night, we went to Pan-ch'iao to visit the market and to buy some pears. We were disappointed, only few of the Tung-li 東梨 or East Pear were seen. This decided us to go on to Yung-ch'ang. Ere we started we visited a pear garden where three varieties were in cultivation; we secured bud wood of the two of them and the P'ang-li 胖梨 or Fat Pear and then made ready for Yung-ch'ang. The well paved road with large flat slabs of rock in the center lead through rice fields lined with willow trees to the village of Pan-ch'iao, Wooden Plank Bridge, one long narrow village. En route we saw large vats buried or built into the ground. These vats are used for storing or mixing a dye produced from the leaves of the acanthaceous plant *Strobilanthes cusia* (Ham.) O. Ktze; it has whitish flowers; the leaves and twigs are thrown into water and left there, a rich dark blue dye is obtained. The lower part of Pan-ch'iao, 20 li from Yung-ch'ang, that is the center street was lined with cloth hanging down 50-60 feet from poles, to dry, it having been just taken from the dyeing vats where they are left to soak. The odor is a peculiar and nauseating one. After passing Pan-ch'iao, and its little village 1 li behind it, called Pan-ch'iao-ho-kai 板橋河街, we entered the broad plain of Yung-ch'ang with its vast rice fields. The people were just harvesting their crops. They had large wooden troughs carried into the fields where the rice had been cut and stacked up in bunches tied together, these bunches they took and with all their might struck against the walls of the trough, four men performing this threshing operation at one trough. This is the only place where I observed this method, but doubtless it is employed elsewhere in Yünnan. We passed several villages, I think three, ere we reached the north gate of the rather low city wall. The gate is also low and leads into the usual fore-court and to the main gate, which is about 35 feet thick. This part of the town is full of vegetables and fruit gardens and there is one continuation of houses as in other Chinese (Yünnanese) towns. We soon came to the main street with its unappetizing restaurants, stalls of all sorts, and the lively market; the streets were crowded thousands of men (and women) moving to and fro. Passing caravans shoved people into the vendors and their ware. In search of quarters we went to the west gate of Yung-ch'ang and there found a large temple at the foot of the hill called Pao Shan, on which by the way, there are many temples. We are located in a practically new one, not on the slopes but at the very foot of the hill. The temple were astonished when they saw us come with such huge quantities of pears, and more so when we invited everybody to eat, as we only wanted the seeds. They thought we were crazy, when we began picking out the seeds and giving away the expensive pears, 6 ¢ a piece of the large ones, or 3 ¢ gold, few people can afford to buy more than one or two pears which cost 120 cash a piece or 3 ¢ gold, and here I was giving away over a hundred. Of course we ate what we wanted cooked others and the rest we gave to whoever wanted them. Everybody looked on in wonder and munching pears.

Yung-ch'ang is surrounded by a low wall which encloses the hill at the foot of which it is situated. The walls form a triangle on the top of the hill, part of which is densely wooded with *Pinus sinensis*, it is on this hill called Pao Shan 保山 which is also the present name of the town, that all the numerous temples are situated. It is a quiet place, away from the noise and filth of the town. The whole place appears very prosperous, and certainly the Yung-ch'ang plain, about 18 miles long is one of the largest and richest of Yünnan. The hills surrounding it are bare not a tree anywhere, grassy slopes, and eroded hillsides. Fire wood is very expensive and scarce, it has to be brought from quite a distance. Vegetables of all kinds are sold, among the many, one new to me, apparently the basal part of a liliaceous plant which is white, and of the consistency of bamboo shoots, it is sliced thin and cut lengthwise and fried in butter. Spinach, enormous squashes, turnips and the tubers of a leguminous vine *Pueraria* are sold, it is especially plentiful.

October 7, 1922

We left our comfortable temple called Pai-ho-kuan 白鶴觀, View of the White Crane, at the foot of the hills very early. Passing through the dirty, filthy streets to the main street on the corner of which is a nauseating open public latrine, everything accumulating under the rocks of the street, this for decades if not centuries. Filthy mangy dogs were roaming the rather empty street picking up odds and ends from the market sweepings of the day before. My horse is lame so I have to walk or ride a baby horse bought by Ma Ting-an at Li-chiang.

As we passed near the south gate the odor became absolutely nauseating and this continued for some time. I nearly lost my breakfast and had to run. In such places people sat and talked as if it were the purest air imaginable, some lousing each other; the stench was abominable, open latrines neither limed nor cleaned and swarming with myriads of maggots. How people can live under such conditions is a riddle to me. We followed the filthy road for some distance and I gave a sigh of relief when we struck the rice fields. The road, partly paved, partly muddy, follows the plain southwest skirting the hillsides till we strike the village of A-shih-wo 阿石窩. Everybody gazes with mouth wide open. Soon the place is left behind as we turn to the right, through pear groves and wild persimmons and ascend the limestone mountain range bare to a degree. One third up on the right side there is a cave called Shih-hua-tung 石花洞 or Stone Flower Cave. The cave is pure limestone, and full of huge stalagmites, the walls forming broad sheets of stalagmites. At the entrance sit three gods the one in the center being Lung-wang or Dragon king. There are inscriptions cut into the limestone above the altar; every Chinese First moon thousands of people come to this place to worship. It is said that the exit of this cave, which is about 20 feet high and inhabited by hundreds of bats, giving a monotonous concert of *chi, chi, chi*, is near Pan-ch'iao about 30 li away. This of course may be a story. We ascend higher over a very rocky and muddy road till we reach the top of this lower range. The trail leads to the left, leaving a valley and the right hillside is covered with pines. We also enter pine scrub, many cut-over *Alnus* trees, *Hypericum patulum, Prunus* trees without flower or fruit, many wild pears, *Desmodium* with rich purple flowers, *Malus* trees with large apples and a handsome *Osbeckia crinita*, a melastomaceous shrub. Wild *Diospyros*, the shrubby *Crataegus*, here about 10 feet tall, loaded with the transparent crimson fruits, a shrubby *Corylus*, yellow composites and others. We climb the slopes diagonally of the limestone hills, yellowish walls being exposed, many blocks of limestone jutting out and many covering the ground below having tumbled from the heights above. To the left there is a pretty valley or rather deep undulated hilly depression. In the distance blue mountain ranges, quite low, and looking back one can still see the Yung-ch'ang plain but only a small part of it. The mountains become partly bare again till we reach the village of Leng-shui-ts'un 冷水村 or Cold Water Village, where the hillsides are covered with scrub, bamboo and small trees as the Shan-li-kuo 山梨果 a species of wild pear. We stopped for lunch in a miserable shed in which an old woman sold tea to passing coolies and the cook. After a quick repast and paying the woman 20 ¢ for letting us use her fire and place, we moved on.

The trail ascends steeply for awhile till we reach a gap, and lo and behold, there is a descent before us of about 4,000 feet. The paved road is in parts broken off as if cut with a knife leaving the traveler stranded in mid air. It leads down the steep grade in a tortuosity of a huge corkscrew, resembling a spiral circus chute. In addition to the miserableness of the road it began to pour, making the smooth worn stones so slippery that it was next to impossible to descend. Ma Ting-an fell several times; how the caravan managed to negotiate that trail I don't know, for they were behind. The hills here are absolutely bare, only grass, huge eroded areas, the roads washed out, and mud everywhere; the soil is a stiff yellowish-red clay. Here and there were outcroppings of slate and then traveling became a bit better. There is nothing whatsoever of botanical interest on today's journey. It began to rain so hard that we took shelter in a grass-covered shed where a man sold hot tea to travelers. We sat down and waited. The whole narrow valley of P'u-p'iao 蒲縹 was filled with black mist and rain. When we reached the gap above we had a magnificent view of the valley which is dotted with many villages. Rice and corn being the crops.

We finally descended over the most miserable road possible – washouts everywhere – to the plain and the first village reached is P'u-p'iao. The sun shone again but the road was still a quagmire. To the right, in the distance, seems to be a deep valley and a high mountain range, I suspect it to be the Salwin valley and river which we shall cross tomorrow. We followed the main road through the village, passed a large inn to the right with a new loft but were told that it was full of grass or hay, so we moved on to an old temple on the banks of the Hua-ch'iao Ho 花橋河 close to the bridge where we found shelter. There are two courtyards and in the last building, in the second yard, a small square where we made our beds. In the room I am occupying there are three shrines one large, center one, and two lesser ones, filled with a peculiar mud-clay or limestone mass which the keeper says has grown up around the image of which only the head protrudes, it is a curious mass passing from the base over the frame of the shrine and completely covering the image save the head. The people removed it once but it grew again and so they left it thinking the god wanted to be hidden from view. I examined the formidable mass some 6 feet high and four feet broad and found it full of chambers and channels, which makes me think that it is the work of some insect or termite. But I could detect no life whatever. Having a bad cold and tomorrow being a long stage I shall stop for tonight.

October 9, 1922
We left the dirty village of Hung-mu-shu 紅木樹 shortly after the sun had risen over the hills of the Salwin valley; yet the valley itself was still dark and misty. The trail leads steeply through the filthy village out into cut-over forest and hillsides covered with young growth of *Castanopsis delavayi*? and *Erythrina arborescens*. The vegetation up here is quite different from that of the Salwin valley below, which is purely tropical. Up here the temperature is much cooler, although it was quite warm climbing the steep hillsides. Zig zag leads the trail up a steep spur covered with *Castanopsis, Quercus*, several species, one with dense foliage and a beautiful round crown (*Lithocarpus spicata* var. *chittagonga,* 6998) the acorns being born in spikes densely packed. The soil is a loamy clay and the road was one big morass. Near the top of the spur we found the beautiful pink-flowered shrub *Luculia intermedia* Hutch. (neg. 30313) with delicious jasmin-like color. It was not common and we only saw it near the top of the spur. There are many *Quercus*, but only the small fruited *Castanopsis delavayi* and pine trees. Here the forest had been cut down and grasses have come in. We had a wonderful view of the valley to our left with a higher mountain spur on the other side, densely covered with virgin forest. Straight below us lay the Salwin Valley with the winding river, while to our right was a deep narrow valley partly filled with clouds. We climbed to the very top of this spur only to descend into a sort of depression in which there was situated a small village called *Elephant's neck* – Hsiang-po-tzu 象膊子; just before reaching that village we found again the large shrub with beautiful corymbs of large pink fragrant flowers *Luculia intermedia* Hutch. (7032) which I photographed with Lau-loo standing beneath. The trail rises again and climbs the adjoining spur, densely covered with virgin forest. *Castanopsis concolor* becomes very common, large pines, trees 70-80 feet tall cover the very steep slopes; large *Andromeda* trees are common. Rhododendrons make their appearance, several species of *Lithocarpus, L. dealbata* (7038), *L. fenestrata* (7039), *L. cleistocarpa* (7040), and large theaceous trees

(*Schima argentea*, 7041), the trunks moss-covered. After a climb of over two hours we reached the summit of that range along the top of which the trail continues. There is no more mud, for the whole summit of this magnificently forested range is mica rock and pure mica sand. The whole vegetation had the aspect of a rain forest, ferns abound in great numbers, especially epiphytic species covering the mossy trunks. There are almost pure stands of *Castanopsis* thickly covered with moss, ferns and orchids. Large trees of *Rhododendron delavayi* and *Rh. bullatum* occur and also what appears to be *Rhododendron decorum* but large specimens. Theaceous trees (*Schima*) are also common and a great many others without flower or fruit. *Lithocarpus* was most numerous. *Polypodium* ferns were especially abundant climbing the trees; a large shrubby spineless *Rubus* with digitate leaves, silvery beneath, was also common with melastomaceous shrubs, Spiraeas, and many others. We reached what is known as the *Wind Pass*, Feng-k'ou 風口, also written Feng-k'ou 峰 or Peak Pass. On both sides of the pass there still stand the remains of a wall which once was arched over, forming the boundary between the Yung-ch'ang and T'eng-yüeh districts. A stone tablet lies about inscribed on both sides in Chinese. To the left, a little beyond, is a small meadow, a camping place and to the left of it a small knoll on the top of which there is a telegraph pole, this is evidently the knoll mentioned by Davies, which he climbed to take sights.

The trail now descends through dense jungle, araliaceous trees occur, a semi-climbing *Eodia*-like spiny shrub with red fruits, a shrubby *Rhododendron* with very wooly leaves and many other plants. We continue to descend somewhat to the village of Hsiao-p'ing-ho 小平河 where we had lunch. From here on the trail climbs up a hillside only to descend again into dense forest with tall trees festooned with vines; *Gleichenia* ferns with huge fronds are common, three species of *Hedychium*, ferns and many other plants. The road from here on is exceptionally good, it was laid out by an English engineer; we descend to a dirty village with a clearing only to climb again to a densely forested knoll, and finally descend over cut-over hillsides into the Shweli valley to the Lu-chiang or Shweli River. The hills are now bare again with here and there a pine, *Alnus* tree, *Rhicinus* bushes with smooth capsules instead of the ordinary spiny ones. A melastomaceous shrub occurs and lower down a tree with dark red, small, ovoid drupes called Hsiang-kuo-shu 香果樹 by the Chinese. Of the fruit pulp or rather thick skin an oil is expressed used for burning in lamps, and for keeping leather soft, as well as for many other purposes. We now descend rapidly through terraced rice fields to the Shweli River which we cross by an iron chain suspension bridge. We now climb again slightly over a badly paved stoney road, through paddy fields to the miserable village of Kan-lan-chai 橄欖寨 or Olive Village with a long street absolutely filthy. I decided to spend the last night of this journey elsewhere than in this miserable hole so I climbed 2 li beyond to the top of a hill where there is a not altogether clean temple and here we decided to stop. It was long after sunset ere our coolies and caravan arrived. The next day we entered T'eng-yüeh 騰越 now called T'eng-ch'ung 騰衝. For information about T'eng-yüeh see the small diary.

Journey from T'eng-yüeh to Lung-ling, October 19, 1922 to October 25, 1922

October 19, 1922
It was a cloudy day when I awakened ready to start on the trip towards Lung-ling. En route I hope to find the desired chestnuts. We left rather late the mulemen who had promised that they would come early to tie on the loads were as usually late, their early means about half past eight. Of the seven mules hired I hoped at least to have three without loads reserving them for the chestnuts, but as usual there is so much to take along that there was even luggage left over.

The road led us through the same gate by which we entered T'eng-yüeh, through the same villages outside, but instead of crossing the plain diagonally to the hills, we turn to the right following a narrow stone path through paddy fields to the south western end of the plain up its narrowest part, to the village of Ta-tung 大董 crossed the Ta-tung Ho, and gradually ascended the low range encircling the plain at this end. The ridges here are pure sand, the trail being white and visible for long distances ahead. The hills are bare only here and there covered with scrub vegetation. We collected a few ferns growing along the roadside. The trail at this time of the year one would expect to be extremely muddy but for the fact that it follows the crest of the sandy ridge. Water was standing in places and some parts of the road were a rushing streambed. The road at the top of the ridge is about 1,000 feet higher than the T'eng-yüeh plain; in places the gradient is considerable. As we ascend higher there is more vegetation than near the T'eng-yüeh plain where every tree has been cut for firewood. We encountered the pink flowered shrub *Luculia intermedia*, the same species I photographed on the Salwin-Shweli divide, several oaks, I saw three or four species, a few leguminous shrubs, such as *Desmodium*, a lovely *Rhododendron* still in flower, bearing small pink flowers, the lower petal spotted purplish red. Here and there we saw a pine tree. The descent is rather gradual and less pleasant as an intervening small ridge which leads down to the Meng-lien 猛連 plain is not sand but yellowish red clay. In consequence the trail was extremely muddy, in places almost impassable. The roads in this part of Yünnan are much frequented by bullock caravans; the short steps that bullocks make and all stepping in the same place has made deep furrows across the road at intervals of a foot with a ridge between; these furrows are filled with water and mud in the rainy season. If one has to travel over a long stretch of these corrugated muddy and slippery roads it becomes exceedingly trying, walking for a horse is as difficult as for a man walking on railroad ties, with the additional discomfort for the horse of having to step into pools of mud, and for pedestrians on the muddy slippery ridges which often cause one to slide down into the deep muddy trench between them. We halted for lunch, it is really a joke to call it such, as my so-called cook a half Tibetan, had again prepared nothing and said all he had was some pears and greasy cake. So our stop was a very short one. In fact had I known that Meng-lien was so close, only 8 li distant, I would have continued the journey. We reached Meng-lien in about half an hour after leaving the river or rather small stream which is crossed at the foot of the ridge. Our eyes met bright colored paraphernalia on the one street in the village near the school house-temple. White-clad children and grown ups indicated a funeral. There were almost life-sized paper horses, the frame of bamboo, covered with paper and painted, paper hoa-gans or mountain chairs with pictures of females, a paper tomb stone, carefully made with long inscriptions, towers of paper, paper banners and leaned against the wall of the house there was an army of paper men and women 3 1/2 feet tall, representing servants. All these things were made for the comfort of the departed in the spirit world. His relatives saw that he had horses to ride, chairs to be carried in with chair coolies and servants, all of paper. In addition they burn imitation money to bribe the devils and make life comfortable in the world beyond. The Chinese are extremely materialistic and their religion does not go beyond material comfort in the next world. They would be extremely stingy while their relatives are alive, and weigh every cash before expending it for his benefit while ill, although when dead, a coffin is purchased costing about 200 dollars, plus all the above described paraphernalia for his comforts in the spirit world. Feasting of relatives etc., goes on for weeks afterwards; much noise is made in the evening to drive the spirits away, while probably a little attention as to food and cleanliness and some treatment at a hundredth part of the cost of the funeral, would in all probability saved his or her life.

We left them with all their foolishness and passed on to the Kai-tzu or Market of Meng-lien 3 li beyond, and took up quarters in a temple to the right on a small hill; the temple was in the process of repair. There are several chestnut trees around the temple, all diseased, in all probability with the chestnut fungus. I took specimens which I hope to send home for identification.

October 20, 1922

We left our temple at Meng-lien Market, about 7 a.m. expecting to reach Man-ch'i the same day. The road was abominable although the white sandy ridges are everywhere in evidence. We followed the plain for some distance and near two small villages on the narrow plain below, turned to the right in an easterly direction, crossing the hills. We found many tall straight pine trees, which looked more like *Pinus insularis* than *Pinus sinensis*. The needles are much longer, the trunks straight and tall, the branching habit is exactly that of *Pinus insularis* which I really think it is. Associated with it is what looks like *Pyrus pashia*, the leaves being broad and cordate. The trail ascends a ridge, winding about, with a deep valley to the left and right. Here we found *Castanopsis concolor* or a variety of it, the leaves differing considerably from it, while the fruits are identical. There were large *Schima Wallichii* trees, many flowering *Quercus* or *Lithocarpus*, the large shrubby *Hydrangea strigosa* Rehd. var. *macrophylla* (Hemsl.) Rehd. with cream colored bracts, the pink flowered shrub with smaller flowers is much in evidence, also huge *Betula luminifera* H. Winkl. trees, 80-90 feet tall, with a brown flaky bark, many *Prunus* but neither in flower or fruit. The trail descends rather steeply; on the slopes of this ridge which is well wooded in its upper part, nearly every tree was festooned with orchids as *Dendrobium heterocarpum* (7115), *Dendrob. strongilanthum* (7116), *Oberonia latipetala* which proved to be new, a *Bulbophyllum* etc.

There are many lovely spots for a camp but the slopes of the ridges are extremely steep and it would be a considerable job to bring water from the deep narrow valleys below. Oaks as *Quercus griffithii* (7123), with *Castanopsis tribuloides* A.DC. (7108) abound in great numbers, but strange to say are here all flowering; in that respect the whole flora resembles that of Keng-tung State and northern Laos. There are many tall bamboos here and there a *Schefflera delavayi* tree with digitate leaves. Ferns abound especially epiphytic *Polypodium,* together with orchids on the moss covered trees. We crossed the paddy plain and ascended another sandy ridge, the banks of the trail are lined with tall white-plumed grasses and a tall blue-flowered *Lobelia* and a smaller purple flowered one; *Hedychium* is common as are *Musa* and *Ficus congesta* (7049) but of the latter smaller trees only; the one occuring in Keng-tung with asymmetrical leaves is found together with oaks and pines. In fact the whole vegetation is that of the Meh Len valley of Keng-tung. The large-leaved tree-like *Vaccinium sprengelii* with the commonly found tree in Keng-tung with globose fruits on pendant spikes occurs here also.

We passed a village called Huang-ni-kou 黃泥垢, Yellow Dirt, filthy beyond description and so we went on, the caravan stopping a little beyond in an open grassy spot. We collected some of the large yellow-flowered orchids *Bulbophyllum* with large angular pendant fruits, a pink fragrant *Pedicularis* (7126) which turned out to be new and was called *Pedicularis lunglingensis* by Bonati. We then climbed higher over a gentle sloping ridge and there found large chestnut trees loaded with fruits. I looked about for water and a camping place and finding both, decided to stop. We pitched both tents near some graves on a lovely grassy spot surrounded by large trees.

October 21, 1922

Rain all night and this morning. The place was soaked, breakfast was cooked under an umbrella to keep the fires going. After the rain stopped, unfortunately only for a short time, we went about looking for and examining the large chestnut trees we spotted yesterday. There is one species here with small fruits. The trees are huge some 80-90 feet in height, with a mighty crown and huge straight trunks, 50-60 feet boles 2-3 feet in diameter. The trees were loaded with fruit but after climbing them and examining the fruits we found them all empty, there were a few ripe brown ones but they were all worm eaten. We examined tree after tree which proved to be *Castanopsis hystrix* A.DC. (6716) but all had empty fruits. It was exceedingly disappointing. Some villagers came with antiquated fire arms out for deer shooting. I employed two picking the chestnuts gathered from a tree near the camp which resembled much *Castanopsis concolor* but the fruits were much smaller, the seeds black and shining (6714) –it turned out to be *Castanopsis Wattii* (King) A. Camus. The leaves were lanceolate and fawn-colored beneath; this tree was not so large as the one with the empty fruits. We went out on the opposite ridge in torrential rain, but found no more chestnuts. One of the boys told us there was a large fruited species at the foot of the mountain. This decided us to climb down part ways; the slopes are nearly vertical and at least 1,000 feet deep, and there we found tall huge trees with spreading crowns filling the narrow valley to the right. Here were two *Castanopsis* – one *Castanopsis tribuloides* A.DC. (2360), the other *Castanopsis diversifolia* King (2405; neg. 30064), huge trees 90 feet tall, 3 feet in diameter, with burs 2 inches in diameter, such as we found in the extreme south of Yünnan near the British border. The boys said that last year there was a great abundance of fruit but now we could find none, save old ones on the ground from last year. As there are no more to be collected in this neighborhood we are going on tomorrow to Man-ch'i 蠻岐, hoping to find others on the way. It is a cold night the stars are out and we hope no rain will come.

The scenery from this camp is lovely, large trees everywhere. From north to south stretches for quite a distance a long regularly formed mountain chain, while numerous spurs like the one we are on drop abruptly and steeply into the narrow valleys below. A deep valley is to the back of us extending in a westerly direction.

October 22, 1922

We left our camp at about 8 a.m., with less chestnuts than I hoped to get. There were many tall trees but all had empty burs, and many were worm eaten. The trees were magnificent with straight boles and a huge wonderful crown. There are so many mountain trails and as the main road is not any broader than the trails it is very difficult to find one's way. This caused us to get off the right track; we had gone for a considerable distance ere we found out that the trail to Man-ch'i led due south. We had to retrace our steps and go on to another ridge following the deep valley to our right. We climbed another mountain spur, densely wooded with *Schima wallichiana* and *Betula luminifera* majestic tall trees, but deciduous at this time of the year. Wild persimmons (*Diospyros kaki* L. var. *silvestris* Mak. (6717)) with pubescent fruits velvety outside, 2 inches in diameter, gradation 7,000 feet. Of this we collected seed. The trail descended rather steeply into the valley and here we found *Castanopsis Wattii* (King) A. Camus (6714) such as we collected near our camp, the one with small black shiny fruits. From now on the trail winding about, led through uninteresting country mostly open grassland actually in the valley of the Shweli river which was plainly visible below. This river the Chinese call the Lung-ch'uan 龍川. We descended on to the right hand slope of the Shweli valley till we reached the dirty filthy piggery called Man-ch'i. I stopped for lunch under a huge *Ficus* tree, on a hill overlooking the Shweli valley. We soon were on the road again, the trail leading down on the side of the ridge through rice fields, and grass land with scattered pine groves. Following a narrow valley within the Shweli valley, the grassland gave way to pine forest and *Schima* trees with open spots of lovely meadows. The opposite mountain range forming the Shweli valley is naturally terraced, sloping towards the river. The last terrace is broad and ends abruptly, falling straight into the river bed. Waterfalls descend the steep sides. By the way, Man-ch'i is situated in the Shweli valley on the upper slopes of the western ridge and not on the other side of the ridge, also not in the valley, as is indicated on Davies's map. We followed now through lovely groves of pine trees, with scattered *Schima* trees and shrubs, till we reached cultivated (rice) land on the hillside; and below, on a narrow spur, we saw a conglomeration of huts, this was Man-lo. Market was on; we rode through the narrow main street everybody sitting on the dirty ground. There were pears, corn, and other delicacies only relished by Chinese. We retreated from this nauseating place to a small promontory overlooking the village and near some graves I decided to camp. Graves and temples are always located in the choicest spots.

I sent the interpreter down to the market to inquire about wild chestnuts and he returned with a small boy who said there were several trees 2 li distant and we decided to see them at once. Two were in a deep gulch, huge trees, their enormous trunks diappearing in the shrubs below. We went further on and came to a beautiful flat meadow not large, but sufficiently so to camp. Tall pine trees were plentiful. Here we found three large chestnut trees, the same species I found south of Szemao – *Castanopsis tribuloides* A. DC. (2663), neg. 30094) and which I photographed at that time. I decided to camp right here and sent men back to fetch the others. Our caravan came just along on the road above us and we hailed them to come down. There was a lovely brook quite near, plenty of firewood and below the hill a small village hidden by bamboo groves and tall trees. The village near our camp, consisting of only a few houses is called Lien-chien 連肩. Hardly had we pitched the tents, when the sky became full of dense thunderclouds, lightning flashed, the thunder rolled and it rained in earnest. Soon however all was over and the stars peeped here and there through quickly vanishing clouds. The night was cold.

October 23, 1922

Today we cleaned up all the three trees near here the result being two big loads of chestnut. The nuts are very small but all good, the trees are 80 feet tall with trunks 3 feet in diameter; perfectly sound and most prolific bearers. One of the boys climbed to the tops of these mighty trees and their huge limbs crashed to the ground. Everybody picked and basket after basket was taken to the camp. This is really a lovely spot. A mountain range stretches from north to south wooded with here and there a terraced rice field sandwiched in between ridges. Due south extends a gently-sloping broad range densely wooded.

The people while all speaking Chinese and not admitting any knowledge of Tai or Shan, all chew betelnut. Undoubtedly they are of Shan origin. An old woman of 80 years, came to see us and watch the collecting of the chestnuts. She was kind and gentle, I gave her a small amount of money with which she was delighted; she soon returned with a present of forty Tofu (bean curd) large fat bugs like locusts called feng in Chinese, 3 eggs, and another delicacy which nearly turned my stomach, we accepted all in good grace. She afterwards came and helped us picking chestnuts through the whole afternoon. I washed her eyes with Boric acid, as they were much inflamed and infected; she was very thankful and left smiling.

It is a glorious evening, all the stars are out and also the new moon which is about to disappear behind the western hills.

October 24, 1922

We stopped all day at our camp cleaning chestnuts. A glorious day of sunshine and cool breezes under the pines.

October 25, 1922
Dense fog settled over our camp and the valley of Shweli. Drip drip came the dew on to my tent from the branches of the large pine tree. We left late about 8 a.m., the fog still enshrowding everything. The trail led through the now deserted place of Man-lo 蠻邏 down to the banks of the Shweli. The trail was abominable ploughed in places by care-for-nothing peasants. The bank on the east side of the Shweli is very steep and wooded, a nice, well constructed suspension bridge called Kuang-chi t'eng-lung t'ieh so-ch'iao 廣濟騰龍鐵索橋, The rising dragon of Kuang-chi iron chain bridge. It was built in the 28th year of Kuang-hsü (1902 A.D.) spans the deep green waters of the Lung-ch'uan chiang 龍川江 as the river is called in Chinese. The trail ascends steeply through forests of *Schima wallichiana*, a tall tree, *Quercus, Castanopsis histrix* (7136, neg. 22977) which is here quite common. There is also to be found the birch *Betula luminifera* and the tall and stately *Altingia excelsa* Noronha (7174). Ferns abound on the banks in the shade of bushes. Deep ravines are to our right filled with tall trees, mostly *Castanopsis hystrix*, tall *Ficus* trees with epiphytic ferns, and orchids. When we reached the divide the trail lead down hill winding about through the hills, with terraced rice fields, bananas are plentiful, even *Pandanus* occurs and the whole has a very tropical aspect. We met a carpenter on the road, at least we surmised that he followed that profession as he carried all the implements of that trade on his back. We asked him if he considered the wood of this *Castanopsis histrix* any good, whereupon he replied that it was better than that of any chestnut found in this region. We stopped and cut down a few more branches of two trees gathering the ripe nuts.

We passed a hamlet called Pang-la-chang 邦蜡掌, rather clean with substantially built houses and broad well paved road. From here we climbed again through rather open country then down again a deep ravine, crossed a rushing stream over an arched bridge and followed it for a short distance; a smaller stream joined it here, and it was at that spot where we found numerous ferns, a climbing one, intermixed with *Trichomanes, Polypodium* and orchids with large yellowish flowers, the *Liparis bootanensis* Griffith (7157) originally found in Bhutan, which festooned the boulders. There were various shrubs, *Smilax*, the tea shrub *Thea sinensis* L. (7175), perhaps identical with the one found in the south at Muang-hai. From now on the country was more or less grassy and finally we found ourselves among bare hills through which the tortuous trail led up and down, real Chinese fashion, till we descended to the plain of Lung-ling 龍陵 or Dragon Mound; it is a rather small plain. We followed a narrow path through rice fields and through two gates about 100 yards apart, the latter the real entrance to the town. As usual the front of this old stone gate was pasted over with advertisements of the B.A.T. Tobacco Company, ugly, and absolutely out of place. Lung-ling is as filthy as all the other Chinese towns, but there are some large fine dwellings, at least they appear substantial and decent from the outside, if their owners dwell like pigs and with pigs in the interior recesses of these mansions I am unable to say, but very likely they do.

We went to an old temple on the hill outside the town, not the Kuan-yin temple but further east from it. It is a Buddhist temple of the Tibetan-like Mahâyâna corrupted faith. The priest a jolly good fellow was a native of Li-chiang; there also are two boys here, novices who are also Li-chiang boys one of them still speaking Na-khi. The name of this temple is Wen-ch'ang Kung 文昌宮, or The Palace of the God of Learning.

We left with the intention of going to Hsiang-ta determined to find the large fruited *Castanopsis*. The trail leads through the long and narrow village of Lung-ling between bare hills. On the right there are a few low hills in the center of the plain; on the top of one being a temple. The trail turns to the left by a small village, climbing around a ridge with *Alnus* and a few pine trees, and descends steeply into a narrow ravine following a very rocky streambed. The trail itself is very rocky and in places resembles a narrow ditch. Much granite is in evidence. The trees are mostly *Alnus*, the same as occurs beyond Yang-pi. The common asymmetrically-leaved *Ficus* found also in Keng-tung State and another species resembling *Ficus hispida* are much in evidence. The beautiful melastomaceous shrub *Lasianthus henryi* Hutch. (7187) with *Bauhinia variegata* are also common, and so are the small fruited *Ficus, Mussaenda frondosa* with large white bracts and yellow flowers, the shrubby species *Polygala congesta* with yellow flowers and reddish brown fruits. Leguminous vines several species belonging to the Papilionatae, *Phyllanthus emblica, Callicarpa arborea* Roxb. (7197) tree 25-30 feet are to be found in the rain forest with here and there *Saurauja napaulensis* DC. shrubs with pink flowers. This streambed is enclosed by bare hills, the vegetation being mostly along the actual bed. The stream is known as Pa-chiu Ho; the trail follows it to the bottom of the valley through rice fields, where it enters the Pan Ho 半河, coming from the left hand valley. The soil is almost pure granite sand, mixed with quartz. We cross the Pan Ho over an arched bridge and soon came to another stream without a bridge. The trail now leads steeply zigzag up a hillside. On this lower central ridge we find the common *Alnus* and *Albizzia* with a flat spreading crown and small leaflets, *Schima wallichii*, very large trees become very common, also the *Betula luminifera* trees with fawn colored flaky bark, several species of *Ficus* and tall trees of *Altingia excelsa* collected previously is also found together with a species of *Castanopsis*. From this ridge we descend in to a very sandy ravine of exceedingly tropical aspect, but not rich in species. There are quite a number of ferns, begonias, aroids and urticaceous shrubs, etc. We climb another ridge with large trees very tropical in appearance covered with huge leguminous lianas (Papilionatae). *Alnus* trees are still common as are *Ficus roxburghii*, *Engelhardtia spicata*, araliaceous trees, *Macaranga*, blue-flowered *Hydrangea* shrubs, *Debregeasia* trees with yellow fruits, *Albizzias, Glochidion, Phyllanthus emblica,* an occasional *Castanopsis histrix.* The *Lobelia* with bluish purple flowers adorns the banks of the trail, reaching a height of 10 feet.

We now are in the central part of the valley where we cross a stream known as Pa-chiu Ho; there is a small village of the same name, the houses are of Shan type. Not far from this village we found a large tree of the large fruited *Castanopsis concolor* but without fruits, old ones littered the ground beneath, and we collected specimens of fruits and leaves. The trail ascends again higher and higher following the side of a ridge. The valley is very narrow, terraced in places and planted to rice.

Trees are numerous, especially *Alnus* and *Ficus.* There are no pines until we come to the top of the ridge beyond, overlooking the small plain to the left and the semi-bare hills around Hsiang-ta. There is a tiliaceous tree, a similar to one found in Siam, *Erythrina arborescens, Quercus, Pieris,* 15 feet tall, a sapindaceous tree with huge leaves resembling those of *Melia acedarach* but of the genus *Radermachera* (7230) with long narrow curved pods. *Polygonum* is common on the banks, also *Rubus flavus* and a large coarse-leaved *Rubus, Indigofera,* numerous weeds, a pink flowered melastomaceous shrub with narrow leaves, *Dioscorea,* small singly-stemmed aralioids but none in flower, *Hedychium* sp.? and numerous ferns.

We decided to pitch camp on one of the grassy side spurs of the ridge we were following. Unfortunately the cook and coolie carrying

victuals passed by unnoticed and not seeing us went to Hsiang-ta 象達. Boon Mah fetched them, all arriving in camp at midnight after I had gone to bed with a cheerless supper of rice given me by the boys and improvised bamboo spoon as the only utensil.
The night for this region was bitterly cold.

The following day we went exploring the ridge beyond and I decided to climb a ridge way in the distance and south-west from Hsiang-ta, over which the road leads to Mang-shih 忙市 – Busy Market. The vegetation from the camp on, about 6,000 feet elevation, is the same as previously mentioned, with additional trees and shrubs such as the bright red-berried tree found also at Hsiao-shui chi, the blue-fruited *Gaultheria pyroloides* Hook. f. & Th. (7238), the very spiny (but only on trunks) spreading *Dalbergia mimosoides* Fr. (7236), *Pinus sinensis*, and *Alnus* trees which are here extremely common. The trail to Hsiang-ta leads on the left, down hills, but we continued straight ahead to the end of a narrow valley, crossed it and followed the ridge up the big knoll which was densely forested. There we found *Castanopsis histrix* (7136) very common on the steep slopes. At the foot of the slope there was a small village called Hsiao-shih-fang 小石房. We found the *Castanopsis* tree with fruits resembling *Castanopsis concolor* but with small black shining nuts as previously collected. *Anaphalis* were common as was a species of *Gentiana rigescens* Fr. We found a very large tree with huge spreading crown of the large fruited *Castanopsis* (7229) (later described as a new species as *Castanopsis Rockii* A. Camus), with a trunk 8 feet in diameter, but no fresh fruits, only fallen ones on the ground. There were large fruited, huge *Schima wallichii* trees, *Glochidion*, a few graves, *Phyllanthus* shrubs, Euphorbiaceae like Crotons, a *Diospyros* with large fruits (seeds collected), a large fruited *Elaeocarpus* (7226) (tree 30-40 feet), a *Pyrus* of which we collected seeds, the leaves being oblong and bronze-colored, the bignoniaceous *Thunbergia salweenensis* W. W. Sm. (7217), a vine with white flowers in pinkish sack-like inflated calices. Two species of *Smilax*, both large, *Smilax macrophylla* Roxb. (7231) and small leaved species *Smilax ferox* Wall. (7223). Near the top, on huge limestone boulders we found, creeping *Ficus* and long huge bunches of *Vittaria* fern. Not finding any ripe fruits of *Castanopsis* we retraced our steps to the camp. A splendid view could be had from here. Way in the distance to the right, but almost straight ahead, there was a high mountain in blue haze; the height is given on the map as

10,000 feet beyond which the Salwin winds its way southward. On the top or near it meanders a road which leads to Mang-shih in the plains. There are many trails; we followed up the path leading straight to the knoll but encountered too many spiny bushes of *Berberis sublevis* W. W. Sm. (7244). Returned to camp at 4 p.m.

Journey from T'eng-yüeh to Sadon in Burma
November 7, 1922 to November 22, 1922

November 7, 1922
Left T'eng-yüeh 騰越 about 9 a.m. From my house we follow the road back of the highschool due west towards the pagoda on the hill which like all the hills around T'eng-yüeh are bare. We cross the stone bridge over the Tieh-shui Ho 疊水河, which a hundred yards further on, drops into a deep gorge, forming the T'eng-yüeh water fall, all this tremendous water power goes to waste. We turn to the left, pass the waterfall the spray of which rises to the top of the gorge and spreads like a veil over the road. In front there is a picturesque little hill with a couple of trees on it, between these trees they are building a temple, the front facing the waterfall. We go through a narrow gap or cut made into the quartz-sandy hillside and descend to a small plain. We follow the stream bed for a short distance in the paddy plain. This stream flows into the Ta-ping Ho near Kan-ai. To the left, at the foot and slopes of the hills is the largest village in the T'eng-yüeh district called Ho-shun-hsiang 河順鄉. We keep to the right till we reach the village of Ta-chuang 大莊, here I saw a fruit sold by a woman, resembling a small persimmon, it is ovoid-oblong in outline, greenish-yellow outside, skin chartaceous, the fruitflesh is white slimy and sour, of the flavor of a pickle. This surrounds a rather large stone. The Chinese call it Nien-tsao 年草, and eat it with salt and chilies; it is said to grow on a large tree.

At the end of this village we leave the paved road and follow across the slopes of a grassy hill otherwise bare; the soil is red clay, but lava rocks crop out everywhere. Here are numerous graves built of lava rock but the roofs of these tombs are covered with weeds and look like the head of the „shock-headed Peter“ of the days of my youth. To the right and left are tall, bare (grass-covered) volcanic cones, with huge lava beds of the black vesicular type (*aa*) surrounding their bases. The trail leads now due west again between two ancient flows. These lava beds harbor some vegetation; next to grasses a shrubby *Elsholtzia* is exceedingly common, also a prostrate *Potentilla.* The lava bed to the left is older than that on the right which is covered with red clay. Due west stretches a mountain range from north to south at the foot of this range, semi-bare, is the valley of Mien-ching, in the center of which lies the village of Mien-ching proper. All told there are about 1,000 houses on this plain forming Mien-ching. The trail descends now into the valley and we cross the paddy fields where they are threshing rice with a wooden flail the size and exact shape of a clothes-hanger, only the handle is longer. To the left is a quarry near the top of a hill. Rock is taken out of the interior of the mountain as a sort of tunnel leads into the hill. We passed through the village where they had market; they were selling small cultivated chestnuts which they said came from along the Bhamo road. For lunch we stopped at a school-temple which was unoccupied. From here we cross the paddy fields again due north, follow a stream which makes a bend where another stream joins it and flows south; this latter stream we follow up into the hills. We now enter a narrow gorge filled with vegetation. At first only weeds, *Leontopodium Andersonii* C. B. Clarke (7273), *Anaphalis, Melastoma* and *Artemisia*; as we enter further a large grove of *Alnus* trees, *Schima wallichii*, *Quercus augustinii* var. *rockiana* A. Camus, *Castanopsis delavayi* the common species with brownish silver leaves, and small fruits, many tall araliaceous trees, *Schefflera minutistellata* Merr., a branching tree and others with single stems and a crown of leaves on the top, *Ternstroemia japonica* Thunb. (7289), *Lithocarpus spicata* (Sm.) var. *chittagonga* (King) Rehd. (7302), etc. The trail first leads down to the streambed, crosses it, and then climbs and climbs through jungle. There are many ferns, begonias, several orchids, a small terrestrial one with large white flowers, *Pleione grandiflora* (7281) also an epiphytic one *Coelogyne gardneriae* (7280) with drooping racemes of white flowers; *Smilax* and leguminous vines being common. Many trees of *Castanopsis* are to be found; there was one large tree of the large fruited species but without fruits and many of the species of which I collected seed near Man-lo, *Castanopsis hystrix* (6716). Another one was observed resembling *Castanopsis concolor* as found back of Fu-kuo Ssu near Yang-pi. There were many other trees especially common being *Alnus* the same as found beyond Yang-pi, and the *Betula luminifera* tree with cherry like bark, it is just now coming into flower and the young leaves are also just appearing.

Compositae are common but were not collected. *Engelhardtia spicata* is also met with but is not common. There are a species of *Litsea*, *Litsea cubeba* (Laur.) Pers. (7262), and a *Lindera* with very fragrant leaves, also the small-leaved, pink-flowered *Rhododendron* (7272). As we ascend higher we find *Rhododendron delavayi*, *Rhododendron decorum* Fr., an *Acer,* quite a large tree but without fruit, also the lovely *Prunus majestica* Koehne (7265) with rich pink flowers. *Polypodium* adorn the trunks of trees, canebrake grows near the streambed. The other side of the streambed is very steep and densely covered with forest while the one on which the trail is, is mostly cut over in places, but not a level spot can be found. If such there is, it is cultivated by some one. We are camping along the streambed in a mass of bracken fern and *Artemisia* weeds evidently an abandoned field. The banks of the stream are full of *Rubus*, both *Rubus flavus* and a broad-leaved sinuately lobed species. *Mahonia lomariifolia* (7260) is here very common and an araliaceous tree with very spiny trunk and huge much divided leaves resembling those of *Leea;* this species, if I do not err, I found also in the south, south of Chieng-rung. It is however not in flower and so I cannot say definitely if it is the same species. It is quite cold and I am forced to go to bed to keep warm.

November 8, 1922

Left our camp along the Shih-t'ou Ho 石頭河 at 7 a.m. The trail still ascends following the slopes of the bare hills; the other side is wooded, but seems to be secondary growth; two or three li more and we shall be in Hsin-ch'i. We descend slightly and at a turn of the trail we come to a circular temple on the top of a small hill. To the right there is a small paddy plain and on it, and on the western slopes of the hills, is situated the funniest conglomeration of a village called Hsin-ch'i 新齊. Palms, *Trachycarpus fortunei*, bamboo, and *Pinus armandi* are on the hillsides. Most of the trees in the neighborhood are in their autumn foliage, yellow, red, carmin, brown, etc. There are *Castanopsis, Quercus augustini* Skan. var. *Rockiana* A. Cam. (7296), *Lithocarpus spicata* (Sm.) var. *chittagonga* (King) Rehd. (7302), *Rhus*, a rose climbing over shrubs, *Hypericum* bushes, a monkshood, *Aconitum venatorium* var. *ecalcaratum* Airy-Shaw (7288), tall erect plants, etc. The village itself is awful, there are houses of wood or mud, covered with grass, walls made of bamboo, built on posts like Shan houses; they are leaning in all directions and it is really dangerous to pass through there as some of these walls may collapse any time for they lean half across the stony trail. Pools of water are everywhere; water standing, water running and along the brooks there is one rice pounder after another worked by water which fills a trough and empties it, and the wooden pounder falls into the rice in a hole in the ground. This performance repeats itself every two minutes or so. We climb up through this funny village, situated at the foot of the hills which are wooded with *Pinus armandi*, trees 80 feet or more in height. In the pools of standing water logs are soaking, this is to make them white ant proof. The hills beyond the pine-covered knolls are bare, save in the ravines and on their margins which are wooded, we find a theaceous tree *Gordonia axillaris* (7299) with large white flowers about 4 inches across, and having woody capsules with winged seeds, also a *Dalbergia* with scandent branches; lilies abound (*Lilium giganteum*) some, 12-15 feet tall, with long white tubular flowers, we collected the bulbs. There are several epiphytic orchids and ferns on the trees, also bignoniaceous vines. We climb higher over the otherwise bare hills to the summit of the ridge. Along the trail we find two species of *Diplycosia, Diplycosia muscicola* (W. W. Sm.) Merr. (7293), with large glaucous capsules, the other unidentified, also *Gaultheria pyroloides* Hook. f. & Th. (7238) with blue berries as we found on the Lung-ling Hsiang-ta trail. *Cornus* trees were common, also *Rhododendron delavayi, Rubus pentagona, Rosa, Buddleya* sp.? with long racemes of orange-pinkish flowers; otherwise the hills are bare. We descend now rapidly following a water course, lined with *Alnus* here called Ch'i-mu-shu 期木樹, *Quercus*, willows, bamboo, *Polygonum, Elsholtzia* shrubs, *Anaphalis* and composite weeds. The trail winds among hills still following the streamlet on the banks of which we now find a species of *Malus* with carmin fruits, very small, the tree being deciduous. We now cross a few paddy fields and find ourselves on the edge of the village of Hsiao-ti-fang 小地方, consisting of a few houses built in Shan style. Many tall fan palms protrude beyond the houses. From here the trail ascends again through narrow gaps, in part it is a brook, lined with oak trees with very large acorns. Ferns abound. We climb higher till we reach a narrow gap and lo and behold a wonderful view, a panorama of mountains is opened out before our eyes. Deep to our left is a wooded gorge, also to our right but not wooded; ahead in the distance we see a broad band of water winding through the brown paddy fields, it is the Ta-p'ing Ho 大平河. We descend zigzag a spur which leads to the bottom of the valley. This spur is grass-covered with the exception of a few trees which clothe the precipitous parts of it. There are many *Anneslea fragrans* Wall. (7314) trees in flower, *Alnus* trees, *Macropanax undulatus* (7364), and *Litsea eusoma* W. W. Sm. (7361) while the grass is 6-7 feet high. We intended to camp but could find no water save deep below in the gorge. It began to rain in earnest and so we moved on trying to find a place to pitch a tent. We are now camping near some paddy fields in the narrow valley along a streamlet which flows into the Ta-p'ing Ho. Mosquitos are very busy, so I must close.

November 9, 1922

Left our camp on the paddy fields called Sung-p'o-chio 松坡脚 rather late. The trail descended more or less steeply down to the streambed. The vegetation was composed of tall bamboo, *Ficus* with semi-sagitate leaves, a shrubby species of wild lemon, about 4-5 feet high, branching from the base; the fruit is spherical, compressed, but with a distinct knob at the apex. None of the seeds were good and the fruits were full of maggots, evidently attached by a fruit fly. This shrub was very common, also plentiful along the Ta-p'ing Ho. There were silk cotton trees, *Aleurites fordii*, guava bushes, leguminous shrubs and plenty of *Euphorbia antiquorum,* planted as a hedge. We shortly reached a small village inhabited by Shans, it is called Muang-lung 猛龍 and is situated at the bottom of the narrow gorge. A little beyond to the right is a very narrow gorge from which the Ta-p'ing Ho issues. The stream is quite swift, full of boulders at least in this stretch. We follow the streambed near paddy fields. The banks on the left are flat, that on the right steep; tall bamboo grow on the right bank. The whole stream from this village on is full of fishing weirs and naked men are seen in the swift waters arranging their fishing traps. We follow along the left bank, which is now steep and covered with shrubs of various kind as *Callicarpa,* various ferns, even tree ferns. We now climb the hillside which is covered with *Lindera* oil trees full of red berries. We turn to the left. The plain stretches from south to north, and is one large paddy plain. There are several villages on this plain, inhabited mostly by Shan. A large village on the right bank of the river is called Chiu-ch'eng 舊城 or Old City. On the left we pass through the small village of Man-leh. Next to Chiu-ch'eng is another small village Kuan-shang which extends to the bank of the river where a ferry plies from that village. We now turn to the right to the river bank and cross the stream in a dugout canoe, the horses being taken across a little further up. From here we cross the paddy fields in a northerly direction to the main village of the Chan-hsi plain called Tuan-po, the residence of the sawbwa[6] or t'ou-mu 頭目 – Eyes in the Head, whose name is Mong-tzu-tsai. The heat on the plain, only 3,200 feet elevation, is stifling and causes throbbing in the ears. The air is laden with humidity and not a breath of wind is stirring. We now approach the village, a three storey roof is prominent, to the right on a low hill the sawbwa has his house surrounded by a wall. We decided to call on him so I sent my card in and he promptly appeared in the courtyard ushering us into his reception room, a poor affair open to the roof, the walls covered with red paper inscribed in Chinese, old dusty paper flowers standing in vases on a shelf. He sat down on a campstool and meekly answered all our questions, always rising a bit after having answered. He is quite civil. He owns about half of the rice fields on his side of the plain, and has a household of about 30 people. His son, a spry fellow well dressed and somewhat sporty, appeared and he ordered him to get two men to escort us to the Kachin village of Hpun Kaw. We took leave of him to lunch under a big *Ficus* tree near by. The trail from here on winds through the paddy plain in a northerly direction crossing the stream continuously. We follow to the extreme end of the plain, over a most miserable trail full of boulders of immense size which are strewn about everywhere. The hills are bare, undoubtedly due to the Shan and Kachins; we saw the latter bringing in bundles of fire wood to the village; they look dirty, unkempt and well deserve the name Yeh-jen 野人 or Yi-jen 夷人 both meaning wild men which the Chinese give them. We cross streams, boulders, mud holes, in fact the trail is abominable, finally we ascend, the trail leading steeply up a ridge. There are many *Schima wallichii* trees, tree ferns of which we collected specimens, the trunks about 12 feet tall and nearly one foot thick, the frond having 14 pair of pinnae. There are many large trees of *Castanopsis*, also tall trees, 100 feet in height, with trunks of 5-6 feet in diameter of the nien-ts'ao, apparently a meliaceous tree. We met a number of spirit trees on the way, four bamboo poles tied together about 1 foot

---

6 Vgl. H. R. Davies: *Yun-nan. The link between India and the Yangtze.* Cambridge: Univ. Pr. 1909, 20: Sawbwa, the Burmese corruption of the Shan word Sow-hpa, Lord of the Sky, which is the usual title of a Shan chief.» (Chinesisch: auch *t'u-szu* 土司).

apart with a small seat on top for the spirit. This is erected by the Kachins. A steep trail leads to the edge of the ridge on which is situated Hpun Kaw, the Kachin village where we are going to stop for the night. Hpun Kaw is called P'eng-chio 朋角 in Chinese. The Kachins are really a very curious people. They wear the Chinese blue cotton pants and coats but the women wear home spun stuff of many colors, around the neck they wear silver rings, beads and necklaces, while below the knees they have numerous rings made of cane; these thin rings are dark brown to black in color. Their houses are often 150 feet long, and pitch dark inside even in the daytime. The roof (a big gable) is made of bamboo and thatched with grass, the edges of the roof reaching the ground. In front of their house they keep their caraboas in pools of stinking mud. The floor of the house is only about 1 1/2 feet above ground and pigs and chickens live beneath it. The room itself is not divided except where the fire is kept. There is also a square affair with sticks and offerings where the spirits dwell. In front of their long barn-like houses they have thick posts in the ground, close together, where they kill their cattle with spears. They are certainly a primitive people and remind me much on the Miao I found on Doi Chang in Siam. Their women are however not at all shy as are those of the Miao. They seem to have no conception of the value of money as they ask prohibitive prices for things as chicken, fire wood or their services. One asked 5 Rupees a day as a guide to the next village. They are dirty and filthy, but seem to have a keen wit about them, quite independent and indifferent. From here to Meng-ka 猛嘎 is half day's stage. Our escort sent by the sawbwa, armed with knives, spears and match-lock guns, will return tomorrow and two Kachins will go with us as far as the old boundary gates.

November 10, 1922

We spent a most miserable day at this village, rain all day, mud everywhere. Notwithstanding the rain the Kachins came to my tent to stare and ask for medicines. Nearly all are afflicted with huge goiters which hang in big lumps on to their chests, the men as well as the women are afflicted with them; besides they have all sore eyes and their teeth are black from betel chewing and encrusted with lime. The dirt of as many years as they are old adheres to them. The hair of the women is never kempt or cleaned but swarms with vermin; they tie a small knot right above their forehead with a piece of grass. For a skirt they wear a thick cloth reaching to the knees, it is fastened on the right side, but only begins below the navel. they do wear a jacket but that is all. The men all have their knives about 3 feet long, 2 inches broad, and placed in a sheath of one wall only, it is held in place by 3-4 bamboo cords at the point.

November 11, 1922
Early in the morning I learned that a mule belonging to our caravan had been stolen during the night. I went to see the Kachin Chief who pleaded ignorance. I gave him 15 minutes for the mule to be returned and in the meantime took possession of his horse, had it saddled and ready to start for Meng-ka. On requesting him to accompany us to Meng-ka he disappeared in the jungle. I left a soldier behind in the village who would stay till they brought the mule, and so we left for Meng Ka. From Hpun Kaw the ridge became narrow and densely wooded with huge trees, of *Castanopsis*, several species of *Quercus*, a species of *Malus* with large fruits (seeds collected). The araliaceous *Macropanax undulatus* (Wall.) Seem. trees (7364) (specimens collected) were also to be found, but the most common were *Quercus* and *Castanopsis*, with *Schima* trees.

A little higher up the forest became thinner, and consisted of small trees of *Quercus* mixed with a slender bamboo. We passed abandoned clearings full of tall grasses which hang over the trail, and, being wet from the rain, made walking very disagreeable. We now entered magnificent forests huge trees, moss-covered and one mass of lianas. There were many ferns and orchids but only two species were in flower, both of which we had collected previously. The most conspicuous was the large yellow-flowered one so common on the Lung-ling road. The trees were mostly a narrow-leaved *Quercus*, lauraceous trees not in fruit, *Engelhardtia spicata* and a tree with an *Engelhardtia*-like inflorescence but simple leaves. There was the lovely *Prunus majestica* with deep pink drooping flowers, araliaceous and theaceous trees, and a huge fruited *Castanopsis* with the nut completely enclosed. Bamboo was common, a tall slender species; ferns abounded mainly Polypodiaceae and numerous others (see collection), the whole was a typical virgin rain forest. We passed an occasional clearing full of weeds, secondary scrub and vines, and after some more climbing reached the top of the ridge. Here there is a double gate called Lao-kuan-ch‘eng 老關城 or Old Pass Town, now all tumbled down, the roots of a huge *Ficus* (banyan) penetrating the masonry, this gate which forms now the boundary between Chan-hsi

盞西 and Meng-ka was formerly the Burmese-Chinese border. From here on the trail descends steeply; tall slender timber bamboo formed the main vegetation, huge *Ficus* trees towering above it here and there. The large fruited *Castanopsis* was still common; many lianas, one with pandanus-like fruit, the phalanges were fleshy and born on naked trunks. Many *Pyrus* were in evidence, and a few orchids grew epiphytically on the moss-covered trunks. We passed a few clearings, entering forest again with smaller trees, one a member of the rose family, *Photinia serrulata* Lindl. (7373) with orange red fruits. We followed a stream which we had to cross many times it is called Ku-lien-p'i Ho 苦建皮河 and flows into Meng-ka Ho on the Meng-ka plain. From now on the vegetation changes, huge trees of a *Quercus* with small fruits like *Quercus delavayi*, also resembling it in the leaves, grew scattered on the otherwise grassy slopes. The trunks are almost black, the branches very tortuous. We descend only to ascend some small hills, turn to the right and finally descend to the Meng-ka plain.

On the left on reaching the plain proper there is a temple on the left hill but hidden among the trees. We now follow the Meng-ka Ho for a distance and then follow in the bed of a brook which joins the Meng-ka Ho further down. This brook the trail follows till we reach the village of Meng-ka, which consists of about 200 houses, 150 of which are Chinese, the remainder Li-su. We went to the so-called school-house, but it was such an abominable place full of carabaos, filth etc., that I decided to see the headman of the village, a Chinese by the name of Yang, 33 years old. He told us that he remembered Major Davies staying with his father, he was then 5 years old, also said that Davies gave his father 10 Rupees, as much as to say, I hope you will give me the same. The weather is still bad, raining again. We are now fairly comfortably housed in a loft of the headman's house in Meng-ka.

November 12, 1922

We stopped all day at Meng-ka. In the afternoon I went out to gather seeds of two varieties of wild pear and also seeds of *Photinia serrulata* Lindl. We climbed up the hills back of Meng-ka following a small brook, where I collected numerous ferns and a few flowering plants. There were several scrub oaks and a species of *Malus* with small red fruits.

November 13, 1922

The night was cold and perfectly clear. We started quite early with two Li-su guides. The trail follows the Meng-ka paddy plain in a northwesterly direction among scrub composed of *Rhus* with its red leaves and waxy fruiting panicles, roses, *Rosa moschata* subsp. *longicuspis* (Bertol. Herr.)(7374), all in fruit, the flowers are said to be white, we collected seeds of it. We now turned to the right climbing the hillside to the left where there were a number of graves among huge oak trees such as we found on the slopes coming from Hpun Kaw to Meng-ka. This region has been much cut over. The only interesting tree left is the ubiquitous *Alnus*. There were several *Quercus*, but *Alnus* held sway. At the head of the valley there is a Li-su village which we visited. We also met quite a number of Li-su women fully dressed and covered with numerous ornaments, rings around their necks, heads, and bells tied in bunches to the margins of their jackets and sleeves. They wore short narrow trousers reaching to the knees. Also a long narrow apron decorated with heads and what appeared to be cowrie shells. The trail leads to the left leaving the Li-su village to the right. A steep trail ascends to what appears to be the summit range which enclosed the valley. Vegetation becomes scarce only a few *Alnus* trees remained till we reach the grassy slopes of the hilltop. From here a picture of desolation meets our eyes. As far as I can see bare hills, with dead trees and grasses. We descend steeply on the other side of the range into a valley between canebrake and weeds. On the banks of the trail we see an occasional *Rhododendron decorum* Fr. and *Rhododendron delavayi* Fr. The earth bank itself is covered with the prostrate *Gaultheria nummularioides* D. Don (7384). From here on we climb over absolutely bare hills, only grasses, one very tall species *Eulalia quadrinervis* (Hack.) Kuntze (7385) taking possession of the hillsides. There is an *Anaphalis* and other composite weeds as *Lactuca hastata* (Wall.) D.C. (7383), this is the only vegetation covering these hills up to 8,500 feet. The soil is sandy, the trail miserable, regular ditches 10-15 feet deep filled with blocks of quartz. Every now and then we pass through thickets of a slender canebrake with a occasional *Rhododendron sidereum* Balf. f. (7376) and a short shrubby one with rusty wooly leaves, *Rhododendron bullatum* Fr. (7377) similar to the one found on the Hung-mu-shu ridge. To the left and right are deep valleys both bare except the upper slopes of the main watershed which forms the boundary between China and Burma. We pass through a narrow gap and through a bit of mossy forest, but soon we are out on the bare hillside. Here the slopes are completely covered with a short bamboo 1-2 feet high (1790). All were sterile and afflicted with galls. It grew at 9,000 feet elevation on both the Burmah and China side.

The upper slopes are covered with moss forest, the trees are tinted with all the autumn colors, red, yellow, brown, and the rich green of the evergreen. We pass another narrow gap; to the left is a pile of rocks at first almost appearing like a grave. Back of it is a square block of stone or cement. On the west face of it is engraved the figure 33 both in English and Chinese. This is the actual boundary. We now enter a sort of basin with pools of water, and streamlets running into this basin from all directions. The round hillsides are covered with lovely forest of *Quercus* and many species of *Rhododendron,* one a tree 35-40 feet with a trunk of nearly 2 feet in diameter, *Rhododendron sino-grande* Balf. f. & Sem. (7381) is especially conspicuous on account of its huge leaves which are silvery beneath. Every tree is litterally clothed with moss and hepaticae, lichens are scarce. Ferns abound on the moss covered tree trunks, especially Polypodia, *Trichomanes* and *Hymenophyllum*, also *Vittaria* (two species) one with extremely narrow fronds. There are many peculiar trees here which were new to me as an araliaceous plant *Macropanax undulatus* (Wall.) Seem. (7433), 30-40 feet, the trees were in fruit. There are lauraceous and rutaceous plants and one species of oak, *Lithocarpus variolosa* (Fr.) Chin (7447) is especially common, also one or two other species occur but without fruit. A *Primula* lines the path also an *Aconitum* but passed the fruiting stage. There are several species of raspberries, *Rubus calycinus* Wall. (7391) is prostrate in *Rhododendron* forest. We found a huge stout lily (*Lilium giganteum* var. *yünnanense*) in ripe fruit with deltoid large leaves but dead. The stem was about 5-7 feet tall and 2 inches thick. Fruits collected. The undergrowth in this interesting *Rhododendron* forest is mainly one species of fern, fronds simply pinnate, the fertile frond with very narrow pinnae. This fern is terrestrial only and covers the ground to the exclusion of everything else. One other tree must be mentioned a species of *Camellia wardii* Kob. (7446) but apparently not common. The trail leads from this basin up a small valley densely wooded with forest as described above only to descend into another basin with sufficient flat space for a camp. Here we decided to pitch our tents. The whole basin is covered with the short bamboo previously mentioned while the hillsides are densely wooded with *Rhododendron* forest. The coloring of the autumn foliage in the glow of a setting sun was perfectly glorious, brilliant tints of yellow, the red of the maples, the rich green of the *Rhododendron*, and the whitish silvery leaves of another species of tree unknown to me, made a wonderful color scheme. All these trees have beautiful dense round crowns, and thus the hillsides have an appearance of a billowy sea of magnificent color. The interior of these forests are weird, moss hangs from the trees several feet long, the trunks are enclosed in moss and the ground covered with ferns; brooks are everywhere. When we first arrived we found 5 wolves of a reddish yellow color; they sat down and looked at us, I fired a shot and they all disappeared. One of our Li-su men said that tigers frequented this place also. I however think that it is much to high 9,000 feet and too cold for tigers.

The sky is clear and it is bitterly cold, my fingers are numb from the cold and I can't write any more. The place where we camp is called Ch'ang-ti-fang 長地方.

November 14, 1922

The night was bitterly cold and during the early part of the morning it rained heavily. At about sunrise the wind changed to the east and the sky was clear again. We left at about 8 a.m. The trail leads through the small bamboo 1-2 feet high. Pools of water everywhere and brooks are descending from the hillsides from all directions. We soon entered the mossy *Rhododendron* forest. There were large theaceous trees, *Camellia wardii* Kobuski (7446) with white flowers, *Eurya cerasifolia* (D. Don) Kobuski (7448), many Rhododendrons, but neither in flower nor in fruit. The large heavy silvery *Rhododendron sino-grande* was the predominating one. The *Quercus* with dark green, stiff leaves was common, the acorns are born in spikes and could be found on the ground beneath. Maple trees were very common. Tall araliaceous shrubs were a species of the *Daphne papyracea* Wall. var. *crassiuscula* Rehd. (7412), an *Osmanthus* (7394) in fruit, peculiar pink flowered small trees, *Smilax lebrunii* with small red berries, an *Arisaema* (7429) with dark green divided leaves; an aconite, many ferns, etc. We descended rapidly over a stony trail following a rushing brook its downward course. We now entered dense stands of slender bamboo (*Arundinaria*) here and there a maple, and bushes which were neither in flower nor in fruit, but most of them had dark green leathery leaves. *Lindera thomsonii* Allen (7409) grew with the aralioids higher up. From the bamboo jungle we gradually entered magnificent forest of trees 100-150 feet in height with straight boles 70-80 without a branch and 2-3 feet in diameter. It was of course impossible to ascertain what they were the immense height, the huge moss covered trunks made climbing out of the question, no fruits could be found on the ground. It was a glorious morning and as desolate as the country is on the Chinese side as magnificent is it on the Burmah side, absolutely untouched virgin forest. There were many climbers especially *Pothos* and *Philodendron*, many epiphytic ferns; *Asplenium, Hymenophyllum* and *Trichomanes* ferns growing on the trunks together with milky asclepiaceous creepers with thick fleshy leaves. There were spiny small-leaved shrubs with red berries, tall *Asplenium* ferns, myrsinaceous shrubs, small araliaceous trees and a tall *Arisaema* with huge (5) foliate leaves; but the tall immense trees could neither be collected nor could they be identified. One huge tree, *Litsea khasiana* Meissn. var. *hookeri* Meissn. (7439) we found of which the native from Meng-ka said that the wood is used as incense and is burned in brass bowls before their idols. Steeply we descended a zigzag trail, the forest becoming more and more tropical, tree ferns *Alsophila khasiana* (7431) appeared 40 feet tall with trunks 1 foot in diameter, the fronds 10 feet long, having 16 pairs of alternate pinnae (specimen collected). Here also we found a huge *Castanopsis* with large burs as we found in the Lung-ling Hsiang-ta road, but only old fruits could be found. There were no oaks nor other *Castanopsis* to be seen. The undergrowth became dense, a tangle of ferns, vines, buttressed trees with dense masses of Philodendrons climbing their trunks, no palms were seen nor Pandani. Araliaceous trees 60 to 80 feet tall were encountered as *Schefflera impressa* Harms and a much branched species which formed a tree 60 feet tall with huge terminal panicles, and 5-digitate leaves which proved to be new and was described as *Brassaiopsis Rockii* Merr. (type, 7432). There were simple stemmed species with a crown of leaves at the apex, resembling a *Carica papaya.* Steeper and steeper became the trail till we finally reached the stream called Chieh-ma Ho 疥痲河 – Nettle Stream, on account of the large stinging nettle (also found on Doi Sotop in Siam) found along the banks. There were other Urticaceae, many vines, and huge masses of a species of *Musa* with large truncate leaves, the apex being as broad as near the center. Tree ferns loomed up high out of the forest on the steep slopes on the other side of the streambed. We lunched near the clear, cold waters of the Chieh-ma Ho, at about 5,500 feet elevation, having descended nearly 4,000 feet. Here the forest is of course evergreen, and the rich somber green contrasts exceedingly from the temperate forest above, where the leaves have taken on all the colors of an European or American temperate forest in autumn, the line of demarcation is very striking. Five li more and we shall arrive at the village of Chieh-ma-ho. The trail ascends only to descend again; the vegetation is now more or less uninteresting, consisting mainly of *Alnus* trees, with *Engelhardtia* trees, probably *Engelhardtia spicata*; the *Ficus* with semi-sagittate leaves, the *Saurauja* with brownish leaves and pink flowers; bamboo is also common. I must mention a very broad-leaved, large scandent bamboo which occurs in the large magnificent forest half way down. We arrived at the foot of a bare hill; a narrow valley leads to the left, one trail descends to a few houses, another leads us the bare hill into the small valley, in which Chieh-ma-ho-chai 疥痳河寨 is situated. We pass a fort-like house with masonry gate, and wall, and are lead to the headman's house. The entrance is a massive low gate; the whole compound is surrounded by a brick wall, with rock foundation. As the place reeked of filth from pigs etc., I decided to pitch the tents on the top of a bare hill to the left of the village. It began to rain again in earnest, and although the sky is dark and gloomy it has now stopped. From my camp I can see the fort and white house of Sadon or Hsi-tung 昔董 as the Chinese call it. From below the summit on the Burmese side we had a splendid view over the plain of northern Burmah and its distant mountain border both west and north. Range upon range was visible in the northern part, sharply outlined, as the valleys were filled with clouds.

November 15, 1922

We rose early and broke camp; the sky was clear and the day promised to be a fine one. The trail from this small place was rather bad, narrow and muddy, but soon we struck a broad road well trimmed, not like in China where one wipes the dew of the long overhanging grasses and bushes along the banks of the road. The forest was dense again in parts, mostly Araliaceae, *Eugenia* sp.?, *Callicarpa* sp.?, *Alnus nepalensis*, a *Symplocos* and many nettles or Chieh-ma, Zingiberaceae, ferns, begonias with beautifully marked leaves but not in flower, there were also *Castanopsis indica* (7451) trees but all with empty fruits; a species of *Castanea* about 50 feet tall in flower with empty burs still adhering. We passed through several Kachin villages with their typical houses and near their villages the usual altars made of bamboo under trees for their nats[7] or spirits. The Kachins relate that when the great Spirit called up all the people of the earth to give them presents, everybody went up with large baskets and they returned with wealth. The Kachins had however only little bags which did not hold much, hence their poverty. When God called the people up again, the Kachins thought they would bring large baskets. At that time God gave out all the nats and spirits and the Kachins having brought large baskets got most of the nats which are still with them. This is the reason why they have more than any other people. They also relate how the book of knowledge written on sheep skin was lost to them at the time God distributed it; when they left from on high they got lost and had no food and being hungry they boiled the sheepskin and ate it, and thus was lost the book of knowledge hence their ignorance. After passing Kachin villages and paddy fields we descended to a small paddy plain at the foot of the Sadon hill. There we met a caravan which was going to buy salt to take to China free of duty over the Chang-ti fang pass where there is no custom house. We arrived at the Sadon military post at about 1 p.m. There is a lovely clean Dak Bungalow near Mr. Wilson's house. The fort built of stone and surrounded by barbed wire is on the very top of the hill; in the fort is the post and telegraph office. There is a fine drill ground, and just below the fort is a small village, or rather the cook houses of the Indians. It turned out a wonderful day and we took advantage of it and dried our plants, packed seeds etc.

November 16, 1922

A glorious morning; Tabu Pum is to the left and Matu Pum on the right, sky absolutely clear, not a cloud to be seen, the air was cold but very refreshing and pleasant. To the west the broad bend of the Irrawady is visible, the banks are rather high. On the nearest ridge the heliograph station is plainly visible and beyond, on the plain, the outline of Myitkyina. The last range on the horizon extends clear into Assam and is known as the Pukun? Range, while the one between and beyond the heliograph station harbors on the west the Hukong valley. There is a good P. W. D. road to Khamti 230 miles north of Myitkyina now quite safe. Stopped all day at Sadon; had dinner with

7 Burmesisch nât, «a term applied to all spiritual beings, angels, elfs, demons, or what not ...» H. Yule, A. C. Burnell: *Hobson-Jobson.* London: Routledge & K. Paul 1969, 619.

Mr. Wilson, the Deputy Commissioner, the lone white man in charge of this part of Burmah.

November 17, 1922
Left Sadon at half past six a.m., Mr. Wilson came to the Bungalow and said good bye. The road leads by the fort on the left side and down the hill, passing the target shed. We descend to the little paddy plain, crossed the small stream, and followed the trail on the other side of the valley up into the hills. The vegetation consisted mainly of *Alnus* trees, here and there a *Bauhinia* probably *Bauhinia variegata* a tree 30 feet tall, *Albizzia* sp.? with reddish pods and fine feathery foliage, *Ficus* sp.? the one with asymmetrical clasping leaves; we passed Kachin villages and small clearings only to enter dense forest; there were several species of *Ficus (Ficus roxburghii)* with large leaves and brown hairy fruits as collected in the Keng-tung State, *Castanopsis*, Albizzias, many araliaceous trees, Hedychiums, *Clerodendron nutans* Wall. (7466), a bush with long drooping racemes of white flowers, broad-leaved scandent bamboo, truncate leaved Musas, *Asplenium nidus*, the bird's nest fern of trees grew on the trunks as well as *Philodendron* vines. Here and there a tree fern with a 25 feet tall trunk, dark blackish-brown, the crown of fronds almost horizontal but somewhat spreading upwards, leaves 12 feet long, pinnae 19 pair, stipe prickly, of this we collected specimens. The broad leaved dark-green *Polypodium* was common also *Saurauja* trees with pink flowers and leaves brown beneath. There are many vines and ferns in this somber forest through which an excellent broad graded road winds along the hillside. To left (west) a wonderful panorama of the fantastic peaks unfolds itself. They are densely wooded and rise from the plain of Burmah. In the distance on the horizon is a long range of mountains which extend to Assam. We now descend and reach the village of Lu-chi 10 miles from Sadon; here is a weather-board rest house, very comfortable indeed; we had lunch on the verandah, Yow Bowm the Kachin caretaker bringing water and fire-wood. From here the road is not quite so good. We pass the Kachin village of La-pai of which I took several photos, the houses seem to be a bit better constructed than those of Hpun Kaw. Before reaching the village there is a new road which, instead of passing through the village, goes around the hillside down to the Hou-chiang 猴江 as the Chinese call it: Monkey River or Waw Kyang as it is spelled on the British maps. This is the newly constructed road. We were however advised to take

the old road which we did, passing over a mountain spur instead of going around it. We passed the Hou-chiang or Monkey River by a wire suspension bridge and ascended the hillside on the other side of the valley. There is mostly *Alnus nepalensis*, *Entada* vines, and the ordinary underscrub, as bracken fern etc., a short ride (horseback) of about 4 miles and we reached the stone bungalow on the left of the road, it is surrounded by a stone wall. The caretaker, a Kachin nicknamed *Punjabi*, on account of a few hair on his chin, lives in the Kachin village on the slopes of the hill below the bungalow. My horse took violently sick evidently from having eaten some poisonous weeds; its whole body was convulsed, all the nerves were twitching. It recovered however during the night after having been given it a large dose of epsom salt. The night was clear cool and pleasant.

November 18, 1922

We left our bungalow early. The broad road led through *Alnus nepalensis* forest, and passed Kachin clearings planted to *Coix lachrima jobii*, or Job's tears, the grass being about 6 feet tall; small *Phyllanthus* bushes bordered the road side, *Alnus* forest extended way up into the valley. We crossed a stream by another wire suspension bridge and followed the road through the same type of forest while here and there an araliaceous tree occured; grasses were common. Higher up the valley we crossed the same stream over a log bridge, a narrow cane bridge to the left of it having been abandoned as being too frail. From here we enter moss forest. There are many orchids, a lovely pink one, *Pleione grandiflora* (7512) very fragrant occurs on rocks with many ferns and other orchids. There are several Rubi, *Rubus rufus* Focke (7479), *Rubus ampelinus* Focke (7482) a tall slender bamboo, *Quercus* trees, and all through the forest are huge round boulders covered with moss and ferns, the type of trees now changes, there many Araliaceae, *Schefflera venulosa* (Seem.) Harms (7489) different species from those found lower down, lauraceous trees, *Litsea brachypoda* Allen (7480; paratype) and a host of others as large-flowered theaceous trees *Gordonia axillaris* and *Ternstroemia japonica* which we collected back of Hsin-ch'i, the second day out of T'eng-yüeh. A short walk brought us to the small Li-su village of Kham-bai-ti below the summit range, the border of Burma and Yünnan. North of the camp is a high hill or mountain, semi-bare, but full of huge rocks and pinnacles, southeast of it is a low gap partly wooded. The trail leads however near the bare hill higher up and nearer the big mountain to the left. We are comfortably housed in a bamboo rest-house.

November 19, 1922
We left our Li-su camp at Kham-bai-ti at dawn. A broad road leads through mossy forest; the underbrush alongside of the trail or rather road had been recently cut and there was not much to be collected. There were a number of araliaceous trees, *Quercus*, Theaceae, one a small araliaceous tree with a cauliflourous inflorescence (7530).

The forest was quite different from that found on the Ch'ang-ti-fang Range to Chieh-ma-ho. *Mahonia lomariifolia* Takeda was not uncommon, a small pink flowered melastomaceous shrub grew as underbrush while *Philodendron* infested the tree trunks, there were also *Schima* trees, *Polygonum*, numerous ferns, orchids, the latter not in flower. It appears that the forest on this side has been much disturbed by the Li-su tribe; the upper slopes of the ridge form the Burmah-China boundary; it is here bare on both sides. The peak to the left called Chi-chiao Shan 雞叫山 or The Call of the Chicken Mountain is also bare save in little ravines where there appears to be *Rhododendron* forest. The mountain is very rocky and ragged at the apex, in contrast to those found on the Chang-ti-fang pass which are rounded knobs covered with luxurious *Rhododendron* forest. A splendid road leads to the upper slopes out of the mossy forest into small bamboo, canebrake, with occasional *Cornus* trees, *Rubus pentagona*, the yellow *Senecio scandens* next a climber, *Anaphalis, Mahonia lomariifolia* Takeda, weed and grasses. To the right is a singular peak, pudding-shaped, which is called Hai-ho Shan or the Thin Shoe Hill, while Tabu-bum which is 11187 in height is called Ta-hsüeh Shan 大雪山. The ridge to the summit is bare, weeds to the left, canebrake on the right, infinitely less interesting than the Chang-ti-fang pass. The road leads zigzag over the steep slopes to the top of the ridge. A splendid road leads to the very boundary mark, a square block surrounded by a rock wall. From here the rotten Chinese road commences. The other side is just as disappointing, bare grassy slopes over which a miserable trail full of washouts and boulders leads down into the valley of the Niu-ch'üan Ho 牛圈河. It is a very narrow valley which was filled with clouds. No trees only weeds and grasses, with occasional shrubbery along a brook, till we reach the Chinese Custom house at Niu-ch'üan Ho. Here the hillsides are covered with what appears to be *Pinus insularis*. We follow a central spur down into the valley. Pines everywhere also *Alnus nepalensis*. The hillsides are without underbrush but brown from the dead pine needles; the pine trees here are tall straight trees, with an oblong crown, and reflexed drooping branches on old trees and with ascending branches on young trees; they belong apparently to the same species as found on the Lung-ling road.

A fairly good trail now leads down to a stream bed, the Ta Ho 大河 which is really the Ta-p‘ing Ho 大平河. Here grows *Aster rockianus* H.-M. We skirt a hill to the left follow up stream and reach the small poor Li-su hamlet of Ta-ho. Across the stream in front of the hamlet is a hill with an old broken fort. The Ta Ho is here a terrific torrent tumbling over huge boulders. A Li-su semicircular bridge spans it, a few feet up stream a more solid bridge of two strong logs covered cross-wise with branches and filled in with dirt makes the crossing a much safer venture. This bridge which is kept up by the China Maritime Customs, had a few days ago been washed away by a terrific flood, caused by three days of constant downpours. Here I found Mr. Houston the Customs Commissioner of T‘eng-yüeh and the English Consul (both dead now). The Consul's name was Coles. They were expecting their caravan and were going to camp on the banks of the Ta-ho, inviting me to do likewise. My caravan being late I accepted the invitation and spent a pleasant Sunday afternoon in civilized company.

In the late afternoon, at about sunset, we went up stream collecting, among the boulders I found a purplish daisy, *Aster rockianus* H.-M. (7533), several shrubs, ferns, while the pink *Pleione grandiflora* grew on boulders. Little Li-su boys came with their crossbows and arrows and exhibited their splendid markmanship by hitting a copper piece stuck in the sand. There is quite a lot of vegetation on the west bank of the river, there are *Thea* shrubs, *Mahonia lomariifolia* Takeda, *Quercus*, small shrubbery small-leaved *Rhododendron*, also larger leaved species and a holly-like tree but without flower or fruit. On the east bank are small terraced paddy fields. Ferns festoon the tree trunks.

November 20, 1922

Left our camp at 6:30 a.m. We ascend the hill by the Li-su village, then follow up a broad stream the Ku-yung Ho 古勇河 which flows into the Ta Ho near the village. The pines, described previously, cover the hillsides. Along the bank which is densely wooded are

Quercus, Thea, and many other shrubs and trees but mainly *Castanopsis concolor, Cornus capitata* Wall. (7551), small *Schima* trees with small fruits, the trunks are moss-covered and festooned with ferns of various kinds mostly *Polypodium*. This little forest along the Ku-yung Ho was a pleasant surprise; there are many cascades in this stream which drops very suddenly in places forming many rapids and almost waterfalls. We follow along the stream through jungle till we reach a paddy plain where we cross the stream over a miserable shaky one plank bridge, but some of the men prefer to ford the stream. The plain of Ku-yung is quite broad, surrounded by hills, partly wooded, partly bare, and is almost 10 miles long. We cross the paddy fields keeping to the right and pass through an old stone tower-like gate into the dirty village of Ku-yung. Turning to the left we cross again over paddy fields to Ku-yung kai-tzu 古勇街子. The place was practically empty, there being no market. From here we again cross paddy fields to another village called Shang-ma-ch'iao 上馬橋.

Here along the trail we found an old shed and I decided to stop for lunch. On the whole Ku-yung plain there are about 1,200 houses and about 50 villages, 25 of which are Li-su villages; there also some Shan and on the hills Kachin villages. The town of Ku-yung has no Chinese official but only a headman. The trail now follows the left side of the paddy plain skirting the hillside going west. The trail which we missed at first due to its being a streambed and not a trail, passes by other villages just as filthy and smelly as any other of millions found in China. On the hillsides we now see *Pinus armandi,* some *Catalpa Douclouxii*, the latter apparently planted, some wild small fruited *Diospyros* trees, *Trachycarpus fortunei* the bast of which is used for raincoats and as a sleeve for straining honey, etc. *Schima* trees are also met with. The whole stretch over this plain is most uninteresting and unprofitable from a botanical standpoint. The trail leads through the center of the paddy plain then again skirting the hillside. When passing through the paddy plain it is always a streambed as the farmers when not requiring water to flood the paddy fields divert the water into the Ta-lu or main road. This at a time when they dry their rice in the fields after having been harvested. This so-called highway is at best a miserable trail paved several hundred years ago and not repaired since. Hence the proverb *a road is good for ten years and bad for ten thousand.* As Pan-ch'iao is still some 15 li distant I decided to stop in an old forsaken school house called Chao-chung-tz'u 昭忠祠, on the lonely road near a village called Ma-li-chai 麻力寨. The doors being barricaded from the inside we climbed through an open space in the wall after throwing down a log meant for an obstruction. The whole structure is about to collapse. White ants have eaten the wooden post, the masonry is cracked and the walls leaning in different directions but we hope that it will keep together for the night. Opposite this school house is a hill on the top of which are the remains of an old temple, now without roofs. The school house really an old ancestral temple with the name of hundreds of deads on its walls, is soon going to follow suit. Over the school benches and on the few tables cut paddy was drying.

November 21, 1922

We left this miserable place, continued down another small paddy plain, really a continuation of the Ku-yung plain – to the small hamlet on the other side, called Ching-k'ou 箐口. The trail leads through an old tumbled down gate with the posts supporting the mud bricks, all rotten and ready to fall someday. In no other but a Chinese community would such a menace to traffic be allowed to remain. From this little filthy village we pass to another paddy plain; the air is bitterly cold, the sky grey, the sun without strength and my fingers too blue to take notes. After this the trail winds between bare hills with here and there a *Pinus armandi* tree, the elevation being about 7,600 feet or whereabouts. An occasional *Rhododendron delavayi*, scrubby oaks, araliaceous vines, *Acanthopanax trifoliatus* (L.) Merr. (7570), *Craibiodendron henryi* W.W. Sm., *Bucklandia populnea* R. Br. ferns etc. are met with. We pass thence out on the small paddy plain of Pan-ch'iao; the village is situated on the right side the trail passes through it and up the hillside. Here we find *Castanopsis concolor* trees, *Lithocarpus spicata* trees about 30-40 feet tall with spreading crown, *Populus tibetica* shrubs, a species of *Pyrus*, a tree 50 feet tall with rather large pears, two inches in diameter, and long large shiny leaves turning red. We collected the fruits. The trail now winds around these small hills, through dark narrow gaps where I found two interesting ferns, and out into more open country but between a confusion of small hills, the little spaces intervening between these closely set hills, are cultivated by the Li-su. The Ku-yung Ho flows to the left about 1,000 feet below. The trail now descends steeply a narrow ravine filled with trees of *Castanopsis concolor, Ficus roxburghii* with large leaves and large figs as found on the Hsiang-ta – Lung-ling road, *Alnus* trees, *Prunus majestica* now in flower, a rich dark pink. Zigzag we wind down to the narrow valley with terraced paddy fields to Ch'üan-ma-ching 荃麻箐. There, under a *Zizyphus* tree I had lunch. A miserable trail, part quagmire and streambed full of rocks leads down hill over paddy fields cultivated by Li-su; we enter scrub vegetation again, *Schima* trees, a species of *Pyrus* with small fruits and a tree probably an *Osmanthus* make their appearance. We are now on a central ridge, well wooded with tall trees, such as huge *Schima* trees, the *Betula luminifera* tree with the cherry bark, *Rhododendron stamineum* Fr., rosaceous trees with red berries, and a large peculiar dioecious tree with broad leaves and yellow globular male flowers. There are several species of oak, the nien-ts'ao trees, and an *Osmanthus* tree 40 feet tall or more loaded with fruits which we collected. We decided to camp on a hill surrounded by large old *Schima wallichii* trees, back of us, and a valley below us.

November 22, 1922

From our camp at the Kao-t'ien 高田 village on the small hill with the large *Schima* trees we descend to the paddy plain. At the very foot of the hill the trail leads to a wooden, covered bridge called Kao-ch'iao 高橋, over the Lu-k'ou Ho 路口河. A small village is to the right; the trail is a regular streambed and leads from the paddy plain up a small spur or ridge partly wooded, the Lu-k'ou Ho flowing in a valley to the right, below; we reach the village of Hsiao-chai 小寨. Hsiao-chai is partly divided consisting of two villages; to the left of the road is a paddy plain, much higher than the plain through which the Lu-k'ou Ho flows, but both are in the same valley. The paddy plain near Hsiao-chai is terraced; the trees around the place are mostly *Catalpa*, *Erythrina arborescens*, a *Ficus* etc. We pass through paddy fields over a miserably paved road to a rather large temple called Ya-tung Ssu; this temple is surrounded by a wall and is situated on a low flat hill, in the enclosure is a regular forest of *Catalpa* trees; the entrance is over a small stone bridge which spans a lotus pond. The place is, however, much neglected like all temples in Yünnan at this age of atheism even in China. About half a li ahead to the left is a small village called Fan-chia-ying 范家營 if anyone would wish to halt for the night the temple mentioned above can be recommended. Near the village of Fan-chia-ying the road divides, one trail to the right leads to Mien-chin, the other to the left to Wo-lung 窩龍 and T'eng-yüeh. The trail leads through scrub and then through dense stands of young

Alnus nepalensis trees called Chi-mu-shu by the Chinese, it looks as if it were planted. Mien-chin lies to the right in the plain below.

We pass now close to the quarry I mentioned the first day out of T'eng-yüeh to Mien-chin. There is no tunnel but the rock is just dug out from the side of the mountain. The only trees are *Alnus*, the soil is red clay smooth and very slippery after a heavy dew. We skirt at the foot of a spur to the left is Hsiao-wo-lung and a little further on Ta-wo-lung 大窩龍 or the Large Wo-lung. There are many pine trees and the place is surrounded by a lava-rock wall. The trees found here are *Lindera* sp.?, *Pinus armandi*, *Pinus sinensis*, etc. The road ascends the hill ahead to a flat small plateau, between bare hills which form the walls of an ancient crater. It leads directly through the center of the crater floor. Ahead is a little narrow gap, (artificial) through which the trail passes, it descends rapidly or rather steeply, the soil being whitish coarse sand. At the foot of the hills the road joins the road leading from T'eng-yüeh to Mien-chin. We now follow this road to T'eng-yüeh.

ZWEI WEITERE BRIEFE

Der erste Brief ist an Loy Marks gerichtet, in deren Haus Rock später wohnte, wenn er nicht auf Reisen war. Er hatte das Ehepaar Marks in China kennengelernt, als dieses sich dort kurze Zeit aufhielt. Der Brief ist von Bedeutung, weil er die Frage eines botanischen Gartens diskutiert. Heute gibt es mehrere botanische Gärten auf Oahu; in Honolulu selbst den Foster Garden, dessen Ehrenvorstand Rock später war, sowie das Lyon Arboretum, das der University of Hawaii angeschlossen und nach Rocks altem Freund Lyon benannt ist.

Seattle 5, Wash., April 9, '54

Mrs. Lester Marks
3860 Old Pali Road
Honolulu, T. H.

Dear Mrs. Marks:

In pursuance of our conversation re the establishing of a real Botanic Garden in Hawaii I would like to make the following suggestions:
That there is a real need for a Botanic Garden in Hawaii is obvious. It is high time that such a Garden be established for the following reasons:

Hawaii is part of the United States possessing a climate unsurpassed where both tropical and semi-temperate plants can be grown, and there should be no need for institutions like the Arnold Arboretum of Harvard to support a Botanic Garden in Cuba, or Cornell in Trinidad, countries outside the United States. Honolulu was considered one large garden, but with the increase of population and high taxes the beautiful homes and large private gardens are being done away with, old and rare trees and palms are being cut down and the land cut up for the building of small homes on whose grounds it will not be possible to plant more than a shrub or two. Many rare palms and trees have disappeared as I had opportunity during my recent visit to verify. Mr. Jordan's place, where there were unique specimens of palms, has been cut up beyond recognition and so are many others. Bulldozers are in evidence everywhere.

The native flora is also gradually disappearing. On my recent trips on the various islands I learned to my regret the passing of many rare trees which formerly were plentiful. I refer for instance to the beautiful dry forests on the land of Auahi – Kahikinui on the southern slopes of Haleakala. I explored the region in 1909 and 1910 and paid visits to the area in 1953 and this year; I found that 13 genera not species, have disappeared. Cattle has again been let into the area after the disappearance of Pamakani. It is only a matter of very few years, perhaps even months, before that land will be entirely denuded of trees.

The slopes of Hualalai (Puuwaawaa) are another area where the trees are fast disappearing; where in 1910 a lovely forest existed at Waihau, one finds now single trees scattered over the landscape. And of certain species belonging to endemic genera only single trees are in existence. Some have already become extinct. The wet areas, although set aside as forest reserves, are not what they were. The building of ditches, the draining of the land by means of tunnels and flumes to get every possible drop of water for the irrigation of cane fields, etc., has wrought many changes. Even the summit of the Kohala Mountains has changed beyond recognition. The area was once one vast bog covered with sphagnum moss (in 1909), today the moss is found only in small patches while pests like Pamakani[1] (a species different from the one found on Maui) has taken a hold besides a still worse pest, a Composite shrub of the genus *Pluchea*, which can thrive in the driest as well as wettest areas in the islands.

The flora on the top of Kohala has completely changed. Where formerly there were beautiful species of Lobelias one finds introduced grasses and weeds.

The same condition exists on Molokai.

To help drain these reservoirs, *Eucalyptus* has been planted on the top of the Kohala Mountains to denote a boundary line as I was told.

I speak from experience and a knowledge of Hawaiian forests covering a period of over 45 years.

To preserve this forest flora, one of the most interesting on the globe, with an endemism of over 80%, which means that 80% of the native flora is not known from any other part in the world, is an urgent reason to establish a botanic garden in Hawaii.

1 Wohl *Hibiscus arnottianus*.

Of all the places visited on Oahu, I have come to the conclusion that Kahana Valley on the windward side of Oahu is the most suitable. The valley has been little disturbed, its lateral ridges extend almost to the sea, on the drier areas such native plants (trees) peculiar to the drier slopes can be grown; such plants can be grown from seeds gathered on all the various islands and accomodated in similar locations in Kahana.

The Hawaiian Islands are rich in Lobelias, most of them are restricted to the wet forests of the various islands but of the 160 odd species only very few occur on more than one island. It is easy to propagate them and plant them in the wet forest at the head of the valley on the central ridge.

Exotics and palms can be established in the valley floor. The latter on account of its stream can also harbor aquatic plants, like *Victoria regia*, etc., while on the shore which is a part of the valley, a typical strand forest can be cultivated.

Kahana has great possibilities for a Botanic Garden and is to my mind the most suitable area where such a Garden can be established. It would serve both for the transplanting of Hawaiian species which are gradually becoming extinct, and also for the establishing of an exotic garden of tropical trees and shrubs. Such a garden would attract scholars and students from the various botanical institutions of the United States and abroad. Like the „Buitenzorg“[2] in Java, it would become a Mecca not only for Botanists but also plant lovers and horticulturists from all over the world.

This is just a rough outline, but the time has come, if we want to preserve what still can be preserved, for the setting aside of an area like Kahana and the establishing of a Botanic Garden which will be the pride of Hawaii.

Other areas on Hawaii or rather Oahu are not suitable. It is best to start properly with a comprehensive plan, then with an area like Foster Garden with many old trees many of which have reached their age limit, and also on account of lack of possibility to expand. The Manoa Arboretum[3] is confined to the very head of that valley and is not suitable for the growing of dry district plants.

2 Der historische botanische Garten in Bogor.

3 Das heutige Lyon Arboretum.

I would, therefore, urge that, if the people of Hawaii are really desirous of finally establishing a Botanic Garden, Kahana Valley be selected for the purpose.

Although I am on in years, I shall be only to glad to help to the best of my ability towards guiding and planning in the layout of such a garden.

I wish you all success in convincing the people of Hawaii that the creation of such a Botanic Garden is a must. It will preserve not only the most interesting plants of our native flora, but will be a haven for exotic plants from all over the world.

Very sincerely yours,
Joseph F. Rock

Der zweite Brief ist von Peter Goullart an Rock gerichtet. Beide kannten sich gut aus Li-chiang, und wie das folgende Schreiben zeigt, nutzte Rock Goullarts Hilfe und Beziehungen.

Peter Goullart (1896? Kiev – †1963 in Deutschland) studierte in Moskau und Paris; sein Vater war ein reicher Kaufmann, starb aber, als Peter 2-3 Jahre alt war. Nach der Revolution ging Goullart über Moskau nach Turkestan, von dort nach Vladivostok, dann nach Harbin und Shanghai. Er war als Führer für reiche Amerikaner, dann als Mitarbeiter der gewerblichen Kooperativen tätig. Nach dem Umschwung in China ging er nach Singapore. Er ist u.a. durch sein Buch *Forgotten kingdom* (London 1956) bekannt geworden, das inzwischen ins Chinesische übersetzt ist, und auf diese Weise wurde man in Rußland auf ihn aufmerksam. Vgl. D. Voskresenskij: V kraju dunba. *Vostočnaja kollekcija.* Sommer 2002,84-96

Piter Gullart: Sjuńcin – obrjad „žertvennoj ljubvi“. Fragment iz knigi Zabytoe carstvo. Ebenda 97-103 (Hrsg. und übers. [aus d. Chin.] von D. Voskresenskij)

Im Gegensatz zu Voskresenskij gibt Wei Chien (vgl. Einleitung), S. 160 als Lebensdaten 1901-1975; als Geburtsort ist Moskau, als Sterbeort ist Singapore angegeben.

Kunming, April 4th 1947

Dear Dr. Rock,

I arrived here on March 30 at midnight after a five-day trip from Hsiakwan. Travel conditions are very difficult these days. Most of the trucks are worn out. We had accident after accident and had to change three trucks. I had to leave one of the trucks behind. I do not know when it can arrive, if at all. Big bands of bandits roam between Hsiakwan and Yunnanyi and trucks are constantly stopped and robbed. Four trucks behind us were robbed and a driver killed. Transportation, owing to doubling of price of gasoline, became expensive too. The ticket from Hsiakwan to Kunming or vice-versa costs 80.000 dollars and baggage is now 50.000 per each 100 catties.

Prices here are about treble since your departure from Kunming. Carbide lamps were selling at 3000 during your stay here. Now they are 13.000 each. The city is flooded with them and I cannot understand why you could not find them. I got your two wires and most of the drugs you want, except Penicillin 100.000 which is unobtainable. But I bought two boxes of 200.000 instead. Each box has six bottles at 50.000 per bottle. Nobody knows what hypo is and there are no magnesium tablets. I walked my feet off trying to get all the items you want and at as cheap prices as possible. Mr. Arnold changed your drafts at official rate and is keeping the two banknotes of 100 each which I turned over to him. I shall get all those heavy items for you such as flour, nails, some carbide etc. but it will cost a fortune to bring them to Likiang at present rates. And I am very apprehensive of the bandit situation.

I am planning to leave on the 13th or 14th. On Monday I am going to Yuki on business visit.

I bought the rubber things for Ho Chi Hui but I am unable to buy the bycycle [!] for Chao Chuk Tien as it costs 2.500.000 and he did not give me that money.

*I have not yet gone to the black market but will do so this afternoon. Please tell Lao Wong that nobody is expected to come with me from the Head Office in Shanghai.

Hoping to see you soon.

Very sincerely,

P. Goullart

*U.S. army food supplies are sold there as coffee, dried milk, jam, baking powder. J. F. Rock

REGISTER DER PFLANZENNAMEN

Offensichtliche Verschreibungen sind im Text korrigiert. Im Register ist die Schreibung standardisiert.

REGISTER DER PFLANZENNAMEN (NACH NUMMERN)

1551 – *Bridelia monoica* (Laur.) Merr.
1618 – *Engelhardtia spicata* Bl.
1646 – *Strychnos Nux-vomica* L.
1645 – *Spatholobus Roxburghii* Benth.
1681 – *Oxytenanthera albociliata* Munroe
1721 – *Dioscorea rockii* Prain & Burk.
1722 – *Bauhinia tenuiflora* Watt.
1744 – *Celastrus hindsii* Benth.
1755 – *Quercus daichangensis* Camus
1757 – *Hymenophyllum exsertum* Wall.
1791 – *Castanopsis armata* Spach
1793 –*Senecio acuminatus* Wall. forma *breviligulata* H.-M.
1995 – (*Quercus*) *Lithocarpus fenestrata* (Roxb.) Rehd.
2002 – *Syzygium abortivum* (Gagnep.) Merr. & L. M. Perry
2004 – *Cibotium barometz* (L.) Thn.
2031 – *Pteris ensiformis* Burm.
2034 – *Ficus gibbosa* Bl.
2035 – *Ficus comosa* Roxb.
2037 – *Ficus pyriformis* Hook. & Arn.
2040 – *Laportea crenulata* Gaudich. var.
2041 – *Alsophila glabra* (Bl.) Hook.
2051 – *Pittosporopsis kerrii* Craib.
2053 – *Melocalamus compactiflorus* B. & H.
2066 – *Dendrocalamus birmanicus* A. Camus nov. spec.
2097 – *Alsophila glauca* (Bl.) J. Sm.
2154 – *Dendrocalamus strictus* Nees
2155 – *Pinus Merkusii* Junghuhn & de Vriese
2172 – *Lithocarpus spicata* (Sm.) R. & W.
2174 – *Engelhardtia spicata* Bl.
2177 – *Prunus salicina* Lindl.
2192 – *Pyrus pashia* Buch.-Ham.
2195 – *Prunus puddum* Roxb.
2199 – *Prunus persica* (L.) Batsch.
2201 – *Castanopsis hystrix* A. DC.
2214 – *Quercus brandisiana* Kurz
2221 – *Quercus brandisiana* Kurz
2218 – *Lithocarpus fenestrata* (Roxb.) Rehd. & Wils.
2219 – *Castanopsis indica* A. DC.
2220 – *Schima wallichii* Choisy (det. Rehd.)

2226 – *Rubus ellipticus* Sm. (ellipticum)
2236 – *Clematis acuminata* D.C.
2237 – *Celastrus monospermus* Roxb.
2240 – *Prunus persica* (L.) Batsch
2248 – *Quercus acutissima* Carruth.
2252 – *Castanopsis hystrix* D.C.
2253 – *Pyrus pashia* Buch.-Ham.
2269 – *Wendlandia glabrata* D.C.
2270 – *Castanopsis tcheponensis* Hickel & Camus
2272 – *Bauhinia variegata* L., neg. no. 30048
2274 – *Lithocarpus spicata* (Sm.) R. & W.
2280 – *Dendrocalamus strictus* Nees
2295 – *Lithocarpus truncata* (King) Rehd.
2296 – *Vaccinium sprengelii* (G. Don) Sleumer
2288 – *Oxytenanthera albociliata* Munroe
2289 – *Oxytenanthera albociliata* Munroe
2298 – *Cibotium barometz* (L.) Thn.
2304 – *Dendrocalamus hookeri* Munro
2322 – *Dendrocalamus hookeri* Munro
2325 – *Vitex trifolia* L.
2332 – *Ficus altissima* Bl.
2335 – *Ficus obtusifolia* Roxb.
2341 – *Alsophila glabra* (Bl.) Hook.
2347 – *Justicia adhatoda* L.
2348 – *Justicia adhatoda* L.
2360 – *Castanopsis tribuloides* A.DC.
2363 – *Castanopsis tribuloides* A. DC.
2382 – *Engelhardtia spicata* Bl.
2389 – *Mucuna*
2405 – *Castanopsis diversifolia* King
2406 – *Vernonia Volkameriaefolia* D.C.
2407 – *Quercus acutissima* Carruth.
2408 – *Clematis heracleifolia* D.C.?
2414 – *Vernonia papillosa* Fr.
2424 – *Senecio scandens* Buch.-Ham., ex D. Don
2434 – *Mallotus philippinensis* Muell.
2436 – *Senecio scandens* L.
2439 – *Castanopsis hystrix* D.C.
2443 – *Alnus nepalensis* Don.
2453 – *Quercus acutissima* Carruth.
2455 – *Vaccinium sprengelii* (G. Don) Sleumer
2462 – *Oxytenanthera felix* Keng. Det. Keng
2467 – *Dendrocalamus giganteus* Munro
2469 – *Thea sinensis* L., neg. no. 30074
2474 – *Salix araeostachya* Schneider
2477 – *Lithocarpus spicata* var. *brevipetiolata* (A. DC.) Rehd. & Wils.
2479 – *Eugenia* L.

2494 – *Podocarpus macrophyllus* Don.
2495 – *Podocarpus neriifolius* D. Don
2502 – *Boehmeria platyphylla* Don.
2507 – *Jasminum* sp.?
2511 – *Streblus asper* Laur.
2512 – *Clematis kerriana* Craib.
2514 – *Salix wilsonii* Seem.
2520 – *Homonoia riparia* Lour.
2531 – *Phoenix roebelenii* O'Brien
2540 – *Deutzia corymbosa* R. Br.
2541 – *Campanula mekongensis* Diels
2556 – *Mayodendron igneum* Turcz.
2558 – *Lithocarpus cornea*?
2566 – *Bauhinia variegata* Linn.
2568 – *Toxocarpus* ?
2588 – *Lithocarpus spicata* (Sm.) R. & W.
2591 – *Mussaenda hossei* Craib.
2606 – *Sterculia fulgens* Wall.
2615 – *Castanopsis indica* A. DC.
2622 – *Schima Wallichii* Choisy
2631 – *Ficus altissima* Bl., neg. 30088
2633 – *Erythrina stricta* Roxb.
2654 – *Cycas pectinata* Griff.? (perhaps)
2662 – *Quercus helferiana* DC., or 2659
2663 – *Castanopsis tribuloides* A. DC.
2671 – *Vaccinium sprengelii* (G. Don) Sleumer
2680 – *Cycas* sp.?, neg. no. 30092, 30093
2689 – *Benettiodendron brevipes* Merrill, n.sp.
2696 – *Quercus acutissima* Carr.
2704 – *Schima wallichii* Choisy
2707 – *Symplocos paniculata* (Thunb.) Mig.
2728 – *Engelhardtia spicata* Bl.
2735 – *Primula malacoides* Fr., det. W. W. S.
2742 – *Syzygium rockii* sp. n. type Merr & Perry
2743 – *Rubus elongatus* Sm.
2749 – *Olea densiflora* H. L. Li
2751 – *Randia wallichii* Hook. f. ?
2757 – *Tupidanthus calyptratus* Hook. f. & Thoms.
2762 – *Rhododendron stamineum* Franch.
2791 – *Rubus alceaefolius* Poir.?
2803 – *Docynia delavayi* (Fr.) Schneid.
2804 – *Symplocos chinensis* (Laur.) Drude
2806 – *Milletia pulchra* Benth.
2808 – *Leucomeris decora* Kurz
2812 – *Vernonia volkameriaefolia* D. C.
2813 – *Leontopodium andersonii* C. D. Cl.
2816 – *Symplocos paniculata* (Thunb.)

2818 – *Melia azedarach* L.
2825 – *Lobelia pyramidalis* Wall.
2827 – *Cassia javanica* L. var. *indo-chinensis* Gagnep.
2836 – *Prunus puddum* Roxb.
2849 – *Mezoneurum sinense* Hemsl.
2869 – *Dracaena*
2876 – *Symplocos crataegoides* Buch.-Ham.
2877 – *Rhododendron mackenzianum* Forr.
2884 – *Prunus puddum* Roxb.
2885 – *Gordonia chrysandra*?
2888 – *Keteleeria Davidiana* Beiss.
2889 – *Myrica esculenta* Buch.-Ham.
2930 – *Quercus mongolica* Fish.
2933 – *Caesalpinia sepiaria* Roxb.
2934 – *Jasminum lanceolaria* Roxb.
2937 – *Clematis armandii* Fr.
2939 – *Ficus altissima* Bl.
2943 – *Wendlandia paniculata* Roxb,
2945 – *Woodfordia floribunda* Salisb.
2949 – *Trevesia palmata* Vis.
2950 – *Ehretia dicksoni* Hance
2952 – *Diospyros yünnanensis* Rehd. & Wils.
2953 – *Ficus ischnopoda* Miq.?
2957 – *Ficus nitida* Thunb.
2959 – *Quercus mespilifolia* Wall.
2960 – *Albizzia julibrissin* Durazz.
2962 – *Neyraudia madagascariensis* Hook. f.
2964 – *Cedrela Toona* Roxb.
2994 – *Pinus armandii*
2995 – *Osteomeles schwerinae* (Schneid.)
2997 – *Pyrus pashia* Buch.-Ham.
2998 – *Celtis* cf. *sinensis* Pers.
3000 – *Diospyros lotus* Linn.
3004 – *Docynia delavayi* (Fr.) Schneid.
3006 – *Pieris ovalifolia* Don.
3007 – *Rhododendron dendricola* Hutch.
3008 – *Rhododendron microphyton* Fr.
3012 – *Rhododendron pachypodum* Balf. f. S.
3012a – *Rhododendron delavayi* Fr.
3015 – *Rosa odorata* var. *gigantea* (Call.) Rehd. & W.
3026 – *Jasminum diversifolium* var. *glabricosum* (W. W. Sm.) Kobuski
3030 – *Hypericum patulum* Thunb.
3043 – *Rhododendron decorum* Fr.
3045 – *Thea pitardii* (C. Stuart) Rehd.
3072 – *Pistacia chinensis* Bunge
3075 – *Vaccinium fragile* Fr.
3076 – *Castanopsis delavayi* Fr.,

3077 – *Hypericum patulum* Thunb.
3081 – *Rhododendron delavayi* Fr.
3081a – *Rhododendron decorum* Fr.
3084 – *Campylotropis polyantha* (Fr.) Schindl.
3085 – *Leycesteria formosa* Wall. var. *stenosepala* Rehd.
3093 – *Quercus delavayi* Fr.
3094 – *Rosa multiflora* Thunb. var. *legitima* Regel
3097 – *Catalpa duclouxii* Dode
3100 – *Pieris formosa* (Wall.) D. Don.
3105 – *Thea pitardii* (C. Stuart) Rehd.
3112 – *Rosa omeiensis* Rolfe
3113 – *Rosa sericea* subsp. *eusericea* var. *Mairei* (Leid.) Herr.
3118 – *Rhododendron serotinum* Hutch.
3119 – *Pleione yünnanensis* Rolfe
3121 – *Elaeagnus umbellata* Thunb.
3122 – *Salix tenella* Schneid. var. *luctuosa* Lév.
3123 – *Rhododendron oleifolium* Fr.
3125 – *Euphorbia chrysocoma glaucophylla* Lév.?
3127 – *Indigofera mairei* Pamp.
3128 – *Corylopsis yünnanensis* Diels
3129 – *Rhododendron cyanocarpum* Fr.
3131 – *Rhododendron rubiginosum* Fr.
3132 – *Morina betonicoides* Benth
3133 – *Primula sonchifolia* Franch.
3137 – *Rhododendron trichocladum* Fr.
3139 – *Piptanthus concolor* Craib
3142 – *Rhododendron sulphureum* Fr.
3146 – *Rhododendron neriiflorum* Fr.
3147 – *Juniperus squamata* Buch, & Hamil.
3148 – *Abies delavayi* Fr.
3149 – *Rhododendron haematodes* Fr.
3150 – *Juniperus squamata* Buch & Hamil.
3152 – *Rhododendron fastigiatum* Fr.
3156 – *Vaccinium delavayi* Fr.
3161 – *Leptodermis carissa [scissa?]*
3162 – *Stellera Chamaejasme* Linn.
3164 – *Polygala sibirica* L.
3166 – *Rhododendron triflorum* Hook. f.
3180 – *Salix variegata* Franch.
3184 – *Alnus ferdinandi-coburgi* Schneid.
3186 – *Trachelospermum donieri* (Leid.) Woods.
3189 – *Bauhinia densiflora* Fr.
3191 – *Campylotropis polyantha* (Fr.) Schindl.
3193 – *Salix cavaleriei* Leid.
3194 – *Paliurus orientalis* (Fr.) Hemsley
3197 – *Rosa banksiae* Art. var. *normalis* Regel
3198 – *Styrax langkongensis* W. W. Sm.

3199 – *Antiotrema dunnianum* (Diels) H.-M.
3202 – *Pyracantha crenulata* (D. Don) Roem.
3203 – *Pistacia chinensis* Bunge
3205 – *Photinia prionophylla* (Fr.) Schneid.
3206 – *Pyrus calleryana* Decne
3207 – *Sophora flavescens* Ait.
3215 – *Docynia delavayi* (Fr.) Schneid.
3216 – *Jasminum beesianum* Forrest & Diels
3217 – *Catalpa duclouxii* Dode
3220 – *Myosotis sylvatica* (Elrh.) Hoffm.
3221 – *Cynoglossum amabile* Stapf & Dramm.
3222 – *Rosa sericea* subsp. *eusericea* var. *mairei* (Lév.) Herr.
3224 – *Campylotropis polyantha*
3227 – *Quercus semecarpifolia*
3232 – *Androsace spinulifera* Knuth.
3235 – *Salvia digitaliodes* Diels
3238 – *Rosa moschata* subsp. *brunonii* (Lindl.) Herr.
3241 – *Rhododendron delavayi* Fr.
3248 – *Clematis chrysocoma* Fr.
3257 – *Rhododendron decorum*
3258 – *Ligustrum sempervirens* (Fr.) Mansf.
3269 – *Rhododendron chartophyllum praecox*
3274 – *Rhododendron racemosum* Fr.
3281 – *Prinsepia utilis* Royle
3284 – *Clematis montana* Buch.-Ham.
3286 – *Prunus conadenia* Koehne
3291 – *Quercus semecarpifolia* Sm. var. *glabra* (Fr.) H.-H.
3293 – *Jasminum humile* L.
3294 – *Chionanthus retusa* Lindl. & Paxt.
3298 – *Buddleia heliophila* W. W. Sm.
3299 – *Primula begoniiformis* Petetitm.
3301 – *Decaisnea fargesii* Fr.
6566 – *Colquhounia coccinea* Wall. *var. mollis* Prain
6584 – *Castanopsis delavayi* Fr.
6587 – *Hibiscus venustus* Bl.
6594 – *Hibiscus forrestii* Diels
6598 – *Aconitum volubile* Pall.
6605 – *Begonia labordei* Lévl.
6617 – *Indigofera scabrida* Dunn.
6618 – *Gordonia axillaris* (Don) Diels
6631 – *Osbeckia chinensis* L.
6635 – *Corylus yünnanensis* (Fr.) A. Camus
6636 – *Pyrus pashia* Buch.-Ham.
6637 – *Pyrus pashia* Buch.-Ham.
6639 – *Inula cappa* DC.
6640 – *Leptodermis*
6645 – *Castanea mollissima* Bl.

6646 – *Castanopsis concolor* Rehd. & Wils.
6637 – *Castanopsis concolor* Rehd. & Wils.
6647 – *Castanopsis concolor* Rehd. & Wils.
6648 – *Anaphalis margaritacea* Benth & Hook. f. *cinnamomea* (Wall.) H.-M.
6653 – *Dumasia villosa* DC
6657 – *Alnus nepalensis* D. Don
6658 – *Quercus griffithii* Hook. f. & Th.
6659 – *Quercus glauca* Thunb.
6660 – *Indula nervosa* Wall.
6662 – *Leontopodium andersoni* C. B. Clarke
6663 – *Pueraria peduncularis* Graham
6668 – *Rosa moschata* subsp. *longicuspis* (Bertol.) Herr.
6669 – *Apios carnea* (Wall.) Benth.
6670 – *Diospyros lotus* L.
6674 – *Dollinera cinerascens*
6681 – *Rhododendron decorum* Fr.
6683 – *Castanopsis concolor* Rehd. & Wils.
6689 – *Anisadenia pubescens*
6694 – *Juglans regia* L.
6695 – *Zizyphus yünnanensis* Schneid.
6714 – *Castanopsis Wattii* (King) A. Camus.
6716 – *Castanopsis hystrix* A.DC.
6717 – *Diospyros kaki* L. var. *silvestris* Mak.
6836 – *Meliosma pungens* Wall.
6839 – *Lobelia erecta* Hook. f. & Thoms.
6846 – *Thea pitardii* (C. Stuart) Rehd.
6888 – *Ilex polyneura* (H.-M.) S. Y. Ha
6890 – *Impatiens delavayi* Fr.
6911 – *Desmodium griffithianum* Benth.
6913 – *Dalbergia mimosoides* Fr.
6915 – *Senecio clivorum* Maxim.
6917 – *Indigofera gerardiana* Wall.
6918 – *Indigofera scabrida* Dunn.
6919 – *Michelia lanceolata* Wils.
6929 – *Salix* sp.?
6931 – *Quercus griffithii* Hook. f. & Th.
6932 – *Callicarpa rubella* forma *angustata* P'ei
6935 – *Elsholtzia rugulosa* Hemsl.
6936 – *Thea yünnanensis Pitardi*
6939 – *Lithocarpus spicata*
6998 – *Lithocarpus spicata* var. *chittagonga*
7038 – *Lithocarpus dealbata*
7039 – *Lithocarpus fenestrata (Roxb.) Rehd.*
7040 – *Lithocarpus cleistocarpa* Rehd.
7041 – *Schima argentea* E. Pritz
7049 – *Ficus congesta* Roxb.

7108 – *Castanopsis tribuloides* A.DC.
7115 – *Dendrobium heterocarpum* Wall.
7116 – *Dendrobium strongylanthum* Rchb. f.
7123 – *Quercus griffithii*
7126 – *Pedicularis*
7136 – *Castanopsis hystrix*
7157 – *Liparis bootanensis* Griffith
7174 – *Altingia excelsa* Noronha
7175 – *Thea sinensis* L.
7187 – *Lasianthus henryi* Hutch.
7197 – *Callicarpa arborea* Roxb.
7217 – *Thunbergia salweenensis* W. W. Sm.
7223 – *Smilax ferox* Wall.
7226 – *Elaeocarpus*
7229 – *Castanopsis Rockii* A. Camus
7230 – *Radermachera*
7231 – *Smilax macrophylla* Roxb.
7236 – *Dalbergia mimosoides* Fr.
7238 – *Gaultheria pyroloides* Hook. f. & Th.
7244 – *Berberis sublevis* W. W. Sm.
7260 – *Mahonia lomariifolia*
7262 – *Litsea cubeba* (Laur.) Pers.
7265 – *Prunus majestica* Koehne
7272 – *Rhododendron*
7273 – *Leontopodium Andersoni* C. B. Clarke
7280 – *Coelogyne gardneriana* Lindl.
7281 – *Pleione grandiflora*
7288 – *Aconitum venatorium* var. *ecalcaratum* Airy-Shaw
7289 – *Ternstroemia japonica* Thunb.
7293 – *Diplycosia muscicola* (W. W. Sm.) Merr.
7296 – *Quercus augustini* Skan. var. *Rockiana* A. Cam.
7299 – *Gordonia axillaris*
7302 – *Lithocarpus spicata* (Sm.) var. *chittagonga* (King) Rehd.
7314 – *Anneslea fragrans* Wall.
7361 – *Litsea eusoma* W. W. Sm.
7364 – *Macropanax undulatum* (Wall.) Seem.
7373 – *Photinia serrulata* Lindl.
7374 – *Rosa moschata* subsp. *longicuspis* (Bertol.) Herr.
7376 – *Rhododendron sidereum* Balf. f.
7377 – *Rhododendron bullatum* Fr.
7381 – *Rhododendron sino-grande* Balf. f. & Sem.
7383 – *Lactuca hastata* (Wall.) D.C.
7384 – *Gaultheria nummularioides* D. Don
7385 – *Eulalia quadrinervis* (Hack.) Kuntze
7391 – *Rubus calycinus* Wall.
7394 – *Osmanthus*
7409 – *Lindera thomsonii* Allen

7412 – *Daphne papyracea* Wall. var. *crassiuscula* Rehd.
7429 – *Arisaema*
7431 – *Alsophila khasiana*
7432 – *Brassaiopsis Rockii* Merr.
7433 – *Macropanax undulatum* (Wall.) Seem.
7439 – *Litsea khasiana* Meissn. var. *hookeri* Meissn.
7446 – *Camellia wardii* Kob.
7447 – *Lithocarpus cariolosa* (Fr.) Chin
7448 – *Eurya cerasifolia* (D. Don) Kobuski
7451 – *Castanopsis indica*
7466 – *Clerodendron nutans* Wall.
7479 – *Rubus rufus* Focke
7480 – *Litsea brachypoda* Allen
7482 – *Rubus ampelinus* Focke
7489 – *Schefflera venulosa* (Seem.) Harms
7512 – *Pleione grandiflora*
7533 – *Aster rockianus* H.-M.
7551 – *Cornus capitata* Wall.
7570 – *Acanthopanax trifoliatum* (L.) Merr.

NAMENREGISTER

Die Schreibungen der Namen sind teils uneinheitlich. Da eine Standardisierung anhand der Texte nicht möglich war, wurden die Namen in der vorliegenden Form aufgenommen.